T0201783

Option Pricing and Estimation
of Financial Models with R

Option Pricing and Estimation of Financial Models with R

Stefano M. Iacus

Department of Economics, Business and Statistics
University of Milan, Italy

A John Wiley & Sons, Ltd., Publication

Library of Congress Cataloging-in-Publication Data

Iacus, Stefano M. (Stefano Maria)
  Option pricing and estimation of financial models with r / Stefano M. Iacus.
     p. cm.
  Includes bibliographical references and index.
  ISBN 978-0-470-74584-7
  1. Options (Finance) – Prices. 2. Probabilities. 3. Stochastic processes. 4. Time-series analysis. I. Title.
  HG6024.A3.I23 2011
  332.64'53 – dc22

                                                                                          2010045655

A catalogue record for this book is available from the British Library.

Print ISBN: 978-0-470-74584-7
ePDF ISBN: 978-1-119-99008-6
oBook ISBN: 978-1-119-99007-9
ePub ISBN: 978-1-119-99020-8

Set in 10/12 Times by Laserwords Private Limited, Chennai, India
Printed in Malaysia by Ho Printing (M) Sdn Bhd

Contents

Preface   **xiii**

1 A synthetic view   **1**
  1.1 The world of derivatives   2
    1.1.1 Different kinds of contracts   2
    1.1.2 Vanilla options   3
    1.1.3 Why options?   6
    1.1.4 A variety of options   7
    1.1.5 How to model asset prices   8
    1.1.6 One step beyond   9
  1.2 Bibliographical notes   10
    References   10

2 Probability, random variables and statistics   **13**
  2.1 Probability   13
    2.1.1 Conditional probability   15
  2.2 Bayes' rule   16
  2.3 Random variables   18
    2.3.1 Characteristic function   23
    2.3.2 Moment generating function   24
    2.3.3 Examples of random variables   24
    2.3.4 Sum of random variables   35
    2.3.5 Infinitely divisible distributions   37
    2.3.6 Stable laws   38
    2.3.7 Fast Fourier Transform   42
    2.3.8 Inequalities   46
  2.4 Asymptotics   48
    2.4.1 Types of convergences   48
    2.4.2 Law of large numbers   50
    2.4.3 Central limit theorem   52
  2.5 Conditional expectation   54
  2.6 Statistics   57
    2.6.1 Properties of estimators   57
    2.6.2 The likelihood function   61

	2.6.3	Efficiency of estimators	63
	2.6.4	Maximum likelihood estimation	64
	2.6.5	Moment type estimators	65
	2.6.6	Least squares method	65
	2.6.7	Estimating functions	66
	2.6.8	Confidence intervals	66
	2.6.9	Numerical maximization of the likelihood	68
	2.6.10	The δ-method	70
2.7	Solution to exercises	71
2.8	Bibliographical notes	77
	References	77

3	**Stochastic processes**	**79**
3.1	Definition and first properties	79
	3.1.1	Measurability and filtrations	81
	3.1.2	Simple and quadratic variation of a process	83
	3.1.3	Moments, covariance, and increments of stochastic processes	84
3.2	Martingales	84
	3.2.1	Examples of martingales	85
	3.2.2	Inequalities for martingales	88
3.3	Stopping times	89
3.4	Markov property	91
	3.4.1	Discrete time Markov chains	91
	3.4.2	Continuous time Markov processes	98
	3.4.3	Continuous time Markov chains	99
3.5	Mixing property	101
3.6	Stable convergence	103
3.7	Brownian motion	104
	3.7.1	Brownian motion and random walks	106
	3.7.2	Brownian motion is a martingale	107
	3.7.3	Brownian motion and partial differential equations	107
3.8	Counting and marked processes	108
3.9	Poisson process	109
3.10	Compound Poisson process	110
3.11	Compensated Poisson processes	113
3.12	Telegraph process	113
	3.12.1	Telegraph process and partial differential equations	115
	3.12.2	Moments of the telegraph process	117
	3.12.3	Telegraph process and Brownian motion	118
3.13	Stochastic integrals	118
	3.13.1	Properties of the stochastic integral	122
	3.13.2	Itô formula	124

3.14 More properties and inequalities for the Itô integral   127
3.15 Stochastic differential equations   128
    3.15.1 Existence and uniqueness of solutions   128
3.16 Girsanov's theorem for diffusion processes   130
3.17 Local martingales and semimartingales   131
3.18 Lévy processes   132
    3.18.1 Lévy-Khintchine formula   134
    3.18.2 Lévy jumps and random measures   135
    3.18.3 Itô-Lévy decomposition of a Lévy process   137
    3.18.4 More on the Lévy measure   138
    3.18.5 The Itô formula for Lévy processes   139
    3.18.6 Lévy processes and martingales   140
    3.18.7 Stochastic differential equations with jumps   143
    3.18.8 Itô formula for Lévy driven stochastic differential
           equations   144
3.19 Stochastic differential equations in $\mathbb{R}^n$ 145
3.20 Markov switching diffusions   147
3.21 Solution to exercises   148
3.22 Bibliographical notes   155
    References   155

4 Numerical methods   159
4.1 Monte Carlo method   159
    4.1.1 An application   160
4.2 Numerical differentiation   162
4.3 Root finding   165
4.4 Numerical optimization   167
4.5 Simulation of stochastic processes   169
    4.5.1 Poisson processes   169
    4.5.2 Telegraph process   172
    4.5.3 One-dimensional diffusion processes   174
    4.5.4 Multidimensional diffusion processes   177
    4.5.5 Lévy processes   178
    4.5.6 Simulation of stochastic differential equations
           with jumps   181
    4.5.7 Simulation of Markov switching diffusion processes   183
4.6 Solution to exercises   187
4.7 Bibliographical notes   187
    References   187

5 Estimation of stochastic models for finance   191
5.1 Geometric Brownian motion   191
    5.1.1 Properties of the increments   193

		5.1.2	Estimation of the parameters	194
5.2	Quasi-maximum likelihood estimation			195
5.3	Short-term interest rates models			199
		5.3.1	The special case of the CIR model	201
		5.3.2	Ahn-Gao model	202
		5.3.3	Aït-Sahalia model	202
5.4	Exponential Lévy model			205
		5.4.1	Examples of Lévy models in finance	205
5.5	Telegraph and geometric telegraph process			210
		5.5.1	Filtering of the geometric telegraph process	216
5.6	Solution to exercises			217
5.7	Bibliographical notes			217
	References			218

6 European option pricing — **221**

6.1	Contingent claims			221
		6.1.1	The main ingredients of option pricing	223
		6.1.2	One period market	224
		6.1.3	The Black and Scholes market	227
		6.1.4	Portfolio strategies	228
		6.1.5	Arbitrage and completeness	229
		6.1.6	Derivation of the Black and Scholes equation	229
6.2	Solution of the Black and Scholes equation			232
		6.2.1	European call and put prices	236
		6.2.2	Put-call parity	238
		6.2.3	Option pricing with R	239
		6.2.4	The Monte Carlo approach	242
		6.2.5	Sensitivity of price to parameters	246
6.3	The δ-hedging and the Greeks			249
		6.3.1	The hedge ratio as a function of time	251
		6.3.2	Hedging of generic options	252
		6.3.3	The density method	253
		6.3.4	The numerical approximation	254
		6.3.5	The Monte Carlo approach	255
		6.3.6	Mixing Monte Carlo and numerical approximation	256
		6.3.7	Other Greeks of options	258
		6.3.8	Put and call Greeks with `Rmetrics`	260
6.4	Pricing under the equivalent martingale measure			261
		6.4.1	Pricing of generic claims under the risk neutral measure	264
		6.4.2	Arbitrage and equivalent martingale measure	264
6.5	More on numerical option pricing			265
		6.5.1	Pricing of path-dependent options	266
		6.5.2	Asian option pricing via asymptotic expansion	269
		6.5.3	Exotic option pricing with `Rmetrics`	272

6.6   Implied volatility and volatility smiles   273
     6.6.1   Volatility smiles   276
6.7   Pricing of basket options   278
     6.7.1   Numerical implementation   280
     6.7.2   Completeness and arbitrage   280
     6.7.3   An example with two assets   280
     6.7.4   Numerical pricing   282
6.8   Solution to exercises   282
6.9   Bibliographical notes   283
     References   284

7 American options   285
7.1   Finite difference methods   285
7.2   Explicit finite-difference method   286
     7.2.1   Numerical stability   292
7.3   Implicit finite-difference method   293
7.4   The quadratic approximation   297
7.5   Geske and Johnson and other approximations   300
7.6   Monte Carlo methods   300
     7.6.1   Broadie and Glasserman simulation method   300
     7.6.2   Longstaff and Schwartz Least Squares Method   307
7.7   Bibliographical notes   311
     References   311

8 Pricing outside the standard Black and Scholes model   313
8.1   The Lévy market model   313
     8.1.1   Why the Lévy market is incomplete?   314
     8.1.2   The Esscher transform   315
     8.1.3   The mean-correcting martingale measure   317
     8.1.4   Pricing of European options   318
     8.1.5   Option pricing using Fast Fourier Transform method   318
     8.1.6   The numerical implementation of the FFT pricing   320
8.2   Pricing under the jump telegraph process   325
8.3   Markov switching diffusions   327
     8.3.1   Monte Carlo pricing   335
     8.3.2   Semi-Monte Carlo method   337
     8.3.3   Pricing with the Fast Fourier Transform   339
     8.3.4   Other applications of Markov switching diffusion models   341
8.4   The benchmark approach   341
     8.4.1   Benchmarking of the savings account   344
     8.4.2   Benchmarking of the risky asset   344
     8.4.3   Benchmarking the option price   344
     8.4.4   Martingale representation of the option price process   345
8.5   Bibliographical notes   346
     References   346

9   Miscellanea   **349**
   9.1   Monitoring of the volatility   349
         9.1.1   The least squares approach   350
         9.1.2   Analysis of multiple change points   352
         9.1.3   An example of real-time analysis   354
         9.1.4   More general quasi maximum likelihood
                 approach   355
         9.1.5   Construction of the quasi-MLE   356
         9.1.6   A modified quasi-MLE   357
         9.1.7   First- and second-stage estimators   358
         9.1.8   Numerical example   359
   9.2   Asynchronous covariation estimation   362
         9.2.1   Numerical example   364
   9.3   LASSO model selection   367
         9.3.1   Modified LASSO objective function   369
         9.3.2   Adaptiveness of the method   370
         9.3.3   LASSO identification of the model for term
                 structure of interest rates   370
   9.4   Clustering of financial time series   374
         9.4.1   The Markov operator distance   375
         9.4.2   Application to real data   376
         9.4.3   Sensitivity to misspecification   383
   9.5   Bibliographical notes   387
         References   387

APPENDICES

A   'How to' guide to R   **393**
   A.1   Something to know first about R   393
         A.1.1   The workspace   394
         A.1.2   Graphics   394
         A.1.3   Getting help   394
         A.1.4   Installing packages   395
   A.2   Objects   395
         A.2.1   Assignments   395
         A.2.2   Basic object types   398
         A.2.3   Accessing objects and subsetting   401
         A.2.4   Coercion between data types   405
   A.3   S4 objects   405
   A.4   Functions   408
   A.5   Vectorization   409

A.6   Parallel computing in R   411
      A.6.1   The foreach approach   413
      A.6.2   A note of warning on the multicore package   416
A.7   Bibliographical notes   416
      References   417

B R in finance   **419**
B.1   Overview of existing R frameworks   419
      B.1.1   Rmetrics   420
      B.1.2   RQuantLib   420
      B.1.3   The quantmod package   421
B.2   Summary of main time series objects in R   422
      B.2.1   The ts class   423
      B.2.2   The zoo class   424
      B.2.3   The xts class   426
      B.2.4   The irts class   427
      B.2.5   The timeSeries class   428
B.3   Dates and time handling   428
      B.3.1   Dates manipulation   431
      B.3.2   Using date objects to index time series   433
B.4   Binding of time series   434
      B.4.1   Subsetting of time series   440
B.5   Loading data from financial data servers   442
B.6   Bibliographical notes   445
      References   445

Index   **447**

Preface

Why another book on option pricing and why the choice of R language? The R language is increasingly accepted by so-called 'quants' as the basic infrastructure for financial applications. A growing number of projects, papers and conferences are either R-centric or, at least, deal with R solutions. R itself may not initially be very friendly, but, on the other hand, are stochastic integrals, martingales, and the Lévy process that friendly? In addition to this argument, we should take into account the famous quote from the R community by Greg Snow which describes the correct approach to R but equally applies to the Itô integral and to mathematical finance in general:

> When talking about user friendliness of computer software I like the analogy of cars vs. busses: Busses are very easy to use, you just need to know which bus to get on, where to get on, and where to get off (and you need to pay your fare). Cars on the other hand require much more work, you need to have some type of map or directions (even if the map is in your head), you need to put gas in every now and then, you need to know the rules of the road (have some type of driver's licence). The big advantage of the car is that it can take you a bunch of places that the bus does not go and it is quicker for some trips that would require transferring between busses. R is a 4-wheel drive SUV (though environmentally friendly) with a bike on the back, a kayak on top, good walking and running shoes in the passenger seat, and mountain climbing and spelunking gear in the back. R can take you anywhere you want to go if you take time to learn how to use the equipment, but that is going to take longer than learning where the bus stops are in a point-and-click GUI.

This book aims to present an indication of what is going on in modern finance and how this can be quickly implemented in a general computational framework like R rather than providing extra optimized lower-level programming languages' ad hoc solutions. For example, this book tries to explain how to simulate and calibrate models describing financial assets by general methods and generic functions rather than offering a series of highly specialized functions. Of course, the code in the book tries to be efficient while being generalized and some hints are given in the direction of further optimization when needed.

The choice of the R language is motivated both by the fact that the author is one of the developers of the R Project but also because R, being open source, is transparent in that it is always possible to see how numerical results are obtained without being deterred by a 'black-box' of a commercial product. And the R community, which is made by users and developers who in many cases correspond to researchers in the field, do a continuous referee process on the code. This has been one of the reasons why R has gained so much popularity in the last few years, but this is not without cost (the 'no free lunch' aspect of R), in particular because most R software is given under the disclaimer 'use at your own risk' and because, in general, there is no commercial support for R software, although one can easily experience peer-to-peer support from public mailing lists. This situation is also changing nowadays because an increasing number of companies are selling services and support for R-related software, in particular in finance and genetics.

When passing from the theory of mathematical finance to applied finance, many details should be taken into account such as handling the dates and times, the source of the time series in use, the time spent in running a simulation etc. This books tries to keep this level rather than a very abstract formulation of the problems and solutions, while still trying to present the mathematical models in a proper form to stimulate further reading and study.

The mathematics in this book is necessarily kept to a minimum for reasons of space and to keep the focus on the description and implementation of a wider class of models and estimation techniques. Indeed, while it is true that most mathematical papers contain a section on numerical results and empirical analysis, very few textbooks discuss these topics for models outside the standard Black and Scholes world.

The first chapters of the book provide a more in-depth treatment with exercises and examples from basic probability theory and statistics, because they rely on the basic instruments of calculus an average student (e.g. in economics) should know. They also contain several results without proof (such as inequalities), which will be used to sketch the proofs of the more advanced parts of the book. The second part of the book only touches the surface of mathematical abstraction and provides sketches of the proofs when the mathematical details are too technical, but still tries to give the correct indication of why the level of mathematical abstraction is needed. So the first part can be used by students in finance as a summary and the second part as the main section of the book. It is assumed that readers are familiar with R, but a summary of what they need to know to understand this book is contained in the two appendices as well as some general description of what is available and up-to-date in R in the context of finance.

So, back to Snow's quote: this book is more a car than a bus, but maybe with automatic gears and a solar-power engine, rather than a sport car with completely manual gears that requires continuous refueling and tuning.

A big and long-lasting smile is dedicated to my beloved Ilia, Ludovico and Lucia, for the time I spent away from them during the preparation of this manuscript. As V. Borges said once, 'a smile is the shortest distance between two persons'.

S.M. Iacus
November 2010

1

A synthetic view

Mathematical finance has been an exponentially growing field of research in the last decades and is still impressively active. There are also many directions and subfields under the hat of 'finance' and researchers from very different fields, such as economics (of course), engineering, mathematics, numerical analysis and recently statistics, have been involved in this area.

This chapter is intended to give a guidance on the reading of the book and to provide a better focus on the topics discussed herein. The book is intended to be self-contained in its exposition, introducing all the concepts, including very preliminary ones, which are required to better understand more complex topics and to appreciate the details and the beauty of some of the results.

This book is also very computer-oriented and it often moves from theory to applications and examples. The R statistical environment has been chosen as a basis. All the code presented in this book is free and available as an R statistical package called **opefimor** on CRAN.[1]

There are many good publications on mathematical finance on the market. Some of them consider only mathematical aspects of the matter at different level of complexity. Other books that mix theoretical results and software applications are usually based on copyright protected software. These publications touch upon the problem of model calibration only incidentally and in most cases the focus is on discrete time models mainly (ARCH, GARCH, etc.) with notable exceptions.

The main topics of this book are the description of models for asset dynamics and interest rates along with their statistical calibration. In particular, the attention is on continuous time models observed at discrete times and calibration techniques for them in terms of statistical estimation. Then pricing of derivative contracts on a single underlining asset in the Black and Scholes-Merton framework (Black and Scholes 1973; Merton 1973), pricing of basket options, volatility, covariation and regime switching analysis are considered. At the same

---

[1] CRAN, Comprehensive R Archive Network, http://cran.r-project.org.

Option Pricing and Estimation of Financial Models with R, First Edition. Stefano M. Iacus.
© 2011 John Wiley & Sons, Ltd. Published 2011 by John Wiley & Sons, Ltd.

time, the book considers jump diffusions and telegraph process models and pricing under these dynamics.

1.1   The world of derivatives

There are many kinds of financial markets characterized by the nature of the financial products exchanged rather than their geographical or physical location. Examples of these markets are:

- **stock markets**: this is the familiar notion of stock exchange markets, like New York, London, Tokyo, Milan, etc.;

- **bond markets**: for fixed return financial products, usually issued by central banks, etc.;

- **currency markets or foreign exchange markets**: where currencies are exchanged and their prices are determined;

- **commodity markets**: where prices of commodities like oil, gold, etc. are fixed;

- **futures and options markets**: for derivative products based on one or more other underlying products typical of the previous markets.

The book is divided into two parts (although some natural overlapping occurs). In the first part the modelling and analysis of dynamics of assets prices and interest rates are presented (Chapters 3, 4 and 5). In the second part, the theory and practice on derivatives pricing are presented (Chapters 6 and 7). Chapter 2 and part of Chapter 3 contain basic probabilistic and statistical infrastructure for the subsequent chapters. Chapter 4 introduces the numerical basic tools which, usually in finance, complement the analytical results presented in the other parts. Chapter 8 presents an introduction to recently introduced models which go beyond the standard model of Black and Scholes and the Chapter 9 presents accessory results for the analysis of financial time series which are useful in risk analysis and portfolio choices.

1.1.1   Different kinds of contracts

Derivatives are simply contracts applied to financial products. The most traded and also the object of our interest are the *options*. An option is a contract that gives the right to sell or buy a particular financial product at a given price on a predetermined date. They are clearly asymmetric contracts and what is really sold is the 'option' of exercise of a given right. Other asymmetric contracts are so-called *futures* or *forwards*. Forwards and futures are contracts which oblige one to sell or buy a financial product at a given price on a certain date to another party. Options and futures are similar in that, e.g., prices and dates are prescribed but clearly in one case what is traded is an opportunity of trade and in the other an obligation. We mainly focus on option pricing and we start with an example.

1.1.2   Vanilla options

Vanilla options[2] is a term that indicates the most common form of options. An option is a contract with several ingredients:

- the *holder*: who subscribes the financial contract;

- the *writer*: the seller of the contract;

- the *underlying asset*: the financial product, usually but not necessarily a stock asset, on which the contract is based;

- the *expiry date*: the date on which the right (to sell or buy) the underlying asset can be exercised by the holder;

- the *exercise or strike price*: the predetermined price for the underlying asset at the given date.

Hence, the holder buys a right and not an obligation (to sell or buy), conversely the writer is obliged to honor the contract (sell or buy at a given price) at the expiry date.

The right of this choice has an economical value which has to be paid in advance. At the same time, the writer has to be compensated from the obligation. Hence the problem of fixing a *fair* price for an option contract arises. So, option pricing should answer the following two questions:

- how much should one pay for his right of choice? i.e. how to fix the price of an option in order to be accepted by the holder?

- how to minimize the risk associated with the obligation of the writer? i.e. to which (economical) extent can the writer reasonably support the cost of the contract?

Example 1.1.1 (From Wilmott *et al.* (1995)) *Suppose that there exists an asset on the market which is sold at $25 and assume we want to fix the price of an option on this asset with an expiry date of 8 months and exercise price of buying this asset at $25. Assume there are only two possible scenarios: (i) in 8 months the price of the asset rises to $27 or (ii) in 8 months the price of the asset falls to*

---

[2] From *Free On-Line Dictionary of Computing*, http://foldoc.doc.ic.ac.uk. Vanilla : *(Default flavour of ice cream in the US) Ordinary flavour, standard. When used of food, very often does not mean that the food is flavoured with vanilla extract! For example, 'vanilla wonton soup' means ordinary wonton soup, as opposed to hot-and-sour wonton soup. This word differs from the canonical in that the latter means 'default', whereas vanilla simply means 'ordinary'. For example, when hackers go to a Chinese restaurant, hot-and-sour wonton soup is the canonical wonton soup to get (because that is what most of them usually order) even though it isn't the vanilla wonton soup.*

$23. In case (i) the potential holder of the option can exercise the right, pay $25 to the writer to get the asset, sell it on the market at $27 to get a return of $2, i.e.*

$$\$27 - \$25 = \$2.$$

In scenario (ii), the option will not be exercised, hence the expected return is $0. If both scenarios are likely to happen with the same probability of $\frac{1}{2}$, the expected return for the potential holder of this option will be

$$\frac{1}{2} \times \$0 + \frac{1}{2} \times \$2 = \$1.$$

So, if we assume no transaction costs, no interest rates, etc., the fair value of this option should be $1. If this is the fair price, a holder investing $1 in this contract could gain $-\$1 + \$2 = \$1$, which means 100% of the invested money in scenario (i) and in scenario (ii) $-\$1 + \$0 = -\$1$, i.e. 100% of total loss. Which means that derivatives are extremely risky financial contracts that, even in this simple example, may lead to 100% of gain or 100% of loss.

Now, assume that the potential holder, instead of buying the option, just buys the asset. In case (i) the return from this investment would be $-\$25 + \$27 = \$2$ which means $+2/25 = 0.08 \ (+8\%)$ and in scenario (ii) $-\$25 + \$23 = -\$2$ which equates to a loss of value of $-2/25 = -0.08 \ (-8\%)$.

From the previous example we learn different things:

- the value of an option reacts quickly (instantaneously) to the variation of the underlying asset;

- to fix the fair price of an option we need to know the price of the underlying asset at the expiry date: either we have a crystal ball or a good predictive model. We try the second approach in Chapters 3 and 5;

- the higher the final price of the underlying asset the larger will be the profit; hence the price depends on the present and future values of the asset;

- the value of the option also depends on the strike price: the lower the strike price, the less the loss for the writer;

- clearly, the expiry date of the contract is another key ingredient: the closer the expiry date, the less the uncertainty on future values of the asset and vice versa;

- if the underlying asset has high volatility (variability) this is reflected by the risk (and price) of the contract, because it is less certainty about future values of the asset. The study of volatility and Greeks will be the subject of Chapters 5, 6 and 9.

It is also worth remarking that, in pricing an option (as any other risky contract) there is a need to compare the potential revenue of the investment against fixed return contracts, like bonds, or, at least, interest rates. We will discuss models for the description of interest rates in the second part of Chapter 5. To summarize, the value of an option is a function of roughly the following quantities:

$$\text{option value} = f(\text{current asset price, strike price,}$$

$$\text{final asset price, expiry date,}$$

$$\text{interest rate})$$

Although we can observe the current price of the asset and predict interest rates, and we can fix the strike price and the expiry date, there is still the need to build predictive models for the final price of the asset. In particular, we will not be faced with two simple scenarios as in the previous example, but with a complete range of values with some variability which is different from asset to asset. So not only do we need good predictive models but also some statistical assessment and calibration of the proposed models. In particular we will be interested in calculating the expected value of f mainly as a function of the final value of the asset price, i.e.

$$\text{payoff} = \mathbb{E}\{f(\cdots)\}$$

this is the *payoff* of the contract which will be used to determine the fair value of an option. This payoff is rarely available in closed analytical form and hence simulation and Monte Carlo techniques are needed to estimate or approximate it. The bases of this numerical approach are set in Chapter 4.

The option presented in Example 1.1.1 was in fact a *call* option, where call means the 'right to buy'. An option that gives a right to sell at some price is called a *put* option. In a put option, the writer is obliged to buy from the holder an asset to some given price (clearly, when the underlying asset has a lower value on the market). We will see that the structure of the payoff of a put option is very similar to that of a call, although specular considerations on its value are appropriate, e.g. while the holder of a call option hopes for the rise of the price of the assets, the owner of the put hopes for the decrease of this price, etc. Table 1.1 reports put and call prices for the Roll Royce asset. When the strike price is 130, the cost of a call is higher than the cost of the put. This is because the current price is 134 and even a small increase in the value produces a gain of at least \$4. In the case of the put, the price should fall more than \$4 in order

Table 1.1 *Financial Times*, 4 Feb. 1993. (134): asset price at closing on 3 Feb. 1993. Mar., June, Sep.: expiry date, third Wednesday of each month.

		Calls			Puts		
Option	Ex. Price	Mar	Jun	Sep	Mar	Jun	Sep
R.Royce	130	11	15	19	9	14	17
(134)	140	6	11	16	16	20	23

to exercise the option. Of course all the prices are functions of the expiry dates. This is a similar situation but with smaller prices for options with a higher strike price (140).

1.1.3 Why options?

Usually options are not primary financial contracts in one's portfolio, but they are often used along with assets on which the derivative is based. A traditional investor may decide to buy stocks of a company if he or she believes that the company will increase its value. If right, the investor can sell at a proper time and obtain some gain, if wrong the investor can sell the shares before the price falls too much. If instead of real stocks the investor buys options on that stock, her fall or gain can go up to 100% of the investment as shown in the trivial example. But if one is risk adverse and wants to add a small risk to the portfolio, a good way to do this is to buy regular stocks and some options on the same stock. Also, in a long-term strategy, if one owns shares and options of the same asset and some temporary decrease of value occurs, one can decide to use or buy options to compensate this temporary loss of value instead of selling the stocks. For one reason or another, options are more liquid than the underlying assets, i.e. there are more options on an asset than available shares of that asset.

So options imply high risk for the holder which, in turn, implies complete loss of investment up to doubling. Symmetrically, the writer exposes himself to this obligation for a small initial payment of the contract (see e.g. Table 1.1). So, who on earth may decide to be a writer of one of these contracts of small revenue and high risk? Because an option exists on the market, their price should be fixed in a way that is considered convenient (or fair) for both the holder and the writer. Surely, if writers have more information on the market than a casual holder, then transition costs and other technical aspects may give enough profit to afford the role of writers. The strategy that allows the writer to cover the risk of selling an option to a holder at a given price is called *hedging*. More precisely, the hedging strategy is part of the way option pricing is realized (along with the notion of *non-arbitrage* which will be discussed in details in Chapter 6). Suppose we have an asset with decreasing value. If a portfolio contains only assets of this type, its value will decrease accordingly. If the portfolio contains only put options on that asset, the value of the portfolio will increase. A portfolio which includes both assets and put options in appropriate proportion may reduce the risk to the extent of eliminating the risk (*risk free strategy*). Hedging is a portfolio strategy which balances options and assets in order to reduce the risk. If the writer is able to sell an option at some price slightly higher than its real value, he may construct a hedging strategy which covers the risk of selling the option and eventually gain some money, i.e. obtain a risk-free profit. Risk-free strategies (as defined in Chapter 6) are excluded in the theory of Black and Scholes.

1.1.4 A variety of options

Options like the ones introduced so far are called *European* options. The name European has nothing to do with the market on which they are exchanged but on the typology of the contract itself. European options are contracts for which the right to sell (European call option) or buy (European put option) can be exercised only at a fixed expiry date. These options are the object of Chapter 6.

Options which can be exercised during the whole period of existence of the contract are called *American options*. Surely, the pricing of American options is more complicated than the pricing of European options because instead of a single fixed horizon, the complete evolution of the underlying asset has to be predicted in the most accurate way. In particular, the main point in possessing an American option is to find the optimal time on which exercise the option. This is the object of Chapter 7.

In both cases, options have not only an initial value (the initial fair price) but their value changes with time and options can be exchanged on the market before expiry date. So, the knowledge of the price of an option over the whole life of the contract is interesting in both situations.

Another classification of options is based on the way the payoff is determined. Even in the case of European options, it might happen that the final payoff of the option is determined not only by the last value of the underlying asset but also on the complete evolution of the price of the same asset, for example, via some kind of averaging. These are called *exotic options* (or *path-dependent options*). This is typical of options based on underlying products like commodities, where usually the payoff depends on the distance between the strike price and the average price during the whole life of the contract, the maximal or minimal value, etc.) or interest rates, where some geometric average is considered.

Average is a concept that applies to discrete values as well as to continuous values (think about the expected value of random variables). Observations always come in discrete form as a sequence of numbers, but analytical calculations are made on continuous time processes. The errors due to discretization of continuous time models affect both calibration and estimation of the payoffs. We will discuss this issue throughout the text.

Path-dependent options can be of European or American type and can be further subclassified according to the following categories which actually reflect analytical ways to treat them:

- barrier options: exercised only if the underlying asset reaches (or not) a prescribed value during the whole period (for example, in a simple European option with a given strike price, the option may be exercised by the holder only if the asset does not grow too much in order to contain the risk);

- Asian options: the final payoff is a function of some average of the price of the underlying asset;

- lookback options: the payoff depends on the maximal or minimal value of the asset during the whole period.

Notice that in this brief outlook on option pricing we mention only options on a single asset. Although very pedagogical to explain basic concepts of option pricing, many real options are based on more than one underlying asset. We will refer to these options as *basket options* and consider them in Chapter 6, Section 6.7. As for any portfolio strategy, correlation of financial products is something to take into account and not just the volatility of each single asset included in the portfolio. We will discuss the monitoring of volatility and covariance estimation of multidimensional financial time series in Chapter 9.

1.1.5   How to model asset prices

Modern mathematical finance originated from the doctoral thesis of Bachelier (1900) but was formally proposed in a complete financial perspective by Black and Scholes (1973) and Merton (1973). The basic model to describe asset prices is the *geometric Brownian motion*. Let us denote by $\{S(t), t \geq 0\}$ the price of an asset at time t, for $t \geq 0$. Consider the small time interval dt and the variation of the asset price in the interval $[t, t + dt)$ which we denote by $dS(t) = S(t + dt) - S(t)$. The return for this asset is the ratio between $dS(t)$ and $S(t)$. We can model the returns as the result of two components:

$$\frac{dS(t)}{S(t)} = \text{deterministic contribution} + \text{stochastic contribution}$$

the deterministic contribution is related to interest rates or bonds and is a risk free trend of this model (usually called the *drift*). If we assume a constant return μ, after dt times, the deterministic contribution to the returns of S will be μdt:

$$\text{deterministic contribution} = \mu dt.$$

The stochastic contribution is instead related to exogenous and nonpredictable shocks on the assets or on the whole market. For simplicity, these shocks are assumed to be symmetric, zero mean etc., i.e. typical Gaussian shocks. To separate the contribution of the natural volatility of the asset from the stochastic shocks, we assume further that the stochastic part is the product of $\sigma > 0$ (the volatility) and the variation of stochastic Gaussian noise $dW(t)$:

$$\text{stochastic contribution} = \sigma dW(t).$$

It is further assumed that the stochastic variation $dW(t)$ has a variance proportional to the time increment, i.e. $dW(t) \sim N(0, dt)$. The process $W(t)$, which is such that $dW(t) = W(t + dt) - W(t) \sim N(0, dt)$, is called the Wiener process or Brownian motion. Putting all together, we obtain the following equation:

$$\frac{dS(t)}{S(t)} = \mu dt + \sigma dW(t),$$

which we can rewrite in differential form as follows:

$$dS(t) = \mu S(t)dt + \sigma S(t)dW(t). \qquad (1.1)$$

This is a difference equation, i.e. $S(t + dt) - S(t) = \mu S(t)dt + \sigma S(t)(W(t + dt) - W(t))$ and if we take the limit as $dt \to 0$, the above is a formal writing of what is called a *stochastic differential equation*, which is intuitively very simple but mathematically not well defined as is. Indeed, taking the limit as $dt \to 0$ we obtain the following differential equation:

$$S'(t) = \mu S(t) + \sigma S(t)W'(t)$$

but the $W'(t)$, the derivative of the Wiener process with respect to time, is not well defined in the mathematical sense. But if we rewrite (1.1) in integral form as follows:

$$S(t) = S(0) + \mu \int_0^t S(u)du + \sigma \int_0^t S(u)dW(u)$$

it is well defined. The last integral is called *stochastic integral* or *Itô integral* and will be defined in Chapter 3. The geometric Brownian motion is the process $S(t)$ which solves the stochastic differential equation (1.1) and is at the basis of the Black and Scholes and Merton theory of option pricing. Chapters 2 and 3 contain the necessary building blocks to understand the rest of the book.

1.1.6   One step beyond

Unfortunately, the statistical analysis of financial time series, as described by the geometric Brownian motion, is not always satisfactory in that financial data do not fit very well the hypotheses of this theory (for example the Gaussian assumption on the returns). In Chapter 8 we present other recently introduced models which account for several discrepancies noticed on the analysis of real data and theoretical results where the stochastic noise $W(t)$ is replaced by other stochastic processes. Chapter 9 treats some advanced applied topics like monitoring of the volatility, estimation of covariation for asynchronous time series, model selection for sparse diffusion models and explorative data analysis of financial time series using cluster analysis.

1.2   Bibliographical notes

The principal text on mathematical finance is surely Hull (2000). This is so far the most complete overview of modern finance. Other text may be good companion to enter the arguments because they use different level of formalism. Just to mention a few, we can signal the two books Wilmott (2001) and Wilmott et al. (1995). The more mathematically oriented reader may prefer books like Shreve (2004a,b), Dineen (2005), Williams (2006), Mikosch (1998), Cerný (2004), Grossinho et al. (2006), Korn and Korn (2001) and Musiela and Rutkowski (2005). More numerically oriented publications are Fries (2007), Jäckel (2002), Glasserman (2004), Benth (2004), Ross (2003), Rachev (2004) and Scherer and Martin (2005). Other books more oriented to the statistical analysis of financial times series are Tsay (2005), Carmona (2004), Ruppert (2006) and Franke et al. (2004).

References

Bachelier, L. (1900). Théorie de la spéculation. *Analles Scientifique de l'École Normale Superieure* **17**, 21–86.

Benth, F. (2004). *Option Theory with Stochastic Analysis. An introduction to Mathematical Finance*. Springer-Verlag Berlin, Heidelberg.

Black, F. and Scholes, M. (1973). The pricing of options and corporate liabilities. *Journal of Political Economy* **81**, 637–654.

Carmona, R. (2004). *Statistical Analysis of Financial Data in S-Plus*. Springer, New York.

Cerný, A. (2004). *Mathematical Techniques in Finance. Tools for Incomplete Markets*. Princeton University Press, Princeton, N.J.

Dineen, S. (2005). *Probability Theory in Finance. A Mathematical Guide to the Black-Scholes Formula*. American Mathematical Society, Providence, R.I.

Franke, J., Härdle, W., and Hafner, C. (2004). *Statistics of Financial Markets: An Introduction*. Springer, New York.

Fries, C. (2007). *Mathematical Finance. Theory, Modeling, Implementation*. John Wiley & Sons, Inc., Hoboken, N.J.

Glasserman, P. (2004). *Monte Carlo Methods in Financial Engineering*. Springer, New York.

Grossinho, M., Shiryaev, A., Esqível, M., and Oliveira, P. (2006). *Stochastic Finance*. Springer Science Business Media, Inc., New York.

Hull, J. (2000). *Options, Futures and Other Derivatives*. Prentice-Hall, Englewood Cliffs, N.J.

Jäckel, P. (2002). *Monte Carlo Methods in Finance*. John Wiley & Sons, Ltd, Chichester, England.

Korn, R. and Korn, E. (2001). *Option Pricing and Portfolio Optimization*. American Mathematical Society, Providence, R.I.

Merton, R. C. (1973). Theory of rational option pricing. *Bell Journal of Economics and Management Science* **4**, 1, 141–183.

Mikosch, T. (1998). *Elementary Stochastic Calculus with Finance in View*. World Scientific, Singapore.

Musiela, M. and Rutkowski, M. (2005). *Martingale Methods in Financial Modeling. Second Edition*. Springer-Verlag Berlin, Heidelberg.

Rachev, S. (2004). *Handbook of Computational and Numerical Methods in Finance*. Birkhäuser, Boston.

Ross, S. (2003). *An Elementary Introduction to Mathematical Finance. Options and Other Topics*. Cambridge University Press, New York.

Ruppert, D. (2006). *Statistics and Finance: An Introduction*. Springer, New York.

Scherer, B. and Martin, D. (2005). *Introduction to Modern Portfolio Optimization with NuOPT, S-Plus and S+Bayes*. Springer Science Media Business, Inc., New York.

Shreve, S. (2004a). *Stochastic Calculus for Finance I. The Binomial Asset Pricing Model*. Springer, New York.

Shreve, S. (2004b). *Stochastic Calculus for Finance II. Continuous-Time Models*. Springer, New York.

Tsay, R. S. (2005). *Analysis of Financial Time Series. Second Edition*. John Wiley & Sons, Inc., Hoboken, N.J.

Williams, R. (2006). *Introduction to the Mathematics of Finance*. American Mathematical Society, Providence, R.I.

Wilmott, P. (2001). *Paul Wilmott Introduces Quantitative Finance*. John Wiley & Sons, Ltd, Chichester, England.

Wilmott, P., Howison, S., and Dewynne, J. (1995). *The Mathematics of Financial Derivatives. A Student Introduction*. Cambridge University Press, New York.

2

Probability, random variables and statistics

As seen, the modeling of randomness is one of the building blocks of mathematical finance. We recall here the basic notions on probability and random variables limiting the exposition to what will be really used in the rest of the book. The expert reader may skip this chapter and use it only as a reference and to align his own notation to the one used in this book.

Later on we will also discuss the problem of good calibration strategies of financial models, hence in this chapter we also recall some elementary concepts of statistics.

2.1 Probability

Random variables are functions of random elements with the property of being measurable. To make this subtle statement more precise we need the following preliminary definitions.

Definition 2.1.1 *Let Ω be some set. A family $\mathcal{A}$ of subsets of Ω is called σ-algebra on Ω if it satisfies the following properties:*

(i) $\emptyset \in \mathcal{A}$;

(ii) if $A \in \mathcal{A}$ then its complementary set $\bar{A}$ is in $\mathcal{A}$;

(iii) countable unions of elements of $\mathcal{A}$ belong to $\mathcal{A}$, i.e. let $A_n \in \mathcal{A}$, $n = 1, 2, \ldots$ then $\cup_n A_n \in \mathcal{A}$.

For example, the family $\{\emptyset, \Omega\}$ is a σ-algebra and it is called *trivial σ-algebra*. Let S be some set. We denote by $\sigma(S)$ or $\sigma\text{-}\{A : A \subset S\}$ the σ-algebra *generated* by

---

Option Pricing and Estimation of Financial Models with R, First Edition. Stefano M. Iacus.
© 2011 John Wiley & Sons, Ltd. Published 2011 by John Wiley & Sons, Ltd.

the subsets A of S, i.e. the family of sets which includes the empty set, each single subset of S, their complementary sets and all their possible countable unions.

Definition 2.1.2 *Let $\mathcal{A}$ be the σ-algebra on Ω. A set function P is a probability measure on $\mathcal{A}$ if it is a function $P : \mathcal{A} \mapsto [0, 1]$ which satisfies the three axioms of probability due to Kolmogorov (1933):*

(i) $\forall A \subset \mathcal{A}, P(A) \geq 0$;

(ii) $P(\Omega) = 1$;

(iii) *let $A_1, A_2, \ldots,$ a sequence of elements of $\mathcal{A}$ such that $A_i \cap A_j = \emptyset$ for $i \neq j$, then*

$$P\left(\bigcup_{i=1}^{\infty} A_i\right) = \sum_{i=1}^{\infty} P(A_i).$$

The last formula in axiom (iii) is called *complete additivity*.

Definition 2.1.3 *A probability space is a triplet $(\Omega, \mathcal{A}, P)$, where Ω is a generic set of elements (which can be thought as the collection of all possible outcomes of a random experiment) sometimes called 'sample space', $\mathcal{A}$ is the σ-algebra generated by Ω, i.e. the set of all sets (the events) for which it is possible to calculate a probability, and P is a probability measure.*

Example 2.1.4 *Consider the experiment which consists of throwing a dice. The possible outcomes of the experiment or the sample space is $\Omega = \{1, 2, 3, 4, 5, 6\}$ and $\mathcal{A}$ is the σ-algebra generated by the elementary events $\{1\}, \ldots, \{6\}$ of Ω, i.e. the σ-algebra of all events for which it is possible to evaluate some probability. Consider the events $E = $ 'even number' $= \{2, 4, 6\}$ and $F = $ 'number greater than 4' $= \{5, 6\}$. Both events can be obtained as union of elementary events of Ω and both belong to $\mathcal{A}$. Does the event $G = $ 'number 7' belong to $\mathcal{A}$? Of course not, indeed, it is not possible to obtain $\{7\}$ by unions, complementation, etc., of elements of Ω. In the previous example, if the dice is fair, i.e. each face is equiprobable, then P can be constructed as follows:*

$$P(A) = \frac{\#elementary\ elements\ in\ A}{\#elementary\ elements\ in\ \Omega} = \frac{|A|}{|\Omega|}$$

therefore

$$P(E) = \frac{\#\{2, 4, 6\}}{\#\{1, 2, 3, 4, 5, 6\}} = \frac{3}{6} = \frac{1}{2}$$

$$P(F) = \frac{\#\{5, 6\}}{\#\{1, 2, 3, 4, 5, 6\}} = \frac{2}{6} = \frac{1}{3}$$

where #A stands for: 'number of elements in set A'.

The following properties follow as easy consequences from the axioms of probability and are left as exercise.

Exercise 2.1 *Let $A \in \mathcal{A}$. Prove that $P(\bar{A}) = 1 - P(A)$ and $P(A) \le 1$.*

Theorem 2.1.5 *If A and B are two elements of $\mathcal{A}$, then*

$$P(A \cup B) = P(A) + P(B) - P(A \cap B) \tag{2.1}$$

For any collection of sets $A_1, A_2, \ldots$ in $\mathcal{A}$, Equation (2.1) generalizes to

$$P\left(\bigcup_i A_i\right) \le \sum_i P(A_i).$$

This last property is called sub-additivity. If $A \subset B$ then

$$P(A) \le P(B). \tag{2.2}$$

Exercise 2.2 *Prove (2.1) and (2.2).*

Using set theory one can prove the following relationships among complementary sets called De Morgan's laws

$$\overline{A \cup B} = \bar{A} \cap \bar{B} \quad \text{and} \quad \overline{A \cap B} = \bar{A} \cup \bar{B}$$

and the distributive property of intersection and union operators

$$A \cap (B \cup C) = (A \cap B) \cup (A \cap C).$$

2.1.1   Conditional probability

Definition 2.1.6 *The conditional probability of A given B is defined as*

$$P(A|B) = \frac{P(A \cap B)}{P(B)}$$

for sets B such that $P(B) > 0$.

The conditional probability seen as $P(\cdot|B)$ is a true probability measure, which can be eventually written as $P_B(\cdot) = P(\cdot|B)$. Conditioning only affects the probability of events but not events themselves, which means that the expectation on the realization of some event A may change due to the knowledge about another event B.

Exercise 2.3 *Prove that $P_B(\cdot) = P(\cdot|B)$ is a probability measure.*

Conditional probability is a key aspect of the calculus of probability from which other notions derive like the next one.

Definition 2.1.7 *Two events* $A, B \in \mathcal{A}$ *are said to be independent if and only if*

$$P(A \cap B) = P(A) \cdot P(B)$$

or, alternatively, if and only if $P(A|B) = P(A)$ *and* $P(B|A) = P(B)$.

If A is independent of B also B is independent of A. In fact, given that A is independent of B we have $P(A|B) = P(A \cap B)/P(B) = P(A)$, hence $P(A \cap B) = P(A|B)P(B) = P(A)P(B)$. Then, $P(B|A) = P(A \cap B)/P(A) = P(A)P(B)/P(A) = P(B)$.

Definition 2.1.8 *The events* $A_1, A_2, \ldots,$ *are said to be independent if and only if for all collection of indexes* $j_1 \neq j_2 \neq \ldots \neq j_k$ *and any* $k > 1$ *we have*

$$P\left(\bigcap_{i=j_1}^{j_k} A_i\right) = \prod_{i=j_1}^{j_k} P(A_i).$$

So, from previous definition, events are independent if they are mutually independent in couples, triplets, etc.

2.2 Bayes' rule

Definition 2.2.1 *A family* $\{A_i, i = 1, \ldots, n\}$ *of subsets of* Ω *is called partition of* Ω *if*

$$\bigcup_{i=1}^{n} A_i = \Omega \quad and \quad A_i \cap A_j = \emptyset \quad \forall i \neq j = 1, \ldots, n.$$

The following result is easy to prove.

Exercise 2.4 *Let* $\{A_i, i = 1, \ldots, n\}$ *be a partition of* Ω *and* E *another event in* Ω. *Prove that*

$$P(E) = \sum_{i=1}^{n} P(E|A_i)P(A_i). \tag{2.3}$$

Result (2.3) is sometimes called the *law of total probability* and it is a very useful tool to decompose the probability of some event into the sum of weighted conditional probabilities which are often easier to obtain. The next result comes directly as an application of the previous formula and is called Bayes' rule.

Theorem 2.2.2 (Bayes' rule) *Let* $\{A_i, i = 1, \ldots, n\}$ *be a partition of* Ω *and* $E \subset \Omega$, *then*

$$P(A_j|E) = \frac{P(E|A_j)P(A_j)}{\sum_{i=1}^{n} P(E|A_i)P(A_i)} = \frac{P(E|A_j)P(A_j)}{P(E)}, \quad j = 1, \ldots, n. \tag{2.4}$$

The terms $P(A_i)$ are sometimes called the *prior* probabilities and the terms $P(A_j|E)$ the *posterior* probabilities, where prior and posterior are with reference to the occurrence of (or knowledge about) the event E. This means that the knowledge about event E changes the initial belief on the realization of event A_j, i.e. $P(A_j)$, into $P(A_j|E)$. The proof of (2.4) is easy if one notices that $P(A_j|E) = P(A_j \cap E)/P(E) = P(E|A_j)P(A_j)/P(E)$.

Example 2.2.3 (The Monty Hall problem) *A hypothetical man dies in a hypothetical probabilistic world. After his death he faces three doors: red, green and blue. A desk clerk explains that two doors will bring him to hell, one to paradise. The man chooses the red door. Before he opens the red door, the desk clerk (who knows about the door to paradise) opens the blue door showing that it is the door to hell. Then he asks the man if he wants to keep the red door or change it with the green one. Will you change the door?*

Apparently, given that two doors are left, each door has a 50:50 percent probability of leading to paradise, so there is no point in changing the doors. But this is not the case. Denote by D_R, D_G and D_B the events 'paradise is behind red/green/blue door' respectively and denote by B the event 'the clerk desk opens the blue door'. The three events D_R, D_G and D_B constitute a partition of Ω and, before event B occurs, $P(D_R) = P(D_G) = P(D_B) = 1/3$ because there is no initial clue which one is the door to paradise. Clearly $P(D_B|B) = 0$. We now calculate $P(D_R|B)$ and $P(D_G|B)$. We first need $P(B)$. If paradise is behind red door, the clerk chooses to show either the blue or the green door, so in particular B with 50% probability, hence $P(B|D_R) = 1/2$. If paradise is behind the green door, the clerk is forced to open the blue door, hence $P(B|D_G) = 1$. Clearly, $P(B|D_B) = 0$. By (2.3) we have

$$P(B) = P(B|D_R)P(D_R) + P(B|D_G)P(D_G) + P(B|D_B)P(D_B)$$

$$= \frac{1}{2} \cdot \frac{1}{3} + 1 \cdot \frac{1}{3} = \frac{1}{2}$$

Therefore, by Bayes' rule

$$P(D_R|B) = \frac{P(B|D_R)P(D_R)}{P(B)} = \frac{\frac{1}{2} \cdot \frac{1}{3}}{\frac{1}{2}} = \frac{1}{3}$$

$$P(D_G|B) = \frac{P(B|D_G)P(D_G)}{P(B)} = \frac{1 \cdot \frac{1}{3}}{\frac{1}{2}} = \frac{2}{3}$$

$$P(D_B|B) = 0$$

So the conclusion is that the man should change the red door with the green one because $P(D_G|B) > P(D_R|B)$.

Definition 2.2.4 *Two probability measures P and Q on $(\Omega, \mathcal{A})$ are said to be equivalent if, for each subset, $A \subset \mathcal{A}$, $P(A) > 0$ implies $Q(A) > 0$ and vice versa.*

The above definition means that the two measures P and Q assign positive probability to the same events A (though not necessarily the same numerical values of the probability).

2.3   Random variables

We denote by $\mathcal{B}(\mathbb{R})$ the Borel σ-algebra generated by the open sets of $\mathbb{R}$ of the form $(-\infty, x)$, $x \in \mathbb{R}$, i.e. $\mathcal{B}(\mathbb{R}) = \sigma\text{-}\{(-\infty, x), x \in \mathbb{R}\}$. Consider a function $f : A \to B$ and take $S \subset B$, then the *inverse image* of S by f is denoted by $f^{-1}(S)$ and corresponds to the following subset of A:

$$f^{-1}(B) = \{a \in A : f(a) \in B\}.$$

Definition 2.3.1 *Given a probability space* $(\Omega, \mathcal{A}, P)$, *a real random variable X is a measurable function from* $(\Omega, \mathcal{A})$ *to* $(\mathbb{R}, \mathcal{B}(\mathbb{R}))$.

Hence a random variable transforms the events belonging to Ω into real numbers with the additional requirement of measurability. Measurability of X means the following: let A be a given subset of $\mathcal{B}(\mathbb{R})$, then to the event $X \in A$ it has to correspond a subset $B \in \mathcal{A}$ so that $P(B)$ is well defined. So it means that, whenever we want to calculate the probability of '$X \in A$', we can calculate it from the corresponding $P(B)$. More formally,

$$\forall A \in \mathcal{B}(\mathbb{R}), \ \exists B \in \mathcal{A} : X^{-1}(A) = B$$

and hence

$$Pr(X \in A) = P(\{\omega \in \Omega : \omega \in X^{-1}(A)\}) = P(B), \quad A \subset \mathbb{R}, B \in \mathcal{A},$$

where $X^{-1}(A)$ is the inverse image of A by X. Note that we wrote $Pr(X \in A)$ and not $P(X \in A)$ because the probability measure P works on the elements of $\mathcal{A}$ but X takes values in $\mathbb{R}$. More formally, a probability measure, say Q, on $(\mathbb{R}, \mathcal{B}(\mathbb{R}))$ can be defined as $Q(X \in A) = P(\{\omega \in \Omega : \omega \in X^{-1}(A)\})$ and hence the probability space $(\Omega, \mathcal{A}, P)$ is transformed by X into a new probability space $(\mathbb{R}, \mathcal{B}(\mathbb{R}), Q)$. To avoid too much abstraction, with abuse of notation, we simply write $P(X \in A)$ for $Pr(X \in A) = Q(X \in A)$. That's because once Q is defined, the original measure P can be disregarded.

Example 2.3.2 *Consider again the experiment in Example 2.1.4. Let X be the random variable defined as follows:* $X(\omega) = -1$, *if* $\omega = 1$ *or 2*, $X(\omega) = 0$, *if* $\omega = 3, 4, 5$ *and* $X(\omega) = +1$ *otherwise. We want to calculate* $P(X \geq 0)$.

$$P(X \geq 0) = P(X \in \{0, +1\})$$

$$= P(\omega \in \Omega : \omega \in X^{-1}(\{0, +1\}))$$

$$= P(\{\omega \in \Omega : X(\omega) = 0\} \cup (\{\omega \in \Omega : X(\omega) = +1\})$$

$$= P(\{3, 4, 5\} \cup \{6\}) = P(\{3, 4, 5, 6\}) = \frac{4}{6}$$

so $X^{-1}(\{0, +1\}) = B$, where B must be a subset of $\mathcal{A}$, the σ-algebra of Ω, in order to be able to calculate $P(X \geq 0)$.

Measurability of random variables at the moment appears as a technical require-ment but for stochastic processes it is a more delicate subject because it is strictly related to the notion of *information*, and information in finance is a key aspect of the analysis. We will return to this aspect later on.

As mentioned, in practice with random variables, we never work directly with the probability measure P on $(\Omega, \mathcal{A})$ but it is preferable to define their *cumulative distribution function* or simply their *distribution function* as follows:

$$F(x) = P(X \leq x) = P(X \in (-\infty, x]), \quad x \in \mathbb{R}.$$

The cumulative distribution function is an increasing function of its argument and such that $\lim_{x \to -\infty} F(x) = 0$ and $\lim_{x \to +\infty} F(x) = 1$. Further, for any two given numbers $a < b \in \mathbb{R}$, we have that

$$P(a \leq X \leq b) = P(X \leq b) - P(X \leq a) = F(b) - F(a).$$

When the random variable is *continuous* $F(\cdot)$ may admit the representation[1]

$$F(x) = \int_{-\infty}^{x} f(u) du,$$

where $f(\cdot)$ is called the *density function* or simply density of X and has the following properties: $f(x) \geq 0$ and $\int_{\mathbb{R}} f(x) dx = 1$. If X is a discrete random variable, then $F(\cdot)$ is written as

$$F(x) = P(X \leq x) = \sum_{x_j \leq x} P(X = x_j) = \sum_{x_j \leq x} p(x_j)$$

and $p(\cdot)$ is called *probability density function* or *mass function*. Here and in the following, we denote by x_i the generic value taken by the random variable X. Clearly, $p(x_i) \geq 0$ for all i and $\sum_i p(x_i) = 1$. Notice that, while for a discrete random variable $P(X = k) \geq 0$, for a continuous random variable $P(X = k) = 0$ always, hence the density function $f(\cdot)$ and probability mass function $p(\cdot)$ have different interpretations. For example, while $p(k) = P(X = k)$ in discrete case, $f(x) \neq P(X = x) = 0$ but $f(x) dx \simeq P(X \in [x, x + dx)) \geq 0$.

---

[1] Not all continuous random variables admit a density but, for simplicity, we assume that it is always the case in this chapter.

Definition 2.3.3 *Let* X *be a one-dimensional random variable with distribution function* $F(\cdot)$. *We call* q-*th quantile of* X *the real number* x_q *which satisfies*

$$x_q = \inf_x \{x : F(x) \geq q\}, \quad q \in [0, 1].$$

Definition 2.3.4 *Two random variables* X *and* Y *are said to be independent if*

$$P(X \in A, Y \in B) = P(X \in A)P(Y \in B), \quad \text{for all } A, B \in \mathbb{R},$$

i.e. the probability that X *takes particular values is not influenced by that of* Y.

The couple (X, Y) of random variables has a joint cumulative distribution function which we denote by $F(x, y) = P(X \leq x, Y \leq y)$. If both variables are continuous we denote the joint density function as $f_{XY}(x, y)$ and

$$F_{XY}(x, y) = \int_{-\infty}^{x} \int_{-\infty}^{y} f_{XY}(u, v) \mathrm{d}u \mathrm{d}v.$$

If both are discrete we denote the probability density function by $p_{XY}(x_i, y_j) = P(X = x_i, Y = y_j)$ and

$$F_{XY}(x, y) = \sum_{x_i \leq x} \sum_{y_j \leq y} p_{XY}(x_i, y_j).$$

In case X and Y are independent, we have that $F_{XY}(x, y) = F_X(x)F_Y(y)$ where $F_X(\cdot)$ and $F_Y(\cdot)$ are the distribution functions of X and Y. Similarly for the densities, i.e. $f_{XY}(x, y) = f_X(x)f_Y(y)$ and $p_{XY}(x_i, y_j) = p_X(x_i)p_Y(y_j)$. In general if $X = (X_1, X_2, \ldots, X_n)^\top$ is random vector[2], we denote the distribution function by

$$F_X(x_1, x_2, \ldots, x_n) = P(X_1 \leq x_1, X_2 \leq x_2, \ldots, X_n \leq x_n)$$

and, if all components of X are independent, the usual factorization holds, i.e.

$$F_X(x_1, x_2, \ldots, x_n) = F_{X_1}(x_1)F_{X_2}(x_2) \cdots F_{X_n}(x_n)$$

with obvious notation. Similarly for the joint densities and/or mass functions.

Definition 2.3.5 *Let* P *and* Q *be two equivalent probability measures on* $(\Omega, \mathcal{A})$. *A function* f *such that*

$$Q(A) = \int_A f \mathrm{d}P$$

is called Radon-Nikodým derivative and it is usually denoted as

$$f = \frac{\mathrm{d}Q}{\mathrm{d}P}.$$

---

[2] We denote by $A^\top$ the transpose of A.

The density of a continuous random variable is defined exactly in this way. Indeed

$$P_X((-\infty, x]) = P(X \in (-\infty, x]) = F(x) = \int_{-\infty}^{x} f(x)dx,$$

where $dx = \lambda(dx)$ is nothing but the Lebesgue measure of the interval dx. Thus, the density f is the Radon-Nikodým derivative of P_X with respect to the Lebesgue measure λ, i.e. $f = dP_X/d\lambda$.

Definition 2.3.6 *We define the expected value of a random variable X in the following integral transform*

$$\mathbb{E}(X) = \int_{\Omega} X(\omega)P(d\omega) = \int_{\mathbb{R}} xdF(x)$$

where the last integral is a Riemann-Stieltjes integral. If X is continuous we have $\mathbb{E}(X) = \int_{\mathbb{R}} xdF(x) = \int_{\mathbb{R}} xf(x)dx$ and when X is discrete $\mathbb{E}(X) = \sum_i x_i p_X(x_i)$. The variance of X is defined as

$$\text{Var}(X) = \mathbb{E}(X - \mathbb{E}\{X\})^2 = \int_{\Omega} (X(\omega) - \mathbb{E}\{X\})^2 P(d\omega).$$

The n-th moment of a random variable is defined as $\mu_n = \mathbb{E}\{X^n\}$.

2.3.6.1 Some properties of the expected value operator

The expected value has also several properties we mention without proof. Let X and Y be two random variables and c, M some real constants. Then

(a) $\mathbb{E}(X \pm Y) = \mathbb{E}(X) \pm \mathbb{E}(Y)$;

(b) if X and Y are independent, then $\mathbb{E}(XY) = \mathbb{E}(X)\mathbb{E}(Y)$;

(c) if c is a constant, then $\mathbb{E}(cX) = c\mathbb{E}(X)$ and $\mathbb{E}(c) = c$;

(d) if $X \geq 0$, then $\mathbb{E}(X) \geq 0$;

(e) if $X \geq Y$ (i.e. $X(\omega) - Y(\omega) \geq 0, \forall \omega \in \Omega$), then $\mathbb{E}(X) \geq \mathbb{E}(Y)$;

(f) let $\mathbf{1}_A(\omega)$ be the random variable which takes value 1 if $\omega \in A$ and 0 otherwise; then $\mathbb{E}(\mathbf{1}_A) = P(A)$;

(g) if $|X| \leq M$ (i.e. $|X(\omega)| \leq M, \forall \omega \in \Omega$) then $|\mathbb{E}(X\mathbf{1}_A)| \leq MP(A)$;

(h) $P(X \in B) = \mathbb{E}(\mathbf{1}_{\{X \in B\}}) = \int_A X(\omega)P(d\omega) = \mathbb{E}(X\mathbf{1}_A)$, with $A = X^{-1}(B)$.

When a random variable X is such that $\mathbb{E}(X) < \infty$ we say that X is *integrable*. Clearly X is integrable if $|X|$ is integrable. It is easy to see that $\text{Var}(X) = \mathbb{E}(X^2) - (\mathbb{E}\{X\})^2$ by simple expansion of the binomial $(X - \mathbb{E}(X))^2$. In general, given a transform $Y = g(X)$ of a random variable X, with $g(\cdot)$ a measurable function, it is possible to calculate $\mathbb{E}(Y)$ as

$$\mathbb{E}(Y) = \mathbb{E}\{g(X)\} = \int_{\mathbb{R}} g(x) dF(x).$$

Definition 2.3.7 *The covariance between two random variables X and Y is the quantity*

$$\mathrm{Cov}(X, Y) = \mathbb{E}\{(X - \mathbb{E}(X))(Y - \mathbb{E}(Y))\} = \mathbb{E}(XY) - \mathbb{E}(X)\mathbb{E}(Y)$$

where

$$\mathbb{E}(XY) = \int_{\Omega} X(\omega) Y(\omega) P(d\omega)$$

is the mixed moment of X and Y. When X and Y are both discrete $\mathbb{E}(XY) = \sum_i \sum_j x_i y_j p_{XY}(x_i, y_j)$ and when both are continuous $\mathbb{E}(XY) = \int_{\mathbb{R}} \int_{\mathbb{R}} xy f_{XY}(x, y) dx dy$.

The covariance between two random variables is the notion of joint variability and is a direct extension of the notion of variance in the univariate case. Indeed, $\mathrm{Cov}(X, X) = \mathrm{Var}(X)$. For a random vector $X = (X_1, X_2, \ldots, X_n)$ it is usually worth introducing the variance–covariance matrix between the components (X_i, X_j), $i, j = 1, \ldots, n$, which is the following matrix

$$\mathrm{Var}(X) = \begin{bmatrix} \sigma_1^2 & \sigma_{12} & \cdots & \sigma_{1n} \\ \sigma_{21} & \sigma_2^2 & & \sigma_{2n}^2 \\ \vdots & \vdots & \ddots & \vdots \\ \sigma_{n1} & \sigma_{n2} & \cdots & \sigma_n^2 \end{bmatrix}$$

with $\sigma_i^2 = \mathrm{Var}(X_i)$, $i = 1, \ldots, n$, and $\sigma_{ij} = \mathrm{Cov}(X_i, X_j)$, $i \neq j$.

Exercise 2.5 *Prove that if X and Y are independent then $\mathrm{Cov}(X, Y) = 0$. Provide a counter example to show that the contrary is not true, i.e., $\mathrm{Cov}(X, Y) = 0$ does not imply that X and Y are independent.*

If $\{X_i, i = 1, \ldots, n\}$ is a family of random variables and a_i, $i = 1, \ldots, n$, are some constants, then

$$\mathbb{E}\left\{ \sum_{i=1}^n a_i X_i \right\} = \sum_{i=1}^n a_i \mathbb{E}(X_i). \tag{2.5}$$

Moreover, if the X_i are mutually independent, then

$$\mathrm{Var}\left\{ \sum_{i=1}^n a_i X_i \right\} = \sum_{i=1}^n a_i^2 \mathrm{Var}(X_i). \tag{2.6}$$

If the random variables are not mutually independent the formula for the variance takes the form:

$$\text{Var}\left\{\sum_{i=1}^{n} a_i X_i\right\} = \sum_{i=1}^{n} a_i^2 \text{Var}(X_i) + 2 \sum_{i,j:i<j} a_i a_j \text{Cov}(X_i, X_j).$$

Definition 2.3.8 *A random variable X is L^p integrable, and we write $X \in L^p(\Omega, \mathcal{A}, P)$, or simply $X \in L^p$, if $\int_{\Omega} |X(\omega)|^p P(d\omega) = \int_R |x|^p dF(x) < \infty$. We call $X \in L^2$ a square integrable random variable.*

2.3.1   Characteristic function

The *characteristic function* of a random variable X is the following integral transform

$$\varphi(t) = \mathbb{E}\left\{e^{itX}\right\} = \int_{-\infty}^{\infty} e^{itx} dF(x)$$

where i is the imaginary unit. When X has a density, the characteristic function becomes $\varphi(t) = \int_R e^{itx} f(x) dx$. The characteristic function has the following elementary properties

   (i) $\varphi(0) = 1$;

   (ii) $|\varphi(t)| \leq 1$ for all t, indeed $|\varphi(t)| = \left|\mathbb{E}\left\{e^{itX}\right\}\right| \leq \mathbb{E}\left|e^{itX}\right| \leq 1$ as $|e^{ix}| \leq 1$;

   (iii) $\overline{\varphi(t)} = \varphi(-t)$, with $\bar{z}$ is the complex conjugate of $z = a + ib$, i.e. $\bar{z} = a - ib$;

   (iv) the function $\varphi(t)$ is a continuous function for all real numbers t.

The characteristic function uniquely identifies the probability law of a random variable, so each random variable has one and only one characteristic function and each characteristic function corresponds to a single random variable. We will make use of this property in Section 2.3.7.

Theorem 2.3.9 *If $\{X_i, i = 1, \ldots, n\}$ is a family of mutually independent random variables and a_i, $i = 1, \ldots, n$, a sequence of constants, then the characteristic function of $S_n = \sum_{i=1}^{n} a_i X_i$ satisfies the following equality*

$$\varphi_{S_n}(t) = \varphi_{X_1}(a_1 t)\varphi_{X_2}(a_2 t) \cdots \varphi_{X_n}(a_n t). \tag{2.7}$$

Proof. Indeed

$$\varphi_{S_n}(t) = \mathbb{E}\left\{e^{itS_n}\right\} = \mathbb{E}\left\{e^{it\sum_{i=1}^{n} a_i X_i}\right\} = \mathbb{E}\left\{\prod_{i=1}^{n} e^{ita_i X_i}\right\} = \prod_{i=1}^{n} \mathbb{E}\left\{e^{ita_i X_i}\right\},$$

where independence has been used in the last equality.

A corollary of the last theorem is the following: consider a sequence of *independent and identically distributed* (i.i.d.) random variables $\{X_i, i = 1, \ldots, n\}$ and define $S_n = \sum_{i=1}^{n} X_i$. Then,

$$\phi_{S_n}(t) = \{\phi_X(t)\}^n,$$

where $\phi_X(\cdot)$ is the common characteristic function of the X_i's. When the moment of order n of the random variable exists, it is possible to obtain it by n-times differentiation of the characteristic function with respect to t as follows:

$$\mathbb{E}\left(X^n\right) = i^{-n} \left. \frac{d^n}{dt^n} \varphi(t) \right|_{t=0}. \tag{2.8}$$

This is easy to see: consider $\frac{d}{dt}\varphi(t) = \mathbb{E}\left(iXe^{itX}\right)$, evaluated in $t = 0$ gives $i\mathbb{E}(X)$. Then by induction one proves (2.8).

2.3.2 Moment generating function

Closely related to the characteristic function is the so called *moment generating function* of a random variable X which is defined as follows:

$$M(\alpha) = \mathbb{E}\left\{e^{\alpha X}\right\} = \int_{-\infty}^{\infty} e^{\alpha x} dF(x), \quad \alpha \in \mathbb{R}.$$

It easy to prove that

$$\mathbb{E}\left(X^n\right) = \left. \frac{d^n}{d\alpha^n} M(\alpha) \right|_{\alpha=0}$$

from which its name. Further, under the same conditions for which (2.7) holds, we have that

$$M_{S_n}(\alpha) = M_{X_1}(a_1\alpha) M_{X_2}(a_2\alpha) \cdots M_{X_n}(a_n\alpha). \tag{2.9}$$

2.3.3 Examples of random variables

We will mention here a few random variables which will play a central role in the next chapters.

2.3.3.1 Bernoulli random variable

The Bernoulli random variable takes only the two values 1 and 0, respectively with probability p and $1 - p$, i.e. $P(X = 1) = p$, $P(X = 0) = 1 - p$, with $0 < p < 1$. It is usually interpreted as the indicator variable of some related events (for example, failure-functioning, on/off, etc.).

Exercise 2.6 *Prove that* $\mathbb{E}(X) = p$, $\text{Var}(X) = p(1 - p)$ *and* $\varphi(t) = 1 - p + pe^{it}$.

We denote the Bernoulli random variable X as $X \sim \text{Ber}(p)$. This variable is the building block of the following Binomial random variable.

2.3.3.2 Binomial random variable

Let $X_i, i = 1, \ldots, n$, be a sequence *independent and identically distributed* (i.i.d.) Bernoulli random variables with parameter p, i.e. $X_i \sim \text{Ber}(p)$. The random variable which counts the number of ones (successes) in a sequence of n Bernoulli trials is called the *Binomial* random variable. More precisely

$$Y = \sum_{i=1}^{n} X_i \sim \text{Bin}(n, p)$$

where $\text{Bin}(n, p)$ stands for *Binomial law*[3] *with parameters n and p*, which is the following discrete distribution

$$P(Y = k) = \binom{n}{k} p^k (1 - p)^{n-k}, \quad k = 0, 1, \ldots, n. \tag{2.10}$$

Exercise 2.7 *Prove that* $\mathbb{E}(Y) = np$, $\text{Var}(Y) = np(1 - p)$ *and* $\varphi(t) = \left(1 - p + pe^{it}\right)^n$.

R functions to obtain density, cumulative distribution function, quantiles and random numbers for the Binomial random variable are of the form [dpqr]binom.

2.3.3.3 Poisson random variable

The Poisson random variable is the limit (as $n \to \infty$) of the binomial experiment in the case of rare events, i.e. when the probability of observing one event is close to zero and $n \cdot p = \lambda$ remains fixed as n increases. The Poisson law of parameter λ, $\text{Poi}(\lambda)$, $\lambda > 0$ is the following probability mass function

$$P(X = k) = \frac{\lambda^k e^{-\lambda}}{k!}, \quad k = 0, 1, \ldots \tag{2.11}$$

We denote a Poisson random X by $X \sim \text{Poi}(\lambda)$.

Exercise 2.8 *Prove that X has both mean and variance equal to λ and characteristic function* $\varphi(t) = \exp\left\{\lambda \left(e^{it} - 1\right)\right\}$.

R functions to obtain density, cumulative distribution function, quantiles and random numbers for the Poisson random variable are of the form [dpqr]pois.

---

[3] In Equation (2.10) the term $\binom{n}{k}$ is the binomial coefficient: $\binom{n}{k} = \frac{n!}{k!(n-k)!}$, with $n! = n \cdot (n - 1) \cdot (n - 2) \cdots 2 \cdot 1$.

2.3.3.4   Uniform random variable

Let $[a, b]$ be a finite interval, the Uniform random variable $X \sim U(a, b)$, is a continuous random variable with density

$$f(x) = \begin{cases} \frac{1}{b-a}, & x \in (a, b), \\ 0, & \text{otherwise;} \end{cases}$$

and distribution function

$$F(x) = \begin{cases} 0, & x < a, \\ \frac{x-a}{b-a}, & x \in [a, b], \\ 1, & x > b. \end{cases}$$

Exercise 2.9 *Prove that*

$$\mathbb{E}(X) = \frac{a+b}{2}, \quad \text{Var}(X) = \frac{(b-a)^2}{12} \quad and \quad \varphi(t) = \frac{e^{itb} - e^{ita}}{it(b-a)}.$$

R functions to obtain density, cumulative distribution function, quantiles and random numbers for the Uniform random variable are of the form [dpqr]unif.

2.3.3.5   Exponential random variable

This variable is related to the Poisson random variable. While the Poisson random variable counts the number of rare events, the exponential random variable measures the time between two occurrence of Poisson events, so it is a positive and continuous random variable. The exponential random variable X of parameter $\lambda > 0$, i.e. $X \sim \text{Exp}(\lambda)$, has density

$$f(x) = \begin{cases} \lambda e^{-\lambda x}, & x > 0, \\ 0, & \text{otherwise;} \end{cases}$$

and distribution function

$$F(x) = \begin{cases} 0, & x < 0, \\ 1 - e^{-\lambda x}, & x > 0. \end{cases}$$

Let $Y \sim \text{Poi}(\lambda)$, then

$$P(Y = 0) = \frac{\lambda^0 e^{-\lambda}}{\lambda^0} = e^{-\lambda} = 1 - P(X < 1) = P(X > 1),$$

i.e. the probability of no events in the time unit is equivalent to waiting more than 1 unit of time for the first event to occur. In general, if x is some amount of time, then $P(Y = 0)$ with $Y \sim \text{Poi}(\lambda x)$ is equal to $P(X > x)$, with $X \sim \text{Exp}(\lambda)$.

Exercise 2.10 *Prove that*

$$\mathbb{E}(X) = \frac{1}{\lambda}, \quad \text{Var}(X) = \frac{1}{\lambda^2} \quad and \quad \varphi(t) = \frac{\lambda}{\lambda - it}.$$

An interesting property of the exponential distribution is that it is *memoryless* which means that, if X is an exponential random variable, then

$$P(X > t + s | X > t) = P(X > s), \quad \forall s, t \geq 0,$$

which is easy to prove. Indeed,

$$P(X > t + s | X > t) = \frac{P(X > t + s)}{P(X > t)} = \frac{1 - F(t + s)}{1 - F(s)} = \frac{e^{-\lambda(t+s)}}{e^{-\lambda t}}$$

$$= e^{-\lambda s} = P(X > s).$$

The memoryless property means that given no event occurred before time t, the probability that we need to wait more s instants for the event to occur, is the same as waiting s instants from the origin. For example, if time unit is in seconds s and no event occurred in the first $10s$, then the probability that the event occurs after $30s$, i.e. wait for another $20s$ or more, is $P(X > 30 | X > 10) = P(X > 20)$.

R functions to obtain density, cumulative distribution function, quantiles and random numbers for the Exponential random variable are of the form [dpqr]exp.

2.3.3.6 Gamma random variable

The Gamma random variable, $X \sim \text{Gamma}(\alpha, \beta)$, has a continuous distribution with two parameters $\alpha > 0$ and $\beta > 0$. The density function of its law has the form:

$$f(x) = \frac{\beta^\alpha}{\Gamma(\alpha)} x^{\alpha-1} e^{-\beta x}, \quad x > 0,$$

where the function

$$\Gamma(k) = \int_0^\infty t^{k-1} e^{-t} dt,$$

is called the *gamma function*. The Γ function has several properties. We just list them without proof

- if k is an integer, then $\Gamma(k + 1) = k!$;

- $\Gamma\left(\frac{1}{2}\right) = \sqrt{\pi}$;

- $\dfrac{\Gamma(x + 1)}{\Gamma(x)} = x.$

The Gamma is distribution is such that, if $X \sim \text{Gamma}(\alpha, \beta)$, then

$$\mathbb{E}(X) = \frac{\alpha}{\beta}, \quad \text{Var}(X) = \frac{\alpha}{\beta^2}$$

and its characteristic function is

$$\varphi(t) = \mathbb{E}\left\{e^{itX}\right\} = \left(1 - \frac{it}{\beta}\right)^{-\alpha}.$$

The Gamma distribution includes several special cases, the most important one being the exponential random variable and the χ^2 random variable (see below). Indeed, $\text{Gamma}(1, \beta) = \text{Exp}(\beta)$ and $\text{Gamma}\left(\frac{n}{2}, \frac{1}{2}\right) = \chi_n^2$.

R functions to obtain density, cumulative distribution function, quantiles and random numbers for the Gamma random variable are of the form [dpqr]gamma.

2.3.3.7 Gaussian random variable

The Gaussian or Normal random variable is a continuous distribution with two parameters μ and σ^2 which correspond respectively to its mean and variance. The density function of its law $N(\mu, \sigma^2)$ is

$$f(x) = \frac{1}{\sqrt{2\pi\sigma^2}} e^{-\frac{(x-\mu)^2}{2\sigma^2}}, \quad x \in \mathbb{R}.$$

Its characteristic function is given by

$$\varphi(t) = \mathbb{E}\left\{e^{itX}\right\} = \exp\left\{\mu it - \frac{\sigma^2 t^2}{2}\right\}.$$

Exercise 2.11 *Derive by explicit calculations the mean, the variance and the characteristic function of the Gaussian random variable.*

The very special case of $N(0, 1)$ is called a *standard* normal random variable. This random variable is also symmetric around its mean.

R functions to obtain density, cumulative distribution function, quantiles and random numbers for the Gaussian random variable are of the form [dpqr]norm.

2.3.3.8 Chi-square random variable

The Chi-square distribution has one parameter n, the degrees of freedom and it is a non-negative random continuous random variable denoted as χ_n^2 with density

$$f(x) = \frac{1}{2^{\frac{n}{2}} \Gamma\left(\frac{n}{2}\right)} x^{\frac{n}{2}-1} e^{-\frac{x}{2}}, \quad x > 0,$$

and characteristic function

$$\varphi(t) = \mathbb{E}\left\{e^{itX}\right\} = (1 - 2it)^{-\frac{n}{2}}.$$

If $X \sim \chi_n^2$, then

$$\mathbb{E}(X) = n, \quad \text{Var}(X) = 2n.$$

The square of a Gaussian random variable N(0, 1) is χ_1^2 distributed. Along with the standard Chi-square distribution it is possible to define the noncentral Chi-square random variable with density

$$f(x) = \frac{1}{2} e^{-\frac{x+\delta}{2}} \left(\frac{x}{\delta}\right)^{\frac{n}{4}-\frac{1}{2}} I_{\frac{k}{2}-1}(\sqrt{\delta x}),$$

where n are the degrees of freedom, $\delta > 0$ is the noncentrality parameter and the random variable is denoted by $\chi_n^2(\delta)$. The function $I_k(x)$ is the modified Bessel function of the first kind (see Abramowitz and Stegun 1964). If $X \sim \chi_n^2(\delta)$, then

$$\varphi(t) = \mathbb{E}\left\{e^{itX}\right\} = \frac{e^{\frac{it\delta}{1-2it}}}{(1-2it)^{\frac{n}{2}}}$$

and

$$\mathbb{E}(X) = n + \delta, \quad \text{Var}(X) = 2(n + 2\delta).$$

R functions to obtain density, cumulative distribution function, quantiles and random numbers for the Chi-square random variable are of the form [dpqr]chisq.

2.3.3.9   Student's t random variable

The Student's t random variable is symmetric and continuous with zero mean and a parameter n, the degrees of freedom. It arises in many contexts and in particular in statistics. If a random variable has a Student's t distribution with $n > 0$ degrees of freedom we write $X \sim t^n$. The density of X is

$$f(x) = \frac{\Gamma\left(\frac{n+1}{2}\right)}{\sqrt{n\pi}\,\Gamma\left(\frac{n}{2}\right)} \left(1 + \frac{x^2}{n}\right)^{-\frac{n+1}{2}}, \quad x \in \mathbb{R},$$

and its characteristic function is

$$\varphi(t) = \mathbb{E}\, e^{itX} = \frac{K_{\frac{n}{2}}(\sqrt{n}|t|)(\sqrt{n}|t|)^{\frac{n}{2}}}{\Gamma\left(\frac{n}{2}\right) 2^{\frac{n}{2}-1}}.$$

Moreover,

$$\mathbb{E}(X) = 0, \quad \text{Var}(X) = \begin{cases} \frac{n}{n-2}, & n > 2, \\ \infty, & 1 < n \leq 2, \\ \text{undefined otherwise.} \end{cases}$$

If $Z \sim N(0, 1)$ and $Y \sim \chi_n^2$, with Z and Y independent, then $t = Z/\sqrt{Y/n} \sim t_n$. The noncentral version of the Student's t distribution exists as well.

R functions to obtain density, cumulative distribution function, quantiles and random numbers for the Student's t random variable are of the form [dpqr]t.

2.3.3.10   Cauchy-Lorentz random variable

The Cauchy or Lorentz random variable Cauchy(γ, δ) has a continuous distribution with two parameters γ and δ and density

$$f(x) = \frac{1}{\pi} \frac{\gamma}{\gamma^2 + (x - \delta)^2}, \quad x \in \mathbb{R},$$

and cumulative distribution function

$$F_X(x) = \frac{1}{\pi} \arctan \left(\frac{x - \delta}{\gamma} \right) + \frac{1}{2}.$$

This distribution is characterized by the fact that all its moments are infinite but the mode and the median are both equal to δ. Its characteristic function is given by

$$\varphi(t) = \mathbb{E}\left\{ e^{itX} \right\} = \exp\left\{ \delta it - \gamma |t| \right\}.$$

If X and Y are two independent standard Gaussian random variables, then the ratio X/Y is a Cauchy random variable with parameters $(0, 1)$. The Student's t distribution with $n = 1$ degrees of freedom is again the Cauchy random variable of parameters $(0, 1)$.

R functions to obtain density, cumulative distribution function, quantiles and random numbers for the Cauchy random variable are of the form [dpqr]cauchy.

2.3.3.11   Beta random variable

The Beta random variable, $X \sim \text{Beta}(\alpha, \beta)$, $\alpha, \beta > 0$, has a continuous distribution with support in $[0, 1]$ and density

$$f(x) = \frac{\Gamma(\alpha + \beta)}{\Gamma(\alpha)\Gamma(\beta)} x^{\alpha-1}(1 - x)^{\beta-1}, \quad 0 < x < 1.$$

If $X \sim \text{Beta}(\alpha, \beta)$, then

$$\mathbb{E}(X) = \frac{\alpha}{\alpha + \beta}, \quad \text{Var}(X) = \frac{\alpha\beta}{(\alpha + \beta)^2(\alpha + \beta + 1)},$$

while the characteristic function is expressed using series expansion formulas. The name of this distribution comes from the fact that the normalizing constant is the so-called Beta function

$$\text{Beta}(\alpha, \beta) = \frac{\Gamma(\alpha)\Gamma(\beta)}{\Gamma(\alpha + \beta)}.$$

A special case of this distribution is the Beta$(1, 1)$ which corresponds to the uniform distribution U$(0, 1)$. This distribution has a compact support with a density that can be flat, concave U-shaped or convex U-shaped, symmetric (for $\alpha = \beta$) or asymmetric, depending on the choice of the parameters α and β.

R functions to obtain density, cumulative distribution function, quantiles and random numbers for the Beta random variable are of the form [dpqr]beta.

2.3.3.12   The log-normal random variable

The log-Normal random variable, $X \sim \log N(\mu, \sigma^2)$, has a continuous distribution with two parameters μ and σ^2 and density

$$f(x) = \frac{1}{x\sqrt{2\pi\sigma^2}}e^{-\frac{(\log x - \mu)^2}{2\sigma^2}}, \quad x > 0.$$

Its cumulative distribution function is given by

$$F_X(x) = \Phi\left(\frac{\log x - \mu}{\sigma}\right),$$

where $\Phi(\cdot)$ is the cumulative distribution function of the standard Gaussian random variable. The log-Normal distribution is sometimes called the Galton's distribution. This random variable is called log-Normal because, if $X \sim N(\mu, \sigma^2)$ then $Y = \exp\{X\} \sim \log N(\mu, \sigma^2)$. Its characteristic function $\varphi(t)$ exists if $\text{Im}(t) \leq 0$. The moments of the log-Normal distributions are given by the formula

$$\mathbb{E}(X^k) = e^{k\mu + \frac{1}{2}k^2\sigma^2}$$

and, in particular, we have

$$\mathbb{E}(X) = e^{\mu + \frac{\sigma^2}{2}}, \quad \text{Var}(X) = \left(e^{\sigma^2} - 1\right)e^{2\mu + \sigma^2}$$

and

$$\mu = \log \mathbb{E}(X) - \frac{1}{2}\log\left(1 + \frac{\text{Var}(X)}{\mathbb{E}(X^2)}\right), \quad \sigma^2 = \log\left(1 + \frac{\text{Var}(X)}{\mathbb{E}(X^2)}\right).$$

The log-Normal takes a particular role in the Black and Scholes model presented in Chapter 6. In particular, the next result plays a role in the price formula of European call options (see Section 6.2.1):

$$\mathbb{E}\left(X\mathbf{1}_{\{X > k\}}\right) = \int_k^\infty xf(x)\mathrm{d}x = e^{\mu + \frac{1}{2}\sigma^2}\Phi\left(\frac{\mu + \sigma^2 - \log k}{\sigma}\right).$$

R functions to obtain density, cumulative distribution function, quantiles and random numbers for the log-normal random variable are of the form: [dpqr]lnorm.

2.3.3.13   Normal inverse Gaussian

The Normal Inverse Gaussian distribution, or NIG$(\alpha, \beta, \delta, \mu)$, was introduced in finance by Barndorff-Nielsen (1997). Its density has the form:

$$f(x) = \frac{\alpha}{\pi} \exp\left\{\delta\sqrt{\alpha^2 - \beta^2} + \beta(x - \mu)\right\} \frac{K_1\left(\alpha\delta\sqrt{1 + \left(\frac{x-\mu}{\delta}\right)^2}\right)}{\sqrt{1 + \left(\frac{x-\mu}{\delta}\right)^2}}, \quad x \in \mathbb{R},$$

(2.12)

where K_1 denotes the Bessel function of the third kind with index 1 (Abramowitz and Stegun 1964). In this distribution μ is a location parameter, α represents heaviness of the tails, β the asymmetry and δ the scale parameter. The characteristic function of this distribution has the following simple form:

$$\varphi_X(t) = \mathbb{E}\left\{e^{itX}\right\} = e^{it\mu} \frac{\exp\{\delta\sqrt{\alpha^2 - \beta^2}\}}{\exp\{\delta\sqrt{\alpha^2 - (\beta + it)^2}\}}.$$

If $X \sim \text{NIG}(\alpha, \beta, \delta, \mu)$, then

$$\mathbb{E}(X) = \mu + \frac{\beta\delta}{\sqrt{\alpha^2 - \beta^2}}, \quad \text{Var}(X) = \frac{\delta}{\sqrt{\alpha^2 - \beta^2}} + \frac{\beta^2\delta}{\left(\alpha^2 - \beta^2\right)^{\frac{3}{2}}}.$$

R functions to obtain density, cumulative distribution function, quantiles and random numbers for the normal inverse Gaussian random variable are available in package **fBasics** and are of the form [dpqr]nig.

2.3.3.14   Generalized hyperbolic distribution

The Generalized Hyperbolic distribution, or GH$(\alpha, \beta, \delta, \mu, \lambda)$, was introduced in Eberlein and Prause (2002) as a generalization of the hyperbolic distribution. As special cases, it includes the normal inverse Gaussian law for $\lambda = \frac{1}{2}$ and the hyperbolic distribution for $\lambda = 1$ (see, Barndorff-Nielsen (1977)). Its density has the form:

$$f(x) = c(\lambda, \alpha, \beta, \delta)(\delta^2 + (x - \mu)^2)^{\frac{1}{2}\left(\lambda - \frac{1}{2}\right)} K_{\lambda - \frac{1}{2}}$$

$$\times \left(\alpha\sqrt{\delta^2 + (x - \mu)^2}\right) \exp\{\beta(x - \mu)\}, \quad x \in \mathbb{R},$$

(2.13)

with

$$c(\lambda, \alpha, \beta, \delta) = \frac{(\alpha^2 - \beta^2)^{\frac{\lambda}{2}}}{\sqrt{2\pi}\alpha^{\lambda - \frac{1}{2}} K_\lambda\left(\delta\sqrt{\alpha^2 - \beta^2}\right)}$$

and K_λ is the Bessel function of the third kind of index λ (Abramowitz and Stegun 1964). The parameters α, β, δ and μ have the same interpretation as in the NIG distribution. The characteristic function has the form:

$$\varphi_X(t) = \mathbb{E}\left\{e^{itX}\right\} = e^{it\mu} \left(\frac{\alpha^2 - \beta^2}{\alpha^2 - (\beta + it)^2}\right)^{\frac{\lambda}{2}} \frac{K_\lambda\left(\delta\sqrt{\alpha^2 - (\beta + it)^2}\right)}{K_\lambda\left(\delta\sqrt{\alpha^2 - \beta^2}\right)}.$$

The mean and the variance of $X \sim \text{GH}(\alpha, \beta, \delta, \mu, \lambda)$ are given by the two formulas:

$$\mathbb{E}(X) = \mu + \frac{\beta\delta^2}{\gamma} \frac{K_{\lambda+1}(\gamma)}{K_\lambda(\gamma)}$$

and

$$\text{Var}(X) = \frac{\delta^2}{\gamma} \frac{K_{\lambda+1}(\gamma)}{K_\lambda(\gamma)} + \frac{\beta^2\delta^4}{\gamma^2}\left(\frac{K_{\lambda+2}(\gamma)}{K_\lambda(\gamma)} - \frac{K_{\lambda+1}^2(\gamma)}{K_\lambda^2(\gamma)}\right),$$

with $\gamma = \delta\sqrt{\alpha^2 - \beta^2}$.

R functions to obtain density, cumulative distribution function; quantiles and random numbers for the generalized hyperbolic random variable are available in package **fBasics** and are of the form [dpqr]gh.

2.3.3.15 Meixner distribution

The Meixner distribution originates from the theory of orthogonal polynomials and was suggested as a model for financial data returns in Grigelionis (1999) and Schoutens and Teugels (1998). The density of the Meixner(α, β, δ) distribution is given by

$$f(x) = \frac{\left\{2\cos\left(\frac{\beta}{2}\right)\right\}^{2d}}{2\alpha\pi\,\Gamma(2d)} \exp\left(\frac{\beta x}{\alpha}\right)\left|\Gamma\left(\delta + \frac{ix}{\alpha}\right)\right|^2, \quad x \in \mathbb{R}, \tag{2.14}$$

with $\alpha > 0$, $\beta \in (-\pi, \pi)$ and $\delta > 0$. The characteristic function has the following simple form:

$$\varphi_X(t) = \mathbb{E}\left\{e^{itX}\right\} = \left(\frac{\cos\left(\frac{\beta}{2}\right)}{\cosh\left(\frac{\alpha t - i\beta}{2}\right)}\right)^{2d}.$$

For $X \sim \text{Meixner}(\alpha, \beta, \delta)$ we have

$$\mathbb{E}(X) = \alpha\delta\tan\left(\frac{\beta}{2}\right), \quad \text{Var}(X) = \frac{\alpha^2\delta}{2\cos^2\left(\frac{\beta}{2}\right)}.$$

2.3.3.16 Multivariate Gaussian random variable

A random vector $X = (X_1, X_2, \ldots, X_n)^\top$ follows a multivariate Gaussian distribution with vector mean $\mu = (\mu_1, \mu_2, \ldots, \mu_n)^\top$ and variance-covariance matrix

Σ if the joint density of $X = (X_1, \ldots, X_n)^\top$ has the following form:

$$f(x_1, x_2, \ldots, x_n) = \frac{1}{(2\pi)^{\frac{n}{2}} |\Sigma|^{\frac{1}{2}}} \exp\left\{-\frac{1}{2}(x - \mu)^\top \Sigma^{-1}(x - \mu)\right\},$$

with $x = (x_1, x_2, \ldots, x_n)$. In the above $|\Sigma|$ is the determinant of the $n \times n$ positive semi-definite matrix Σ and Σ^{-1} is the inverse matrix of Σ. For this random variable, $\mathbb{E}(X) = \mu$ and $\text{Var}(X) = \Sigma$ and we write $X \sim \text{N}(\mu, \Sigma)$. Its characteristic function is given by the formula

$$\varphi_X(t) = \exp\left\{i\mu^\top t - \frac{1}{2}t^\top \Sigma t\right\}, \quad t \in \mathbb{R}^n.$$

The following two sets of conditions are equivalent to the above:

(i) if every linear combination $Y = a_1 X_1 + a_2 X_2 + a_n X_n$ of $(X_1, X_2, \ldots, X_n)$ is Gaussian distributed, then X is a multivariate normal;

(ii) let $Z = (Z_1, \ldots, Z_m)^\top$ be a vector of independent normal random variables $\text{N}(0, 1)$ and let $\mu = (\mu_1, \ldots, \mu_n)^\top$ be some vector, with A and $n \times m$ matrix. Then $X = AZ + \mu$ is a multivariate Gaussian random variable.

Consider now a two-dimensional Gaussian vector with mean $\mu = (0, 0)^\top$ and variance-covariance matrix

$$\Sigma = \begin{bmatrix} \sigma_1^2 & \sigma_{12} \\ \sigma_{21} & \sigma_2^2 \end{bmatrix} = \begin{bmatrix} \sigma_1^2 & \rho\sigma_1\sigma_2 \\ \rho\sigma_1\sigma_2 & \sigma_2^2 \end{bmatrix}$$

where $\rho = \text{Cov}(X_1, X_2)/(\sqrt{\text{Var}X_1}\sqrt{\text{Var}X_2})$ is the correlation coefficient between the two components X_1 and X_2. We have

$$f(x_1, x_2) = \frac{1}{2\pi\sigma_1\sigma_2\sqrt{1 - \rho^2}} \exp\left\{-\frac{1}{2(1 - \rho^2)}\left(\frac{x_1^2}{\sigma_1^2} + \frac{x_2^2}{\sigma_2^2} - \frac{2\rho x_1 x_2}{\sigma_1\sigma_2}\right)\right\}.$$

From the above formula is clear that when $\text{Cov}(X_1, X_2) = 0$ (and hence $\rho = 0$) the joint density $f(x_1, x_2)$ factorizes into the product of the two densities of X_1 and X_2, i.e. this condition is enough to deduce that X_1 and X_2 are also independent.

We have seen in Exercise 2.5 that in general null correlation does not imply independence and we should also remark that what we have just shown is different from saying that *any* two Gaussian random variables with null correlation are always independent. What the previous result shows is that, if X and Y are Gaussian and their joint distribution is *also* Gaussian, then null correlation implies independence. Indeed, consider the following example by Melnick and Tenenbein (1982): take $X \sim \text{N}(0, 1)$ and let $Y = -X$ if $|X| < c$ and $Y = X$ if $|X| > c$, where $c > 0$ is some constant. Looking at the definition of Y we see that if c is very small, than Y is essentially equal to X so we expect high positive correlation; the converse is true if c is very large (negative correlation). So there will be some value of c such that the correlation between X and Y is exactly zero. We now show that Y is also Gaussian.

$$P(Y \le x) = P(\{(|X| < c) \cap (-X < x)\} \cup \{(|X| > c) \cap (X < x)\})$$
$$= P((|X| < c) \cap (-X < x)) + P((|X| > c) \cap (X < x))$$

now, by the symmetry of X, $P(-X < x) = P(X > -x) = P(X < x)$. Therefore we conclude that $P(Y \le x) = P(X \le x)$. So Y and X are both Gaussian (actually they have the same law), they may have null correlation but they are not independent.

In general, it is useful to know that the components of a multivariate Gaussian random vector are mutually independent if and only if the variance-covariance matrix is diagonal.

The R function `mvrnorm` in package **MASS** or the function `rmvnorm` in package **mvtnorm** can be used to obtain random numbers from the multivariate normal distribution. Moreover, the package **mvtnorm** also implements `[dpq]mvnorm` functions to obtain cumulative distribution function, density function and quantiles.

2.3.4   Sum of random variables

Although it is easy to derive the expected value of the sum of two random variables X and Y, it is less obvious to derive the distribution of $Z = X + Y$. But if X and Y are independent, then the probability measure of Z is the *convolution* of the probability measures of X and Y. Consider X and Y discrete random variables taking arbitrary integer values, then $Z = X + Y$ can take also integer values. When $X = k$ then $Z = z$ if and only if $Y = z - k$, hence the probability of the event $Z = z$ is the sum, over all possible values of k, of the probabilities of events $(X = k) \cap (Y = z - k)$. Given that X and Y are independent, we have

$$P(Z = z) = \sum_k P(X = k) P(Y = z - k).$$

If X and Y have density respectively $f_X(\cdot)$ and $f_Y(\cdot)$ with support in $\mathbb{R}$, then, by the same reasoning, the density of $Z = X + Y$ is given by

$$f_Z(z) = \int_{\mathbb{R}} f_X(z - y) f_Y(y) \mathrm{d}y = \int_{\mathbb{R}} f_Y(z - x) f_X(z) \mathrm{d}x \qquad (2.15)$$

Sometimes the convolution is denoted by $f_Z(z) = (f_X * f_Y)(z)$.

Example 2.3.10 (Sum of uniform random variables) *Let X and Y be two independent uniform random variables $U(0, 1)$ and consider $Z = X + Y$. We now* *apply* (2.15). *First notice that z can take any value in* $[0, 2]$. *Then*

$$f_Z(z) = \int_{\mathbb{R}} f_X(z-y) f_Y(y) dy = \int_0^1 f_X(z-y) dy.$$

But $f_X(z-y) \neq 0$ *only if* $0 \leq z - y \leq 1$, *i.e.* $z - 1 \leq y \leq z$. *We now split the range of z into* $[0, 1]$ *and* $[1, 2]$. *So, if* $z \in [0, 1]$ *we have*

$$f_Z(z) = \int_{z-1}^z 1 dy = \int_0^z 1 dy = z, \quad 0 \leq z \leq 1$$

and, if $1 < z \leq 2$

$$f_Z(z) = \int_{z-1}^z 1 dy = \int_{z-1}^1 dy = 2 - z, \quad 1 < z \leq 2,$$

and $f_Z(z) = 0$ *if* $z < 0$ *and* $z > 2$. *Then*

$$f_Z(z) = \begin{cases} z, & 0 \leq z \leq 1, \\ 2 - z, & 1 < z \leq 2, \\ 0, & otherwise. \end{cases}$$

Exercise 2.12 (Sum of exponential random variables) *Let* $X \sim \text{Exp}(\lambda)$ *and* $Y \sim \text{Exp}(\lambda)$, $\lambda > 0$, *be two independent random variables. Find the density of* $Z = X + Y$.

Exercise 2.13 (Sum of Gaussian random variables) *Let* $X \sim \text{N}(\mu_1, \sigma_1^2)$ *and* $Y \sim \text{N}(\mu_2, \sigma_2^2)$ *be two independent random variables. Prove that* $Z = X + Y \sim \text{N}(\mu_1 + \mu_2, \sigma_1^2 + \sigma_2^2)$.

Theorem 2.3.11 *Let* $X_1, X_2, \ldots, X_n$ *be n independent Gaussian random variables respectively with means* μ_i *and variances* σ_i^2, $i = 1, \ldots, n$. *Then,*

$$Z = \sum_{i=1}^n X_i \sim \text{N}\left(\sum_{i=1}^n \mu_i, \sum_{i=1}^n \sigma_i^2\right).$$

We now enumerate few special cases of sum of random variables without proof, though in most cases playing with the characteristic function is enough to obtain the results.

  (i) the NIG distribution is closed under convolution in the following sense, i.e. if $X \sim \text{NIG}(\alpha, \beta, \delta_1, \mu_1)$ and $Y \sim \text{NIG}(\alpha, \beta, \delta_2, \mu_2)$ are two independent random variables, then

$$X + Y \sim \text{NIG}(\alpha, \beta, \delta_1 + \delta_2, \mu_1 + \mu_2);$$

(ii) if $X_i \sim \text{Gamma}(\alpha_i, \beta)$, $i = 1, \ldots, n$, is a sequence of independent Gamma random variables, then

$$X_1 + X_2 + \cdots + X_n \sim \text{Gamma}(\alpha_1 + \cdots + \alpha_n, \beta);$$

(iii) if $X_i \sim N(\mu_i, \sigma_i^2)$, $i = 1, \ldots, n$, is a sequence of independent Gaussian random variables, then

$$X_1^2 + X_2^2 + \cdots + X_n^2 \sim \chi_n^2(\delta)$$

where $\chi_n^2(\delta)$ is the noncentral Chi-square random variable with n degrees of freedom and noncentral parameter δ given by

$$\delta = \sum_{i=1}^{n} \left(\frac{\mu_i}{\sigma_i^2} \right)^2.$$

2.3.5 Infinitely divisible distributions

Let X be a random variable with distribution function $F_X(x)$ and characteristic function $\phi_X(u)$. We now introduce the notion of infinitely divisible laws which represent a class of random variables whose distributions are closed with respect to the convolution operator. This property will be particularly useful in financial applications in relation to Lévy processes.

Definition 2.3.12 *The distribution F_X is infinitely divisible if, for all $n \in \mathbb{N}$, there exist i.i.d. random variables, say $X_1^{(1/n)}$, $X_2^{(1/n)}$, $\ldots$, $X_n^{(1/n)}$, such that*

$$X \sim X_1^{(1/n)} + X_2^{(1/n)} + \cdots + X_n^{(1/n)}.$$

Equivalently, F_X is infinitely divisible if, for all $n \in \mathbb{N}$, there exists another law $F_{X^{(1/n)}}$ such that

$$F_X(x) = F_{X^{(1/n)}} * F_{X^{(1/n)}} * \cdots * F_{X^{(1/n)}},$$

i.e. $F_X(x)$ is the n-times convolution of $F_{X^{(1/n)}}$.

Not all random variables are infinitely divisible, but some notable cases are. We show here a few examples because this property come from very easy manipulation of the characteristic functions. Indeed, we will make use of the following characterization: is X has an infinitely divisible law, then

$$\varphi_X(u) = \left(\varphi_{X^{(1/n)}}(u) \right)^n.$$

Example 2.3.13 (Gaussian case) *The law of $X \sim N(\mu, \sigma^2)$ is infinitely divisible. Indeed,*

$$\varphi_X(u) = \exp\left\{iu\mu - \frac{1}{2}u^2\sigma^2\right\} = \exp\left\{n\left(\frac{iu\mu}{n} - \frac{1}{2}\frac{u^2\sigma^2}{n}\right)\right\}$$

$$= \left(\exp\left\{\frac{iu\mu}{n} - \frac{1}{2}\frac{u^2\sigma^2}{n}\right\}\right)^n = \left(\varphi_{X^{(1/n)}}(u)\right)^n$$

with $X^{(1/n)} \sim N\left(\frac{\mu}{n}, \frac{\sigma^2}{n}\right)$.

Exercise 2.14 (Poisson case) *Prove that the law of $X \sim \text{Poi}(\lambda)$ is infinitely divisible.*

We have seen in Example 2.3.10 that the sum of two uniform random variables, is not a rescaled uniform distribution (rather a triangular shaped distribution). Similarly, it is possible to prove that the Binomial distribution is not infinitely divisible, while the Gamma, the exponential, the Meixner and many others are.

The following result (see e.g. Sato (1999), Lemma 7.8) characterizes infinitely divisible laws in terms of the so-called characteristic triplet.

Theorem 2.3.14 *The law of a random variable X is infinitely divisible if and only if there exists a triplet (b, c, v), with $b \in \mathbb{R}$, $c \geq 0$ and a measure $v(\cdot)$ satisfying $v(\{0\}) = 0$ and $\int_{\mathbb{R}} (1 \wedge |x|^2) v(dx) < \infty$, such that*

$$\mathbb{E}\left\{e^{iuX}\right\} = \exp\left\{ibu - \frac{u^2 c}{2} + \int_{\mathbb{R}} \left(e^{iux} - 1 - iux\mathbf{1}_{\{|x|<1\}}\right) v(dx)\right\}. \tag{2.16}$$

Equation (2.16) is also known as the Lévy-Khintchine formula. Notice that the *characteristic triplet* for $X \sim N(\mu, \sigma^2)$ is $(b = \mu, c = \sigma^2, v)$ where $v(A) = 0$ for all $A \subset \mathbb{R}$. For $X \sim \text{Poi}(\lambda)$, we can take $v(\{x\}) = \delta(x - 1)\lambda$, where δ is the Dirac delta function. Therefore, the characteristic triplet is $(b = 0, c = 0, v)$. The exponent in Equation (2.16)

$$\psi(u) = ibu - \frac{u^2 c}{2} + \int_{\mathbb{R}} \left(e^{iux} - 1 - iux\mathbf{1}_{\{|x|<1\}}\right) v(dx),$$

is called *Lévy* or *characteristic exponent*.

2.3.6 Stable laws

Also random variables with stable law are often used in finance jointly with Lévy processes to model asset prices. The notion of stable law emerged in the study of the distribution of the sum of random variables and indeed, stability preserves the family of distributions with respect to sum. A very recent account on stable distribution and their applications is Nolan (2010).

Definition 2.3.15 *A random variable X is stable or stable in the broad sense if for X_1 and X_2 independent copies of X and positive constants a and b we have*

$$aX_1 + bX_2 \sim cX + d \tag{2.17}$$

for some positive c and some $d \in \mathbb{R}$. The random variable is strictly stable or stable in the narrow sense if (2.17) holds with $d = 0$. A random variable is symmetrically stable if it is stable and symmetrically distributed around 0, i.e. $-X \sim X$.

Clearly, the Gaussian law is stable.

Example 2.3.16 (Gaussian case) *let X_1 and X_2 be two independent copies of $X \sim N(\mu, \sigma^2)$, then*

$$aX_1 + bX_2 \sim N((a+b)\mu, (a^2+b^2)\sigma^2) \sim cX + d$$

with $c^2 = a^2 + b^2$ and $d = \mu(a + b - c)$.

There are other two distributions which admit a closed form of the convolution which are also stable, the Cauchy distribution and the Lévy distribution Lévy(γ, δ) with two parameters $\gamma > 0$ and δ and density

$$f(x) = \sqrt{\frac{\gamma}{2\pi}} \frac{1}{(x-\delta)^{\frac{3}{2}}} \exp\left(-\frac{\gamma}{2(x-\delta)}\right), \quad \delta < x < \infty.$$

There are other equivalent definitions of stable laws which are useful. The first one extends the definition to the sum of n random variables.

Definition 2.3.17 *A nondegenerate random variable X is stable if and only if for all $n > 1$, there exist constants $c_n > 0$ and $d_n \in \mathbb{R}$ such that*

$$X_1 + \cdots + X_n \sim c_n X + d_n,$$

where $X_1, \ldots, X_n$ are i.i.d. as X. X is strictly stable if and only if, it is stable and for all $n > 1$, $d_n = 0$.

The only possible choice of the scaling constant c_n is $c_n = n^{\frac{1}{\alpha}}$, for some $\alpha \in (0, 2]$, see Nolan (2010). We have already seen the central role of the characteristic function in the analysis of the sum of independent random variables. An equivalent definition is based indeed on the following construction.

Definition 2.3.18 *A random variable X is stable if and only if $X \sim aZ + b$, where $\gamma > 0$, $\delta \in \mathbb{R}$ and Z is a random variable with parameters (α, β), $-1 \leq \beta \leq 1$, $\alpha \in (0, 2]$, such that*

$$\varphi_Z(t) = \begin{cases} \exp\left\{-|t|^\alpha \left(1 - i\beta \tan\left(\alpha\frac{\pi}{2}\right) \operatorname{sgn}(u)\right)\right\}, & \alpha \neq 1, \\ \exp\left\{-|t| \left(1 + i\beta\frac{2}{\pi}\operatorname{sgn}(u) \log|u|\right)\right\}, & \alpha = 1 \end{cases} \tag{2.18}$$

and hence

$$\varphi_X(t) = \begin{cases} \exp\left\{itb - |at|^\alpha\left(1 - i\beta\tan\left(\alpha\frac{\pi}{2}\right)\mathrm{sgn}(u)\right)\right\}, & \alpha \neq 1, \\ \exp\left\{itb - |at|\left(1 + i\beta\frac{2}{\pi}\mathrm{sgn}(u)\log|u|\right)\right\}, & \alpha = 1. \end{cases}$$

In the above, when $\alpha = 1$, $0\log(0)$ is taken as 0 and $\mathrm{sgn}(\cdot)$ is the sign function, i.e.

$$\mathrm{sgn}(x) = \begin{cases} -1, & x < 0, \\ 0, & x = 0, \\ 1, & x > 0. \end{cases}$$

These random variables will be denoted by $S(\alpha, \beta, \gamma, \delta)$, and they have a symmetric distribution when $\beta = 0$ and $\delta = 0$, so that the characteristic function of Z in the definition takes the form:

$$\varphi_z(t) = e^{-\gamma^\alpha|u|^\alpha}.$$

It is possible to check that

$$S\left(\alpha = 2, \beta = 0, \gamma = \frac{\sigma}{\sqrt{2}}, \delta = \mu\right) = N(\mu, \sigma^2)$$

$$S\left(\alpha = \frac{1}{2}, \beta = 1, \gamma = \gamma, \delta = \delta\right) = \text{Lévy}(\gamma, \delta)$$

$$S(\alpha = 1, \beta = 0, \gamma = \gamma, \delta = \delta) = \text{Cauchy}(\gamma, \delta)$$

Stable distributions are hence characterized by four parameters, where α is called *index of stability* or *characteristic exponent*, β is the *skewness* parameter, $\gamma > 0$ is the *scale* parameter and $\delta \in \mathbb{R}$ is the *location* parameter. Notice that γ is called scale parameter and not standard deviation because even in the Gaussian case $\gamma = \frac{\sigma}{\sqrt{2}} \neq \sigma$, and the same for δ which is only a location parameter and not the mean (indeed, the Cauchy distribution has no mean). The notation $S(\alpha, \beta, \gamma, \delta)$ is sometimes written as $S(\alpha, \beta, \gamma, \delta; 1)$ where the last '1' denotes one kind of parametrization. This means that stable laws can be parametrized in several ways, and one should always check which version is used. In particular, the parametrization of this book, $S(\alpha, \beta, \gamma, \delta) = S(\alpha, \beta, \gamma, \delta; 1)$, coincides with

$$X \sim \begin{cases} \gamma Z + \delta, & \alpha \neq 1, \\ \gamma Z + \delta + \beta\frac{2}{\pi}\gamma\log\gamma, & \alpha = 1 \end{cases}$$

and the parametrization $S(\alpha, \beta, \gamma, \delta; 0)$ with

$$X \sim \begin{cases} \gamma\left(Z - \beta\tan\left(\alpha\frac{\pi}{2}\right)\right) + \delta, & \alpha \neq 1, \\ \gamma Z + \delta, & \alpha = 1, \end{cases}$$

with Z as in (2.18).

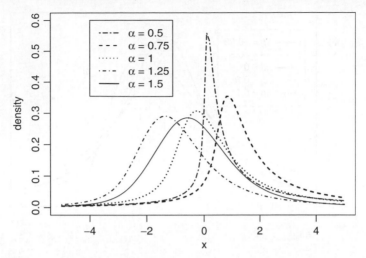

Figure 2.1   Shape of stable distributions $S(\alpha, \beta = 0.5, \gamma = 1, \delta = 0)$, *for* $\alpha = 0.5, 0.75, 1, 1.25, 1.5$.

R functions to obtain density and cumulative distribution function; quantiles and random numbers for stable random variables are of the form `[dpqr]stable` and available in the contributed R package called `fBasics`. The functions allow for several parametrizations of stable distributions including the previous ones. The next code produces the plot in Figure 2.1 using the function `dstable` which by default sets `gamma = 1` and `delta = 0`. The argument `pm` denotes the parametrization in use. The plot shows the different shapes assumed by the stable distribution as a function of α for the stable law $S(\alpha, \beta = 0.5, \gamma = 1, \delta = 0)$.

```
R> require(fBasics)
R> x <- seq(-10, 10, length = 500)
R> y1 <- function(x) dstable(x, alpha = 0.5, beta = 0.5, pm = 1)
R> y2 <- function(x) dstable(x, alpha = 0.75, beta = 0.5, pm = 1)
R> y3 <- function(x) dstable(x, alpha = 1, beta = 0.5, pm = 1)
R> y4 <- function(x) dstable(x, alpha = 1.25, beta = 0.5, pm = 1)
R> y5 <- function(x) dstable(x, alpha = 1.5, beta = 0.5, pm = 1)
R> curve(y1, -5, 5, lty = 6, ylim = c(0, 0.6), ylab = "density")
R> curve(y2, -5, 5, lty = 2, add = TRUE)
R> curve(y3, -5, 5, lty = 3, add = TRUE)
R> curve(y4, -5, 5, lty = 4, add = TRUE)
R> curve(y5, -5, 5, lty = 1, add = TRUE)
R> legend(-4, 0.6, legend = c(expression(alpha == 0.5),
    expression(alpha ==
+       0.75), expression(alpha == 1), expression(alpha == 1.25),
+       expression(alpha == 1.5)), lty = c(6, 2, 3, 4, 1))
```

Similarly, next code shows the skewness of the stable distribution $S(\alpha = 1, \beta, \gamma = 1, \delta = 0)$ as a function of $\beta = -0.99, -0.3, 0, 0.5, 0.7$. The resulting plot is given in Figure 2.2.

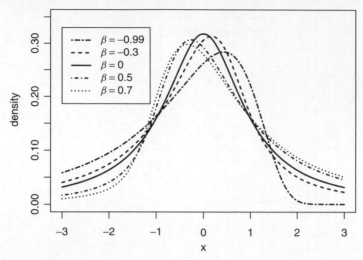

Figure 2.2 Shape of stable distributions $S(\alpha = 1, \beta, \gamma = 1, \delta = 0)$, for $\beta = -0.99, -0.3, 0, 0.5, 0.7$.

```
R> require(fBasics)
R> x <- seq(-10, 10, length = 500)
R> y1 <- function(x) dstable(x, alpha = 1, beta = -0.99, pm = 1)
R> y2 <- function(x) dstable(x, alpha = 1, beta = -0.3, pm = 1)
R> y3 <- function(x) dstable(x, alpha = 1, beta = 0, pm = 1)
R> y4 <- function(x) dstable(x, alpha = 1, beta = 0.5, pm = 1)
R> y5 <- function(x) dstable(x, alpha = 1, beta = 0.7, pm = 1)
R> curve(y1, -3, 3, lty = 6, ylim = c(0, 0.35), ylab = "density")
R> curve(y2, -3, 3, lty = 2, add = TRUE)
R> curve(y3, -3, 3, lty = 1, add = TRUE)
R> curve(y4, -3, 3, lty = 4, add = TRUE)
R> curve(y5, -3, 3, lty = 3, add = TRUE)
R> legend(-3, 0.33, legend = c(expression(beta == -0.99),
    expression(beta ==
+      -0.3), expression(beta == 0), expression(beta == 0.5),
    expression(beta ==
+      0.7)), lty = c(6, 2, 1, 4, 3))
```

2.3.7 Fast Fourier Transform

We have seen the central role of characteristic functions and we know that each random variable has one and only one characteristic function and vice versa. The characteristic function of random variables X with density $f(\cdot)$ can be expressed as

$$\varphi(t) = \mathbb{E}\left\{e^{itX}\right\} = \int_{-\infty}^{\infty} f(x)e^{itx}\,\mathrm{d}x. \qquad (2.19)$$

It is always possible to recover the density or the distribution function of a random variable from its characteristic function as the following theorem shows (see e.g. Kendall and Stuart 1977).

Theorem 2.3.19 (Inversion theorem) *Let $\varphi(t)$ be the characteristic function of a random variable with distribution function $F(x)$ and density function $f(x)$. Then,*

$$F(x) = \frac{1}{2} - \frac{1}{2\pi} \int_{-\infty}^{\infty} \frac{e^{-itx}\varphi(t)}{it}\,dt = F(0) - \frac{1}{2\pi} \int_{-\infty}^{\infty} \frac{e^{-itx} - 1}{it}\varphi(t)\,dt$$

and

$$f(x) = \frac{1}{2\pi} \int_{-\infty}^{\infty} e^{-itx}\varphi(t)\,dt. \tag{2.20}$$

Although the previous theorem established a direct link between $\varphi(\cdot)$, $F(\cdot)$ and $f(\cdot)$, closed form results are rarely obtained and in most cases the solution is obtained by numerical methods. In particular, numerical approximation of the integral (2.20) has to be calculated by some quadrature formula which is based on the discretization along the integration variable t, for each given x. So to obtain the shape of the function $f(x)$ in (2.20) one also needs to discretize the x axis. Assuming that N points are chosen for the x grid and N for the t grid, this numerical problem requires at least N^2 operations to be performed. In this respect, one of the most important advances in numerical analysis was the Fast Fourier Transform (FFT) algorithm by Cooley and Tukey (1965). Although there is no need to know the details, it is worth understanding that the merit of this algorithm is to reduce the computational burden from order N^2 to order $N \log_2(N)$, which is the reason for the adjective 'fast'. Indeed, if $N = 1000$, then $N^2 = 1,000,000$ but $N \log_2(N) = 9,965.784$. This algorithm allows us to calculate a discrete version of the Fourier transform or the inverse of a given discretized Fourier transform. For a given vector of complex numbers x_n, $n = 1, \ldots, N$, the FFT algorithm efficiently calculates the discrete Fourier transform

$$X_k = \sum_{n=1}^{N} x_n e^{-i\frac{2\pi}{N}(k-1)(n-1)}, \quad \text{for} \quad k = 1, \ldots, N \tag{2.21}$$

and, starting from the sequence X_k, the values of the inverse of the discrete Fourier transform

$$x_n = \frac{1}{N} \sum_{k=1}^{N} X_k e^{i\frac{2\pi}{N}(k-1)(n-1)}, \quad \text{for} \quad n = 1, \ldots, N. \tag{2.22}$$

The expression in (2.22) is close to a quadrature formula for the integral (2.19) and (2.21) is close to the quadrature of (2.20). In order to see exact correspondence between the formulas, we need to manipulate them, but first we notice the (direct) FFT corresponds to the inversion formula for the characteristic function (2.20), while inverse FFT corresponds to the characteristic function (2.19). Let $g(t) = e^{-itx}\varphi(t)$, then

$$g(-t) = e^{itx}\varphi(-t) = \overline{e^{itx}\varphi(t)} \quad \text{and} \quad g(t) + g(-t) = 2\mathrm{Re}\left\{e^{-itx}\varphi(t)\right\}.$$

We can rewrite (2.20) as follows:

$$f(x) = \frac{1}{2\pi}\int_{-\infty}^{\infty} e^{-itx}\varphi(t)\mathrm{d}t = \frac{1}{2\pi}\int_{-\infty}^{0} g(t)\mathrm{d}t + \frac{1}{2\pi}\int_{0}^{\infty} g(t)\mathrm{d}t$$

$$= \mathrm{Re}\left\{\frac{1}{\pi}\int_{0}^{\infty} e^{-itx}\varphi(t)\mathrm{d}t\right\}.$$

We now discretize the last integral using a grid of points $t_n = \Delta_t(n-1)$, $n = 1,\ldots,N$, so that in practice we evaluate the integral on the interval $[0, T = t_N = (N-1)\Delta_t]$ instead of the interval $(0, +\infty)$. If N is relatively big, then the truncation of the integral will not affect the approximation too much. We obtain the following approximation:

$$f(x) \approx \mathrm{Re}\left\{\frac{1}{\pi}\sum_{n=1}^{N} e^{-it_n x}\varphi(t_n)\Delta_t\right\}.$$

The value of Δ_t, which is crucial, will be specified later. Now, assume that the function $f(x)$ has a finite support $[x_{\min}, x_{\max}]$ and we set $x_{\min} = 0$ to simplify the exposition. Then take $\Delta_x = (x_{\max} - x_{\min})/(N-1)$ and set $x_k = x_{\min} + \Delta_x(k-1) = \Delta_x(k-1)$, for $k = 1,\ldots,N$. Therefore, for each x_k we have

$$f(x_k) \approx \mathrm{Re}\left\{\frac{1}{\pi}\sum_{n=1}^{N} e^{-i(n-1)(k-1)\Delta_t \Delta_x}\varphi(t_n)\Delta_t\right\}.$$

The crucial position is now to impose the condition $\Delta_x\Delta_t = \frac{2\pi}{N}$, which implies

$$\Delta_t = \frac{2\pi}{N}\frac{N-1}{x_{\max} - x_{\min}} \approx \frac{2\pi}{x_{\max} - x_{\min}}$$

for large N. So finally we have

$$f(x_k) \approx \mathrm{Re}\left\{\frac{2}{x_{\max}}\sum_{n=1}^{N} e^{-i\frac{2\pi}{N}(k-1)(n-1)}\varphi(t_n)\right\}.$$

Taking $X_k = f(x_k)$ and $x_n = \varphi(t_n)$ we obtain, up to normalizing constants, a version of (2.21). Similar manipulation allows to express (2.22). All those manipulations are made by the software interface, so now we explain how to execute FFT (2.22) and its inverse (2.21) with R. The algorithm for the FFT is designed to efficiently calculate expressions of the form (2.21). So, if we want to obtain the characteristic function from the FFT transform we need to use the inverse of the FFT. R is not special in this sense because the function fft calculates exactly (2.21), so to obtain the characteristic function we need to use the argument inverse = TRUE (the default being FALSE) in the function fft. Indeed, R just executes the same algorithm fft plugging a sign '+' in the exponential. Notice that the normalizing factor $1/N$ is missing. Assume we take the density $f(\cdot)$ of the standard Gaussian density $N(0, 1)$ and let us calculate the characteristic function on a grid of points over the interval $(-3, 3)$.

```
R> x <- seq(-3, 3, length = 20)
R> f <- function(x) dnorm(x)
R> f(x)
```

```
 [1] 0.004431848 0.010873446 0.024145731 0.048529339 0.088279375
     0.145346632
 [7] 0.216591572 0.292125176 0.356604876 0.394000182 0.394000182
     0.356604876
[13] 0.292125176 0.216591572 0.145346632 0.088279375 0.048529339
     0.024145731
[19] 0.010873446 0.004431848
```

Then, we calculate the characteristic function using the R function ftt

```
R> y <- fft(f(x), inverse = TRUE)
R> y
```

```
 [1]   3.161856e+00+0.000000e+00i -1.911250e+00+3.027123e-01i
 [3]   4.125760e-01-1.340541e-01i -3.528574e-02+1.797898e-02i
 [5]  -6.619828e-04+4.809586e-04i -9.869481e-04+9.869481e-04i
 [7]  -5.609033e-04+7.720172e-04i -2.895818e-04+5.683363e-04i
 [9]  -1.211024e-04+3.727148e-04i -2.920041e-05+1.843641e-04i
[11]  -2.220446e-16+0.000000e+00i -2.920041e-05-1.843641e-04i
[13]  -1.211024e-04-3.727148e-04i -2.895818e-04-5.683363e-04i
[15]  -5.609033e-04-7.720172e-04i -9.869481e-04-9.869481e-04i
[17]  -6.619828e-04-4.809586e-04i -3.528574e-02-1.797898e-02i
[19]   4.125760e-01+1.340541e-01i -1.911250e+00-3.027123e-01i
```

Now y is the output of the FFT algorithm in (2.22) without the normalizing factor $1/N$. Therefore, to get density back from the FFT we proceed as follows:

```
R> invFFT <- as.numeric(fft(y)/length(y))
R> invFFT
```

```
 [1] 0.004431848 0.010873446 0.024145731 0.048529339 0.088279375
     0.145346632
 [7] 0.216591572 0.292125176 0.356604876 0.394000182 0.394000182
     0.356604876
[13] 0.292125176 0.216591572 0.145346632 0.088279375 0.048529339
     0.024145731
[19] 0.010873446 0.004431848
```

where `as.numeric` transform the `complex` vector into one of real numbers by dropping the imaginary part. We will discuss in more depth the use of the inverse Fourier transform as an alternative to the Monte Carlo method, or the exact formulae, for the calculation of option prices, in Section 8.1.5.

2.3.8   Inequalities

There are some fundamental inequalities in the calculus of probability which are often used in proofs so we collect them here as a reference. We will review some of them, providing only hints for the proof of some of them.

Theorem 2.3.20 (Chebyshev's inequality) *Let X be a random variable with expected value $\mathbb{E}(X) = \mu$ and let $\epsilon > 0$ be any positive real number. Then*

$$P(|X - \mu| \geq \epsilon) \leq \frac{\text{Var}(X)}{\epsilon^2}. \tag{2.23}$$

Proof. Remember that we can always write $\text{Var}(X) = \int_{\mathbb{R}} |x - \mu|^2 \mathrm{d}F(x)$. Because this is a sum of positive terms, if we restrict the summation to the set on which $|X - \mu| \geq \epsilon$ we have that

$$\text{Var}(X) \geq \int_{\mathbb{R}} \mathbf{1}_{\{|x-\mu| \geq \epsilon\}} |x - \mu|^2 \mathrm{d}F(x).$$

The proof ends by noting that, on this subset of $\mathbb{R}$, all terms are at least ϵ and we have that

$$\int_{\mathbb{R}} \mathbf{1}_{\{|x-\mu| \geq \epsilon\}} |x - \mu|^2 \mathrm{d}F(x) \geq \epsilon^2 \int_{\mathbb{R}} \mathbf{1}_{\{|x-\mu| \geq \epsilon\}} \mathrm{d}F(x) = \epsilon^2 P(|X - \mu| \geq \epsilon).$$

A generalization of this inequality, called Chebyshev-Markov inequality, is the following:

$$P(|X| \geq \epsilon) \leq \frac{\mathbb{E}|X|^p}{\epsilon^p} \tag{2.24}$$

which is well defined when all quantities exist. Similarly we have the Chebyshev-Cantelli inequality:

$$P(X - \mathbb{E}(X) > \epsilon) \leq \frac{\text{Var}(X)}{\text{Var}(X) + \epsilon^2}.$$

Theorem 2.3.21 (Lyapunov's inequality) *Let X be a random variable with finite moments up to order r. Then*

$$(\mathbb{E}|X|^p)^{\frac{1}{p}} \leq (\mathbb{E}|X|^r)^{\frac{1}{r}}, \quad for \ \ 0 < p < r.$$

Theorem 2.3.22 (Cauchy-Schwarz-Bunyakovsky inequality) *Let X and Y be two square integrable random variables, then*

$$|\mathbb{E}(XY)| \leq \sqrt{\mathbb{E}(X^2)\mathbb{E}(Y^2)}.$$

Theorem 2.3.23 (Hölder inequality) *Let $p, q \in (1, \infty)$ and such that $\frac{1}{p} + \frac{1}{q} = 1$. Let X and Y be two random variables such that $(\mathbb{E}|X|^p)^{\frac{1}{p}} < \infty$ and $(\mathbb{E}|X|^q)^{\frac{1}{q}} < \infty$, then*

$$|\mathbb{E}(XY)| \leq (\mathbb{E}|X|^p)^{\frac{1}{p}} (\mathbb{E}|Y|^q)^{\frac{1}{q}}.$$

Definition 2.3.24 *A function $f(\cdot)$, $f : \mathbb{R} \to \mathbb{R}$, is said to be convex if for any $(x_1, x_2, \ldots, x_n) \in \mathbb{R}^n$ and non-negative constants $a_1, a_2, \ldots, a_n$ then*

$$f\left(\sum_{i=1}^{n} a_i x_i\right) \leq \sum_{i=1}^{n} a_i f(x_i).$$

Theorem 2.3.25 (Jensen's inequality) *Let $f(\cdot)$ be any real-valued convex function on $\mathbb{R}$ and X a random variable with finite expectation. Then*

$$f(\mathbb{E}\{X\}) \leq \mathbb{E}\{f(X)\}.$$

Proof. We present the proof for discrete random variables. Let X be a discrete random variable taking a finite number of values, then by definition of convex function, with $a_i = P(X = x_i) = p_X(x_i)$, we have

$$f(\mathbb{E}\{X\}) = f\left(\sum_{i=1}^{n} x_i p_X(x_i)\right) \leq \sum_{i=1}^{n} f(x_i) p_X(x_i) = \mathbb{E}\{f(X)\}.$$

From Jensen's inequality it immediately follows that $|\mathbb{E}(X)| \leq \mathbb{E}|X|$.

Theorem 2.3.26 *Let X be a random variable and $f(\cdot)$, $g(\cdot)$ monotone nondecreasing measurable functions. Then*

$$\mathbb{E}(f(X)g(X)) \geq \mathbb{E}\{f(X)\mathbb{E}g(X)\},$$

provided all expectations are finite. If $f(\cdot)$ is monotone increasing and $g(\cdot)$ monotone decreasing, then

$$\mathbb{E}(f(X)g(X)) \leq \mathbb{E}\{f(X)\}\mathbb{E}\{g(X)\}.$$

Theorem 2.3.27 (Kolmogorov inequality) *Let X_i, $i = 1, 2, \ldots, n$ be independent random variables, with $\mathbb{E}(X_i) = 0$ and $\mathbb{E}(X_i^2) < \infty$, then*

$$P\left(\max_{1 \leq i \leq n} |X_1 + X_2 + \cdots + X_i| \geq \epsilon\right) \leq \frac{1}{\epsilon^2} \sum_{i=1}^{n} \mathbb{E}(X_i^2),$$

for all $\epsilon > 0$.

2.4   Asymptotics

Sequences of random variables can be defined in a natural way, but because of their structure, convergence is intended in a slightly different way from what is usually the case in basic calculus courses. In particular, measurability and distributions of these random objects define different types of convergence. Finally, for what matters to statistics and finance, some particular sequences have very peculiar limits as we will see.

2.4.1   Types of convergences

Definition 2.4.1 (Convergence in distribution) *Let* $\{F_n, n \in \mathbb{N}\}$ *be a sequence of distribution functions for the sequence of random variables* $\{X_n, n \in \mathbb{N}\}$. *Assume that*

$$\lim_{n \to \infty} F_n(x) = F_X(x)$$

for all $x \in \mathbb{R}$ *such that* $F_X(\cdot)$ *is continuous in* x, *where* F_X *is the distribution function of some random variable* X. *Then, the sequence* X_n *is said to converge in distribution to the random variable* X, *and this is denoted by* $X_n \xrightarrow{d} X$.

This convergence only means that the distributions F_n of the random variables converge to another distribution F, but nothing is said about the random variables. So this convergence is only about the probabilistic behavior of the random variables on some intervals $(-\infty, x]$, $x \in \mathbb{R}$.

Definition 2.4.2 (weak convergence) *A sequence of random variables* X_n *weakly converges to* X *if, for all smooth functions* $f(\cdot)$, *we have*

$$\lim_{n \to \infty} \int_{\mathbb{R}} f(x) \mathrm{d}F_n(x) = \int_{\mathbb{R}} f(x) \mathrm{d}F_X(x)$$

and we write $X_n \xrightarrow{w} X$.

Theorem 2.4.3 *A sequence of random variables* X_n *weakly converges to* X *if and only if it also converge in distribution.*

So previous results say that there is no difference between weak convergence and convergences in distribution, so one of the two can be used as a criterion to prove the other.

Definition 2.4.4 (Convergence in probability) *A sequence of random variables* X_n *is said to converge in probability to a random variable* X *if, for any* $\epsilon > 0$, *the following limit holds true*

$$\lim_{n \to \infty} P(|X_n - X| \geq \epsilon) = 0.$$

This is denoted by $X_n \xrightarrow{p} X$ *and it is the pointwise convergence of probabilities.*

This convergence implies the convergence in distribution. Sometimes we use the notation

$$p\!-\!\lim_{n\to\infty} |X_n - X| = 0$$

for the convergence in probability. A stronger type of convergence is defined as the probability of the limit in the sense $P(\lim_{n\to\infty} X_n = X) = 1$ or, more precisely,

Definition 2.4.5 (Almost sure convergence) *The sequence X_n is said to converge almost surely to X if*

$$P\left(\left\{\omega \in \Omega : \lim_{n\to\infty} X_n(\omega) = X(\omega)\right\}\right) = 1.$$

When this happens we write $X_n \overset{a.s.}{\to} X$.

Almost sure convergence implies convergence in probability.

Definition 2.4.6 (r-th mean convergence) *A sequence of random variables X_n is said to converge in the r-th mean to a random variable X if*

$$\lim_{n\to\infty} \mathbb{E}|X_n - X|^r = 0, r \geq 1.$$

and we write $X_n \overset{L^r}{\to} X$.

The convergence in the r-th mean implies the convergence in probability thanks to Chebyshev's inequality, and if X_n converges to X in the r-th mean, then it also converges in the s-th mean for all $r > s \geq 1$. *Mean square convergence* is a particular case of interest for stochastic analysis and corresponds to the case $r = 2$.

Theorem 2.4.7 *If for each $\epsilon > 0$ we have $\sum_{n=1}^{\infty} P(|X_n - X| \geq \epsilon) < \infty$, then $X_n \overset{a.s.}{\to} X$.*

Example 2.4.8 *Let $X \sim U(0, 1)$ and consider the sequence*

$$X_n(\omega) = \begin{cases} 0, & 0 \leq X(\omega) \leq \frac{1}{n^2}, \\ X(\omega), & \frac{1}{n^2} < X(\omega) \leq 1, \end{cases}$$

for $n = 1, 2, \ldots$. Then

$$\sum_{n=1}^{\infty} P(|X_n - X| \geq \epsilon) \leq \sum_{n=1}^{\infty} \frac{1}{n^2} < \infty$$

for any $\epsilon > 0$, because $|X_n - X| > 0$ on the interval $(0, 1/n^2)$, hence $P(|X_n - X| \geq \epsilon) = P(X \leq 1/n^2) = 1/n^2$. Therefore, $X_n \overset{a.s.}{\to} X$ by Theorem 2.4.7. Moreover,

$$\mathbb{E}|X_n - X|^2 = \int_0^{\frac{1}{n^2}} x^2 \mathrm{d}x = \frac{1}{3n^6} \to 0.$$

Then X_n converges also in quadratic mean to X.

The following implications $\Rightarrow$ hold in general:

$$\overset{a.s.}{\rightarrow} \Rightarrow \overset{p}{\rightarrow} \Rightarrow \overset{d/w}{\rightarrow}$$

$$\forall r > 0: \quad \overset{L^r}{\rightarrow} \Rightarrow \overset{p}{\rightarrow}$$

$$\forall r > s \geq 1: \quad \overset{L^r}{\rightarrow} \Rightarrow \overset{L^s}{\rightarrow}$$

Further, if a sequence of random variables X_n converges in distribution to some constant $c < \infty$, i.e. $X_n \overset{d}{\rightarrow} c$, then it also converges in probability to the same constant, i.e. $X_n \overset{p}{\rightarrow} c$.

Theorem 2.4.9 (Slutsky's) *Let X_n and Y_n be two sequences of random variables such that $X_n \overset{d}{\rightarrow} X$ and $Y_n \overset{d}{\rightarrow} c$, where X is a random variable and c a constant. Then*

$$X_n + Y_n \overset{d}{\rightarrow} X + c \quad and \quad X_n Y_n \overset{d}{\rightarrow} cX.$$

Theorem 2.4.10 (Continuous mapping theorem) *If $X_n \overset{d}{\rightarrow} X$ and $h(\cdot)$ is continuous function such that the probability of X taking values where $h(\cdot)$ is discontinuous is zero, then $h(X_n) \overset{d}{\rightarrow} h(X)$.*

2.4.2 Law of large numbers

Theorem 2.4.11 *Let $\{X_n, n = 1, \ldots\}$ be a sequence of independent and identically distributed random variables with $\mathbb{E}(X_n) = \mu < \infty$ and $\mathrm{Var}(X_n) = \sigma^2 < \infty$ for all n. Let $S_n = \sum_{i=1}^{n} X_i$. Then*

$$\frac{S_n}{n} \overset{p}{\rightarrow} \mu.$$

Proof. We simply need to use Chebyshev's inequality (2.23). In fact, take $\epsilon > 0$

$$P\left(\left|\frac{S_n}{n} - \mu\right| \geq \epsilon\right) \leq \frac{\mathrm{Var}\left(\frac{S_n}{n}\right)}{\epsilon^2} = \frac{\sigma^2}{n\epsilon^2} \rightarrow 0 \quad as \quad n \rightarrow \infty.$$

It is also possible to remove the requirement on the finiteness of the second moment using a different proof based on characteristic functions.

Theorem 2.4.12 *Let $\{X_n, n = 1, \ldots\}$ be a sequence of independent and identically distributed random variables with $\mathbb{E}(X_n) = \mu < \infty$. Then $\frac{S_n}{n} \overset{d}{\rightarrow} \mu$.*

Proof. We make use of the properties of the characteristic functions. In particular we use (2.7) with $a_i = 1/n$. Let $\varphi_X(t)$ denote the characteristic function of X_i and $\varphi_n(t)$ the characteristic function of S_n/n. Then

$$\varphi_n(t) = \left[\varphi\left(\frac{t}{n}\right)\right]^n.$$

We now use Taylor expansion and obtain $\varphi(t) = 1 + i\varphi'(t)t + o(t)$, hence

$$\varphi\left(\frac{t}{n}\right) = 1 + \frac{i\mu t}{n} + o\left(\frac{1}{n}\right).$$

Therefore, remembering that $\lim_{n\to\infty}(1 - a/n)^n = e^a$, we get

$$\lim_{n\to\infty}\varphi_n(t) = e^{i\mu t},$$

which is the characteristic function of degenerate random variable taking only the value μ. Then we have proved that $\frac{S_n}{n} \xrightarrow{d} \mu$ and hence also $\frac{S_n}{n} \xrightarrow{P} \mu$.

Previous results concern only the distributional properties of the arithmetic mean of independent random variables so they are called *weak law of large numbers* but, we can notice that, for any n we have

$$\mathbb{E}\left(\frac{S_n}{n}\right) = \frac{n\mathbb{E}(X_i)}{n} = \mu,$$

so we can expect a stronger result. This is indeed the case and the limit theorem is called *strong law of large numbers*.

Theorem 2.4.13 (Strong L.L.N.) *Let $\{X_n, n = 1, \ldots\}$ be a sequence of independent and identically distributed random variables with $\mathbb{E}|X_n|^4 = M < \infty$. Then $\frac{S_n}{n} \xrightarrow{a.s.} \mu$.*

To prove the strong law of large numbers we need the following Lemma without proof.

Lemma 2.4.14 (Borel-Cantelli) *Let A_n be a sequence of events. If*

$$\sum_n P(A_n) < \infty \quad then \quad P\left\{\omega : \lim_{n\to\infty} \mathbf{1}_{A_n}(\omega) = 0\right\} = 1.$$

If the events A_n are mutually independent, the converse is also true.

Loosely speaking, Borel-Cantelli Lemma says that if, for a sequence of events A_n, the series of probability is convergent, then in the limit the events A_n will not occur with probability one.

Proof. [of Theorem 2.4.13] We assume $\mathbb{E}(X_i) = 0$, otherwise we just consider $Y_i = X_i - \mu$ and then proceed with the same proof. Simple calculation gives

$$\mathbb{E}\left\{S_n^4\right\} = n\mathbb{E}\left\{X_i^4\right\} + 3n(n-1)\left(\mathbb{E}\left\{X^2\right\}\right)^2 \leq nM + 3n^2\sigma^4.$$

Now we use Chebyshev-Markov's inequality (2.24) with power four and obtain, for any $\epsilon > 0$, that

$$P\left(\left|\frac{S_n}{n}\right| \geq \epsilon\right) = P\left(|S_n| \geq n\epsilon\right) \leq \frac{nM + 3n^2\sigma^4}{n^4\epsilon^4}.$$

So, the events $A_n = \left\{\omega : \left|\frac{S_n(\omega)}{n}\right| \geq \epsilon\right\}$ are such that

$$\sum_{n=1}^{\infty} P(A_n) < \infty$$

and applying Borel-Cantelli Lemma we get almost user convergence.

2.4.3   Central limit theorem

Theorem 2.4.15 *Let $\{X_n, n = 1, \ldots\}$ be a sequence of independent and identically distributed random variables with $\mathbb{E}(X_n) = \mu < \infty$ and $\mathrm{Var}(X_n) = \sigma^2 < \infty$ for all n. Let $Y_n = \frac{1}{n}\sum_{i=1}^{n} X_i$ be a new sequence defined for each n. Then*

$$\frac{Y_n - \mu}{\frac{\sigma}{\sqrt{n}}} = \frac{\sum_{i=1}^{n} X_i - n\mu}{\sigma\sqrt{n}} \xrightarrow{d} N(0, 1).$$

Proof. We prove the result by using the convergence of the characteristic functions which corresponds to convergence in distribution. It is easier to obtain the result if we notice that $Z_i = X_i - \mu$ are such that $\mathbb{E}(Z_i) = 0$ and $\mathrm{Var}(Z_i) = \mathbb{E}(Z_i^2) = \sigma^2$. Then

$$S_n = \sum_{i=1}^{n} \frac{Z_i}{\sigma\sqrt{n}} = \frac{Y_n - \mu}{\frac{\sigma}{\sqrt{n}}}.$$

Let us denote by $\varphi_Z(t)$ the characteristic function of the random variables Z_i, then by (2.7) with $a_i = \frac{1}{\sigma\sqrt{n}}$, we have

$$\varphi_{S_n}(t) = \prod_{i=1}^{n} \varphi_Z\left(\frac{t}{\sigma\sqrt{n}}\right) = \left[\varphi_Z\left(\frac{t}{\sigma\sqrt{n}}\right)\right]^n.$$

Now we expand $\varphi(t)$ in powers of t

$$\varphi_Z(t) = \varphi_Z(0) + \varphi_Z'(t)t + \varphi_Z''(t)\frac{t^2}{2!} + o(t^2) = 1 + it\mathbb{E}Z_i + i^2\frac{t^2}{2}\mathbb{E}Z_i^2 + o(t^2)$$

$$= 1 - \frac{t^2}{2}\sigma^2 + o(t^2).$$

Hence

$$\varphi_Z\left(\frac{t}{\sigma\sqrt{n}}\right) = 1 - \frac{t^2}{2\sigma^2 n}\sigma^2 + o\left(\frac{t^2}{\sigma^2 n}\right) = 1 - \frac{t^2}{2n} + o\left(\frac{1}{n}\right).$$

Finally, we obtain

$$\lim_{n\to\infty}\left[1 - \frac{t^2}{2n} + o\left(\frac{1}{n}\right)\right]^n = e^{-\frac{t^2}{2}},$$

which is the characteristic function of the random variable $N(0, 1)$.

Next result, given without proof, involves the infinitely divisible distributions.

Theorem 2.4.16 *Let $\{F_n, n \geq 0\}$ a sequence of infinitely divisible distributions and assume that $F_n(x) \to F(x)$. Then $F(\cdot)$ is also infinitely divisible.*

This result ensures that the limit keeps the same property of being infinitely divisible as the elements of the sequence.

Theorem 2.4.17 (Lindeberg's condition) *Let $\{X_n, n = 1, \ldots\}$ be a sequence of independent random variables such that $\mathbb{E}(X_n) = \mu_n < \infty$, $\mathrm{Var}(X_n) = \sigma_n^2 < \infty$ and let $s_n^2 = \sum_{k=1}^n \sigma_k^2$. If, for any $\epsilon > 0$,*

$$\lim_{n\to\infty}\frac{1}{s_n^2}\sum_{k=1}^n\mathbb{E}\left\{\mathbf{1}_{\{|X_k-\mu_k|>\epsilon s_n\}}(X_k - \mu_k)^2\right\} = 0$$

then the Central Limit Theorem holds, i.e.

$$Z_n = \frac{\sum_{k=1}^n(X_k - \mu_k)}{s_n} \xrightarrow{d} N(0, 1).$$

Lindeberg's condition is sufficient but not necessary and it is a very useful tool to prove asymptotic normality of, e.g., estimators in statistics. This condition guarantees that, for large n, the contribution of each random variable to the total variance is small.

2.5   Conditional expectation

As seen in Section 2.1.1, the conditional probability of A given B is defined as $P(A|B) = P(A \cap B)/P(B)$. In the same way, it is possible to introduce the conditional distribution of a random variable X with respect to the event B as

$$F_X(x|B) = \frac{P((X \le x) \cap B)}{P(B)}, \quad x \in \mathbb{R},$$

and the expectation with respect to this conditional distribution is naturally introduced as (see Mikosch (1998) for a similar treatise)

$$\mathbb{E}\{X|B\} = \frac{\mathbb{E}(X\mathbf{1}_B)}{P(B)} = \frac{1}{P(B)} \int_\Omega X(\omega)\mathbf{1}_B(\omega)P(d\omega)$$

$$= \frac{1}{P(B)} \int_{B \cup \bar{B}} X(\omega)\mathbf{1}_B(\omega)P(d\omega) = \frac{1}{P(B)} \int_B X(\omega)P(d\omega),$$

where $\mathbf{1}_B$ is the indicator function of the set $B \subset \Omega$, which means $\mathbf{1}_B(\omega) = 1$ if $\omega \in B$ and 0 otherwise. For discrete random variables, the conditional expectation takes the form:

$$\mathbb{E}\{X|B\} = \sum_i x_i \frac{P(\{\omega : X(\omega) = x_i\} \cap B)}{P(B)} = \sum_i x_i P(X = x_i|B).$$

For continuous random variables with density $f_X(\cdot)$, if we denote $X(B) = \{x = X(\omega) : \omega \in B\}$ we can write

$$\mathbb{E}\{X|B\} = \frac{\mathbb{E}(X\mathbf{1}_B)}{P(B)} = \frac{1}{P(B)} \int_\mathbb{R} x\mathbf{1}_{\{X(B)\}}(x)f_X(x)dx = \frac{1}{P(B)} \int_{X(B)} xf_X(x)dx,$$

where $\mathbf{1}_{\{X(B)\}}(x) = 1$ if $x \in X(B)$ and 0 otherwise. Consider now a discrete random variable Y that takes distinct values $y_1, y_2, \ldots, y_k$ and define $A_i = \{\omega : Y(\omega) = y_i\} = Y^{-1}(y_i)$, $i = 1, \ldots, k$. Assume that all $P(A_i)$ are positive. Let $\mathbb{E}|X| < \infty$. Then a new random variable Z can be defined as follows:

$$Z(\omega) = \mathbb{E}\{X|Y\}(\omega) \quad \text{where} \quad \mathbb{E}\{X|Y\}(\omega) = \mathbb{E}\{X|Y(\omega) = y_i\}$$

$$= \mathbb{E}\{X|A_i\} \quad \text{for} \quad \omega \in A_i.$$

For all $\omega \in A_i$ the conditional expectation $\mathbb{E}\{X|Y\}(\omega)$ coincides with $\mathbb{E}\{X|A_i\} = \mathbb{E}(X\mathbf{1}_{A_i})$, but, as a function of $\omega \in \Omega$, it is a random variable itself because it depends on the events generated by $Y(\omega)$ and each value taken by the conditional expectation $\mathbb{E}\{X|Y\}(\omega)$, i.e. each $\mathbb{E}\{X|Y\}(\omega) = \mathbb{E}\{X|A_i\}$ has its own probability $P(Y = y_i)$.

If instead of a single set B or a finite number of sets A_i, $i = 1, \ldots, k$, we consider a complete σ-algebra of events (for example, the one generated by a generic random variable Y), we arrive at the general definition of conditional expectation which is given in implicit form as follows.

Definition 2.5.1 *Let X be a random variable such that $\mathbb{E}|X| < \infty$. A random variable Z is called the conditional expectation of X with respect to the σ-algebra $\mathcal{F}$ if:*

(i) *Z is $\mathcal{F}$-measurable and*

(ii) *Z is such that $\mathbb{E}(Z\mathbf{1}_A) = \mathbb{E}(X\mathbf{1}_A)$ for every $A \in \mathcal{F}$.*

The conditional expectation is unique and will be denoted as $Z = \mathbb{E}\{X|\mathcal{F}\}$. With this notation, the equivalence above can be written as

$$\mathbb{E}(\mathbb{E}\{X|\mathcal{F}\}\mathbf{1}_A) = \mathbb{E}(X\mathbf{1}_A) \text{ for every } A \in \mathcal{F}. \qquad (2.25)$$

By definition the conditional expectation is a random variable and the above equality is only true up to null-measure sets. Among the properties of the conditional expectation, we note only the following. Let X and Y be random variables and a, b two constants. Then, provided that each quantity exists, we have

(i) linearity

$$\mathbb{E}\{a \cdot X + b \cdot Y|\mathcal{F}\} = a \cdot \mathbb{E}\{X|\mathcal{F}\} + b \cdot \mathbb{E}\{Y|\mathcal{F}\};$$

(ii) if we condition with respect to the trivial σ-algebra, $\mathcal{F}_0 = \{\Omega, \emptyset\}$, we have

$$\mathbb{E}\{X|\mathcal{F}_0\} = \mathbb{E}(X);$$

(iii) if Y is $\mathcal{F}$-measurable, then

$$\mathbb{E}\{Y \cdot X|\mathcal{F}\} = Y \cdot \mathbb{E}\{X|\mathcal{F}\};$$

(iv) choose $X = 1$ in iii), then it follows that

$$\mathbb{E}\{Y|\mathcal{F}\} = Y;$$

(v) choose $A = \Omega$ in (2.25), then it follows that

$$\mathbb{E}(\mathbb{E}\{X|\mathcal{F}\}) = \mathbb{E}(X); \qquad (2.26)$$

(vi) if X is independent of $\mathcal{F}$, then it follows that

$$\mathbb{E}\{X|\mathcal{F}\} = \mathbb{E}(X)$$

and, in particular, if X and Y are independent, we have $\mathbb{E}\{X|Y\} = \mathbb{E}\{X|\sigma(Y)\} = \mathbb{E}(X)$, where $\sigma(Y)$ is the σ-algebra generated by the random variable Y.

Another interesting property of the conditional expectation which is often used in statistics is the next one which we give with detailed proof because we will make use most of the above properties.

Theorem 2.5.2 *Let X be a square integrable random variable. Then*

$$\text{Var}(X) = \text{Var}(\mathbb{E}\{X|Y\}) + \mathbb{E}(\text{Var}\{X|Y\}).$$

Proof.

$$\text{Var}(X) = \mathbb{E}(X - \mathbb{E}(X))^2 = \mathbb{E}\{\mathbb{E}(X - \mathbb{E}(X))^2|Y\}$$

$$= \mathbb{E}\left\{\mathbb{E}(X - \mathbb{E}\{X|Y\} + \mathbb{E}\{X|Y\} - \mathbb{E}(X))^2 \,\middle|\, Y\right\}$$

$$= \mathbb{E}\left\{\mathbb{E}(X - \mathbb{E}[X|Y])^2 + \mathbb{E}(\mathbb{E}\{X|Y\} - \mathbb{E}(X))^2\right.$$

$$\left. + 2\mathbb{E}[(X - \mathbb{E}\{X|Y\})(\mathbb{E}\{X|Y\} - \mathbb{E}(X))] \,\middle|\, Y\right\}$$

$$= \mathbb{E}\{\mathbb{E}(X - \mathbb{E}\{X|Y\})^2|Y\} + \mathbb{E}\{\mathbb{E}(\mathbb{E}\{X|Y\} - \mathbb{E}(X))^2|Y\}$$

$$+ 2\mathbb{E}\{\mathbb{E}[(X - \mathbb{E}\{X|Y\})(\mathbb{E}\{X|Y\} - \mathbb{E}(X))]|Y\}$$

$$= a + b + c.$$

Consider the first term and notice that

$$\mathbb{E}\{(X - \mathbb{E}\{X|Y\})^2|Y\} = \text{Var}\{X|Y\}$$

where $\text{Var}\{X|Y\}$ is the variance of X calculated using the conditional distribution function of X given Y. Hence $a = \mathbb{E}(\text{Var}\{X|Y\})$. Now, consider the second term. By measurability of $\mathbb{E}\{X|Y\} - \mathbb{E}(X)$ w.r.t. Y we have

$$b = \mathbb{E}\{\mathbb{E}(\mathbb{E}\{X|Y\} - \mathbb{E}(X))^2|Y\} = \mathbb{E}(\mathbb{E}\{X|Y\} - \mathbb{E}(X))^2$$

and using (2.26) we get

$$b = \mathbb{E}(\mathbb{E}\{X|Y\} - \mathbb{E}(X))^2 = \mathbb{E}\{\mathbb{E}\{X|Y\} - \mathbb{E}(\mathbb{E}\{X|Y\})\}^2 = \text{Var}(\mathbb{E}\{X|Y\}).$$

For the last term, using again the measurability of $\mathbb{E}\{X|Y\} - \mathbb{E}(X)$, we get

$$c = 2\mathbb{E}\{\mathbb{E}[(X - \mathbb{E}\{X|Y\})(\mathbb{E}\{X|Y\} - \mathbb{E}(X))]|Y\}$$

$$= 2(\mathbb{E}\{X|Y\} - \mathbb{E}(X))\mathbb{E}\{\mathbb{E}(X - \mathbb{E}\{X|Y\})|Y\}$$

$$= 2(\mathbb{E}\{X|Y\} - \mathbb{E}(X))(\mathbb{E}\{X|Y\} - \mathbb{E}\{X|Y\}) = 0.$$

Finally, we present a version of Jensen's inequality for the conditional expectation, which is given without proof.

Theorem 2.5.3 (Conditional Jensen's inequality) *Let X be a random variable on the probability space $(\Omega, \mathcal{F}, P)$, and let $f(\cdot)$ be any real-valued convex function on $\mathbb{R}$. Assume that X is such that $\mathbb{E}|X| < \infty$ and $\mathbb{E}|f(X)| < \infty$. Then*

$$f(\mathbb{E}\{X|\mathcal{F}\}) \leq \mathbb{E}\{f(X)|\mathcal{F}\}. \tag{2.27}$$

2.6   Statistics

Suppose there is a population of individuals on which we want to measure some quantity of interest. Let us denote by X the random variable which describes this characteristic in the population. Assume that the distribution of X is characterized by some parameter $\theta \subset \Theta \subset \mathbb{R}^k$. The object of statistical inference is to recover the value of θ from a random sample of data extracted from the population. We denote the random sample with the random vector $(X_1, X_2, \ldots, X_n)$ where X_i is the i-th potential observed value on individual i of the random sample. Each X_i is supposed to be extracted independently from the population X so that all the X_i's are copies of X, i.e. they have all the same distribution of X. In this case we say that $(X_1, X_2, \ldots, X_n)$ is a random sample of *independent and identically distributed* (*i.i.d.*) random variables.

Definition 2.6.1 *An estimator T_n of θ is a function of the random sample which takes values on Θ but it is not a function of $\theta \in \Theta$ itself. We write*

$$T_n(\omega) = T_n(\omega, (X_1, X_2, \ldots, X_n)) : \Omega \times \mathbb{R}^n \to \Theta.$$

Example 2.6.2 *Let X_i, $i = 1, \ldots, n$ be an i.i.d. random sample with common distribution with mean $\mathbb{E}(X) = \theta$. The so-called plug-in estimator of θ is the sample mean, i.e.*

$$T_n = \bar{X}_n = \frac{1}{n} \sum_{i=1}^{n} X_i,$$

which we expect to be a 'good' estimator of θ.

2.6.1   Properties of estimators

Before judging estimators we need to discuss about quality measures. We denote by $\mathbb{E}_\theta(X)$ the expected value of X under the distribution P_θ, i.e. $\mathbb{E}_\theta(X) = \int_\Omega X(\omega) P_\theta(d\omega)$. The probability measure P_θ is supposed to be a member of the parametric family of models $\{P_\theta, \theta \in \Theta\}$. More precisely, to each sample size $n \geq 1$ we should consider a parametric family of models $\{P_\theta^n, \theta \in \Theta\}$ and then introduce the so-called *family of experiments* $\{\Omega, \mathcal{A}, \{P_\theta^n, \theta \in \Theta\}\}$, but for simplicity we drop index n from the probability measures and we assume this family of experiments is defined somehow.

Consider again Example 2.6.2. To each θ in the family P_θ we can associate, e.g., the distribution of the Gaussian random variable $N(\mu, \sigma^2)$ with, e.g. $\theta = (\mu, \sigma^2)$ or $\theta = \mu$ for a given value of σ^2, or the one of the Bernoulli random variable Ber(θ) with $\theta = p$.

Definition 2.6.3 *An estimator of θ is said to be unbiased if*

$$\mathbb{E}_\theta(T_n) = \theta, \quad \text{for all} \quad n$$

and asymptotically unbiased if

$$\lim_{n\to\infty} \mathbb{E}_\theta(T_n) = \theta.$$

So, an unbiased estimator T_n recovers the true value θ on average. On average means that, from random sample to random sample, we can get different values of our estimator, but on the average of all possible random samples, it has a nice behaviour. If this fact happens independently of the sample size n, then we call these estimators unbiased, but if this is true only for *large* samples, we have only asymptotically unbiased estimators.

Knowing that, on average, an estimator correctly recovers the true value is not enough if the variability from sample to sample is too high. So we also need to control for the variability.

Definition 2.6.4 *The* mean square error *(MSE) of an estimator T_n of θ is defined as*

$$\text{MSE}_\theta(T_n) = \mathbb{E}_\theta(T_n - \theta)^2.$$

By adding and subtracting the quantity $\mathbb{E}_\theta(T_n)$ in the formula of MSE_θ, we obtain an expression of the MSE which involves both the *bias*, defined as $\text{Bias}_\theta(T_n) = \mathbb{E}_\theta(T_n - \theta)$, and the variance $\text{Var}_\theta(T_n)$ of the estimator

$$\begin{aligned}
\text{MSE}_\theta(T_n) &= \mathbb{E}_\theta(T_n \pm \mathbb{E}_\theta(T_n) - \theta)^2 \\
&= \mathbb{E}_\theta(T_n - \mathbb{E}_\theta(T_n))^2 + \mathbb{E}_\theta(\mathbb{E}_\theta(T_n) - \theta)^2 \\
&\quad + 2\mathbb{E}_\theta(T_n - \mathbb{E}_\theta(T_n))(\mathbb{E}_\theta(T_n) - \theta) \\
&= \text{Var}_\theta(T_n) + (\text{Bias}_\theta(T_n))^2 + 2(\mathbb{E}_\theta(T_n) - \theta)\mathbb{E}_\theta(T_n - \mathbb{E}_\theta(T_n)) \\
&= \text{Var}_\theta(T_n) + (\text{Bias}_\theta(T_n))^2.
\end{aligned}$$

So, given two estimators T_n and S_n, if we want to compare them we need to evaluate both bias and variance of each at the same time and then: choose T_n if $\text{MSE}(T_n) < \text{MSE}(S_n)$ or choose S_n otherwise.

Example 2.6.5 *Let us consider again $\bar{X}_n$ from Example 2.6.2.*

$$\text{MSE}(\bar{X}_n) = \text{Var}_\theta(\bar{X}_n) + (\text{Bias}_\theta(\bar{X}_n))^2$$

but

$$\mathbb{E}_\theta(\bar{X}_n) = \mathbb{E}_\theta\left(\frac{1}{n}\sum_{i=1}^n X_i\right) = \frac{1}{n}\sum_{i=1}^n \mathbb{E}_\theta(X_i) = \frac{1}{n}\sum_{i=1}^n \theta = \frac{n\theta}{n} = \theta.$$

Hence, $\bar{X}_n$ is an unbiased estimator of θ and the mean square error is then just its variance. So we calculate it now.

$$\text{Var}_\theta(\bar{X}_n) = \text{Var}_\theta\left(\frac{1}{n}\bar{X}_n\right) = \frac{1}{n^2}\sum_{i=1}^n \text{Var}_\theta(X_i) = \frac{n\sigma^2}{n^2} = \frac{\sigma^2}{n}.$$

Therefore we have $\text{MSE}_\theta(\bar{X}_n) = \frac{\sigma^2}{n}$. Let us consider now an estimator T_n of the following form:

$$T_n = \frac{3X_1 + X_2 + \cdots + X_{n-1} - X_n}{n}.$$

We have that T_n is also unbiased. Indeed,

$$\mathbb{E}_\theta(T_n) = \frac{3\theta + (n-2)\theta - \theta}{n} = \frac{n\theta}{n} = \theta.$$

Let us calculate the variance of T_n

$$\text{Var}_\theta(T_n) = \text{Var}_\theta\left(\frac{3X_1 + X_2 + \cdots + X_{n-1} - X_n}{n}\right)$$

$$= \frac{3^2\sigma^2 + (n-2)\sigma^2 + \sigma^2}{n^2} = \frac{(8+n)\sigma^2}{n^2}$$

$$= \frac{8\sigma^2}{n^2} + \frac{\sigma^2}{n} > \frac{\sigma^2}{n} = \text{Var}_\theta(\bar{X}_n).$$

Therefore, $\text{MSE}_\theta(T_n) > \text{MSE}_\theta(\bar{X}_n)$ and we should prefer $\bar{X}_n$ to T_n.

Exercise 2.15 *Let $T_n = \frac{1}{n}\sum_{i=1}^n a_i X_i$, with $\sum_{i=1}^n a_i = n$, with X_i i.i.d. random variables with mean θ and variance σ^2. Prove that T_n is an unbiased estimator of θ and that the minimal variance of T_n is obtained for $a_i = 1$, $i = 1, \ldots, n$, i.e. $T_n = \bar{X}_n$.*

The previous exercise shows that the sample mean $\bar{X}_n$ is, in terms of the mean square error, the best estimator of the mean of the population θ among all estimators which are linear combinations of the X_i and it is also unbiased. This is a special case of estimators called *best linear unbiased estimator* (BLUE).

We have also used the plug-in approach to define an estimator of θ, i.e. to estimate the mean of the population we used the empirical mean. But plug-in estimators are not necessarily the best ones. Suppose now we also want to estimate the variance σ^2 and consider the empirical variance as an estimator of σ^2. Because we also estimate μ we write $\mathbb{E}_\theta$ with $\theta = (\mu, \sigma^2)$ to denote expectation under the true unknown model. The empirical variance is defined as

$$S_n^2 = \frac{1}{n}\sum_{i=1}^n (X_i - \bar{X}_n)^2.$$

Then

$$\mathbb{E}_\theta(S_n^2) = \frac{1}{n} \sum_{i=1}^{n} \mathbb{E}_\theta(X_i - \bar{X}_n)^2$$

but

$$\mathbb{E}_\theta(X_i - \bar{X}_n)^2 = \mathbb{E}_\theta(X_i^2) + \mathbb{E}_\theta(\bar{X}_n)^2 - 2\mathbb{E}_\theta(X_i \bar{X}_n)$$

$$= \mu^2 + \sigma^2 + \mu^2 + \frac{\sigma^2}{n} - 2\mathbb{E}_\theta \left\{ \frac{X_i^2}{n} + \frac{X_i}{n} \sum_{j \neq i} X_j \right\}$$

$$= 2\mu^2 + \sigma^2 \frac{n+1}{n} - 2 \left\{ \frac{\mu^2 + \sigma^2}{n} + \mu^2 \frac{n-1}{n} \right\}$$

$$= 2\mu^2 + \sigma^2 \frac{n+1}{n} - 2\mu^2 - 2\frac{\sigma^2}{n} = \frac{n-1}{n}\sigma^2.$$

So $\mathbb{E}_\theta(S_n^2) = \frac{n-1}{n}\sigma^2 < \sigma^2$ and S_n^2 is a biased estimator of σ^2. We can correct the estimator as follows:

$$\bar{S}_n^2 = \frac{1}{n-1} \sum_{i=1}^{n} (X_i - \bar{X}_n)^2 = \frac{n}{n-1} S_n^2 \quad \text{and hence} \quad \mathbb{E}_\theta(\bar{S}_n^2) = \sigma^2.$$

For completeness we mention that

$$\mathrm{Var}_\theta(\bar{S}_n^2) = \frac{2\sigma^4}{n-1} \quad \text{and} \quad \mathrm{Var}_\theta(S_n^2) = \frac{2(n-1)\sigma^4}{n^2}.$$

The proof of this fact is simple algebra but very lengthy and we omit it. But, we remark that, if we compare the mean square error of the two estimators we obtain

$$\frac{\mathrm{MSE}_\theta(\bar{S}_n^2)}{\mathrm{MSE}_\theta(S_n^2)} = \frac{\frac{2}{n-1}\sigma^4}{\frac{2n-1}{n^2}\sigma^4} = \frac{2n^2}{2n^2 - 3n + 1} > 1,$$

which shows that, in the trade-off between bias and variance, the estimator S_n^2 is better than its uncorrected version. Of course, those differences vanish asymptotically.

Definition 2.6.6 *An estimator T_n of θ is said to be* consistent *if, for all $\epsilon > 0$, we have that*

$$\lim_{n \to \infty} P_\theta(|T_n - \theta| \geq \epsilon) = 0.$$

Clearly, by Chebyshev's inequality (2.23) one can usually prove consistency using mean square error

$$P_\theta(|T_n - \theta| \geq \epsilon) \leq \frac{\mathbb{E}_\theta(T_n - \theta)^2}{\epsilon^2}$$

and for unbiased estimators, one just need to check that the variance of the estimator converges to zero. Consistent estimators are convenient also because, by using the continuous mapping theorem 2.4.10, one has that, if $g(\cdot)$ is a continuous function and T_n is a consistent estimator of θ, then $g(T_n)$ is a consistent estimator of $g(\theta)$.

In i.i.d. sampling when all conditions are fulfilled, one can immediately prove that the sample mean $\bar{X}_n$ is a consistent estimator of the mean of the population using either the law of large numbers or, alternatively, Chebyshev's inequality recalling that $\mathbb{E}_\theta(\bar{X}_n - \theta)^2 = \sigma^2/n$.

2.6.2   The likelihood function

Suppose we have a sample of i.i.d. observations X_i, $i = 1, \ldots, n$, with common distribution indexed by some parameter $\theta \in \Theta$. Seen as a random vector, the sample $(X_1, X_2, \ldots, X_n)$ has its own probability. So, for a given set of observed values $(x_1, x_2, \ldots, x_n)$ from the random vector $(X_1, X_2, \ldots, X_n)$, we might wonder what is the probability that these data come from a given model specified by θ. Assume that the X_i's are discrete random variables with probability mass function $p(x; \theta) = P_\theta(X = x)$. Let us construct the probability of the observed sample as

$$P_\theta(X_1 = x_1, X_2 = x_2, \ldots, X_n = x_n) = \prod_{i=1}^{n} p(x_i; \theta).$$

Seen as a function of $\theta \in \Theta$ and given the observed values $(X_1 = x_1, X_2 = x_2, \ldots, X_n = x_n)$, this quantity if called the '*likelihood* of θ given the sample data' and we write

$$L_n(\theta) = L_n(\theta|x_1, \ldots, x_n) = \prod_{i=1}^{n} p(x_i; \theta).$$

In case of continuous random variables with density function $f(x; \theta)$ we denote the likelihood as

$$L_n(\theta) = L_n(\theta|x_1, \ldots, x_n) = \prod_{i=1}^{n} f(x_i; \theta).$$

Now recall that $f(x) \neq P(X = x) = 0$ for continuous random variables, so it is important to interpret $L_n(\theta)$ as the likelihood of θ, rather than the probability of the sample. Indeed, $L_n(\theta)$ weights different values of $\theta \in \Theta$ on the basis of the observed (and given) data. This allows us to define a general approach in the search of estimators of the unknown parameter θ as we will see shortly.

The likelihood function has several derived quantities of interest. The log-likelihood $\ell_n(\theta) = \log L_n(\theta)$ plays some role and it is such that, under some regularity conditions, i.e. when the order of integration and derivation can be exchanged, we have that

$$
\begin{aligned}
\mathbb{E}_\theta \left\{ \frac{\partial}{\partial \theta} \ell_n(\theta) \right\} &= \mathbb{E}_\theta \left\{ \frac{1}{L_n(\theta)} \frac{\partial}{\partial \theta} L_n(\theta) \right\} \\
&= \int_{R^n} \left\{ \frac{1}{L_n(\theta|x_1, \ldots, x_n)} \frac{\partial}{\partial \theta} L_n(\theta|x_1, \ldots, x_n) \right\} \\
&\quad \times L_n(\theta|x_1, \ldots, x_n) \mathrm{d}x_1, \ldots \mathrm{d}x_n \\
&= \int_{R^n} \frac{\partial}{\partial \theta} L_n(\theta|x_1, \ldots, x_n) \mathrm{d}x_1, \ldots \mathrm{d}x_n \\
&= \frac{\partial}{\partial \theta} \int_{R^n} L_n(\theta|x_1, \ldots, x_n) \mathrm{d}x_1, \ldots \mathrm{d}x_n \\
&= \frac{\partial}{\partial \theta} 1 = 0.
\end{aligned}
$$

The function $\frac{\partial}{\partial \theta} \ell_n(\theta)$ is called *score function* and the variance of the score function is called *Fisher information*

$$
\mathcal{I}_n(\theta) = \mathbb{E}_\theta \left\{ \frac{\partial}{\partial \theta} \ell_n(\theta) \right\}^2.
$$

Further, under the same regularity conditions, it is possible to show that the variance of the score function can be obtained by the second derivative (Hessian matrix in the multidimensional case) of the log-likelihood

$$
\mathcal{I}_n(\theta) = \mathbb{E}_\theta \left\{ \frac{\partial}{\partial \theta} \ell_n(\theta) \right\}^2 = -\mathbb{E}_\theta \left\{ \frac{\partial^2}{\partial \theta^2} \ell_n(\theta) \right\}. \tag{2.28}
$$

Indeed, by differentiating twice $\ell(\theta)$ we obtain

$$
\frac{\partial^2}{\partial \theta^2} \ell_n(\theta) = \frac{1}{L_n(\theta)} \frac{\partial^2}{\partial \theta^2} L_n(\theta) - \left\{ \frac{\partial}{\partial \theta} \ell_n(\theta) \right\}^2
$$

hence, reorganizing the terms and taking the expectation, we obtain

$$
\begin{aligned}
\mathbb{E}_\theta &\left\{ \frac{\partial^2}{\partial \theta^2} \ell_n(\theta) \right\} + \mathbb{E}_\theta \left\{ \frac{\partial}{\partial \theta} \ell_n(\theta) \right\}^2 \\
&= \mathbb{E}_\theta \left\{ \frac{1}{L_n(\theta)} \frac{\partial^2}{\partial \theta^2} L_n(\theta) \right\} \\
&= \int_{R^n} \frac{1}{L_n(\theta|x1, \ldots, x_n)} \frac{\partial^2}{\partial \theta^2} L_n(\theta) L_n(\theta|x1, \ldots, x_n) \mathrm{d}x_1, \ldots \mathrm{d}x_n
\end{aligned}
$$

$$= \frac{\partial^2}{\partial \theta^2} \int_{R^n} L_n(\theta) \mathrm{d}x_1, \dots \mathrm{d}x_n$$

$$= 0$$

which proves the fact. In analogy with $\mathcal{I}_n(\theta)$, we can define the Fisher information for the single random variable (or for the model). For a random variable X, we can introduce $\ell(\theta) = \log f(x; \theta)$ is X is continuous or $\ell(\theta) = \log p(x; \theta)$ if X is discrete. Then the Fisher information is defined as follows:

$$\mathcal{I}(\theta) = \mathbb{E}_\theta \left\{ \frac{\partial}{\partial \theta} \ell(\theta) \right\}^2 = -\mathbb{E}_\theta \left\{ \frac{\partial^2}{\partial \theta^2} \ell(\theta) \right\}.$$

For i.i.d. samples, the likelihood $L_n(\theta)$ factorizes in the product of the single densities (or probability mass functions) and hence, in formula (2.28), the log-likelihood $\ell_n(\theta)$ takes the form of the sum of the log-densities (log-probabilities) and then the Fisher information $\mathcal{I}_n(\theta)$ can be rewritten as

$$\mathcal{I}_n(\theta) = \sum_{i=1}^n \mathcal{I}(\theta) = n\mathcal{I}(\theta).$$

Why the quantities $\mathcal{I}(\theta)$ and $\mathcal{I}_n(\theta)$ are called *information* will be clear in the next section.

2.6.3 Efficiency of estimators

Another way to compare estimators is to consider *efficiency* in terms of ratio of their variances. In particular for unbiased or consistent estimators, this approach is equivalent to consider comparison using the mean square error. The following general result states that under regularity conditions, the variance of any estimator of θ in a statistical model satisfies this inequality.

Theorem 2.6.7 (Cramér-Rao)

$$\mathrm{Var}_\theta(T_n) \geq \frac{\left(1 + \frac{\partial}{\partial \theta} \mathrm{Bias}_\theta(T_n)\right)^2}{\mathcal{I}(\theta)}.$$

So if we restrict the attention to unbiased estimators we can reread the above result as follows: no matter which estimator one chooses, the best one can do is to obtain an estimator whose variance is not less that the inverse of the Fisher information. So the Fisher information, and necessarily the likelihood of a model, describe the maximal information which can be extracted from the data coming from that particular model.

2.6.4 Maximum likelihood estimation

If we study the likelihood $L_n(\theta)$ as a function of θ given the n numbers $(X_1 = x_1, \ldots, X_n = x_n)$ and we find that this function has a maximum, we can use this maximum value as an estimate of θ. In general we define *maximum likelihood estimator* of θ, and we abbreviate this with *MLE*, the following estimator

$$\hat{\theta}_n = \arg\max_{\theta \in \Theta} L_n(\theta)$$

$$= \arg\max_{\theta \in \Theta} L_n(\theta | X_1, X_2, \ldots, X_n)$$

provided that the maximum exists. The estimator $\hat{\theta}_n$ is a real estimator because in its definition it depends on the random vector $(X_1, \ldots, X_n)$ and, of course, it does not depend on θ.

Example 2.6.8 *Let X_i, $i = 1, \ldots, n$, be an i.i.d. sample extracted from the Gaussian distribution $N(\mu, \sigma^2)$. For simplicity, assume σ^2 is known. We want to find the MLE of μ. Hence*

$$L_n(\mu) = \prod_{i=1}^{n} \frac{1}{\sqrt{2\pi\sigma^2}} e^{-\frac{(X_i - \mu)^2}{2\sigma^2}} = \left(\frac{1}{\sqrt{2\pi\sigma^2}}\right)^n \prod_{i=1}^{n} e^{-\frac{(X_i - \mu)^2}{2\sigma^2}}.$$

Instead of maximizing $L_n(\mu)$ we maximize the log-likelihood $\ell_n(\mu) = \log L_n(\mu)$

$$\ell_n(\mu) = n \log\left(\frac{1}{\sqrt{2\pi\sigma^2}}\right) - \sum_{i=1}^{n} \frac{(X_i - \mu)^2}{2\sigma^2}$$

but, maximizing $\ell_n(\mu)$ is equivalent to minimizing $-\ell_n(\mu)$. Moreover, the maximum in μ does not depend on the first term of $\ell_n(\mu)$ which contains only constants, hence we just need to solve

$$\hat{\mu}_n = \arg\min_{\mu} \frac{1}{2\sigma^2} \sum_{i=1}^{n} (X_i - \mu)^2$$

but this minimum is exactly $\bar{X}_n = \frac{1}{n} \sum_{i=1}^{n} X_i$ by the properties of the arithmetic mean[4]. Hence the ML estimator of μ is $\hat{\mu}_n = \bar{X}_n$.

Exercise 2.16 *Consider the setup of Example 2.6.8. Find the maximum likelihood estimator of $\theta = (\mu, \sigma^2)$.*

Exercise 2.17 *Let X_i, $i = 1, \ldots, n$, be i.i.d. random variables distributed as Ber(p). Find the maximum likelihood estimator of p.*

The maximum likelihood estimators have the following properties under very mild conditions in the i.i.d. case. We do not state these conditions here because

---

[4] It is easy to show that: $\min_a \sum_{i=1}^{n} (x_i - a)^2 = \sum_{i=1}^{n} (x_i - \bar{x}_n)^2$.

in what follows we will discuss more complicated settings, but we mention these properties because they are the typical result for ML estimators.

(i) these estimators are usually biased but asymptotically unbiased and hence consistent;

(ii) their limiting variance attain the Cramér-Rao bound for unbiased estimators, i.e. the limiting variance is the reciprocal (or inverse in the multidimensional case) of the Fisher information $\mathcal{I}(\theta)$;

(iii) they are asymptotically normal distributed, i.e. if T_n is MLE of θ, then
$$\sqrt{n}(T_n - \theta) \xrightarrow{d} N(0, \mathcal{I}^{-1}(\theta));$$

(iv) these estimators are also invariant to reparametrization. For example, if $\eta = g(\theta)$ and $g(\cdot)$ is some transform, then the MLE for the new parameter η can be obtained as $\hat{\eta}_n = g(\hat{\theta}_n)$ where $\hat{\theta}_n$ is the MLE for θ.

2.6.5   Moment type estimators

Another approach to derive estimators is to use the *method of moments* (for a complete treatment see, e.g. Durham and Gallant (2002); Gallant and Tauchen (1996); Hall (2005).) The idea is to match the moments of the population with the empirical moments in order to get estimators of the parameters. In practice, the estimators are obtained after solving a system of equations like these

$$\mu_j = \mathbb{E}_\theta(X^j) = \frac{1}{n} \sum_{i=1}^{n} X_i^j, \quad j = 1, \ldots, k$$

where k is the number of parameters to estimate.

Example 2.6.9 *Consider the setup of Example 2.6.8 with σ^2 unknown, and let us find moment-type estimators of μ and σ^2. We know that $\sigma^2 = \mathrm{Var}_\theta(X) = \mathbb{E}_\theta(X^2) - \mu^2 = \mu_2 - \mu^2$, hence we need to set*

$$\begin{cases} \mu = \frac{1}{n} \sum_{i=1}^{n} X_i \\ \mu_2 = \frac{1}{n} \sum_{i=1}^{n} X_i^2 \end{cases}$$

and then $\hat{\mu}_n = \bar{X}_n$ and $\hat{\sigma}_n^2 = \frac{1}{n} \sum_{i=1}^{n} X_i^2 - (\bar{X}_n)^2 = \frac{1}{n} \sum_{i=1}^{n} (X_i - \bar{X}_n)^2$.

2.6.6   Least squares method

This method is one of the oldest in statistics and it is based on the minimization of the quadratic norm between some theoretical function of the random variable and the corresponding function on the observed data. A typical application is regression analysis but the general setup is to solve a minimization problem which involves the residuals in some model. Assume (X_i, Y_i), $i = 1, \ldots, n$, are the observations and assume some statistical relationship between the variables

X and Y like $Y = f(X; \theta)$. Define the residuals as $R_i = Y_i - f(X_i; \theta)$, then the *least squares method* consists in finding the solution to

$$\hat{\theta}_n = \arg\min_\theta \sum_{i=1}^n (f(X_i; \theta) - Y_i)^2 = \arg\min_\theta \sum_{i=1}^n R_i^2.$$

The residual R_i is assumed to be related to the error due to randomness in the sample. If $f(\cdot)$ is a linear function, the resulting estimator has several properties and can be obtained in an explicit form, in the nonlinear case some numerical method is needed to find the solution to this quadratic problem.

2.6.7   Estimating functions

Similarly to the least squares or moment-type methods one can consider *estimating functions* as some form of distance between the true parameter and the sample counterpart. Estimating functions are functions of both the data and the parameter, i.e. functions of the form $H(X_1, \ldots, X_n, \theta)$ with the property that $\mathbb{E}_\theta\{H(X_1, \ldots, X_n, \theta)\} = 0$ if θ is the true value. The estimator is then obtained in implicit form as the solution to

$$\hat{\theta}_n : H(X_1, \ldots, X_n; \hat{\theta}_n) = 0.$$

For example, if we take as $H(X_1, \ldots, X_n; \theta) =$ minus the derivative of the log-likelihood function, i.e.

$$H(X_1, \ldots, X_n; \theta) = -\frac{\partial}{\partial \theta} \ell_n(\theta)$$

the value of θ which makes $H(; \theta) = 0$ is nothing but the maximum likelihood estimator of θ when the conditions for which the score function has zero mean are satisfied.

2.6.8   Confidence intervals

Along with point estimates produced by some estimator T_n it is sometimes convenient to get an indication of an interval of plausible values centered around the estimate which is likely (or supposed) to contain the unknown parameter θ. Although, for any deterministic interval $[a, b]$ the probability of the event '$\theta \in [a, b]$' is either 1 or 0, it is still possible to obtain some information using both the mean and the variance of the estimator. In particular, if an estimator is unbiased (or consistent) and we know its variance, we can use a version of the central limit theorem[5] which gives as a result

$$\frac{T_n - \mathbb{E}_\theta(T_n)}{\sqrt{\mathrm{Var}_\theta(T_n)}} \xrightarrow{d} Z \sim N(0, 1)$$

---

[5] To be proved case by case.

and so we can approximate the distribution of T_n for large n as

$$T_n \sim \mathbb{E}_\theta T_n + Z\sqrt{\mathrm{Var}_\theta(T_n)} \sim \theta + Z\sqrt{\mathrm{Var}_\theta(T_n)}.$$

In this case, the quantity

$$\frac{T_n - \theta}{\sqrt{\mathrm{Var}_\theta(T_n)}} \xrightarrow{d} Z \sim \mathrm{N}(0, 1)$$

is the standardized distance of the estimator T_n from the target unknown value θ. We can try to control this distance in probability in the following way

$$P_\theta \left\{ z_{\frac{\alpha}{2}} < \frac{T_n - \theta}{\sqrt{\mathrm{Var}_\theta(T_n)}} < z_{1-\frac{\alpha}{2}} \right\} = 1 - \alpha$$

where $\alpha \in (0, 1)$ is interpreted as an error and $z_q = P(Z \leq q)$ is the q-th quantile of the standard normal distribution. If we want to interpret this probability of 'an interval for the random variable T_n' in 'an interval for the unknown θ' we need to reorganize the writing as follows:

$$P_\theta \left\{ \left(T_n - z_{\frac{\alpha}{2}}\sqrt{\mathrm{Var}_\theta(T_n)} > \theta \right) \cap \left(T_n - z_{1-\frac{\alpha}{2}}\sqrt{\mathrm{Var}_\theta(T_n)} < \theta \right) \right\} = 1 - \alpha$$

or

$$P_\theta \left\{ T_n - z_{1-\frac{\alpha}{2}}\sqrt{\mathrm{Var}_\theta(T_n)} < \theta < T_n - z_{\frac{\alpha}{2}}\sqrt{\mathrm{Var}_\theta(T_n)} \right\} = 1 - \alpha.$$

Now, remember that, by symmetry of the Gaussian distribution, we have

$$z_{\frac{\alpha}{2}} = -z_{1-\frac{\alpha}{2}},$$

thus, finally

$$P_\theta \left\{ T_n - z_{1-\frac{\alpha}{2}}\sqrt{\mathrm{Var}_\theta(T_n)} < \theta < T_n + z_{1-\frac{\alpha}{2}}\sqrt{\mathrm{Var}_\theta(T_n)} \right\} = 1 - \alpha.$$

The above interval is usually rewritten as $\theta \in [T_n \pm z_{1-\frac{\alpha}{2}}\sqrt{\mathrm{Var}_\theta(T_n)}]$ but when we see the writing

$$P_\theta \left\{ \theta \in \left[T_n \pm z_{1-\frac{\alpha}{2}}\sqrt{\mathrm{Var}_\theta(T_n)} \right] \right\} = 1 - \alpha$$

it is important to notice that θ is a given and unknown *fixed* constant, but the extremes of the interval vary because T_n varies. This is interpreted as $(1 - \alpha)\%$ of the times, the above interval contains the true value θ and the remaining $\alpha\%$ of times produce intervals which do not contain the true value.

In some cases, for example $T_n = \bar{X}_n$, with known variance σ^2 and Gaussian random sample, the above interval is exact, i.e. we don't need the central limit theorem. In most cases, the interval is only approximate, so the interval should be regarded as an approximation. Moreover, the calculation of the interval depends on the variance of the estimator which usually needs to be estimated. In such a case, a consistent estimator is needed to estimate $\mathrm{Var}_\theta(T_n)$ in order to justify the approximation of the asymptotic distribution of T_n and hence obtain the confidence interval.

Example 2.6.10 *Consider an i.i.d. sample from the Bernoulli distribution with parameter p. The MLE of p is the sample mean $\bar{X}_n$ which we denote by $\hat{p}_n$ (see Exercise 2.17). The estimator is unbiased $\mathbb{E}_p(\hat{p}_n) = p$ because it is just the sample mean and using the central limit theorem we get:*

$$\hat{p}_n \sim p + \sqrt{\mathrm{Var}_p(\hat{p}_n)}Z = p + \sqrt{\frac{p(1-p)}{n}}Z;$$

clearly we cannot calculate the approximate interval as

$$p \in \left[\hat{p}_n \pm z_{1-\frac{\alpha}{2}} \sqrt{\frac{p(1-p)}{n}} \right]$$

because p is unknown. Then we need to estimate the variance of $\hat{p}_n$. By the law of large numbers, the estimator $\hat{p}_n$ is consistent and hence we can propose the following approximated interval

$$p \in \left[\hat{p}_n \pm z_{1-\frac{\alpha}{2}} \sqrt{\frac{\hat{p}_n(1-\hat{p}_n)}{n}} \right]$$

but this is only asymptotically of level α. Another conservative approach is to consider that $f(x) = x(1-x)$, for $x \in (0, 1)$, has its maximum at $x = 0.5$. Thus, we can use $p = 0.5$ to estimate the largest variance of $\hat{p}_n$ and obtain the following confidence interval

$$p \in \left[\hat{p}_n \pm \frac{z_{1-\frac{\alpha}{2}}}{2\sqrt{n}} \right]$$

whose length is independent of the actual value of $\hat{p}_n$.

For asymptotically efficient estimators (like the maximum likelihood estimator in most of the cases) the asymptotic variance coincides with the inverse of the Fisher information and it is possible to obtain it numerically as part of the estimation procedure. In particular, the `mle` function of R produces both estimates and confidence intervals based on this strategy.

2.6.9   Numerical maximization of the likelihood

It is not always the case, that maximum likelihood estimators can be obtained in explicit form. For what concerns applications to real data, it is important to know if mathematical results about optimality of MLE estimators exist and then find the estimators numerically. R offers a prebuilt generic function called `mle` in the package **stats4** which can be used to maximize a likelihood. The mle function actually minimizes the negative log-likelihood $-\ell(\theta)$ as a function of the parameter θ. For example, consider a sample of $n = 1000$ observations from a Gaussian law with $N(\mu = 5, \sigma^2 = 4)$, and let us estimate the parameters numerically:

```
R> set.seed(123)
R> library("stats4")
R> x <- rnorm(1000, mean = 5, sd = 2)
R> log.lik <- function(mu = 1, sigma = 1) -sum(dnorm(x, mean = mu,
+       sd = sigma, log = TRUE))
R> fit <- mle(log.lik, lower = c(0, 0), method = "L-BFGS-B")
R> fit

Call:
mle(minuslogl = log.lik, method = "L-BFGS-B", lower = c(0, 0))

Coefficients:
      mu     sigma
5.032256 1.982398
```

and, using explicit estimators for μ and σ^2 we get:

```
R> mean(x)

[1] 5.032256

R> sd(x)

[1] 1.98339
```

which almost coincides numerically (remember from Exercise 2.16 that the MLE estimator of σ^2 is S_n^2 while sd calculates $\bar{S}_n^2$). What is worth knowing is that the output of the mle function is an object which contains several informations, including the value of $\ell(\theta)$ at the point of its maximum

```
R> logLik(fit)

'log Lik.' -2103.246 (df=2)
```

the variance-covariance matrix of the estimators, which is obtained inverting the Hessian matrix at the point θ corresponding to the maximum likelihood estimate assuming that, under the regularity conditions for the data generating model, (2.28) holds and that MLE are asymptotically efficient

```
R> vcov(fit)

              mu           sigma
mu     3.929901e-03 8.067853e-10
sigma  8.067853e-10 1.964946e-03
```

Similarly, approximate confidence intervals and the complete summary of the estimated parameters can be obtained using respectively the functions `confint` and `summary`:

```
R> confint(fit)

Profiling...
          2.5 %    97.5 %
mu      4.909269 5.155242
sigma   1.898595 2.072562

R> summary(fit)

Maximum likelihood estimation

Call:
mle(minuslogl = log.lik, method = "L-BFGS-B", lower = c(0, 0))

Coefficients:
       Estimate Std. Error
mu     5.032256 0.06268893
sigma  1.982398 0.04432771

-2 log L: 4206.492
```

In our example above, we have specified the option `method`, which tells R to maximize the likelihood when parameters are subject to constraints. In our example, we specified a vector of lower bounds for the two parameters using the argument `lower`. We will return to the general problem of function minimization and numerical optimization in Appendix A.

2.6.10   The δ-method

The so-called δ-method is a technique to derive the approximate distribution of an estimator which is a function of another consistent estimator. Assume that T_n is a consistent estimator for θ such that

$$\sqrt{n}(T_n - \theta) \xrightarrow{d} N(0, \sigma^2).$$

Now let $g(\theta)$ be a differentiable and non-null function of θ. Then

$$\sqrt{n}(g(T_n) - g(\theta)) \xrightarrow{d} N\left(0, \sigma^2 \left[\frac{\partial}{\partial \theta} g(\theta)\right]^2\right). \tag{2.29}$$

We now prove this result under the assumption that $g(\cdot)$ is a continuous function.

Proof. Using Taylor expansion, we can write

$$g(T_n) = g(\theta) + \frac{\partial}{\partial \theta} g(\tilde{\theta})(T_n - \theta)$$

for some $\tilde{\theta}$ between T_n and θ. Since T_n is a consistent estimator of θ then also $\tilde{\theta} \overset{p}{\to} \theta$. Since $g(\cdot)$ is continuous, by continuously mapping Theorem 2.4.10 we also have that

$$\frac{\partial}{\partial \theta} g(\tilde{\theta}) \overset{p}{\to} \frac{\partial}{\partial \theta} g(\theta).$$

Therefore,

$$\sqrt{n}(g(T_n) - g(\theta)) = \frac{\partial}{\partial \theta} \sqrt{n} g(\tilde{\theta})(T_n - \theta).$$

By assumption $\sqrt{n}(T_n - \theta) \overset{d}{\to} N(0, \sigma^2)$, hence applying Slutsky's Theorem 2.4.9, we end up with the result (2.29).

2.7   Solution to exercises

Solution 2.1 (Exercise 2.1) *Notice that by axiom (ii)* $1 = P(\Omega)$. *By definition of complementary set, we have that* $\Omega = A \cup \bar{A}$ *because A and $\bar{A}$ are disjoint sets. Then, by axiom (iii)* $1 = P(\Omega) = P(A \cup \bar{A}) = P(A) + P(\bar{A})$. *By the same decomposition* $P(A) \leq 1$ *because $P(\bar{A}) \geq 0$ by axiom (i).*

Solution 2.2 (Exercise 2.2) *We only prove (2.1) and (2.2) because the proof of sub-additivity requires additional preliminary results. Notice that* $A \cup B = A \cup (\bar{A} \cap B)$ *and then* $P(A \cup B) = P(A) + P(\bar{A} \cap B)$. *Further,* $B = (A \cap B) \cup (\bar{A} \cap B)$, *hence* $P(B) = P(A \cap B) + P(\bar{A} \cap B) = P(A \cap B) + P(A \cup B) - P(A)$, *which proves (2.1). We now observe that* $A \subset B$ *implies* $A \cap B = A$, *then* $P(B) = P(B \cap \Omega) = P(B \cap (A \cup \bar{A})) = P(B \cap A) + P(B \cap \bar{A}) = P(A) + P(B \cap \bar{A})$, *so* $P(B) \geq P(A)$ *because* $P(B \cap \bar{A}) \geq 0$.

Solution 2.3 (Exercise 2.3) *We need to show (i) to (iii) of Definition 2.1.2. Property (i) is trivial because for all A,* $P_B(A) = P(A|B)$ *is a ratio of a non-negative quantity $P(A \cap B)$ and a positive quantity $P(B)$. For (ii) we have that* $P_B(\Omega) = P(\Omega \cap B)/P(B) = P(B)/P(B) = 1$; *For (iii)*

$$P_B(\cup_i A_i) = P(\cup_i A_i | B) = \frac{P((\cup_i A_i) \cap B)}{P(B)} = \frac{P(\cup_i (A_i \cap B))}{P(B)}$$

$$= \frac{1}{P(B)} \sum_i P(A_i \cap B) = \sum_i P(A_i | B) = \sum_i P_B(A_i).$$

Solution 2.4 (Exercise 2.4) *Given that $E \cap \Omega = E$ and that $\Omega = \bigcup_{i=1}^{n} A_i$ we can rewrite $E = E \cap (\bigcup_{i=1}^{n} A_i)$. Applying distributional properties of $\cap$ and $\cup$ operators, we obtain*

$$E = E \cap \left(\bigcup_{i=1}^{n} A_i \right) = \bigcup_{i=1}^{n} (E \cap A_i).$$

Due to the fact that the sets A_i in the partition are disjoint, so are the events $E \cap A_i$. Finally, we notice that $P(E \cap A_i) = P(E|A_i)P(A_i)$ from Definition 2.1.7 and the proof is complete.

Solution 2.5 (Exercise 2.5) *If X and Y are independent we have that $\mathbb{E}(XY) = \mathbb{E}(X)\mathbb{E}(Y)$, hence $\mathrm{Cov}(X, Y) = \mathbb{E}(XY) - \mathbb{E}(X)\mathbb{E}(Y) = 0$. We now derive a counter example. Let X be a continuous random variable defined in $(-1, 1)$ with density $f(x) = \frac{1}{2}$. Define $Y = X^2$. We now calculate the covariance. $\mathrm{Cov}(X, Y) = \mathbb{E}(XY) - \mathbb{E}(X)\mathbb{E}(Y) = \mathbb{E}(X^3) - \mathbb{E}(X)\mathbb{E}(X^2) = \frac{1}{2}\int_{-1}^{1} x^3 dx - 0 \cdot \mathbb{E}(X^2) = 0$.*

Solution 2.6 (Exercise 2.6) $\mathbb{E}(X) = 1 \cdot P(X = 1) + 0 \cdot P(X = 0) = p$ *and* $\mathrm{Var}(X) = \mathbb{E}\{X - \mathbb{E}(X)\}^2 = (1 - p)^2 \cdot p + (0 - p)^2 \cdot (1 - p) = (1 - p)p\{(1 - p) + p\} = p(1 - p)$. *Finally,* $\varphi(t) = \mathbb{E}\{e^{itX}\} = e^{it}p + e^0(1 - p)$.

Solution 2.7 (Exercise 2.7) *Noticing that the Binomial random variable can be represented as $Y = \sum_{i=1}^{n} X_i$, where the X_i are i.i.d. Ber(p), applying (2.5), (2.6) and (2.7), one obtains respectively mean, variance and characteristic function of Y.*

Solution 2.8 (Exercise 2.8)

$$\mathbb{E}(X) = \sum_{k=0}^{\infty} k \frac{\lambda^k e^{-\lambda}}{k!} = e^{-\lambda} \sum_{k=1}^{\infty} k \frac{\lambda^k}{k!} = \lambda e^{-\lambda} \sum_{k=1}^{\infty} \frac{\lambda^{k-1}}{(k-1)!}$$

$$= \lambda e^{-\lambda} \sum_{j=0}^{\infty} \frac{\lambda^j}{j!} = \lambda e^{-\lambda} e^{\lambda}.$$

$$\mathbb{E}(X^2) = \sum_{k=0}^{\infty} k^2 \frac{\lambda^k e^{-\lambda}}{k!} = e^{-\lambda} \lambda \sum_{k=1}^{\infty} k \frac{\lambda^{k-1}}{(k-1)!}$$

$$= e^{-\lambda} \lambda \left\{ \sum_{k=1}^{\infty} (k-1) \frac{\lambda^{k-1}}{(k-1)!} + \sum_{k=1}^{\infty} \frac{\lambda^{k-1}}{(k-1)!} \right\}$$

$$= e^{-\lambda} \lambda \left\{ \frac{\mathbb{E}(X)}{e^{-\lambda}} + e^{\lambda} \right\} = e^{-\lambda} \lambda \left\{ \frac{\lambda}{e^{-\lambda}} + e^{\lambda} \right\} = \lambda^2 + \lambda$$

Hence $\mathrm{Var}(X) = \mathbb{E}(X^2) - \{\mathbb{E}(X)\}^2 = \lambda$. *Finally*

$$\phi(t) = \sum_{k=0}^{\infty} e^{itk} \frac{\lambda^k e^{-\lambda}}{k!} = e^{-\lambda} \sum_{k=0}^{\infty} \frac{(\lambda e^{it})^k}{k!} = e^{-\lambda} \exp\{\lambda e^{it}\} = \exp\left\{\lambda \left(e^{it} - 1\right)\right\}.$$

Solution 2.9 (Exercise 2.9) *This is a simple exercise of calculus.* $\mathbb{E}(X) = \int_a^b \frac{x}{b-a} dx = \frac{x^2}{2(b-a)}\Big|_a^b = (a + b)/2$. *We now calculate* $\mathbb{E}(X^2) = \int_a^b \frac{x^2}{b-a} dx$

PROBABILITY, RANDOM VARIABLES AND STATISTICS   73

$$= \frac{x^3}{3(b-a)}\Big|_a^b = \frac{b^3-a^3}{3(b-a)}, \text{ therefore } \mathrm{Var}(X) = \mathbb{E}(X^2) - \{\mathbb{E}(X)\}^2 = \frac{b^3-a^3}{3(b-a)} - \frac{(b+a)^2}{4} =$$
$\frac{(b-a)^3}{12(b-a)}$. *Finally,* $\phi(t) = \int_a^b \frac{e^{itx}}{b-a}dx = \frac{e^{itb}-e^{ita}}{it(b-a)}.$

Solution 2.10 (Exercise 2.10)

$$\mathbb{E}(X) = \int_0^\infty x\lambda e^{-\lambda x}dx = -\int_0^\infty x\frac{d}{dx}e^{-\lambda x} = -xe^{-\lambda x}\Big|_0^\infty + \int_0^\infty e^{-\lambda x}dx$$
$$= -\frac{1}{\lambda}e^{-\lambda x}\Big|_0^\infty = \frac{1}{\lambda}.$$

Similarly

$$\mathbb{E}(X^2) = \int_0^\infty x^2\lambda e^{-\lambda x}dx = 2\int_0^\infty x\frac{d}{dx}e^{-\lambda x} = -\frac{2}{\lambda}\int_0^\infty x\frac{d}{dx}e^{-\lambda x} = \frac{2}{\lambda^2}.$$

Therefore, $\mathrm{Var}(X) = \mathbb{E}(X^2) - \{\mathbb{E}(X)\}^2 = \lambda^{-2}$. *For the characteristic function we remember that, by Euler's formula* $e^{ix} = \cos(x) + i\sin(x)$, *then* e^{ix} *is always limited, hence*

$$\phi(t) = \int_0^\infty e^{itx}\lambda e^{-\lambda x}dx = \frac{\lambda}{it-\lambda}\left(e^{-\lambda x}e^{itx}\right)\Big|_0^\infty = \frac{\lambda}{\lambda-it}.$$

Solution 2.11 (Exercise 2.11) *Notice that* $\frac{d}{dx}f(x) = \frac{\mu-x}{\sigma^2}f(x)$, *hence*

$$\mathbb{E}(X) = \int_\mathbb{R} xf(x)dx = -\sigma^2\int_\mathbb{R}\frac{(\mu-x)-\mu}{\sigma^2}f(x)dx = -\sigma^2\int_\mathbb{R}\frac{d}{dx}f(x)dx + \mu$$
$$= -\sigma^2 f(x)\big|_{-\infty}^\infty + \mu = \mu.$$

Similarly, for the variance we notice that $\frac{d^2}{dx^2}f(x) = \left(\frac{(\mu-x)^2}{\sigma^4} - \frac{1}{\sigma^2}\right)f(x)$, *then*

$$\mathbb{E}(X-\mu)^2 = \int_\mathbb{R}(x-\mu)^2 f(x)dx = \sigma^4\int_\mathbb{R}\left(\frac{(\mu-x)^2}{\sigma^4} - \frac{1}{\sigma^2}\right)f(x)dx + \sigma^2$$
$$= \sigma^4\int_\mathbb{R}\frac{d^2}{dx^2}f(x)dx + \sigma^2$$
$$= \sigma^4\frac{d}{dx}f(x)\Big|_{-\infty}^\infty + \sigma^2 = \sigma^2(x-\mu)f(x)\big|_{-\infty}^\infty + \sigma^2 = \sigma^2.$$

Further, for the characteristic function we have

$$\mathbb{E}\{e^{itX}\} = \int_\mathbb{R}\frac{1}{\sqrt{2\pi\sigma^2}}\exp\left\{-\frac{(x-\mu)^2 - 2itx\sigma^2}{2\sigma^2}\right\}dx$$
$$= \exp\frac{(\sigma^2 it+\mu)^2 - \mu^2}{2\sigma^2}\int_\mathbb{R}\frac{1}{\sqrt{2\pi\sigma^2}}\exp\left\{-\frac{(x-(\mu+\sigma^2 it))^2}{2\sigma^2}\right\}dx$$
$$= \exp\frac{(\sigma^2 it+\mu)^2 - \mu^2}{2\sigma^2} = \exp\left\{\mu it - \frac{\sigma^2 t^2}{2}\right\}.$$

Solution 2.12 (Exercise 2.12) *We apply* (2.15) *with the exponential densities* $f_X(x) = f_Y(x) = \lambda e^{\lambda x}$. *We only consider the case* $z > 0$ *because* $f_Z(z) = 0$ *for* $z < 0$.

$$f_Z(z) = \int_{\mathbb{R}} f_X(z - y) f_Y(y) dy = \int_0^z \lambda e^{-\lambda(z-y)} \lambda e^{-\lambda y} dy$$

$$= \int_0^z \lambda^2 e^{-\lambda z} dy = \lambda^2 z e^{-\lambda z}.$$

Solution 2.13 (Exercise 2.13) *We use the convolution formula* (2.15) *only for the particular case* $\mu_1 = \mu_2 = 0$ *and* $\sigma_1^2 = \sigma_2^2 = 1$.

$$f_Z(z) = \int_{\mathbb{R}} \frac{e^{-\frac{(z-y)^2}{2}}}{\sqrt{2\pi}} \frac{e^{-\frac{y^2}{2}}}{\sqrt{2\pi}} dy = \frac{1}{2\pi} \int_{\mathbb{R}} e^{-\frac{z^2}{2} - y^2 + zy} dy = \frac{e^{-\frac{z^2}{4}}}{2\pi} \int_{\mathbb{R}} e^{-(y-\frac{z}{2})^2} dy$$

$$= \frac{e^{-\frac{z^2}{4}}}{2\sqrt{\pi}} \left[\int_{\mathbb{R}} \frac{1}{\sqrt{\pi}} e^{-(y-\frac{z}{2})^2} dy = 1 \right],$$

where the integral in the brackets evaluates to one because it is the integral of the density of a $N\left(\frac{z}{2}, \frac{1}{2}\right)$ *and*

$$f_Z(z) = \frac{e^{-\frac{z^2}{2\cdot 2}}}{\sqrt{2\pi 2}}, \quad hence \quad Z \sim N(0, 2).$$

For the general case, instead of the convolution formula (2.15) *we use the characteristic function*

$$\varphi_Z(t) = \varphi_X(t) \varphi_Y(t) = e^{\mu_1 it - \frac{\sigma_1^2 t^2}{2}} e^{\mu_2 it - \frac{\sigma_2^2 t^2}{2}}$$

$$= \exp\left\{ (\mu_1 + \mu_2) it - \frac{(\sigma_1^2 + \sigma_2^2) t^2}{2} \right\}.$$

Therefore, $Z \sim N(\mu_1 + \mu_2, \sigma_1^2 + \sigma_2^2)$.

Solution 2.14 (Exercise 2.14) *We have that*

$$\varphi_X(u) = \exp\left\{ \lambda \left(e^{iu} - 1 \right) \right\} = \left(\exp\left\{ \frac{\lambda}{n} \left(e^{iu} - 1 \right) \right\} \right)^n = \left(\varphi_{X^{(1/n)}}(u) \right)^n,$$

with $X^{(1/n)} \sim \text{Poi}\left(\frac{\lambda}{n}\right)$.

Solution 2.15 (Exercise 2.15) *Clearly* $\mathbb{E}_\theta(T_n) = \frac{1}{n} \sum_{i=1}^n \{a_i \mathbb{E}_\theta(X_i)\} = \frac{\theta}{n} \sum_{i=1}^n a_i = \theta$. *Now we look for the* a_i *such that we get the minimal variance of* T_n. *So we need to search for the minimum of*

$$\text{Var}_\theta(T_n) = \frac{1}{n^2} \sum_{i=1}^n a_i^2 \text{Var}_\theta(X_i)$$

which corresponds to minimizing the quantity $\sum_{i=1}^{n} a_i^2$ under the constraint $\sum_{i=1}^{n} a_i = n$. We make use of Lagrange multipliers, therefore we construct the function

$$f(\lambda, a_1, \ldots, a_n) = \sum_{i=1}^{n} a_i^2 - \lambda \left(\sum_{i=1}^{n} a_i - n \right)$$

and calculate the derivatives of $f(\lambda, a_i)$ with respect to all variables, i.e.

$$\frac{\partial}{\partial a_i} f(\lambda, a_1, \ldots, a_n) = 2a_i - \lambda = 0, \quad \frac{\partial}{\partial \lambda} f(\lambda, a_1, \ldots, a_n) = \sum_{i=1}^{n} a_i - n = 0.$$

Then we sum up all the equations involving the a_i's and obtain $2 \sum_{i=1}^{n} a_i - n\lambda = 0$ from which in turn we have $\lambda = 2$. Now put back this λ in $2a_i - \lambda = 0$ and obtain the desired result $a_i = 1$.

Solution 2.16 (Exercise 2.16) *We minimize minus the log-likelihood function as a function of μ and σ*

$$h(\mu, \sigma^2) = -\ell_n(\mu, \sigma^2) = \frac{n}{2} \log(2\pi) + \frac{n}{2} \log \sigma^2 + \sum_{i=1}^{n} \frac{(X_i - \mu)^2}{2\sigma^2}$$

$$\frac{\partial}{\partial \mu} h(\mu, \sigma^2) = -\frac{1}{\sigma^2} \sum_{i=1}^{n} (X_i - \mu) = 0$$

$$\frac{\partial}{\partial \sigma^2} h(\mu, \sigma^2) = \frac{n}{2\sigma^2} - \frac{1}{2\sigma^4} \sum_{i=1}^{n} (X_i - \mu)^2 = 0$$

From the first equation we get $\hat{\mu} = \bar{X}_n$ and pluggin-in this value into the second equation we obtain $\hat{\sigma}^2 = S_n^2 = \frac{1}{n} \sum_{i=1}^{n} (X_1 - \bar{X}_n)^2$. We now need to verify that at least one of the two second derivatives at point $(\hat{\mu}, \hat{\sigma}^2)$ is positive and the determinant of the Hessian matrix of second-order partial derivatives of $h(\mu, \sigma^2)$ evaluated at the point $(\hat{\mu}, \hat{\sigma}^2)$ is positive. So we calculate partial derivatives first.

$$\frac{\partial^2}{\partial \mu^2} h(\mu, \sigma^2) = \frac{n}{\sigma^2},$$

$$\frac{\partial^2}{\partial (\sigma^2)^2} h(\mu, \sigma^2) = -\frac{n}{2\sigma^4} + \frac{1}{\sigma^6} \sum_{i=1}^{n} (X_i - \mu)^2,$$

$$\frac{\partial^2}{\partial \mu \sigma^2} h(\mu, \sigma^2) = +\frac{1}{\sigma^4} \sum_{i=1}^{n} (X_i - \mu).$$

Now we recall that $\hat{\sigma}^2 = \frac{1}{n}\sum_{i=1}^{n}(X_i - \hat{\mu})^2$ *and* $\sum_{i=1}^{n}(X_i - \hat{\mu}) = 0$, *hence*

$$\frac{\partial^2}{\partial\mu^2}h(\mu,\sigma^2)\bigg|_{\mu=\hat{\mu},\sigma^2=\hat{\sigma}^2} = \frac{n}{\hat{\sigma}^2} > 0,$$

$$\frac{\partial^2}{\partial(\sigma^2)^2}h(\mu,\sigma^2)\bigg|_{\mu=\hat{\mu},\sigma^2=\hat{\sigma}^2} = -\frac{n}{2\hat{\sigma}^4} + \frac{n}{\hat{\sigma}^4} = \frac{n}{2\hat{\sigma}^4} > 0,$$

$$\frac{\partial^2}{\partial\mu\sigma^2}h(\mu,\sigma^2)\bigg|_{\mu=\hat{\mu},\sigma^2=\hat{\sigma}^2} = \frac{1}{\hat{\sigma}^4}\sum_{i=1}^{n}(X_i - \hat{\mu}) = 0.$$

Finally, we calculate the determinant of the Hessian matrix evaluated at point $(\hat{\mu}, \hat{\sigma}^2)$ *to check if it is positive*

$$H(\hat{\mu},\hat{\sigma}^2) = \begin{vmatrix} \frac{\partial^2}{\partial\mu^2}h(\mu,\sigma^2) & \frac{\partial^2}{\partial\mu\sigma^2}h(\mu,\sigma^2) \\ \frac{\partial^2}{\partial\mu\sigma^2}h(\mu,\sigma^2) & \frac{\partial^2}{\partial(\sigma^2)^2}h(\mu,\sigma^2) \end{vmatrix}_{\mu=\hat{\mu},\sigma^2=\hat{\sigma}^2}$$

$$= \begin{vmatrix} \frac{n}{\hat{\sigma}^2} & 0 \\ 0 & \frac{n}{2\hat{\sigma}^4} \end{vmatrix}$$

$$= \frac{1}{2}\frac{n^2}{\hat{\sigma}^6} > 0.$$

Solution 2.17 (Exercise 2.17) *For each observation we can write* $P(X = x_i) = p^{x_i}(1-p)^{1-x_i}$ *hence*

$$L_n(p) = \prod_{i=1}^{n} p^{X_i}(1-p)^{1-X_i} = p^{\sum_{i=1}^{n} X_i}(1-p)^{n-\sum_{i=1}^{n} X_i}$$

hence

$$\ell_n(p) = \sum_{i=1}^{n} X_i \log p + \left(n - \sum_{i=1}^{n} X_i\right)\log(1-p)$$

and

$$\frac{\partial}{\partial p}\ell_n(p) = \frac{\sum_{i=1}^{n} X_i}{p} - \frac{n - \sum_{i=1}^{n} X_i}{1-p} = 0,$$

from which we obtain

$$\frac{1-p}{p} = \frac{n}{\sum_{i=1}^{n} X_i} - 1 \iff 1-p = \frac{np}{\sum_{i=1}^{n} X_i} - p \iff p = \bar{X}_n.$$

2.8   Bibliographical notes

One of the basic books in probability to study the subject in more detail is, for example, Billingsley (1986). This book also includes a complete treatment on conditional expectation. For an updated account on several probability distribution, one should not miss Johnson *et al.* (1994, 1995). Two textbooks on statistical inference is the classical reference Mood *et al.* (1974) and Casella and Berger (2001).

References

Abramowitz, M. and Stegun, I. (1964). *Handbook of Mathematical Functions*. Dover Publications, New York.

Barndorff-Nielsen, O. E. (1977). Exponentially decreasing distributions for the logarithm of particle size. *Proceedings of the Royal Society of London. Series A, Mathematical and Physical Sciences (The Royal Society)* **353**, 1674, 401–409.

Barndorff-Nielsen, O. E. (1997). Normal inverse gaussian distributions and stochastic volatility modelling. *Scand. J. Statist*. **24**, 113.

Billingsley, P. (1986). *Probability and Measure, Second Edition*. John Wiley & Sons, Inc., New York.

Casella, G. and Berger, R. (2001). *Statistical Inference. Second edition*. Duxbury Press, New York.

Cooley, J. and Tukey, J. (1965). An algorithm for the machine calculation of complex fourier series. *Math. Comput*. **19**, 297–301.

Durham, G. and Gallant, A. (2002). Numerical techniques for maximum likelihood estimation of continuous-time diffusion processes. *J. Bus. Econ. Stat*. **20**, 297–316.

Eberlein, E. and Prause, K. (2002). The generalized hyperbolic model: financial derivatives and risk measures. *In* H. Geman, D. Madan, S. Pliska, and T. Vorst *(Eds.)*, *Mathematical Finance – Bachelier Congress 2000* 245–267.

Gallant, A. and Tauchen, G. (1996). Which moments to match. *Econ. Theory* **12**, 657–681.

Grigelionis, B. (1999). Processes of meixner type. *Lith. Math. J*. **39**, 1, 33–41.

Hall, A. (2005). *Generalized Method of Moments*. Advanced Texts in Econometrics, Oxford University Press, New York.

Johnson, N., Kotz, S., and Balakrishnan, N. (1994). *Continuous Univariate Distributions, Volume 1*. John Wiley & Sons, Inc., New York.

Johnson, N., Kotz, S., and Balakrishnan, N. (1995). *Continuous Univariate Distributions, Volume 2*. John Wiley & Sons, Inc., New York.

Kendall, M. and Stuart, L. (1977). *The Advanced Theory of Statistics, Vol. 1*. Macmillan, New York.

Kolmogorov, A. (1933). *Foundation of Probability*. Wiley, New York.

Melnick, E. and Tenenbein, A. (1982). Misspecifications of the normal distribution. *The American Statistician* **36**, 4, 372–373.

Mikosch, T. (1998). *Elementary Stochastic Calculus with Finance in View*. World Scientific, Singapore.

Mood, A., Graybill, F., and Boes, D. (1974). *Introduction to the Theory of Statistics. Third edition*. McGraw-Hill, New York.

Nolan, J. P. (2010). *Stable Distributions – Models for Heavy Tailed Data*. Birkhäuser, Boston. In progress, Chapter 1 online at academic2.american.edu/~jpnolan.

Sato, K. (1999). *Lévy Processes and Infinitely Divisible Distributions*. Cambridge University Press, Cambridge.

Schoutens, W. and Teugels, J. (1998). Lévy processes, polynomials and martingales. *Commun. Statist.- Stochastic Models* **14**, 1, 335–349.

3

Stochastic processes

When there is the need to model some form of dependence in a sample of observations, stochastic processes arise quite naturally. Dependence is a generic term which must be specified case by case in statistics and probability, but the most common situations include time and/or spatial dependence. In our context, the most interesting form of dependence is time dependence. Still, time dependency can be modelled in a variety of forms as we will discuss in this chapter.

3.1 Definition and first properties

We assume to have a probability space $(\Omega, \mathcal{A}, P)$. A real-valued, one-dimensional, *stochastic process* is a family of random variables $\{X_\gamma, \gamma \in \Gamma\}$ defined on $\Omega \times \Gamma$ taking values in $\mathbb{R}$. The set Γ may be any abstract set, but we will restrict our attention to particular cases. For each $\gamma \in \Gamma$, the random variable $X(\gamma, \omega)$ is a measurable map $X(\gamma, \omega) \mapsto \mathbb{R}$. For a given fixed value of ω, say $\bar{\omega}$, the map $X(\gamma, \bar{\omega})$, seen as a function of $\gamma \in \Gamma$, represents one evolution of the process and the set

$$\{X(\gamma, \bar{\omega}), \gamma \in \Gamma\}$$

is called *trajectory of the process*. In what follows, to keep the notation compact, we will adopt the notation $X_\gamma = X_\gamma(\omega) = X(\gamma, \omega) = X(\gamma)$ whenever needed. We will now consider more concrete cases of stochastic processes.

Example 3.1.1 *If $\Gamma = \mathbb{N}$ and the X_n, $n \in \mathbb{N}$, are independent and identically distributed, the process $\{X_n, n \in \mathbb{N}\}$ represents an i.i.d. sample. This sequence is a discrete time process, usually called Bernoulli or simply random sample and, as seen in Chapter 2, it is the basis of elementary statistical inference.*

Option Pricing and Estimation of Financial Models with R, First Edition. Stefano M. Iacus.
© 2011 John Wiley & Sons, Ltd. Published 2011 by John Wiley & Sons, Ltd.

Example 3.1.2 *If Γ is the axis of positive times $[0, \infty)$, then $\{X_t, t \geq 0\}$ is called continuous time process and each trajectory represents the evolution in time of X. This is the usual case of financial time series.*

Consider for a while $\Gamma = [0, \infty)$. Each value of $\omega \in \Omega$ generates a trajectory $X(t, \omega)$ as a function of $t \in \Gamma$. Each of these trajectories is called *path* of the process. In finance the dynamic of asset prices or their returns are modeled via some kind of stochastic process. For example, the observation of a sequence of quotations of some asset is the one and only one (statistical) observation or path from such models. This means that, unlike the i.i.d. case in which it is assumed to have n replications of the same random variable, in finance we have only one observation for the model which consists in the whole observed path of the process.

For a given $\bar{t}$, the distribution of $X_{\bar{t}}(\omega)$ as a function of $\omega \in \Omega$ is called *finite dimensional distribution* of the process. Clearly, one of the most interesting questions in finance is how to predict future values of a process, say at time $\bar{t} + h$, $h > 0$, given the information (observation) at time $\bar{t}$ (see Figure 3.1.) In order to provide an answer, a correct specification of the statistical model is required, its calibration from the past and actual data and a proper simulation or prediction scheme for the chosen model.

Processes are also classified according to their *state space*. The state space of the process is the set of values assumed by the process X. If the process takes a finite (or countable) number of values (states), the process is called a *discrete* state space process, otherwise the process has *continuous* state space. Combinations of state space and time space identify different classes of processes. Table 3.1 reports some classes of processes.

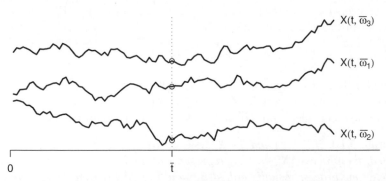

Figure 3.1   The graph represents three different trajectories for three different values of $\bar{\omega}_1, \bar{\omega}_2, \bar{\omega}_3$ for $t \geq 0$. For a given value of t, say $\bar{t}$, the set $\{X_{\bar{t}}(\omega), \omega \in \Omega\}$ represents the set of values of the process X at time $\bar{t}$, from which the finite dimensional distribution of $X_{\bar{t}}$ is obtained.

Table 3.1   Example of processes in terms of state space and time.

| Time | State space | |
	Discrete	Continuous
Discrete	Random walk Markov Chain	ARIMA, GARCH
Continuous	Counting, Poisson	Telegraph, Lévy Wiener, Diffusion

3.1.1   Measurability and filtrations

Consider a stochastic process $\{X(t), t \geq 0\}$. At each time t, it is possible to associate to X_t a σ-algebra denoted by $\mathcal{F}_t = \sigma(X(s); 0 \leq s \leq t)$ (the σ-algebra generated by X up to time t), which is the smallest σ-algebra which makes $X(s, \omega)$ measurable for all $0 \leq s \leq t$. This σ-algebra is the smallest set of subsets of Ω which allows to evaluate probabilities of to events related to $X(t)$. More precisely, we can write

$$\sigma(X(s); 0 \leq s \leq t) = \sigma(X_s^{-1}(B), B \in \mathcal{B}(\mathbb{R}); 0 \leq s \leq t).$$

Definition 3.1.3 *Let $\mathcal{A}_0 = \{\emptyset, \Omega\}$ be the trivial σ-algebra. A family $\mathcal{A} = \{\mathcal{A}_t, t \geq 0\}$ of sub σ-algebras $\mathcal{A}_t \subset \mathcal{A}$, such that $\mathcal{A}_0 \subset \mathcal{A}_s \subset \mathcal{A}_t$, for $0 < s < t$, is called filtration.*

The filtration $\{\mathcal{F}_t, t \geq 0\}$, with $\mathcal{F}_t = \sigma(X(s); 0 \leq s \leq t)$, is called *natural filtration of the process* X_t. Clearly, X_t is $\mathcal{F}_t$-measurable for each $t \geq 0$, but in general stochastic processes can be measurable with respect to generic filtrations.

Definition 3.1.4 *Let $\{\mathcal{A}_t, t \geq 0\}$ be a filtration. A process $\{X_t, t \geq 0\}$ is said to be* adapted *to the filtration $\{\mathcal{A}_t, t \geq 0\}$ if for each t the random variable X_t is $\mathcal{A}_t$-measurable.*

Similarly, for $\mathbb{N}$-indexed stochastic processes, we can construct filtrations $\{\mathcal{F}_n, n \in \mathbb{N}\}$ and natural filtrations, but the main point here is that filtrations are increasing sequences of sub σ-algebras.

Example 3.1.5 (See Billingsley (1986), Th 5.2) *Consider a binary experiment, like throwing a coin. We can associate this experiment with the σ-algebra built on $\Omega = [0, 1]$ in the following way: the random variable $X_1(\omega)$ takes value 0 if $\omega = [0, 1/2)$ and 1 if $\omega = [1/2, 1]$, i.e. $X_1(\omega) = 0$ if $\omega \in [0, 1/2)$ and $X_1(\omega) = 1$ if $\omega \in [1/2, 1]$ (see bottom of Figure 3.2). The probability measure is the Lebesgue*

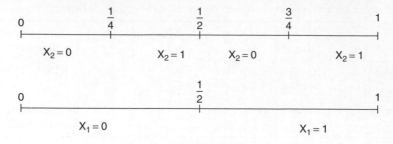

Figure 3.2   Example of building of a filtration. See text of Example 3.1.5.

measure of the corresponding interval, i.e. the length of the interval. Hence $P(X = k) = P(\{\omega \in \Omega : X(\omega) = k\}) = \mu(\{\omega \in \Omega : X(\omega) = k\})$. Hence we have

$$P(X_1 = 0) = P(\{\omega \in \Omega : X(\omega) = 0\}) = \mu\left(\left[0, \frac{1}{2}\right)\right) = \frac{1}{2}$$

and similarly $P(X_1 = 1) = 1/2$. The following σ-algebra

$$\mathcal{F}_1 = \{\emptyset, [0, 1/2), [1/2, 1], [0, 1]\},$$

makes X_1 measurable. We now define X_2 in the following way: $X_2(\omega) = 0$ if $\omega \in [0, 1/4) \cup [1/2, 3/4)$ and $X_2(\omega) = 1$ if $\omega \in [1/4, 1/2) \cup [3/4, 1]$. Hence

$$P(X_2 = 0) = \mu([0, 1/4) \cup [1/2, 3/4)) = 1/4 + 1/4 = 1/2$$

and similarly $P(X_2 = 1) = 1/2$. In this case, the σ-algebra which makes X_2 measurable is the following

$$\mathcal{F}_2 = \{\emptyset, [0, 1/2), [1/2, 1], [0, 1/4), [1/4, 1/2),$$
$$[1/2, 3/4), [3/4, 1], [0, 1], \ldots\}$$

(where by '...' we mean all possible unions of the intervals listed). Of course we have $\mathcal{F}_1 \subset \mathcal{F}_2$. Now $\mathcal{F}_2$ makes both X_2 and X_1 measurable, while $\mathcal{F}_1$ makes measurable X_1 but not X_2. In fact, there is no set in $\mathcal{F}_1$ which makes measurable the event $X_2 = 1$. Similarly one can proceed with X_3. Thinking of the n-th throwing of the coin, we will end with subdividing the interval $[0, 1]$ in subintervals of length $1/2^n$ and obtain the σ-algebra $\mathcal{F}_n$ which includes all previous ones. Hence $\{\mathcal{F}_i, i \geq 1\}$ is a filtration and the process $\{X_i, i \geq n\}$ is adapted to it.

Filtrations are then a way to describe the increasing of information as time pass by. In finance a filtration represents all the information available on the process up to time t. Asking measurability to a process means that for that process it is possible to evaluate probabilities at any given time instant.

Definition 3.1.6 *A stochastic process* $\{X_n, n \geq 1\}$ *is said to be predictable with respect to the filtration* $\{\mathcal{F}_n, n \geq 1\}$ *if* $X_0 \in \mathcal{F}_0$ *and* X_{n+1} *is* $\mathcal{F}_n$*-measurable.*

Hence, for a predictable process, the knowledge of $\mathcal{F}_n$ is sufficient to describe the process at time $n + 1$.

3.1.2   Simple and quadratic variation of a process

The notion of *total variation* or *first order* variation of a process $\{X_t, t \geq 0\}$ is linked to the differentiability of its paths seen as a function of t. Let $\Pi_n = \Pi_n([0, t]) = \{0 = t_0 < t_1 < \cdots < t_i < \cdots < t_n = t\}$ be any partition of the interval $[0, t]$ into n intervals and denote by

$$||\Pi_n|| = \max_{j=0,\dots,n-1} (t_{j+1} - t_j)$$

the maximal step size of the partition Π_n, i.e. the mesh of the partition. The first order variation of X is defined as

$$V_t(X) = p - \lim_{||\Pi_n|| \to 0} \sum_{k=0}^{n-1} |X(t_{k+1}) - X(t_k)|.$$

If X is differentiable, then $V_t(X) = \int_0^t |X'(u)| du$. If $V_t(X) < \infty$, then X is said to be of bounded variation on $[0, t]$. If this is true for all $t \geq 0$, then X is said to have bounded variation. The *quadratic variation* $[X, X]_t$ at time t of a process X is defined as

$$[X, X]_t = p - \lim_{||\Pi_n|| \to 0} \sum_{k=0}^{n-1} |X(t_{k+1}) - X(t_k)|^2.$$

The limit exists for stochastic processes with continuous paths. In this case, the notation $<X, X>_t$ is usually adopted. The quadratic variation can also be introduced as

$$[X, X]_t = p - \lim_{n \to \infty} \sum_{k=1}^{2^n} \left(X_{t \wedge k/2^n} - X_{t \wedge (k-1)/2^n} \right)^2,$$

where $a \wedge b = \min(a, b)$. If a process X is differentiable, then it has quadratic variation equal to zero. Moreover, total and quadratic variation are related by the following inequality

$$\sum_{k=0}^{n-1} |X(t_{k+1}) - X(t_k)| \geq \sum_{k=0}^{n-1} |X(t_{k+1}) - X(t_k)| \frac{|X(t_{k+1}) - X(t_k)|}{\max_{\Pi_n} |X(t_{k+1}) - X(t_k)|}$$

$$= \frac{\sum_{k=0}^{n-1} |X(t_{k+1}) - X(t_k)|^2}{\max_{\Pi_n} |X(t_{k+1}) - X(t_k)|}. \tag{3.1}$$

Therefore, if X is continuous and has finite quadratic variation, then its first order variation is necessarily infinite. Note that $V_t(X)$ and $[X, X]_t$ are stochastic processes as well.

3.1.3   Moments, covariance, and increments of stochastic processes

The expected value and variance of a stochastic process are defined as

$$\mathbb{E}(X_t) = \int_\Omega X(t, \omega) \mathrm{d}P(\omega), \quad t \in [0, T],$$

and

$$\mathrm{Var}(X_t) = \mathbb{E}\{X_t - \mathbb{E}(X_t)\}^2, \quad t \in [0, T].$$

The k-th moment of X_t, $k \geq 1$, is defined, for all $t \in [0, T]$, as $\mathbb{E}\{X_t^k\}$. These quantities are well-defined when the corresponding integrals are finite. The *covariance function* of the process for two time instants s and t is defined as

$$\mathrm{Cov}(X_s, X_t) = \mathbb{E}\left\{(X_s - \mathbb{E}(X_s))(X_t - \mathbb{E}(X_t))\right\}.$$

The quantity $X_t - X_s$ is called the *increment* of the process from s to t, $s < t$.

These quantities are useful in the description of stochastic processes that are usually introduced to model evolution subject to some stochastic shocks. There are different ways to introduce processes based on the characteristics one wants to model. A couple of the most commonly used approaches are the modeling of increments and/or the choice of the covariance function.

3.2   Martingales

Definition 3.2.1 *Given a probability space* $(\Omega, \mathcal{F}, P)$ *and a filtration* $\{\mathcal{F}_t, t \geq 0\}$ *on* $\mathcal{F}$, *a martingale is a stochastic process* $\{X_t, t \geq 0\}$ *such that*

(i) $\mathbb{E}|X_t| < \infty$ *for all* $t \geq 0$

(ii) it is adapted to a filtration $\{\mathcal{F}_t, t \geq 0\}$

(iii) for each $0 \leq s \leq t < \infty$, *it holds true that*

$$\mathbb{E}\{X_t | \mathcal{F}_s\} = X_s,$$

i.e. X_s *is the best predictor of* X_t *given* $\mathcal{F}_s$.

If in the definition above the equality '$=$' is replaced by '$\geq$', the process is called *submartingale*, and if it is replaced by '$\leq$', it is called *supermartingale*.

From the properties of the expected value operator it follows that if X is a martingale, then

$$\mathbb{E}(X_s) = \text{(by definition of martingale)} = \mathbb{E}(\mathbb{E}\{X_t|\mathcal{F}_s\})$$

$$= \text{(by measurability of } X_t \text{ w.r.t. } \mathcal{F}_s \text{ and (2.26))} = \mathbb{E}(X_t),$$

which means that martingales have a constant mean for all $t \geq 0$.

Again, a similar notion can be given for $\mathbb{N}$-indexed stochastic processes. In this case, the martingale property is written as $\mathbb{E}\{X_n|\mathcal{F}_{n-1}\} = X_{n-1}$. We present now some discrete time processes for which it is easy to show their martingale property.

3.2.1   Examples of martingales

Example 3.2.2 (The random walk) *Let* $X_1, X_2, \ldots, X_n$, *be a sequence of independent random variables such that* $\mathbb{E}(X_n) = 0$ *and* $\mathbb{E}|X_n| < \infty$, *for all* $n \in \mathbb{N}$. *Consider the random walk defined as*

$$S_n = \sum_{i=1}^{n} X_i, \quad n \geq 1,$$

and define $\mathcal{F}_n = \sigma(X_i; i \leq n)$. *Then* S_n *is clearly* $\mathcal{F}_n$-*measurable. Moreover,*

$$\mathbb{E}|S_n| \leq \sum_{i=1}^{n} \mathbb{E}|X_i| < \infty.$$

Finally, we check the martingale property (iii):

$$\mathbb{E}\{S_n|\mathcal{F}_{n-1}\} = \mathbb{E}\{X_n + S_{n-1}|\mathcal{F}_{n-1}\} = \mathbb{E}\{X_n|\mathcal{F}_{n-1}\} + \mathbb{E}\{S_{n-1}|\mathcal{F}_{n-1}\}$$

$$= \mathbb{E}(X_n) + S_{n-1} = S_{n-1}.$$

Clearly, the assumption $\mathbb{E}(X_n) = 0$ *is crucial to verify (iii).*

Example 3.2.3 (The likelihood ratio process) *Consider a random sample of i.i.d. random variables* $X_1, X_2, \ldots, X_n$ *distributed as* X *which has density* $f(x)$. *Denote the joint density of* $X_1, X_2, \ldots, X_n$ *by*

$$f_n(X_1, X_2, \ldots, X_n) = f(X_1)f(X_2) \cdots f(X_n)$$

and consider another distribution (i.e. a different statistical model) for X *whose density is denoted by* $g(x)$. *Under the second model, we can consider the joint density* $g_n(\cdots)$ *defined similarly to* $f_n(\cdots)$. *Notice that the assumption is still that*

X_i are actually distributed with density $f(\cdot)$ and not $g(\cdot)$. Consider the stochastic process defined as

$$L_n = L_n(X_1, \ldots, X_n) = \frac{g_n(X_1, X_2, \ldots, X_n)}{f_n(X_1, X_2, \ldots, X_n)}.$$

Then $\{L_n, n \geq 1\}$ is a martingale with respect to the filtration $\{\mathcal{F}_n, n \geq 1\}$ generated by the X_i, i.e. $\mathcal{F}_n = \sigma(X_i; i \leq n)$. Clearly, L_n is the likelihood ratio process (although we don't mind about the parameter θ in this notation) and it is $\mathcal{F}_n$-measurable. We now prove the martingale property

$$\mathbb{E}\{L_n | \mathcal{F}_{n-1}\} = \mathbb{E}\left\{ \frac{g_n(X_1, X_2, \ldots, X_n)}{f_n(X_1, X_2, \ldots, X_n)} \middle| \mathcal{F}_{n-1} \right\}$$

$$= \mathbb{E}\left\{ \frac{g_{n-1}(X_1, X_2, \ldots, X_{n-1})g(X_n)}{f_{n-1}(X_1, X_2, \ldots, X_{n-1})f(X_n)} \middle| \mathcal{F}_{n-1} \right\}$$

$$= L_{n-1}\mathbb{E}\left\{ \frac{g(X_n)}{f(X_n)} \middle| \mathcal{F}_{n-1} \right\}$$

$$= L_{n-1}\mathbb{E}\left(\frac{g(X_n)}{f(X_n)} \right)$$

$$= L_{n-1}$$

by independence of X_n with respect to $\mathcal{F}_{n-1}$ and because

$$\mathbb{E}\left\{ \frac{g(X_n)}{f(X_n)} \right\} = \int_{\mathbb{R}} \frac{g(x_n)}{f(x_n)} f(x_n)\mathrm{d}x_n = \int_{\mathbb{R}} g(x_n)\mathrm{d}x_n = 1.$$

The latter also proves that each L_n is such that $\mathbb{E}|L_n| = 1 < \infty$ because $f(\cdot)$ and $g(\cdot)$ are densities (and hence non-negative).

Exercise 3.1 Consider the random walk $\{S_n, n \geq 1\}$. Prove that $\{Z_n = |S_n|, n \geq 1\}$ is a sub-martingale.

Exercise 3.2 Consider the random walk $\{S_n, n \geq 1\}$ and the moment generating function of X_i $M(\alpha) = \mathbb{E}\{e^{\alpha X_i}\}$. Let

$$Z_n = M(\alpha)^{-n} \exp\{\alpha S_n\}, \quad n \geq 1.$$

Prove that $\{Z_n, n \geq 1\}$ is a martingale with respect to $\{\mathcal{F}_n, n \geq 1\}$. Z_n is called the exponential martingale.

Exercise 3.3 Let $X_1, X_2, \ldots, X_n$, be a sequence of i.i.d. random variables with mean $\mathbb{E}(X_i) = 0$ and $\mathbb{E}(X_i^2) = \sigma^2 < \infty$ for all $i = 1, \ldots, n$. Prove that

$$Z_n = \left(\sum_{i=1}^n X_i \right)^2 - n\sigma^2, \quad n \geq 1,$$

is a martingale with respect to the filtration $\mathcal{F}_n = \sigma(X_i, i \leq n)$.

Theorem 3.2.4 (Doob-Meyer decomposition) *Let $\{X_n, n \geq 1\}$ be a stochastic process adapted to the filtration $\{\mathcal{F}_n, n \geq 1\}$ and such that $\mathbb{E}|X_n| < \infty$ for all $n \geq 1$. Then, the following decomposition exists and it is unique:*

$$X_n = M_n + P_n, \quad n \geq 1,$$

where $\{M_n, n \geq 1\}$ and $\{P_n, n \geq 1\}$ are, respectively, a martingale and a predictable process with respect to $\mathcal{F}_n$, with $P_0 = 0$. P_n is also called the compensator.

Proof. We start proving that the decomposition is unique. Assume that two decompositions exist

$$X_n = M_n + P_n \quad \text{and} \quad X_n = M_n^* + P_n^*.$$

Let us introduce the predictable process $Y_n = M_n - M_n^* = P_n^* - P_n$ which is also a martingale. Then

$$Y_{n+1} - Y_n = \mathbb{E}\{Y_{n+1}|\mathcal{F}_n\} - \mathbb{E}\{Y_n|\mathcal{F}_n\} = \mathbb{E}\{Y_{n+1} - Y_n|\mathcal{F}_n\} = 0$$

from which we obtain that $Y_{n+1} = Y_n = Y_{n-1} = \cdots = Y_1 = Y_0$ but $Y_0 = P_0^* - P_0 = 0$. Therefore, $M_n^* = M_n$ and $P_n^* = P_n$. The existence, is proved by direct construction of the decomposition. Let us start with $M_0 = X_0$ and

$$M_{n+1} = M_n + X_{n+1} - \mathbb{E}\{X_{n+1}|\mathcal{F}_n\}, \quad \text{and hence} \quad P_n = X_n - M_n.$$

First we prove that M_n is a martingale.

$$\mathbb{E}\{M_{n+1}|\mathcal{F}_n\} = \mathbb{E}\{M_n|\mathcal{F}_n\} + \mathbb{E}\{X_{n+1}|\mathcal{F}_n\} - \mathbb{E}\{X_{n+1}|\mathcal{F}_n\} = M_n,$$

and of course $\mathbb{E}|M_n| < \infty$ and M_n is $\mathcal{F}_n$-measurable. Now, let us check that P_n is a predictable compensator. Clearly, with $P_0 = 0$. Moreover,

$$P_n = X_n - M_n = X_n - (M_{n-1} + X_n - \mathbb{E}\{X_n|\mathcal{F}_{n-1}\}) = -M_{n-1} + \mathbb{E}\{X_n|\mathcal{F}_{n-1}\}$$

is $\mathcal{F}_{n-1}$-measurable and hence predictable.

The next couple of properties, whose proof can be found in Section 7 of Klebaner (2005), are useful to describe martingales.

Theorem 3.2.5 *Let M_t be a martingale with finite second moments, i.e. $\mathbb{E}(M_t^2) < \infty$ for all t. Then, its quadratic variation exists and $M_t^2 - [M, M]_t$ is a martingale.*

Theorem 3.2.6 *Let M_t be a martingale with $M_0 = 0$. If, for some t, M_t is not zero, then $[M, M]_t > 0$. Conversely, if $[M, M]_t = 0$, then $M_s = 0$ almost surely for all $s \leq t$.*

3.2.2 Inequalities for martingales

We present here some inequalities for martingales without proof, which may be useful in what follows.

Theorem 3.2.7 (Doob's maximal inequality) *Assume that* $\{X_n, n \geq 1\}$ *is a non-negative submartingale. Then, for any* $\lambda > 0$, *we have*

$$P\left(\max_{k \leq n} X_k \geq \lambda\right) \leq \frac{1}{\lambda}\mathbb{E}\left(X_n \boldsymbol{I}_{\{\max_{k \leq n} X_k \geq \lambda\}}\right) \leq \frac{1}{\lambda}\mathbb{E}\left\{\max(X_n, 0)\right\} \qquad (3.2)$$

and if X_n *is a non-negative supermartingale*

$$P\left(\sup_{k \leq n} X_k \geq \lambda\right) \leq \frac{1}{\lambda}\mathbb{E}\left\{X_0\right\}.$$

For a continuous-time submartingale with continuous paths

$$P\left(\sup_{t \leq T} X_t \geq \lambda\right) \leq \frac{1}{\lambda}\mathbb{E}\left\{\max(X_T, 0)\right\}. \qquad (3.3)$$

Theorem 3.2.8 (Doob's maximal L^2 inequality) *Assume that* $\{X_n, n \geq 1\}$ *is a non-negative submartingale. Then,*

$$\mathbb{E}\left(\max_{k \leq n} X_k^2\right) \leq 4\mathbb{E}|X_n|^2. \qquad (3.4)$$

The L^2 inequality is a particular case of the more general L^p inequality which requires a bit more regularity on the integrability of the process. We present it for continuous-time martingales.

Theorem 3.2.9 (Doob's maximal L^p inequality) *Assume that* $\{X_t, t \geq 0\}$ *is a non-negative submartingale with continuous paths. Then,*

$$\mathbb{E}\left(\sup_{t \leq T} |X_t|^p\right) \leq \left(\frac{p}{p-1}\right)^p \mathbb{E}|X_T|^p, \quad p \geq 1, \qquad (3.5)$$

provided that $\mathbb{E}|X_T|^p < \infty$.

Theorem 3.2.10 (Hájek-Rényi inequality) *Let* $\{S_n = \sum_{i=1}^n X_i, n \geq 1\}$ *be a martingale with* $\mathbb{E}(S_n^2) < \infty$. *Let* $\{b_n, n \geq 1\}$ *a positive nondecreasing sequence. Then, for any* $\lambda > 0$

$$P\left(\max_{1 \leq j \leq n}\left|\frac{S_j}{b_j}\right| \geq \lambda\right) \leq \frac{1}{\lambda^2}\sum_{j=1}^n \mathbb{E}\left\{\frac{X_j^2}{b_j^2}\right\}. \qquad (3.6)$$

For the particular sequence $\{b_n = 1, n \geq 1\}$, the Hájek-Rényi inequality particularizes to the Kolmogorov inequality for martingales, which is interesting in itself.

Theorem 3.2.11 (Kolmogorov inequality for martingales) *Let* $\{S_n = \sum_{i=1}^{n} X_i, n \geq 1\}$ *be a martingale with* $\mathbb{E}(S_n^2) < \infty$. *Then, for any* $\lambda > 0$, *we have*

$$P\left(\max_{1 \leq j \leq n} |S_j| \geq \lambda\right) \leq \frac{1}{\lambda^2} \mathbb{E}(S_n^2). \tag{3.7}$$

Theorem 3.2.12 *Let* $\{X_t, t \geq 0\}$ *be a martingale, then for* $p \geq 1$, *we have*

$$P\left(\sup_{s \leq t} |X_s| \geq \lambda\right) \leq \frac{1}{\lambda^p} \sup_{s \leq t} \mathbb{E}\left(|X_s|^p\right). \tag{3.8}$$

Theorem 3.2.13 (Burkholder-Davis-Gundy inequality) *Let* $\{X_t, t \geq 0\}$ *be a martingale, null at zero. Then, there exist constants* c_p *and* C_p, *depending only on* p, *such that for* $1 < p < \infty$

$$c_p \mathbb{E}\left([X, X]_T^{\frac{p}{2}}\right) \leq \mathbb{E}\left(\sup_{t \leq T} |X_t|^p\right) \leq C_p \mathbb{E}\left([X, X]_T^{\frac{p}{2}}\right). \tag{3.9}$$

Moreover, if X_t *is also continuous, the result holds also for* $0 < p \leq 1$.

3.3 Stopping times

Definition 3.3.1 *A random variable* $T : \Omega \to [0, +\infty)$ *is called stopping time with respect to the filtration* $\{\mathcal{F}_t, t \geq 0\}$, *if the event* $(T \leq t)$ *is* $\mathcal{F}_t$-*measurable, i.e.*

$$\{\omega \in \Omega : T(\omega) \leq t\} = (T \leq t) \in \mathcal{F}_t, \quad \forall t \geq 0.$$

The definition of stopping time essentially says that we can decide whether the event $(T \leq t)$ occurred or not on the basis of the information up to time t. The random variable $T = t$, with t a constant, is a trivial stopping time because $(T = t \leq t) = \Omega \in \mathcal{F}_t$. Moreover, if T is a stopping time, then $T + s$, with $s > 0$ a constant, is also a stopping time. Indeed,

$$(T + s \leq t) = (T \leq t - s) \in \mathcal{F}_{t-s} \subset \mathcal{F}_t.$$

An important fact concerning stopping times is related to the first passage times. Let $\{X_n, n \geq 1\}$ be a sequence of random variables (for example, a random walk) and fix a value β. We denote by $T_\beta = \inf\{n : X_n \geq \beta\}$ the *first passage time* of the process X to the threshold β. By construction, $(T_\beta \leq k) \in \mathcal{F}_k$ where $\mathcal{F}_k = \sigma(X_1, X_2, \ldots, X_k)$. So, the first passage time is a stopping time. Similarly, with the addition of technical bits, one can prove the same for continuous time processes. On the contrary, the last passage time $S_\beta = \sup\{n : X_n \geq \beta\}$ is not a stopping time. Indeed, $(S_\beta \leq k) \notin \mathcal{F}_k$ because we need to observe the trajectory of X up to infinity.

Theorem 3.3.2 *Let T and S be two stopping times with respect to $\{\mathcal{F}_t, t \geq 0\}$. Then*

$$S \vee T, \quad S \wedge T \quad and \quad S + T$$

are also stopping times.

Proof. We need to prove measurability of each event. Indeed

$$S \vee T = (S \leq t) \cap (T \leq T) \in \mathcal{F}_t, \quad and \quad S \wedge T = (S \leq t) \cup (T \leq T) \in \mathcal{F}_t.$$

For $S + T$, we prove that $(S + T > t)^c \in \mathcal{F}_t$. Let us decompose the event $(S + T > t)$ as follows:

$$(S + T > t) = \Omega \cap (S + T > t)$$

$$= ((T = 0) \cup (0 < T < t) \cup (T > t)) \cap (S + T > t)$$

$$= (T = 0, S + T > t) \cup (0 < T < t, S + T > t) \cup (T \geq t, S + T > t)$$

and prove that each term satisfies measurability. For the first term we have

$$(T = 0, S + T > t) = (S > t) = (S \leq t)^c \in \mathcal{F}_t.$$

We decompose further the last term

$$(T \geq t, S + T > t) = ((T > t, S + T > t) \cap (S = 0 \cup S > 0)$$

$$= (T \geq t, S + T > t, S = 0) \cup (T > t, S + T > t, S > 0)$$

$$= (T > t) \cup (T \geq t, S > 0)$$

$$= (T \leq t)^c \cup ((T < t)^c \cap (S > 0))$$

and all terms are $\mathcal{F}_t$-measurable. We need to rewrite the term $(0 < T < t, S + T > t)$ as follows:

$$(0 < T < t, S + T > t) = \bigcup_{s > 0} (s < T < t, S > t - s)$$

$$= \bigcup_{s > 0} ((s < T < t) \cap (S \leq t - s)^c)$$

and notice that

$$(s < T < t) = (T < t) \cap (T \leq s)^c.$$

In a very similar way one can prove the following.

Exercise 3.4 *Let $\{T_n, n \geq 1\}$ be a sequence of stopping times with respect to the filtration $\{\mathcal{F}_t, t \geq 0\}$. Prove that*

$$\inf_{n \geq 1} T_n \quad and \quad \sup_{n \geq 1} T_n$$

are also stopping times.

Definition 3.3.3 (Stopped σ-algebra) *Let T be a stopping time with respect to the filtration $\{\mathcal{F}_t, t \geq 0\}$, we define $\mathcal{F}_T$, the σ-algebra stopped at T the following set*

$$\mathcal{F}_T = \{A \in \mathcal{F}_\infty : A \cap (T \leq t) \in \mathcal{F}_t, t \geq 0\}$$

where $\mathcal{F}_\infty = \sigma\left(\bigcup_t \mathcal{F}_t\right)$.

3.4 Markov property

A discrete time stochastic process $\{X_n, n \geq 1\}$ is said to be *Markovian* if the conditional distribution of X_k given the past $(X_{k-1}, X_{k-2}, \ldots)$ equals the conditional distribution of X_k given X_{k-1} solely, i.e.

$$\mathcal{L}(X_k | X_{k-1}, X_{k-2}, \ldots) = \mathcal{L}(X_k | X_{k-1}).$$

This means that the 'future' of the process X_k depends only on the 'present' X_{k-1} and the knowledge of the 'past' $X_{k-2}, \ldots, X_1$ does not contribute any additional information. Examples of Markov sequences are the i.i.d. sample or the autoregressive models like the following

$$X_n = \theta X_{n-1} + \epsilon_n, \quad X_0 = x_0, \quad \epsilon_i \quad \text{i.i.d} \sim N(0, \sigma^2),$$

with θ some parameter. The class of Markov processes is quite wide and the characterization of the Markov property relies on the state space (discrete versus continuous) and time specification (again, discrete versus continuous). We limit the treatise of Markov processes to what is interesting to the subsequent part of this book.

3.4.1 Discrete time Markov chains

An important class of Markov processes are the so-called *Markov chains*. Consider a discrete time stochastic process $\{X_n, n \geq 1\}$ with discrete (possibly countable) state space S. For simplicity we denote the states of the process with j, but in practical cases one can imagine to have $S = \{0, 1, 2, \ldots\}$ or $S = \{0, \pm 1, \pm 2, \ldots\}$, or, if the state space is finite, $S = \{1, 2, \ldots, d\}$. We denote by $p_{ij}^n = P(X_{n+1} = j | X_n = i)$, i.e. $\{p_{ij}^n, j \in S\}$ is the conditional distribution of the process X at time $n + 1$ given that its position at time n is the state i.

Definition 3.4.1 *The process* $\{X_n, n \geq 1\}$, *with state space* S *is called Markov chain if*

$$P(X_{n+1} = j | X_n = i, X_{n-1} = i_{n-1}, \ldots, X_0 = i_0)$$
$$= P(X_{n+1} = j | X_n = i) = p_{ij}^n, \quad \forall i, j \in S, \tag{3.10}$$

with $i_{n-1}, \ldots, i_1, i_0 \in S$, *when all probabilities are well defined.*

The quantities p_{ij}^n are called one step *transition probabilities*. These transition probabilities also depend on the instant n. Most interesting to us are homogenous Markov chains.

Definition 3.4.2 *A Markov chain such that*

$$P(X_{n+1} = j | X_n = i) = p_{ij},$$

for all $n \geq 0$ *is called a homogeneous Markov chain.*

For homogenous Markov chains, the transition probabilities are independent of the time n and in principle, one can describe the whole structure of the process taking into account only the initial state X_0 and the first transition to X_1, i.e. $p_{ij} = P(X_1 = j | X_0 = i)$. The one step transition probabilities constitute the *transition matrix* P which, for a state space like $S = \{0, 1, \ldots\}$, we can express in the form:

$$P = \begin{bmatrix} p_{00} & p_{01} & p_{02} & \cdots & p_{0j} & \cdots \\ p_{10} & p_{11} & p_{12} & \cdots & p_{1j} & \cdots \\ p_{20} & p_{21} & p_{22} & \cdots & p_{2j} & \cdots \\ \vdots & \vdots & \vdots & \ddots & \vdots & \vdots \\ p_{i0} & p_{i1} & p_{i2} & \cdots & p_{ij} & \cdots \\ \vdots & \vdots & \vdots & \ddots & \vdots & \vdots \end{bmatrix}.$$

Every transition matrix satisfies the following properties

$$p_{ij} \geq 0, \quad i, j \in S, \quad \sum_{j \in S} p_{ij} = 1 \quad \text{for each } i \in S.$$

Example 3.4.3 (Random walk) *Consider a particle which moves on the real line starting from the origin, i.e.* $S_0 = X_0 = 0$. *At each instant n the particle jumps to the right or to the left according to the following distribution:* $P(X_n = +1) = p$, $P(X_n = -1) = q$, *with* $p + q = 1$. *The position of the particle at time n is* $S_n = \sum_{i=1}^{n} X_i$. *Assume that the* X_i *are all independent. The state space of the process* *is $S = \{0, \pm 1, \pm 2, \ldots\}$ and the transition from one state to another is given by $S_n = S_{n-1} + X_n$. Clearly $\{S_n, \geq 1\}$ is a homogenous Markov chain. Indeed,*

$$P(S_{n+1} = j + 1 | S_n = j, S_{n-1}, \ldots, , S_0)$$

$$= P(S_n + X_{n+1} = j + 1 | S_n = j, S_{n-1}, \ldots, S_0)$$

$$= P(X_{n+1} = +1 | S_n = j, S_{n-1}, \ldots, S_0)$$

$$= p$$

because the X_i's are independent and similarly for $P(S_{n+1} = j - 1 | S_n = j, S_{n-1}, \ldots, , S_0)$. The transition matrix for S_n has an infinite number of rows and columns which can be represented as follows:

$$P = \begin{bmatrix} \ddots & \ddots & & \ddots & & & \\ & q & 0 & p & & & \\ & & q & 0 & p & & \\ & & & q & 0 & p & \\ & & & & \ddots & \ddots & \ddots \end{bmatrix},$$

where the matrix is filled with zeros outside the two diagonals of p's and q's.

Example 3.4.4 (Random walk with reflecting barriers) *Assume that the particle of Example 3.4.3 can move only on the finite state space $S = \{a, \ldots, 0, \ldots, b\}$ such that if the particle is in state a at time n, it is pushed at $a + 1$ at instant $n + 1$ with probability 1; if it is in state b at time n it is pushed back at state $b - 1$ at time $n + 1$; otherwise it behaves as in Example 3.4.3. The transition matrix is then represented as follows:*

$$P = \begin{bmatrix} 0 & 1 & & & & & \\ q & 0 & p & & & & \\ & q & 0 & p & & & \\ & & \ddots & \ddots & \ddots & & \\ & & & q & 0 & p & \\ & & & & q & 0 & p \\ & & & & & 1 & 0 \end{bmatrix},$$

and a and b are called reflecting barriers. The first row of P represents the distribution $P(X_n = k, | X_{n-1} = a)$, $k = a, a + 1, \ldots, 0, \ldots, b - 1, b$, and similarly for the subsequent rows.

Example 3.4.5 (Random walk with absorbing barriers) *Assume the same setup of Example 3.4.4, with the difference that when the particle reaches the* *barrier a or b, the random walk is stopped in those states, i.e. if the particle is in state a (or b) at time n, it remains in state a (respectively b) with probability 1 for all times $k \geq n$; otherwise it behaves as in Example 3.4.3. The transition matrix is then represented as follows:*

$$
P = \begin{bmatrix}
1 & 0 & & & & & \\
q & 0 & p & & & & \\
 & q & 0 & p & & & \\
 & & \ddots & \ddots & \ddots & & \\
 & & & q & 0 & p & \\
 & & & & q & 0 & p \\
 & & & & & 0 & 1
\end{bmatrix},
$$

and a and b are called absorbing barriers.

The transition matrix of a homogenous Markov chain represents the short term evolution of the process. In many cases, the long term evolution of the process is more interesting, in particular in asymptotic statistics. Let us introduce the *n*-step transition distribution denoted as $\{p_{ij}^{(n)}, j \in S\}$, which represents the distribution of the process X_{m+n} given that the present state is $X_m = i$. We can write:

$$
p_{ij}^{(n)} = P(X_{m+n} = j | X_m = i).
$$

Notice that $P_{ij}^{(n)}$ does not depend on the time *m* because the chain is homogenous, so we can also write

$$
p_{ij}^{(n)} = P(X_n = j | X_0 = i).
$$

and clearly $p_{ij}^{(1)} = p_{ij}$.

Theorem 3.4.6 *The following relation holds*

$$
p_{ij}^{(n)} = \sum_{h \in S} p_{ih}^{(n-1)} p_{hj}. \tag{3.11}
$$

Proof. Indeed,

$$
p_{ij}^{(n)} = \frac{P(X_n = j, X_0 = i)}{P(X_0 = i)} = \sum_{h \in S} \frac{P(X_n = j, X_{n-1} = h, X_0 = i)}{P(X_0 = i)}
$$

$$
= \sum_{h \in S} P(X_n = j | X_{n-1} = h, X_0 = i) P(X_{n-1} = h | X_0 = i)
$$

$$
= \sum_{h \in S} P(X_n = j | X_{n-1} = h) p_{ih}^{(n-1)} = \sum_{h \in S} P(X_1 = j | X_0 = h) p_{ih}^{(n-1)},
$$

from which we obtain (3.11).

Let us denote by $P^{(n)} = [P_{ij}^{(n)}, i, j, \in S]$ the n-steps transition matrix, then the following relation holds

$$P^{(n)} = P^{(n-1)} \cdot P \qquad (3.12)$$

in the sense of matrix multiplication (matrixes with possibly infinite number of rows and columns). From (3.12) we can see that

$$P^{(2)} = P \cdot P = P^2$$

and, recursively, we obtain

$$P^{(n)} = P^n.$$

Equation (3.11) is a particular case of the Chapman-Kolmogorov equation presented in the next theorem.

Theorem 3.4.7 (Chapman-Kolmogorov equation) *Given a homogenous Markov chain $\{X_n, n \geq 1\}$ with transition probabilities p_{ij}, the following relations hold*

$$p_{ij}^{(m+n)} = \sum_{h \in S} p_{ih}^{(m)} p_{hj}^{(n)},$$

for each $m, n \geq 0$ and for all $i, j \in S$ or in matrix form:

$$P^{(m+n)} = P^{(m)} \cdot P^{(n)}.$$

The proofs follow the same steps as the proof of Equation (3.11). Along with the transition distributions the finite dimensional distribution of the process at time n is of interest. Let us denote by π_k^n the probability of finding X_n in state k at time n

$$\pi_k^n = P(X_n = k), \quad \forall k \in S.$$

Similarly, we denote the law of X_0 by π^0.

Theorem 3.4.8 *Given a homogenous Markov chain with transition matrix P and initial distribution π^0, the distribution of X_n is given by*

$$\pi_k^n = \sum_{h \in S} \pi_h^0 p_{hk}^{(n)},$$

for each $k \in S$, or in matrix form:

$$\pi^n = \pi^0 P^n.$$

Proof. Indeed, we can write

$$P(X_n = k) = \sum_{h \in S} P(X_n = k, X_0 = h) = \sum_{h \in S} P(X_n = k | X_0 = h) P(X_0 = h).$$

In a similar way, we can derive the joint distribution of any n-tuple $(X_{k_1}, \ldots X_{k_n})$.

Theorem 3.4.9 *Given a homogenous Markov chain, with transition matrix P and initial distribution π^0, the joint distribution of $(X_{k_1}, X_{k_2}, \ldots X_{k_n})$, with $0 < k_1 < k_2 < \cdots < k_n$, is given by*

$$P(X_{k_1} = h_1, \ldots X_{k_n} = h_n) = \sum_{h \in S} \pi_h^0 p_{hh_1}^{(k_1)} p_{h_1 h_2}^{(k_2 - k_1)} \cdots p_{h_{n-1} h_n}^{(k_n - k_{n-1})}$$

for each $h_1, h_2, \ldots h_n \in S$.

Proof. Set

$$v_{h_1, \ldots, h_n}^{k_1, \ldots, k_n} = P(X_{k_1} = h_1, \ldots, X_{k_n} = h_n)$$

and write

$$
\begin{aligned}
v_{h_1, \ldots, h_n}^{k_1, \ldots, k_n} &= P(X_{k_1} = h_1, \ldots X_{k_{n-1}} = h_{n-1}) \\
&\quad \times P(X_{k_n} = h_n | X_{k_1} = h_1, \ldots, X_{k_{n-1}} = h_{n-1}) \\
&= P(X_{k_1} = h_1, \ldots X_{k_{n-1}} = h_{n-1}) p_{h_{n-1} h_n}^{(k_n - k_{n-1})}.
\end{aligned}
$$

Then the proof follows iterating the same argument.

Theorem 3.4.10 (Strong Markov property) *Let τ be a stopping time. Given the knowledge of $\tau = n$ and $X_\tau = i$, no other information on $X_0, X_1, \ldots, X_\tau$ is needed to determine the conditional distribution of the event $X_{\tau+1} = j$, i.e.*

$$P(X_{\tau+1} = j | X_\tau = i, \tau = n) = p_{ij}$$

Proof. Denote by V_n the set of vectors $x = (i_0, i_1, \ldots, i_n)$ such that the event $\{X_0 = i_0, X_1 = i_1, X_n = i_n\}$ implies the event $\tau = n$ and $X_\tau = i$ (note that we should ask $i_n = i$). Then, by the law of total probabilities we have

$$P(X_{\tau+1} = j, X_\tau = i, \tau = n) = \sum_{x \in V_n} P(X_{n+1} = j, X_n = x_n, \ldots, X_0 = x_0)$$

$$(3.13)$$

and the second term can be rewritten as follows:

$$\sum_{x \in V_n} P(X_{n+1} = j | X_n = x_n, \ldots, X_0 = x_0) P(X_n = x_n, \ldots, X_0 = x_0).$$

Remember that if $x \in V_n$ then $\tau = n$ and $X_\tau = i$, hence $x_n = i$ and by the Markov property we have that

$$P(X_{n+1} = j | X_n = x_n, \ldots, X_0 = x_0) = p_{ij}$$

and then

$$P(X_{\tau+1} = j | X_\tau = i, \tau = n) = p_{ij} \sum_{x \in V_n} P(X_n = x_n, \ldots, X_0 = x_0)$$

$$= p_{ij} P(X_\tau = i, \tau = n).$$

The statement follows dividing the expression (3.13) by $P(X_\tau = i, \tau = n)$.

The strong Markov property is sometimes written as

$$\mathcal{L}(X_{n+1} | X_n, X_{n-1}, \ldots) = \mathcal{L}(X_{n+1} | X_\tau)$$

and it essentially states that, conditioning on a proper random time, the chain loses its memory. For example, let $\tau_i = \min\{n \geq 1 : X_n = i\}$ be the first passage time of the Markov chain from state i. The random variable τ_i is a stopping time. Then, the new chain $\{\tilde{X}_k\} \overset{d}{=} \{X_{\tau_i+k}\}$, i.e. it behaves as the original Markov chain but started from i.

Let $\tau_i = \min\{n \geq 1 : X_n = i\}$ be the first passage time and define

$$f_{ii} = P(\tau_i < +\infty | X_0 = i)$$

This quantity represents the probability that X_n returns into state i at least once, given that it started from time i. Notice that the initial state X_0 does not enter the definition of the stopping time τ_i.

The strong Markov property implies that the probability that chain returns in state i given that it has visited that state $n - 1$ times, is again f_{ii}. Therefore, the probability of visiting two times the state i is f_{ii}^2, the probability of visiting three times the same state is $f_{ii}^2 \cdot f_{ii} = f_{ii}^3$ and, in general, the probability of n visits is f_{ii}^n.

Definition 3.4.11 *Let S be the state space of a homogeneous Markov chain. Given $i \in S$, the state i is called transient if $f_{ii} < 1$; the state i is called recurrent if $f_{ii} = 1$.*

Notice that if $f_{ii} < 1$ the probability that a chain comes back n times into state i is f_{ii}^n and this converges to 0 as $n \to \infty$, then, from a certain instant the chain will never visit the state i again. This is why the state is called transient. Conversely, is $f_{ij} = 1$, the chain will visit that state infinitely often. Further, a state i such that $p_{ii} = 1$ is called *absorbing state*. We have already encountered absorbing states in Examples 3.4.5.

Definition 3.4.12 *Let P be the transition matrix of a homogenous Markov chain. A probability distribution π on S is called invariant distribution if it satisfies the following equation*

$$\pi = \pi P.$$

The existence of an invariant distribution means that at each instant the marginal law of the chain remains the same, i.e. the chain is stable. Conditions for the existence and unicity of the invariant distribution π can be given, see e.g. the books of Feller (1968), Feller (1971) and Cox and Miller (1965). We only provide the following result.

Theorem 3.4.13 *Assume that there exists an invariant distribution* $\pi = (\pi_1, \pi_2, \ldots)$ *and that the initial distribution of the Markov chain is* π, *i.e.* $\pi_j = P(X_0 = j)$, $j \in \mathcal{S}$. *The, the marginal distribution of* X_n *is* π *for all n.*

Proof. Denote by v^n, the distribution at time n

$$v_j^n = \sum_{i \in \mathcal{S}} \pi_i \, p_{ij}^{(n)},$$

where we have set the initial distribution equal to π. We prove by induction, that if π is an invariant distribution then

$$\sum_{i \in \mathcal{S}} \pi_i \, p_{ij}^{(n)} = \pi_j, \quad \forall n, j.$$

This will imply that

$$v_j^n = \pi_j, \quad \forall n, j.$$

Indeed, by definition of invariant distribution it is true that, for $n = 1$, $\sum_{i \in \mathcal{S}} \pi_i p_{ij} = \pi_j$. Assume that it holds for $n - 1$, that is $\sum_{i \in \mathcal{S}} \pi_i p_{ij}^{(n-1)} = \pi_j$, then

$$\sum_{i \in \mathcal{S}} \pi_i \, p_{ij}^{(n)} = \sum_{i \in \mathcal{S}} \pi_i \sum_{h \in \mathcal{S}} p_{ih}^{(n-1)} p_{hj} = \sum_{h \in \mathcal{S}} \sum_{i \in \mathcal{S}} \pi_i \, p_{ih}^{(n-1)} p_{hj} = \sum_{h \in \mathcal{S}} \pi_h \, p_{hj} = \pi_j.$$

Therefore, it holds for all n.

3.4.2   Continuous time Markov processes

For a complete treatise on continuous time Markov processes we suggest the reading of Revuz and Yor (2004). In this section we just collect the building blocks for subsequent considerations.

For a continuous time stochastic process, we define the *transition probability* the function $P_{s,t}$, $s < t$ such that

$$P\{X_t \in A | \sigma(X_u, u \le s)\} = P_{s,t}(X_s, A) \quad \text{a.s.}$$

where A is some set in state space of X. In the case of continuous time processes, the Chapman-Kolmogorov equation is presented in the following form: for any $s < t < v$, we have

$$\int P_{s,t}(x, \mathrm{d}y) P_{t,v}(y, A) = P_{s,v}(x, A).$$

If the transition probability is homogeneous, i.e. $P_{s,t}$ depends on s and t only through the difference $t - s$, one usually write P_t for $P_{0,t}$ and then the Chapman-Kolmogorov equation becomes

$$P_{t+s}(A) = \int P_s(x, \mathrm{d}y) P_t(y, A) = P_{s,t}(x, A), \quad \forall\, s, t \geq 0$$

from which we see that the family $\{P_t, t \geq 0\}$ forms a semigroup.

Definition 3.4.14 *Let X be a stochastic process adapted to a filtration $\mathcal{F} = \{\mathcal{F}_t, t \geq 0\}$, with transition probability $P_{s,t}$, $s < t$. The process is said to be a Markov process with respect to the filtration $\mathcal{F}$ if, for any non-negative function f and any pair (s, t), $s < t$, we have*

$$\mathbb{E}\{f(X_t)|\mathcal{F}_s\} = P_{s,t}(f(X_s)) \quad a.s.$$

The process is homogeneous if the corresponding transition probability is homogeneous and in this case we write

$$\mathbb{E}\{f(X_t)|\mathcal{F}_s\} = P_{t-s}(f(X_s)) \quad a.s.$$

Definition 3.4.15 (Strong Markov property) *Let X be a stochastic process adapted to the filtration $\mathcal{F} = \{\mathcal{F}_t, t \geq 0\}$, T a stopping time with respect to $\mathcal{F}$ and $\mathcal{F}_T$ the stopped σ-algebra. The process X possesses the strong Markov property if*

$$P(X_t \in A|\mathcal{F}_T) = P(X_t \in A|X_T).$$

The strong Markov process mean that if we 'start afresh', in the terminology of Itô and McKean (1996), the process X at the stopping time T, the new process possesses the same properties as X. So, this is the strong lack of memory of a Markov process.

3.4.3   Continuous time Markov chains

Although we will see that most of the processes considered later are Markov processes, we introduce here a few details about continuous time Markov chains that will be useful later on.

Definition 3.4.16 *A continuous time process $\{X_t, t \geq 0\}$ with discrete state space S is called continuous time Markov chain if*

$$P(X_{t+u} = j|X_u = i, X_s = x_s, 0 \leq s < u) = P(X_{t+u} = j|X_u = u),$$

where $i, j \in S$ and $\{x_s, 0 \leq s < u\} \subset S$.

When the process is stationary, the analysis is much simpler and close to that of discrete time Markov chains. From now on we assume stationarity and hence, the transition probabilities take the form:

$$p_{ij} = P(X_{t+u} = j | X_u = i) = P(X_t = i, X_0 = k).$$

We denote the state vector $s(t)$ with components $s_k(t) = P(X_t = k)$, $k \in S$. This vector obeys

$$s(t + u) = s(u)P(t)$$

with $P(t) = [p_{ij}(t), i, j \in S]$ the transition matrix. From the above we obtain

$$s(t + u + v) = s(v)P(t + u) = s(u + v)P(t) = s(v)P(u)P(t)$$
$$= s(v + t)P(u) = s(v)P(t)P(u)$$

from which we obtain

$$P(t + u) = P(u)P(t) = P(t)P(u), \quad \forall t, u \geq 0,$$

that is, the Chapman-Kolmogorov equation. As usual, the transition matrix satisfies

$$\sum_{j \in S} p_{ij}(t) = 1, \quad \text{for each state } i \in S.$$

We denote with $P(0)$ the initial transition probability matrix

$$P(0) = \lim_{t \downarrow 0} P(t) = I.$$

and it is possible to prove that the transition probability matrix $P(t)$ is continuous for all $t \geq 0$. This allows us to define the next object which plays an important role for continuous time Markov chains.

Definition 3.4.17 (Infinitesimal generator) *The matrix Q defined as*

$$\lim_{t \downarrow 0} \frac{P(t) - I}{t} = P'(0) = Q$$

is called infinitesimal generator of the Markov chain.

For discrete time Markov chain it corresponds to $P - I$. The sum of the rows in $Q = [q_{ij}]$ is zero and such that

$$\sum_{j, i \neq j} q_{ij} = -q_{ii}$$

where

$$q_{ij} = \lim_{t \downarrow 0} \frac{p_{ij}(t)}{t} \geq 0 \quad \text{and} \quad q_{ii} \leq 0.$$

The quantities q_{ij} are called *rates*. They are derivatives of probabilities and reflect a change in the transition probability from state i towards state j. The name 'rates' will become more precise when we discuss the Poisson process. Let $q_i = -q_{ii} > 0$, then $\sum_{j \in S} |q_{ij}| = 2q_i$. For a Markov chain with finite state space, the quantities q_i are always finite, but in general this fact should not be expected. This allows for an additional classification of the states of the Markov chain: a state j is called *instantaneous* if $q_j = \infty$. In this case, when the process enters the state j it immediately leaves it. Assume there is a chain where all states are not instantaneous. Consider a small time Δt. Then,

$$
\begin{aligned}
P(X(t + \Delta t) = j | X(t) = i) &= q_{ij} \Delta t + o(\Delta t), \quad i \neq j, \\
P(X(t + \Delta t) = i | X(t) = i) &= 1 - q_i \Delta t + o(\Delta t).
\end{aligned}
\tag{3.14}
$$

The above is a generalization of the properties of the Poisson process (see (3.16)) which allows for a direct link to the term 'rates' and Poissonian event arrival rates.

Theorem 3.4.18 *Let Q be the infinitesimal generator of a continuous time Markov chain. Then, the transition probability matrix $P(t)$ is differentiable for all $t \geq 0$ and satisfies*

$$
\begin{aligned}
P'(t) &= P(t)Q \quad \text{the forward equation,} \\
&= QP(t) \quad \text{the backward equation.}
\end{aligned}
$$

The solution of $P'(t) = P(t)Q$ with initial condition $P(0) = I$ is

$$
P(t) = \exp(Qt).
$$

In the previous theorem $\exp(A)$ is the matrix exponential of matrix A, which is given by the power series formula

$$
\exp(A) = I + \frac{1}{2!} A^2 + \frac{1}{3!} A^3 + \cdots
$$

with A^k the matrix power, i.e. k-fold matrix multiplication. Package **msm** has a function called `MatrixExp` to calculate this matrix exponential reliably, which is not completely trivial as mentioned in Moler and van Loan (2003). Obviously, the infinitesimal generator and the backward/forward formulas can be defined for general continuous time Markov processes and the most important case is the family of diffusion processes. We will discuss these points later in the book.

3.5   Mixing property

The notion of mixing corresponds to a relaxation of the notion of independence in several ways. We know that two events are independent if $P(A \cap B) = P(A)$

$P(B)$ or, equivalently if $P(A \cap B) - P(A)P(B) = 0$. The mixing property of a sequence of random events $A_1, A_2, \ldots$ is stated as

$$|P(A_i \cap A_j) - P(A_i)P(A_j)| \to 0$$

at some rate.

Definition 3.5.1 *Let* X_i, *be a stochastic process and let* $\mathcal{F}_k = \sigma(X_i, i \leq k)$ *and* $\mathcal{F}_\infty^k = \sigma(X_i, i = k, \ldots, \infty)$. *Let* $A_1 \in \mathcal{F}_k$ *and* $A_2 \in \mathcal{F}_\infty^{k+n}$ *any two events. Then, if*

$$|P(A_1 \cap A_2) - P(A_1)P(A_2)| \leq \varphi(n)P(A_1), \quad \forall n,$$

we say that the process X_i *is* φ*-mixing.*

If $P(A_1) = 0$, the mixing condition trivially holds, but when $P(A_1) > 0$, we can write the φ-mixing condition as follows:

$$|P(A_2|A_1) - P(A_2)| \leq \varphi(n).$$

In these terms, the φ-mixing property says that the conditional probability of n steps in the future, i.e. A_2, with respect to the present, i.e. A_1, is close to the unconditional probability of A_2 up to a function depending on n. When the two quantities match for all n we have independence, but usually, it is required that

$$\lim_{n \to \infty} \varphi(n) = 0$$

at some rate and the rate specifies the strength of dependence between elements of the sequence X_i. Sometimes the condition is stated in a more general form which includes the supremum

$$\sup\{|P(A_2|A_1) - P(A_2)| : A_1 \in \mathcal{F}_k, A_2 \in \mathcal{F}_\infty^{k+n}\} \leq \varphi(n).$$

Example 3.5.2 (Billingsley) *Consider a Markov chain with finite state space* S. *Let*

$$\varphi(n) = \max_{u,v \in S} \left| \frac{p_{uv}^{(n)}}{p_v} - 1 \right|$$

and suppose it is finite. Let

$$A_1 = \{(X_{k-i}, \ldots, X_k) \in H_1\} \in \mathcal{F}_k$$

with H_1 *a subset of* $i + 1$ *states, and*

$$A_2 = \{(X_{k+n}, \ldots, X_{k+n+j}) \in H_2\} \in \mathcal{F}_\infty^{k+n}$$

with H_2 a subset of $j + 1$ states. Then

$$|P(A_1 \cap A_2) - P(A_1)P(A_2)|$$

$$\leq \sum p_{u_0} p_{u_0 u_1} \cdots p_{u_{i-1} u_i} |p^{(n)}_{u_i v_0} - p_{v_0}| p_{v_0 v_1} \cdots p_{v_{j-1} v_j}$$

where the sums extends to the states $(u_0, u_1, \ldots, u_i) \in H_1$ and $(v_0, \ldots, v_j) \in H_2$. Noticing that $\sum_u p_{uv} = 1$ and using the definition of $\varphi(n)$, we obtain the mixing property.

We will discuss different type of mixing sequences in the second part of the book.

3.6 Stable convergence

Definition 3.6.1 *Assume there are two probability spaces $(\Omega, \mathcal{F}, P)$ and $(\Omega', \mathcal{F}', P')$. Assume further that there exists a $Q(\omega, \omega')$, the transition probability from $(\Omega, \mathcal{F})$ into $(\Omega', \mathcal{F}')$. Set*

$$\bar{\Omega} = \Omega \times \Omega', \quad \bar{\mathcal{F}} = \mathcal{F} \otimes \mathcal{F}', \quad \bar{P}(d\omega, d\omega') = P(d\omega)Q(\omega, \omega').$$

The space $(\bar{\Omega}, \bar{\mathcal{F}}, \bar{P})$ is called an extension of $(\Omega, \mathcal{F}, P)$. Denote by $\bar{\mathbb{E}}$ the expected value with respect to $\bar{P}$.

This extension of the original space, includes events which are independent (orthogonal) to the original space. It is a construction used to make the definition of some limits working, like the one presented in the following. Assume to have a sequence of real-valued random variables $\{X_n, n \geq 1\}$ defined on $(\Omega, \mathcal{F}, P)$ and a random variable X defined on $(\bar{\Omega}, \bar{\mathcal{F}}, \bar{P})$ which is an extension of $(\Omega, \mathcal{F}, P)$. Assume that $\mathcal{G}$ is a sub-σ field of $\mathcal{F}$.

Definition 3.6.2 (**$\mathcal{G}$-stable convergence in law**) *The sequence X_n is said to converge in distribution $\mathcal{G}$-stably to X if*

$$\mathbb{E}\{f(X_n)Y\} \to \bar{\mathbb{E}}\{f(X)Y\}$$

as $n \to \infty$ for all bounded continuous functions f on $\mathbb{R}$ and all $\mathcal{G}$-measurable random variables Y. We will write

$$X_n \overset{d_s(\mathcal{G})}{\to} X.$$

When $\mathcal{G} = \mathcal{F}$, we say that X_n stably converges in distribution or converges in distribution stably and we write

$$X_n \overset{d_s}{\to} X.$$

This notion of convergence was introduced in Rényi (1963) and Aldous and Eagleson (1978) and further extended to the class of stochastic processes of interest for this book in Jacod (1997, 2002).

Theorem 3.6.3 *If G_n is a sequence of $\mathcal{G}$-measurable random variables such that $G_n \overset{p}{\to} G$ and if $X_n \overset{d_s(\mathcal{G})}{\to} X$, then $(X_n, G_n) \overset{d_s(\mathcal{G})}{\to} (X, G)$.*

The previous theorem says that joint stable convergence can be obtained by mixing finite dimensional stable convergence with convergence in probability. This kind of result will be useful in the proof of asymptotic properties of estimators for financial models in the high frequency setup.

3.7 Brownian motion

In general, the introduction of a stochastic process is intended to model some aspects of stochastic evolution and, of course, financial time series present such kind of evolutions. Brownian motion or Wiener process is the basic stochastic process in continuous time, like the Gaussian is for continuous random variables. Brownian motion can be introduced using different characterizations and here we present one.

Definition 3.7.1 *Brownian motion is a stochastic process $\{B_t, t \geq 0\}$ starting from zero almost surely, i.e. $P(B_0 = 0) = 1$, with the following properties:*

   (i) is a process with independent increments: $B_t - B_s$ is independent of $B_v - B_u$ when $(s, t) \cap (u, v) = \emptyset$;

   (ii) is a process with stationary increments, i.e. the distribution of $B_t - B_s$, $t > s \geq 0$, depends only on the difference $t - s$ but not on t and s separately;

   (iii) is a process with Gaussian increment, i.e.

$$B_t - B_s \sim \mathcal{N}(0, t - s).$$

Sometimes B_t is denoted by W_t because Brownian motion is also called *Wiener process*. The Brownian motion is a process which runs its trajectories at infinite speed or, more formally, it has infinite first order variation $V_t(B) = +\infty$. It is possible to prove that the trajectories of B_t are nowhere differentiable with probability one but, nevertheless, these trajectories are also continuous. At the same time the quadratic variation of the Brownian motion is t, i.e. $[B, B]_t = t$ as next theorem shows.

Theorem 3.7.2 *The quadratic variation of B is $[B, B]_t = t$ in quadratic mean and almost surely.*

Proof. It is sufficient to prove that

$$\lim_{n \to \infty} \sum_{k=1}^{2^n} \Delta_{n,k}^2 = t$$

almost surely and in quadratic mean, where

$$\Delta_{n,k} = B\left(\frac{kt}{2^n}\right) - B\left(\frac{(k-1)t}{2^n}\right).$$

We start proving the limit in the quadratic mean. Note that the random variables $\Delta_{n,k}^2$ are independent and such that $\mathbb{E}(\Delta_{n,k}^2) = t 2^{-n}$. We have to prove that

$$\lim_{n \to \infty} \mathbb{E}\left\{\sum_{k=1}^{2^n} (\Delta_{n,k}^2 - t 2^{-n})\right\}^2 = 0.$$

Then

$$\mathbb{E}\left\{\sum_{k=1}^{2^n} (\Delta_{n,k}^2 - t 2^{-n})\right\}^2 = \sum_{k=1}^{2^n} \mathbb{E}\left\{\Delta_{n,k}^2 - t 2^{-n}\right\}^2$$

$$= \sum_{k=1}^{2^n} \left\{\Delta_{n,k}^4 + t^2 2^{-2n} - t 2^{-n+1} \mathbb{E}(\Delta_{n,k}^2)\right\}$$

$$= 2t^2 \sum_{k=1}^{2^n} 2^{-2n} = 2t^2 2^{-n} \stackrel{n \to \infty}{\to} 0.$$

To prove almost sure convergence, we first use Chebyshev's inequality (2.23)

$$P\left\{\left|\sum_{k=1}^{2^n} (\Delta_{n,k}^2 - t 2^{-n})\right| > \epsilon\right\} \leq \frac{1}{\epsilon^2} \mathbb{E}\left\{\sum_{k=1}^{2^n} (\Delta_{n,k}^2 - t 2^{-n})\right\}^2$$

$$= \frac{2t^2 2^{-n}}{\epsilon^2}$$

from which we see that

$$\sum_{n=1}^{\infty} P\left\{\left|\sum_{k=1}^{2^n} (\Delta_{n,k}^2 - t 2^{-n})\right| > \epsilon\right\} < \infty.$$

Finally, the Borel-Cantelli Lemma (2.4.14) along with the definition of almost sure convergence provides the result.

Then again, by previous result, the continuity of the trajectories of B_t and formula (3.1) one derives that the simple variation of the Brownian motion is necessarily infinite.

Theorem 3.7.3 *Let* $\{B_t, t \geq 0\}$ *be a Brownian motion. The next processes are all Brownian motions:*

(i) *reflection:* $-B_t$;

(ii) *shift:* $B_{t+a} - B_a$ *for each fixed* $a > 0$;

(iii) *self-similarity:* $\hat{B}_t = cB\left(\frac{t}{c^2}\right)$ *for each fixed* $c \in \mathbb{R}$, $c \neq 0$;

(iv) *time inversion:*

$$\tilde{B}_t = \begin{cases} 0, & \text{if } t = 0, \\ tB\left(\frac{1}{t}\right), & \text{if } t \neq 0. \end{cases}$$

Proof. We only prove the self-similarity property (iii). Clearly $\hat{B}(0) = 0$. Since $B_t \sim N(0, t)$ also $\hat{B}_t \sim N(0, t)$. Consider the increments $B(t/c^2) - B(s/c^2) \sim (B_t - B_s)/c$ and $B(u/c^2) - B(v/c^2) \sim (B_u - B_v)/c$. They are clearly independent if $(s, t) \cap (u, v) = \emptyset$.

3.7.1   Brownian motion and random walks

It is possible to see the Brownian motion as the limit of a particular random walk. Let X_1, X_2, $\ldots$, X_n be a sequence of i.i.d. random variables taking only values -1 and $+1$ with equal probability. Let

$$S_n = X_1 + X_2 + \cdots + X_n$$

be the partial sum up to time n, with $S_0 = 0$. S_i representing the position of a particle (starting at 0 at time 0) after i (leftward or rightward) jumps of length 1. It is possible to show that, for $t \in [0, 1]$,

$$\frac{S_{[nt]}}{\sqrt{n}} \xrightarrow{d} B_t$$

where $[x]$ denotes the integer part of the number x. The next code is an empirical proof of the above convergence. We set $t = 0.3$ and try to approximate the distribution of $B(0.3)$. The graphical output is given in Figure 3.3.

```
R> set.seed(123)
R> n <- 500
R> t <- 0.3
R> sim <- 10000
R> B <- numeric(sim)
R> for(i in 1:sim){
+    X <- sample(c(-1,1), n, replace=TRUE)
+    S <- cumsum(X)
```

```
+  B[i] <- S[n*t]/sqrt(n)
+ }
R> plot(density(B),main="",ylab="distribution",
+  xlab=expression(B(0.3)),lty=3, axes=FALSE)
R> axis(1)
R> g <- function(x) dnorm(x,sd=sqrt(t))
R> curve( g, -3, 3, add=TRUE)
```

3.7.2   Brownian motion is a martingale

It is important to notice that the Brownian motion is also a martingale with respect to its natural filtration. To see this, it is sufficient to show that $\mathbb{E}\{B_t|\mathcal{F}_s\} = B_s$. Indeed, writing $\mathbb{E}\{B_t|\mathcal{F}_s\} = B_s$ is equivalent to write $\mathbb{E}\{B_t - B_s|\mathcal{F}_s\} = 0$ because B_s is $\mathcal{F}_s$-measurable. Now, the increment $B_t - B_s$ is independent of $\mathcal{F}_s$ by definition of Brownian motion, and hence $\mathbb{E}\{B_t - B_s|\mathcal{F}_s\} = \mathbb{E}(B_t - B_s)$ and the latter is zero because $B_t - B_s \sim N(0, t - s)$.

Exercise 3.5 *Prove that* $\{X_t = B_t^2 - t, t \geq 0\}$ *is a martingale with respect to the natural filtration of* B_t.

Exercise 3.6 *Let* μ *and* σ *be real numbers. Prove that* $\{X_t = e^{\mu t + \sigma B_t}, t \geq 0\}$ *is a martingale, with respect to the natural filtration of* B_t, *if and only if* $\mu = -\frac{1}{2}\sigma^2$.

3.7.3   Brownian motion and partial differential equations

Denote by $p(t - s, x, y)$ the transition density of the Brownian motion, i.e. the density of the distribution $P(B_t \in [x, x + \mathrm{d}x]|B_s = y)$. This density is clearly a Gaussian density

$$p(t - s, x, y) = \frac{1}{\sqrt{2\pi(t - s)}} e^{-\frac{(x-y)^2}{2(t-s)}}.$$

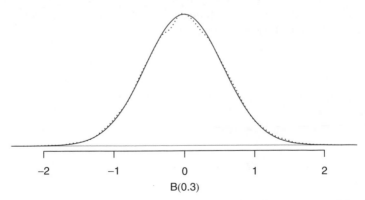

Figure 3.3   Approximation of Brownian motion distribution (dotted line) as the limit of a symmetric random walk.

For simplicity, assume that $s = 0$, then it is easy to show that $p(t, x, y)$ solves the so-called *diffusion* or *heat* partial differential equation:

$$\frac{\partial p}{\partial t} = \frac{1}{2}\frac{\partial^2 p}{\partial x^2}. \tag{3.15}$$

Indeed,

$$\frac{\partial}{\partial t}p(t, x, y) = \frac{(x - y)^2 - t}{2t^2}p(t, x, y)$$

$$\frac{\partial}{\partial x}p(t, x, y) = \frac{y - x}{t}p(t, x, y)$$

$$\frac{\partial^2}{\partial x^2}p(t, x, y) = \frac{(x - y)^2 - t}{2t^2}p(t, x, y)$$

which is enough to get the conclusion.

3.8   Counting and marked processes

A *point process* or a *counting process* $\{N_t, t \geq 0\}$, is a continuous time process which records how many times an event has occurred up to some time t. This process takes values on the integers and can be represented as follows:

$$N_t = \sum_i \mathbf{1}_{\{\tau_i \leq t\}},$$

where τ_i are the random inter-arrival times. A counting process is a *càdlàg* (*continue à droite, limite à gauche*, i.e. right continuous with left limit) process, whose trajectory remains constant up to the time when an event occurs. At the time instant of an arrival, the process N_t jumps upward and continuous with a flat trajectory up to next event. The notion of càdlàg is then equivalent to saying that the points where the process jumps (the arrival times) are included in the upper part of the trajectory (see Figure 3.4).

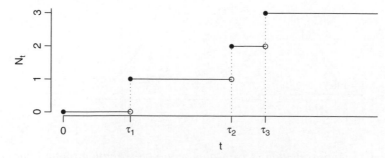

Figure 3.4   Example of a càd-làg process. Empty circles indicate that the points do not belong to that part of the trajectory of the process while filled circles denote the contrary.

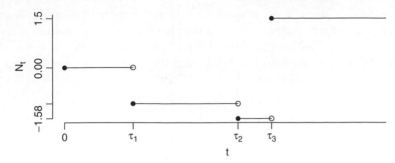

Figure 3.5   Example of marked point process with Gaussian marks.

The random times τ_i, $i = 1, 2, \ldots$ are positive random variables which in most applications are assumed to be independent and with a common law as we will see. When the size of the jump is not of unit length, the process is called *marked point process* and the jumps at time τ_i, say Y_{τ_i}, are called *marks*. Clearly, the process is no longer interpretable as a counting process. The marks Y_{τ_i} can be as well random variables and the resulting marked point process $\{X_t, t \geq 0\}$ is usually written in this form:

$$X_t = \sum_{i=0}^{N_t} Y_{\tau_i}, \quad t \geq 0,$$

where N_t is the counting process for the random times τ_i, $i = 1, 2, \ldots$. Figure 3.5 represents a trajectory of a marked point process with independent Gaussian marks $Y_{\tau_i} \sim N(0, \sigma = 2)$.

3.9   Poisson process

The Poisson process is one of the principal stochastic processes in continuous time with discrete state space. It is a counting process with independent arrival times which are exponentially distributed. This process is used to model the number of rare events in time. In financial markets ruptures or big shocks are usually considered rare events and Poisson process is used to model them. This process can be modeled in several ways as for Brownian motion. We denote by $N_t = N(t) = N([0, t))$ the value of the process up to time t. We denote by $N([a, b]) = N_b - N_a$ the number of events between a and b. Let $[t, t + \Delta t)$ be a small time-interval. The Poisson process is such that in this small interval at most 1 event can occur with non-negligible probability. More precisely

$$\begin{aligned}
P(N([t, t + \Delta t)) = 0) &= 1 - \lambda \Delta t + o(\Delta t) \\
P(N([t, t + \Delta t)) = 1) &= \lambda \Delta t + o(\Delta t) \\
P(N([t, t + \Delta t)) \geq 2) &= o(\Delta t)
\end{aligned} \quad (3.16)$$

where $o(\Delta t)$ is negligible with respect to Δt, as $\Delta t \to 0$. In particular, the second assumption $P(N([t, t + \Delta t)) = 1) = \lambda \Delta t + o(\Delta t)$, is motivated by saying that 'in a small time-interval, the number of arrivals is proportional to the length of that interval'. Notice that the set of equations (3.16) is just a particular case of (3.14) for continuous time Markov chains. It is also assumed that increments of Poisson process are independent of the past. Under these assumptions it is possible to show (see Cox and Miller 1965) that

$$P(N_t = k) = e^{-\lambda t} \frac{(\lambda t)^k}{k!}, \quad k = 0, 1, 2, \ldots$$

which means that N_t is a Poisson random variable with rate λt, i.e. $N_t \sim \text{Poi}(\lambda t)$. Further, let T be the random time between two Poisson events, then $T \sim \text{Exp}(\lambda)$. We will use this aspect in the simulation of the Poisson process.

The above version of the Poisson process is called *homogeneous* because the rate λ at which events occur is constant. It is possible to generalize it to the case in which $\lambda = \lambda(t)$ is a function of time. The nonhomogeneous Poisson process is characterized by its *intensity function*

$$\Lambda(t) = \int_0^t \lambda(s) ds$$

and its distribution has the following form:

$$P(N_t = k) = e^{-\Lambda(t)} \frac{\Lambda(t)^k}{k!}, \quad k = 0, 1, 2, \ldots$$

i.e. $N_t \sim \text{Poi}(\Lambda(t))$. From the fact that N_t is a Poisson random variable we have immediately that $\mathbb{E}(N_t) = \Lambda(t)$. Clearly, when $\lambda(t) = \lambda$ then $\Lambda(t) = \lambda t$ and the process is homogeneous. The statistical analysis and modern treatment of general Poisson processes can be found in Kutoyants (1998).

3.10   Compound Poisson process

The *compound Poisson* process is defined as $X_t = \sum_{i=1}^{N_t} Y_{\tau_i}$, where N_t is a Poisson process and Y_{τ_i} are the jumps (the markers) at random times τ_i. As will be explained later, the compound Poisson process plays an important role in the construction of the Lévy process. It is easy to derive the mean of the compound Poisson process.

Theorem 3.10.1 *Let N_t be a compound Poisson process and let Y_{τ_i} be i.i.d. with common mean μ. Then*

$$\mathbb{E}(X_t) = \mu \Lambda(t).$$

Proof.

$$\mathbb{E}(X_t) = \mathbb{E}\left(\sum_{i=1}^{N_t} Y_{\tau_i}\right) = \mathbb{E}\left(\mathbb{E}\left\{\sum_{i=1}^{N_t} Y_{\tau_i}\,\middle|\, N_t\right\}\right)$$

$$= \sum_{k=1}^{\infty} \mathbb{E}\left\{\sum_{i=1}^{k} Y_{\tau_i}\,\middle|\, N_t = k\right\} P(N_t = k)$$

$$= \sum_{k=1}^{\infty} \mathbb{E}\left(\sum_{i=1}^{k} Y_{\tau_i}\right) P(N_t = k) = \sum_{k=1}^{\infty} \mu k P(N_t = k)$$

$$= \mu \sum_{k=0}^{\infty} k P(N_t = k) = \mu \mathbb{E}(N_t)$$

$$= \mu \Lambda(t)$$

because for $k = 0$, $k P(N_t = k) = 0$.

Although the proof is elementary, it is useful to see it because of the conditioning argument which is often used to prove results involving Poisson processes. When Y_{τ_i} is a sequence of i.i.d. random variables and common characteristic function $\varphi_Y(u)$, also the characteristic function of a compound Poisson process can be easily determined. Indeed

$$\varphi_{X_t}(u) = \mathbb{E}\left(e^{iuX_t}\right) = \mathbb{E}\left(\exp\left\{iu\sum_{i=1}^{N_t} Y_{\tau_i}\right\}\right)$$

$$= \mathbb{E}\left(\mathbb{E}\left\{\exp\left\{iu\sum_{i=1}^{N_t} Y_{\tau_i}\right\}\,\middle|\, N_t\right\}\right) = \mathbb{E}\left(\varphi_Y(u)^{N_t}\right)$$

$$= \sum_{k=0}^{\infty} \frac{\varphi_Y(u)^k (\lambda t)^k}{k!} e^{-\lambda t} = \sum_{k=0}^{\infty} \frac{\varphi_Y(u)^k (\lambda t)^k}{k!} e^{-\lambda t} e^{\pm \varphi_Y(u)\lambda t}$$

$$= e^{-\lambda t + \varphi_Y(u)\lambda t} \sum_{k=0}^{\infty} \frac{(\varphi_Y(u)\lambda t)^k}{k!} e^{-\varphi_Y(u)\lambda t}$$

$$= \exp\{\lambda t(\varphi_Y(u) - 1)\}.$$

Note that, conditionally on the value of N_t, X_t is just a sum of independent random variables with common characteristic function $\varphi_Y(u)$, so we applied Equation (2.7) in the last equality of the first line of the proof. As usual, even if $\varphi_{X_t}(u)$ is a nice formula, an explicit version of the distribution of X_t cannot

always be found and FFT or other methods have to be considered. But there is
at least one special case for which it is possible to derive the distribution of X_t
without using the inverse of the characteristic function. Assume that the markers
are i.i.d. random variables with common law $N(\mu, \sigma^2)$. Then X_t, conditionally on
the value assumed by N_t, is just the sum of N_t i.i.d. Gaussian random variables,
i.e. $X_t|N_t$ is distributed as $N(\mu N_t, \sigma^2 N_t)$. From this fact, summing up with
respect to all possible values of N_t one gets the series-type expression

$$F_{X_t}(x) = P(X_t \le x) = \sum_{k=1}^{\infty} P\left(\sum_{i=1}^{N_t} Y_{\tau_i} \le x \,\bigg|\, N_t = k\right) P(N_t = k)$$

$$= \sum_{k=1}^{\infty} P\left(\sum_{i=1}^{k} Y_{\tau_i} \le x\right) \frac{(\lambda t)^k}{k!} e^{-\lambda t}$$

with $\sum_{i=1}^{k} Y_{\tau_i} \sim N(k\mu, k\sigma^2)$. Therefore, the density of X_t is given by the
following expression

$$f_{X_t}(x) = \sum_{k=1}^{\infty} \frac{e^{-\frac{(x-k\mu)^2}{2k\sigma^2}}}{\sqrt{2\pi k\sigma^2}} \frac{(\lambda t)^k}{k!} e^{-\lambda t}.$$

This is an infinite series where the higher order terms are negligible because
$P(N_t = k)$ converges quite fast to zero as k increases. Therefore, depending on
the values of λ and t a reasonable number of few terms have to be retained in
the approximation. The Poisson and the compound Poisson processes are also
time-invariant, in the following sense:

Theorem 3.10.2 *Let X_t be a compound Poisson process with intensity function
$\Lambda(t)$ and Y_{τ_i} a sequence of i.i.d. random variables. Then*

$$X_t - X_s \sim X_{t-s}, \quad s < t.$$

Proof.

$$P(X_t - X_s \le x) = P\left(\sum_{i=1}^{N_t} Y_{\tau_i} - \sum_{i=1}^{N_s} Y_{\tau_i} \le x\right) = P\left(\sum_{i=N_s+1}^{N_t} Y_{\tau_i} \le x\right)$$

but the above sum contains $N_t - N_s$ terms, and because Y_{τ_i} are i.i.d. random
variables, we have that

$$\sum_{i=N_s+1}^{N_t} Y_{\tau_i} \sim (N_t - N_s) Y_{\tau_i}$$

Now, notice that $N_t - N_s \sim N_{t-s}$ because $N_t = N_s + N_{t-s}$ due to independence of the Poisson increments from the past. Therefore,

$$\sum_{i=1}^{N_t} Y_{\tau_i} - \sum_{i=1}^{N_s} Y_{\tau_i} \sim \sum_{i=1}^{N_{t-s}} Y_{\tau_i}.$$

3.11 Compensated Poisson processes

We have seen that the Poisson process N_t has mean $\Lambda(t)$, and in general $\Lambda(t) > 0$. This implies that N_t is not a martingale, but it is possible to compensate the Poisson process with another process which makes it a martingale. The process A_t which makes

$$M_t = N_t - A_t$$

a martingale is called *compensator*.

Theorem 3.11.1 *Let X_t be a compound Poisson process with intensity function $\Lambda(t) = \int_0^s \lambda(u)\mathrm{d}u$ and let the random variables Y_{τ_i} be i.i.d. with common mean μ, then*

$$M_t = X_t - \mu\Lambda(t)$$

is a martingale.

In short, the compensator A_t of a standard or compound Poisson process is its mean $A_t = \mathbb{E}X_t$. Notice that, for the standard Poisson process, $Y_{\tau_i} = 1$ for all i and hence $\mu = 1$. Therefore, the corresponding martingale process is $M_t = N_t - \Lambda(t)$.

Proof.

$$\mathbb{E}\{M_t|\mathcal{F}_s\} = \mathbb{E}\{X_t - \mu\Lambda(t) \pm X_s|\mathcal{F}_s\} = -\mu\Lambda(t) + \mathbb{E}\{X_t - X_s + X_s|\mathcal{F}_s\}$$

$$= -\mu\Lambda(t) + \mathbb{E}(X_t - X_s) + X_s = -\mu\Lambda(t) + \mathbb{E}(X_{t-s}) + X_s$$

$$= -\mu\Lambda(t) + \mu\Lambda(t-s) + X_s$$

$$= -\mu\int_0^t \lambda(u)\mathrm{d}u + \mu\int_0^{t-s} \lambda(u)\mathrm{d}u + \mu\Lambda(t-s) + X_s$$

$$= -\mu\Lambda(s) + X_s = M_s.$$

3.12 Telegraph process

The telegraph process, studied in (Goldstein 1951) and (Kac 1974), models a random motion with finite velocity and it is usually proposed as an alternative to diffusion models. The process describes the position of a particle moving on the real line, alternatively with constant velocity $+c$ or $-c$. The changes of direction are governed by a Poisson process with intensity function $\Lambda(t)$. The telegraph process or *telegrapher's* process is defined as

$$X_t = x_0 + V_0 \int_0^t (-1)^{N_s} ds, \quad t > 0, \tag{3.17}$$

where V_0 is the initial velocity taking values $\pm c$ with equal probability and independently of the Poisson process $\{N_t, t > 0\}$. For simplicity, we assume that x_0 is not random and we choose it as $x_0 = 0$. Related to the telegraph process is its *velocity process* defined as

$$V_t = V_0(-1)^{N_t}, \quad t > 0,$$

so that $X_t = \int_0^t V_s ds$. The telegraph process is not Markovian, while the coordinate process (X_t, V_t) is. Unfortunately, in applications with discrete time data, only the position X_t is observed but not the velocity. We will return to this problem when we will discuss estimation for this model. We now briefly study the velocity process V_t. The following conditional laws:

$$P(V_t = +c | V_0 = +c) \quad \text{and} \quad P(V_t = -c | V_0 = +c),$$

characterize the velocity process. Their explicit forms are as follows.

Theorem 3.12.1

$$P(V(t) = c | V(0) = c) = 1 - \int_0^t \lambda(s) e^{-2\Lambda(s)} ds = \frac{1 + e^{-2\Lambda(t)}}{2}, \tag{3.18}$$

$$P(V(t) = -c | V(0) = c) = \int_0^t \lambda(s) e^{-2\Lambda(s)} ds = \frac{1 - e^{-2\Lambda(t)}}{2} \tag{3.19}$$

Proof. First notice that, the two probabilities

$$p^{(+c)}(t) = P(V_t = +c) \quad \text{and} \quad p^{(-c)}(t) = P(V_t = -c)$$

are solutions of the following system of partial differential equations

$$\begin{cases} p_t^{(+c)}(t) = \Lambda_t(t)(p^{(-c)}(t) - p^{(+c)}(t)) \\ p_t^{(-c)}(t) = \Lambda_t(t)(p^{(+c)}(t) - p^{(-c)}(t)) = -p_t^{(+c)}(t) \end{cases} \tag{3.20}$$

This can be proved by Taylor expansion of the two functions $p_t^{(+c)}(t)$ and $p_t^{(-c)}(t)$. We give only the derivation of (3.18) because the other follows by symmetry. Conditioning on $V_0 = +c$ implies that: $p^{(+c)}(0) = 1$, $p_t^{(+c)}(0) = -\Lambda_t(0) = -\lambda(0)$, $p^{(-c)}(0) = 0$ and $p_t^{(-c)}(0) = \lambda(0)$. From (3.20) we obtain that

$$\frac{p_{tt}^{(+c)}(t)}{p_t^{(+c)}(t)} = \frac{d}{dt} \log p_t^{(+c)}(t) = -\frac{2\Lambda_t(t)^2 - \Lambda_{tt}(t)}{\Lambda(t)},$$

and by simple integration by parts it emerges that:

$$p_t^{(c)}(t) = \exp\left\{-2\Lambda(t) + \log\left(\frac{\lambda(t)}{\lambda(0)}\right) + \log(-\lambda(0))\right\},$$

from which also follows (3.18). Remark that it is possible to derive (3.18) by simply noting that

$$P(V_t = +c|V_0 = +c) = P\left(\bigcup_{k=0}^{\infty} N_t = 2k\right) = \sum_{k=0}^{\infty}\frac{\Lambda(t)^{2k}}{(2k)!}e^{-\Lambda(t)} = \frac{1 + e^{-2\Lambda(t)}}{2}$$

and then $P(V_t = -c|V_0 = +c) = 1 - P(V_t = +c|V_0 = +c)$.

We conclude the analysis of the velocity process by giving also the covariance function of the couple (V_t, V_s), $s, t > 0$. It can be easily proven that the characteristic function of the couple (V_t, V_s) is, for all $(\alpha, \beta) \in \mathbb{R}^2$,

$$\mathbb{E}\left(e^{i\alpha V_s + i\beta V_t}\right) = \cos(\alpha c)\cos(\beta c) - e^{-2(\Lambda(t)-\Lambda(s))}\sin(\alpha c)\sin(\beta c).$$

The velocity process has zero mean, therefore the covariance function is

$$\text{Cov}(V_s, V_t) = \mathbb{E}(V_s V_t) = -\frac{\partial^2}{\partial\alpha\partial\beta}\mathbb{E}\left(e^{i\alpha V(s) + i\alpha V(t)}\right)\bigg|_{\alpha=\beta=0} = c^2 e^{-2|\Lambda(t)-\Lambda(s)|}.$$

3.12.1 Telegraph process and partial differential equations

The name of this process comes from the fact that the law of the process X_t is a solution of the so-called *telegraph equation* which is a hyperbolic partial differential equation of the following type

$$u_{tt}(t, x) + 2\lambda u_t(t, x) = c^2 u_{xx}(t, x) \qquad (3.21)$$

where $u(t, x)$ is a two-times differentiable function with respect to argument x and one-time differentiable with respect to t, with λ and c two constants. The telegraph equation emerges in the physical analysis of quasi-stationary electric swings propagation along the telegraph cables and it is sometimes called the *wave equation* for this reason. We now give a derivation of the telegraph equation in the nonhomogenous case. We consider the distribution of the position of the particle at time t

$$P(t, x) = P(X_t < x) \qquad (3.22)$$

and we introduce the two distribution functions $F(t, x) = P(X_t < x, V_t = +c)$ and $B(t, x) = P(X_t < x, V_t = -c)$, so that $P(t, x) = F(t, x) + B(t, x)$ and $W(t, x) = F(t, x) - B(t, x)$. The function $W(\cdot, \cdot)$ is usually called the 'flow function'.

Theorem 3.12.2 *Suppose that $F(\cdot, \cdot)$ and $B(\cdot, \cdot)$ are two times differentiable in x and t, then*

$$\begin{cases} F_t(t, x) = -cF_x(t, x) - \lambda(t)(F(t, x) - B(t, x)) \\ B_t(t, x) = cB_x(t, x) + \lambda(t)(F(t, x) - B(t, x)) \end{cases} \tag{3.23}$$

moreover $P(t, x)$ in (3.22) is a solution to the following telegraph equation with nonconstant coefficients

$$\frac{\partial^2}{\partial t^2} u(t, x) + 2\lambda(t) \frac{\partial}{\partial t} u(t, x) = c^2 \frac{\partial^2}{\partial x^2} u(t, x). \tag{3.24}$$

Proof. By Taylor expansion, one gets that $F(\cdot, \cdot)$ and $B(\cdot, \cdot)$ are solutions to (3.23) and rewriting system (3.23) in terms of the functions $W(\cdot, \cdot)$ and $P(\cdot, \cdot)$ it emerges that

$$\begin{cases} P_t(t, x) = -cW_x(t, x) \\ W_t(t, x) = cP_x(t, x) - 2\lambda(t)W(t, x) \end{cases} \tag{3.25}$$

The conclusion arises by direct substitutions. In fact, from the first of system (3.25) we have

$$P_{tt}(t, x) = \frac{\partial}{\partial t} P_t(t, x) = -c \frac{\partial}{\partial t} W_x(t, x).$$

Furthermore,

$$\begin{aligned} W_{tx}(t, x) &= \frac{\partial}{\partial x}(cP_x(t, x) - 2\lambda(t)W(x, t)) \\ &= cP_{xx}(t, x) - 2\lambda(t)W_x(t, x) \\ &= cP_{xx}(t, x) + \frac{2}{c}\lambda(t)P_t(t, x) \end{aligned}$$

by using respectively the second and the first equation of system (3.25).

Consider again the distribution $P(X_t \leq x)$ in (3.22) which can be rewritten as the transition from time 0 to time t of the telegraph process, i.e. $P(X_t \leq x|x_0 = 0)$ (we have initially set $x_0 = 0$, but argument works for any initial nonrandom x_0). This distribution is a mix of a continuous and a discrete component. The discrete component arises when there are no Poisson events in $[0, t)$. In this case, $X_t = \pm ct$ depending on the initial velocity $V_0 = \pm c$, and this happens with probability $\frac{1}{2}e^{-\lambda t}$. We know that (3.22) solves the telegraph equation (3.24), so does the density of its absolute component. By different means Goldstein (1951), Orsingher (1990) and Pinsky (1991) obtained the transition density, which we present as the sum of the continuous and discrete part using the Dirac delta function

$$p(t, x; 0, x_0) = \frac{e^{-\lambda t}}{2c} \left\{ \lambda I_0 \left(\frac{\lambda}{c} \sqrt{c^2 t^2 - (x - x_0)^2} \right) \right.$$

$$+ \frac{\partial}{\partial t} I_0\left(\frac{\lambda}{c}\sqrt{c^2 t^2 - (x - x_0)^2}\right)\bigg\}\mathbf{1}_{\{|x-x_0|<ct\}}$$

$$+ \frac{e^{-\lambda t}}{2}\{\delta(x - x_0 - ct) + \delta(x - x_0 + ct)\} \qquad (3.26)$$

for any $|x - x_0| \le ct$, where $I_\nu(x)$ is the modified Bessel function of order ν (see (Abramowitz and Stegun 1964)) and δ is the Dirac delta function. For the nonhomogenous case, the general solution to (3.24) is not know, but in one case. The proof can be found in Iacus (2001).

Theorem 3.12.3 *Suppose that the intensity function of the Poisson process N_t in (3.24) is*

$$\lambda(t) = \lambda_\theta(t) = \theta \tanh(\theta t), \quad \theta \in \mathbb{R}. \qquad (3.27)$$

Then, the absolutely continuous component $p_\theta(\cdot)$ of distribution (3.22), conditionally on $V_0 = +c$, is given by

$$p_\theta(t, x | V_0 = +c) = \begin{cases} \frac{\theta t}{\cosh(\theta t)} \frac{I_1\left(\frac{\theta}{c}\sqrt{c^2 t^2 - x^2}\right)}{2\sqrt{c^2 t^2 - x^2}}, & |x| < ct, \\ 0, & \text{otherwise.} \end{cases} \qquad (3.28)$$

3.12.2   Moments of the telegraph process

Coming back to the homogenous version of (3.22), we have the following results concerning the moments of the telegraph process. Again we assume $x_0 = 0$. The first two moments of the process are well known and derived in Orsingher (1990)

$$\mathbb{E}(X_t) = 0, \quad \text{and} \quad \mathbb{E}(X_t^2) = \frac{c^2}{\lambda}\left(t - \frac{1 - e^{-2\lambda t}}{2\lambda}\right). \qquad (3.29)$$

Next results generalize (3.29) to any integer q (see, Iacus and Yoshida (2008)).

Theorem 3.12.4 *For every positive integer q,*

$$\mathbb{E}\left\{X_t^{2q}\right\} = (ct)^{2q}\left(\frac{2}{\lambda t}\right)^{q-\frac{1}{2}}\Gamma\left(q + \frac{1}{2}\right)\left\{I_{q+\frac{1}{2}}(\lambda t) + I_{q-\frac{1}{2}}(\lambda t)\right\}e^{-\lambda t}. \qquad (3.30)$$

The modified Bessel functions admit the following expansion

$$I_\nu(x) = \frac{1}{\Gamma(\nu + 1)}\left(\frac{x}{2}\right)^\nu\left(1 + \frac{x^2}{4(\nu + 1)} + \frac{x^4}{32(\nu + 1)(\nu + 2)} + \cdots\right)$$

from which we obtain that $E\left\{X_t^{2q}\right\}$ is of order t^{2q} as $t \to 0$. The following expansion, for $t \to 0$, will be useful in the following

$$\mathbb{E}\{X_t^2\} = c^2 t^2 - \frac{2}{3} c^2 \lambda t^3 + \frac{1}{3} c^2 \lambda^2 t^4 + o(t^4) \tag{3.31}$$

$$\mathbb{E}\{X_t^4\} = c^4 t^4 - \frac{4}{5} c^4 \lambda t^5 + \frac{2}{5} c^4 \lambda^2 t^6 + o(t^6) \tag{3.32}$$

$$\mathbb{E}\{X_t^6\} = c^6 t^6 - \frac{6}{7} c^6 \lambda t^7 + \frac{3}{7} c v^6 \lambda^2 t^8 + o(t^8) \tag{3.33}$$

The moment generating function for X_t was derived in Di Crescenzo and Martinucci (2006).

Theorem 3.12.5 *For all $s \in \mathbb{R}$ and $t \geq 0$*

$$\mathbb{E}\left\{e^{s\,X_t}\right\} = e^{-\lambda t}\left\{\cosh\left(t\sqrt{\lambda^2 + s^2 c^2}\right) + \frac{\lambda}{\sqrt{\lambda^2 + s^2 c^2}}\sinh\left(t\sqrt{\lambda^2 + s^2 c^2}\right)\right\}. \tag{3.34}$$

Many authors analyzed probabilistic properties of the process over the years (see for example Orsingher (1985, 1990); Pinsky (1991); Fong and Kanno (1994); Stadje and Zacks (2004) but here we only mentioned those that will be used in the next chapters.

3.12.3 Telegraph process and Brownian motion

It is quite easy to see that in the limit as $c \to \infty$, $\lambda \to \infty$ and $\lambda/c^2 \to 1$, the telegraph equation (3.21) converges to the heat equation (3.15). Although a rigorous proof can be written, the above result intuitively means that, in the limit when both the velocity and the intensity of the Poisson process diverge, the telegraph process converges in law to a Brownian motion. But this is true at path level. Indeed, the limiting process moves at infinite speed and its trajectory becomes nowhere differentiable like the Brownian motion case.

3.13 Stochastic integrals

In Chapter 1 we introduced stochastic differential equations as a way to model returns of an asset price $\{S_t, t \geq 0\}$ in this form:

$$\frac{\mathrm{d}S_t}{S_t} = \text{deterministic contribution} + \text{stochastic contribution}$$

or, more precisely,

$$\frac{\mathrm{d}S_t}{S_t} = \mu\mathrm{d}t + \sigma\mathrm{d}B_t,$$

where dB_t is the variation of the Brownian motion. This writing is mathematically acceptable if $dt > 0$ is not infinitesimal, otherwise the following limiting equation, as $dt \to 0$,

$$dS_t = \mu S_t dt + \sigma S_t dB_t \qquad (3.35)$$

is not correct because Brownian motion has infinite variation. We also noticed that it is possible to give a precise meaning to (3.35) in integral form:

$$S_t = S_0 + \mu \int_0^t S_u du + \sigma \int_0^t S_u dB_u,$$

where $\int_0^t S_u dB_u$ is the stochastic or Itô integral. We now introduce Itô integral in a more precise way, but before giving the formal definition we offer an intuitive way to construct such an object. In fact, integrals of random processes, say $\{X_t, t \geq 0\}$

$$I(X) = \int_0^t X_u du$$

can be interpreted, ω by ω, as ordinary integrals, i.e.

$$I(X, \omega) = \int_0^t X(u, \omega) du$$

and then, loosely speaking, $I(X)$ can be interpreted as the random 'area' under the trajectory of the process. On the contrary, the stochastic integral involves randomness in the integrator part as well and, even ω by ω, this writing

$$I(X, \omega) = \int_0^t X(u, \omega) dB(u, \omega)$$

needs attention. To understand the following, consider $f(u) = X(u, \omega)$ and $g(u) = B(u, \omega)$, for a given fixed ω as two nonrandom functions and forget for a while the link between $g(\cdot)$ and Brownian motion. When $g(\cdot)$ is differentiable, it is possible to define integrals of the form $\int_0^t f(u) dg(u)$ as

$$\int_0^t f(u) dg(u) = \int_0^t f(u) g'(u) du.$$

If $g(\cdot)$ is not differentiable but has at list finite variation (e.g. $g(x) = |x|$) it is still possible to define the integral of $f(\cdot)$ with respect to the variation of $g(\cdot)$ as follows:

$$\lim_{\|\Pi_n\| \to 0} \sum_{i=1}^{n-1} f(s_i)(g(s_{i+1}) - g(s_i)),$$

where $\Pi_n = \Pi_n([0, t]) = \{0 = s_0 \le s_1 \le \cdots \le s_n = t\}$ and the limit is as $n \to \infty$. As mentioned, we cannot introduce the stochastic integral as follows:

$$\int_0^t X(u)dB(u) = \lim_{n \to \infty} \sum_{i=1}^{n-1} X(s_i)(B(s_{i+1}) - B(s_i)) \tag{3.36}$$

because $V_t(B) = +\infty$. We need to consider the limit in a different metric instead. Assume that $X(s_i)$ is independent of the Brownian increment $B(s_{i+1}) - B(s_i)$ in (3.36). We have that

$$\mathbb{E}\{X(s_i)(B(s_{i+1}) - B(s_i))\}^2 = \mathbb{E}(X^2(s_i))\mathbb{E}(B(s_{i+1}) - B(s_i))^2$$
$$= \mathbb{E}(X^2(s_i)) \cdot (s_{i+1} - s_i)$$

and hence

$$\mathbb{E}\left\{\sum_{i=1}^{n-1} X(s_i)(B(s_{i+1}) - B(s_i))\right\}^2 = \mathbb{E}\left\{\sum_{i=1}^{n-1} X(s_i)^2(B(s_{i+1}) - B(s_i))^2\right\} + A$$
$$= \sum_{i=1}^{n-1} \mathbb{E}(X(s_i)^2)(s_{i+1} - s_i)$$

where the term A contains cross products terms of the form:

$$X(s_i)(B(s_{i+1}) - B(s_i))X(s_j)(B(s_{j+1}) - B(s_j))$$

with $i \ne j$. By the properties of the increments of Brownian motion and the assumption of independence of $X(s_i)$ and $B(s_{i+1}) - B(s_i)$, the expected value of those terms are all zero. Therefore, we have that

$$\lim_{||\Pi_n|| \to 0} \mathbb{E}\left\{\sum_{i=1}^{n-1} X(s_i)(B(s_{i+1}) - B(s_i))\right\}^2$$
$$= \lim_{||\Pi_n|| \to 0} = \sum_{i=1}^{n-1} \mathbb{E}(X(s_i)^2)(s_{i+1} - s_i) = \int_0^t \mathbb{E}(X_u^2)du.$$

So we can define the stochastic integral as follows:

$$I(X) = \lim_{||\Pi_n|| \to 0} \mathbb{E}\left\{\sum_{i=1}^{n-1} X(s_i)(B(s_{i+1}) - B(s_i))\right\}^2. \tag{3.37}$$

Now we need to notice two facts:

(i) the limit in (3.37) exists only if $\int_0^t \mathbb{E}(X_u^2)du < \infty$;

(ii) we assumed independence of $X(s_i)$ from $B(s_{i+1}) - B(s_i)$. This assumption, more precisely, should be stated as X being adapted to the filtration generated by the Brownian motion.

The above are two elementary properties we need to ask to the integrand X in order to have a well-defined stochastic integral.

Definition 3.13.1 (Stochastic integral) *Let $\{X_t, t \geq 0\}$ be a stochastic process adapted to the filtration generated by the Brownian motion and such that $\int_0^t \mathbb{E}(X_u^2)du < \infty$. The stochastic integral of X is defined as*

$$I(X) = \int_0^t X_u dB_u = \lim_{||\Pi_n|| \to 0} \mathbb{E}\left\{\sum_{i=1}^{n-1} X(s_i)(B(s_{i+1}) - B(s_i))\right\}^2.$$

It is not always evident that the stochastic integral exists.

Example 3.13.2 *Consider the stochastic process X defined as $X_s = s^{-1}B_s$. This process is clearly adapted to the natural filtration of B. Let us check the integrability condition*

$$\int_0^t \mathbb{E}(X_u^2)du = \int_0^t u^{-2}\mathbb{E}(B_u^2)du$$

$$= \int_0^t u^{-2}u\,du = \int_0^t u^{-1}du = \log t - \log 0 = \infty.$$

Thus, for this process the stochastic integral is not well defined. If instead we take $X(s) = B(s)$, then the integrability condition is verified

$$\int_0^t \mathbb{E}(B_u^2)du = \int_0^t u\,du = \frac{t^2}{2} < \infty.$$

Example 3.13.3 *Consider the stochastic process $X_s = B(s+1)$. The integrability condition is clearly satisfied, but this process is not adapted to the natural filtration of B, i.e. $X_s = B(s+1)$ is not $\mathcal{F}_s$-measurable, because $\mathcal{F}_s = \sigma(B_u, u \leq s)$ and events concerning $B(s+1)$ are not contained in $\mathcal{F}_s$.*

Notice that, in Definition 3.13.1, the summands are of the following form $X(s_i)((B(s_{i+1}) - B(s_i))$ and we ask for adaptedness in order to be able to calculate expected value of the square of these terms. It is possible to define the stochastic integral in the Stratonovich sense as follows:

$$\int_0^t X_u dB_u = \lim_{||\Pi_n|| \to 0} \sum_{i=1}^{n-1} X\left(\frac{s_i + s_{i+1}}{2}\right)(B(s_{i+1}) - B(s_i))$$

i.e. considering the value of X at the mid-point of the interval $[s_i, s_{i+1}]$. In this case, adaptedness of X is not enough and the stochastic integral has not all the properties which are very useful in stochastic calculus.[1]

Definition 3.13.4 *A process X such that $I(X)$ exists is called an Itô integrable process.*

Notice that the stochastic integral is itself a stochastic process.

3.13.1   Properties of the stochastic integral

Let X and Y be Itô integrable and a and b two real constants. Then

- zero-mean property

$$\mathbb{E}\left(\int_0^t X_u dB_u\right) = 0;\tag{3.38}$$

- Itô isometry

$$\text{Var}\left(\int_0^t X_u dB_u\right) = \mathbb{E}\left(\int_0^t X_u dB_u\right)^2 = \int_0^t \mathbb{E}(X_u^2) du;\tag{3.39}$$

- linearity

$$I(aX + bY) = aI(X) + bI(Y);\tag{3.40}$$

- integration of a constant

$$I(a) = \int_0^t a\, dB_u = a\int_0^t dB_u = aB_t;\tag{3.41}$$

- martingality: let $M_t = M_0 + \int_0^t X_u dB_u$ then M_t is a martingale.

Using the definition of $I(X)$ is not always very easy as the next example shows.

Example 3.13.5 *Let us calculate the stochastic integral of B, i.e. $\int_0^t B_s dB_s$. Some trivial but lengthy algebraic steps show that the following equivalence holds true*

$$\frac{1}{2}\sum_{j=0}^{n-1}(B(s_{i+1}) - B(s_i))^2 = \frac{1}{2}B(t_n)^2 - \sum_{j=0}^{n-1}B(s_i)(B(s_{i+1}) - B(s_i)),$$

so, let us start from the above and evaluate the limit as $n \to \infty$ of

$$\sum_{i=1}^{n-1}B(s_i)(B(s_{i+1}) - B(s_i)) = \frac{1}{2}B(t_n)^2 - \frac{1}{2}\sum_{j=0}^{n-1}(B(s_{i+1}) - B(s_i))^2.$$

---

[1] Although the Stratonovich integral shares similar properties with ordinary integrals.

This limit yealds

$$\frac{1}{2}B_t^2 - \frac{1}{2}[B, B]_t = \frac{1}{2}B_t^2 - \frac{1}{2}t,$$

therefore,

$$\int_0^t B_s \mathrm{d}B_s = \frac{1}{2}B_t^2 - \frac{1}{2}t. \qquad (3.42)$$

From the previous example we obtained a new property for the Itô integral expressed by Equation (3.42). Looking closely to (3.42) we also notice an uncommon property of $I(X)$. Indeed, consider the deterministic and differentiable function $f(\cdot)$, $f(0) = 0$, and calculate $\int_0^t f(s)\mathrm{d}f(s)$. Using integration by parts formula we obtain

$$\int_0^t f(s)\mathrm{d}f(s) = \int_0^t f(s)f'(s)\mathrm{d}s = [f(u)^2]_0^t - \int_0^t f(s)f'(s)\mathrm{d}s$$

and hence

$$\int_0^t f(s)\mathrm{d}f(s) = \frac{1}{2}f^2(t).$$

The last formula above and (3.42) differ in the additional term $-t/2$ present in the stochastic version. So we can conclude two things: *(i)* if available, the formula of integration by parts for the stochastic integral should look different from the usual formulas; *(ii)* calculate the stochastic integral applying the definition is not an easy task. Itô's formula will help in both cases but first we give a formal definition of an Itô process.

Definition 3.13.6 (Itô process) *Let $\{X_t, t \geq 0\}$ be a stochastic process that can be written as follows:*

$$X_t = X_0 + \int_0^t g_s \mathrm{d}s + \int_0^t h_s \mathrm{d}B_s \qquad (3.43)$$

with $g(s, \omega)$ and $h(s, \omega)$ two adapted and progressively measurable random functions such that

$$P\left\{\int_0^t |g(s, \omega)|\mathrm{d}s < \infty\right\} = 1 \quad \textit{and} \quad P\left\{\int_0^t h^2(s, \omega)\mathrm{d}s < \infty\right\} = 1$$

Then X is called an Itô process.

Essentially an Itô process is a process which can be written in the form of a stochastic integral, which is indeed a stochastic process, plus some random variable. Conditions on $g(\cdot)$ an $h(\cdot)$ only require the existence of the integrals apart from null measure sets. Clearly Brownian motion is an Itô process because it can be rewritten as in (3.43) with $g(x, \omega) = 0$ and $h(s, \omega) = 1$. A particular class of Itô processes, which will be defined later, is the class of diffusion process.

3.13.2   Itô formula

Calculating a stochastic integral means essentially to rewrite it in a form in which the integral with respect to Brownian motion disappears. One way to obtain this result very quickly is the famous Itô formula which we are going to introduce. Itô formula (sometimes called Itô Lemma) is a sort of Taylor formula for Itô integrable processes. We are not going to prove the Itô formula, but try to give an intuitive derivation. As usual, we start with two nonrandom functions $f(\cdot)$ and $g(\cdot)$ assuming all regularity conditions needed without mentioning them explicitly. Let us write

$$\frac{\mathrm{d}}{\mathrm{d}t} f(g(t)) = f'(g(t))g'(t),$$

then

$$f(g(t)) = f(g(0)) + \int_0^t f'(g(s))g'(s)\mathrm{d}s$$

and hence

$$f(g(t)) = f(g(0)) + \int_0^t f'(g(s))\mathrm{d}g(s).$$

Now, if we replace $g(\cdot)$ with the Brownian motion, we obtain the following formal writing:

$$f(B_t) = f(0) + \int_0^t f'(B_s)\mathrm{d}B_s,$$

which is actually false (remind that we use differentiability of $g(\cdot)$ in the above). The right formula is in fact the Itô formula.

Lemma 3.13.7 (Itô formula) *Let $f(\cdot)$ be two times differentiable with respect to its argument and measurable. Then*

$$f(B_t) = f(0) + \int_0^t f'(B_s)\mathrm{d}B_s + \frac{1}{2}\int_0^t f''(B_s)\mathrm{d}s. \qquad (3.44)$$

We now show how to obtain (3.42) via Itô formula.

Example 3.13.8 *Consider $f(x) = x^2$ and let us calculate $f(B_t)$ using (3.44). Clearly $f'(x) = 2x$ and $f''(x) = 2$ and in our case x is the Brownian motion. We have*

$$f(B_t) = f(B_0) + \int_0^t f'(B_s)\mathrm{d}B_s + \frac{1}{2}\int_0^t f''(B_s)\mathrm{d}s$$

and

$$B_t^2 = 0^2 + \int_0^t 2B_s\mathrm{d}B_s + \frac{1}{2}\int_0^t 2\mathrm{d}s,$$

from which we easy obtain the (3.42)

$$\int_0^t B_s \mathrm{d}B_s = \frac{1}{2}B_t^2 - \frac{1}{2}t.$$

If we imagine to expand in Taylor series $f(B(t))$ up to the second order, we obtain the following formal expression

$$\mathrm{d}f(B_t) = f'(B_t)\mathrm{d}B_t + \frac{1}{2}f''(B_t)(\mathrm{d}B_t)^2 + \text{rest}$$

for B_t we know from Section 3.7 that the quadratic variation $(\mathrm{d}B_t)^2 = \mathrm{d}t$. Itô's result shows that what is of order greater than $\mathrm{d}t$ goes to zero, hence the rest is indeed negligible and we obtain again (3.44). The general rule of taking into account the order of the differentials is that all cross products of the form '$\mathrm{d}B_t \cdot \mathrm{d}t$' or their powers are negligible, the same is for the powers of $(\mathrm{d}t)^k$ for $k \geq 2$ and, of course, $(\mathrm{d}B_t)^2 = \mathrm{d}t$. So, after the application of the Itô formula, all the differential terms different from $\mathrm{d}t = (\mathrm{d}B_t)^2$ and $\mathrm{d}B_t$ can be neglected.

A general trick to successfully calculate a stochastic integral via Itô formula is to identify the function $f(\cdot)$ or, better, its derivative. Suppose we want to calculate

$$\int_0^t g(B_s)\mathrm{d}B_s.$$

Itô formula is instead given as

$$f(B_t) = f(0) + \int_0^t f'(B_s)\mathrm{d}B_s + \frac{1}{2}\int_0^t f''(B_s)\mathrm{d}s$$

which can be rewritten as

$$\int_0^t f'(B_s)\mathrm{d}B_s = f(B_t) - f(0) - \frac{1}{2}\int_0^t f''(B_s)\mathrm{d}s$$

so, we need to consider the identity $f'(B_s) = g(B_s)$ and try to recognize $f(\cdot)$ from its derivative $f'(\cdot)$. In the above we had $g(x) = x$ and hence $\int_0^t B_s \mathrm{d}B_s$ and we need to set the identity

$$\int_0^t f'(B_s)\mathrm{d}B_s = \int_0^t B_s \mathrm{d}B_s$$

i.e. $f'(x) = x$ from which it is clear that $f(x) = \frac{1}{2}x^2$. Next exercises are examples of use of the Itô formula.

Exercise 3.7 *Calculate*

$$\int_0^t B_s^2 \mathrm{d}B_s.$$

Sometimes, stochastic integrals involve more complex functions of the Brownian motion, for example functions of both t and x. In this case, we have this version of the Itô lemma.

Lemma 3.13.9 (Itô formula) *Let $f(t, x)$ be a measurable function, two times differentiable with respect to x and one time differentiable in t. Then*

$$f(t, B_t) = f(0, 0) + \int_0^t f_t(s, B_s)\mathrm{d}s + \int_0^t f_x(s, B_s)\mathrm{d}B_s + \frac{1}{2}\int_0^t f_{xx}(s, B_s)\mathrm{d}s$$

(3.45)

where

$$f_t(t, x) = \frac{\partial f(t, x)}{\partial t}, \quad f_x(t, x) = \frac{\partial f(t, x)}{\partial x}, \quad f_{xx}(t, x) = \frac{\partial^2 f(t, x)}{\partial x^2}.$$

Exercise 3.8 *Prove that*

$$\int_0^t s\mathrm{d}B_s = tB_t - \int_0^t B_s\mathrm{d}s.$$

It is also possible to prove the following formula of integration by parts. Let $f = f(s)$ a nonstochastic function, continuous and with finite variation on $[0, t]$. Then

$$\int_0^t f(s)\mathrm{d}B_s = f(t)B_t - \int_0^t B_s\mathrm{d}f(s)$$

We now present a version of the Itô formula which holds for Itô processes.

Lemma 3.13.10 (Itô formula) *Let $f(t, x)$ be a measurable function, two times differentiable with respect to x and one time differentiable in t, and let X be an Itô process. Then*

$$f(t, X_t) = f(0, X_0) + \int_0^t f_t(s, X_s)\mathrm{d}s + \int_0^t f_x(s, X_s)\mathrm{d}X_s$$

$$+ \frac{1}{2}\int_0^t f_{xx}(s, X_s)(\mathrm{d}X_s)^2.$$

(3.46)

Clearly (3.46) reduces to (3.45) if X_t is B_t and (3.45) reduces to (3.44) is $f(t, x) = f(x)$. Sometimes, it is useful to present Itô formula using its differential form, which looks as follows:

$$\mathrm{d}f(t, X_t) = f_t(t, X_t)\mathrm{d}t + f_x(t, X_t)\mathrm{d}X_t + \frac{1}{2}f_{xx}(t, X_t)(\mathrm{d}X_t)^2.$$

(3.47)

This version is useful to express the dynamics of stochastic process which are transformation of Itô processes as we will see in details later.

Exercise 3.9 *Calculate $\int_0^t (2B_s^n - s^2)\mathrm{d}B_s$, $n \geq 1$.*

3.14   More properties and inequalities for the Itô integral

Due to the fact that the stochastic integral is a martingale, we can derive some inequalities. Throughout this section we assume that X_t and Y_t are two Itô-integrable process.

Theorem 3.14.1

$$\mathbb{E}\left\{\int_0^T X_t dB_t \int_0^T Y_t dB_t\right\} = \mathbb{E}\left\{\int_0^T X_t Y_t dt\right\}.$$

Theorem 3.14.2 *For any $\delta > 0$ and $\gamma > 0$*

$$P\left\{\sup_{0\le t\le T}\left|\int_0^T X_s dB_s\right| > \delta\right\} \le \frac{\gamma}{\delta} + P\left\{\int_0^T X_t^2 dt > \gamma\right\}.$$

Theorem 3.14.3 *Assume that for some $p \ge 1$ we have*

$$\mathbb{E}\left\{\int_0^T |X_t|^{2p} dt\right\} < \infty.$$

Then,

$$\mathbb{E}\left\{\int_0^T X_t dB_t\right\}^{2p} \le (p(2p-1))^p T^{p-1}\mathbb{E}\left\{\int_0^T |X_t|^{2p} dt\right\}.$$

Theorem 3.14.4 (Burkholder-Davis-Gundy) *There exists a constant $C_p \ge 0$ depending only on p such that*

$$\mathbb{E}\left\{\left|\sup_{0\le t\le T}\int_0^t X_s dB_s\right|^{2p}\right\} \le C_p\mathbb{E}\left\{\int_0^T X_t^2 dt\right\}^p.$$

In particular,

$$\mathbb{E}\left\{\left|\sup_{0\le t\le T}\int_0^t X_s dB_s\right|^2\right\} \le 4\mathbb{E}\left\{\int_0^T X_t^2 dt\right\}.$$

Theorem 3.14.5 *For any Itô integrable process X we have that*

$$\mathbb{E}\left(\exp\left\{\int_0^T X_t dB_t - \frac{1}{2}\int_0^T X_t^2 dt\right\}\right) \le 1.$$

3.15   Stochastic differential equations

In the previous section we have seen that stochastic differential equations of the form:

$$dS_t = \mu S_t dt + \sigma S_t dB_t$$

have a mathematical meaning in the sense of Itô integrals

$$S_t = S_0 + \mu \int_0^t S_u du + \sigma \int_0^t S_u dB_u.$$

We have also introduced Itô processes of the form:

$$X_t = X_0 + \int_0^t g_u du + \int_0^t h_u dB_u$$

where g_t and h_t are stochastic processes like in the geometric Brownian motion above, i.e. $g_t = \mu S_t$ and $h_t = \sigma S_t$. We now introduce the class of diffusion processes which are solutions to stochastic differential equations of the following form:

$$dX_t = b(t, X_t)dt + \sigma(t, X_t)dB_t, \tag{3.48}$$

with some initial condition X_0. The initial condition can be random or not. If random, say $X_0 = Z$, it should be independent of the σ-algebra generated by B and satisfy the condition $\mathbb{E}|Z|^2 < \infty$. The two deterministic functions $b(\cdot, \cdot)$ and $\sigma^2(\cdot, \cdot)$ are called respectively the *drift* and the *diffusion* coefficients of the stochastic differential equation (3.48). In order to have a well defined stochastic differential equation the drift and diffusion coefficients should be measurable functions and satisfy

$$P\left\{\int_0^T \sup_{|x|\leq R} (|b(t, x)| + \sigma^2(t, x))dt < \infty\right\} = 1, \quad \text{for all} \quad T, R \in [0, \infty). \tag{3.49}$$

The above condition is enough to satisfy Definition 3.13.6 of Itô processes. Geometric Brownian motion is clearly an example of diffusion process.

3.15.1   Existence and uniqueness of solutions

In Chapter 1 we introduced stochastic differential equations as a way to model asset dynamics, but writing down a stochastic differential equation does not necessarily mean that a stochastic process solution to it exists.

Moreover, a solution to a stochastic differential equation can be of two different types: *weak* and *strong*. Two weak solutions of a stochastic differential equation are stochastic processes which are equal only in distribution but are not pathwise identical. Strong solutions are on the contrary pathwise determined. A strong solution is also a weak solution, but the contrary is not necessarily true. Statistics usually require weak solutions because it is based on distributional properties of the data. Some numerical analysis may also require strong solutions. We will now introduce a set of conditions which are easier to verify than (3.49), which also qualify the type of solution. The first condition is called *global Lipschitz condition*.

Assumption 1 *For all* $x, y \in \mathbb{R}$ *and* $t \in [0, T]$, *there exists a constant* $K < +\infty$ *such that*

$$|b(t, x) - b(t, y)| + |\sigma(t, x) - \sigma(t, y)| < K|x - y|. \qquad (3.50)$$

Next is the *linear growth condition*, which implies that the solution X_t does not explode in a finite time.

Assumption 2 *For all* $x, y \in \mathbb{R}$ *and* $t \in [0, T]$, *there exists a constant* $C < +\infty$ *such that*

$$|b(t, x)| + |\sigma(t, x)| < C(1 + |x|). \qquad (3.51)$$

Theorem 5.2.1 in Øksendal (1998) states that under conditions (3.50) and (3.51), the stochastic differential equation (3.48) has a unique, continuous, and adapted strong solution such that $\mathbb{E}\left(\int_0^T |X_t|^2 dt\right) < \infty$.

Exercise 3.10 *Write the stochastic differential equation of the process* $Y_t = B_t^2$.

Exercise 3.11 *Write the stochastic differential equation of the process* $Y_t = 2 + t + e^{B_t}$.

Exercise 3.12 *Write the stochastic differential equation of the process* $U_t = \int_0^t e^{-\lambda(t-s)} dB_s$, *with* $\lambda > 0$.

Exercise 3.13 *Let* X *be a solution to* $dX_t = X_t dt + \sigma dB_t$ *and let* $Y_t = f(X_t)$, *with* $f(x) = \sin(x)$. *Write the stochastic differential equation for* Y_t.

Exercise 3.14 *Find the solution of the following stochastic differential equation:* $dX_t = b(t)X_t dt + \sigma(t)X_t dB_t$. *Assume that all necessary conditions on* $b(t)$ *and* $\sigma(t)$ *are fulfilled.*

Exercise 3.15 *Show that* $X_t = \frac{B_t}{1+t}$ *is a solution to* $dX_t = -\frac{1}{1+t}X_t dt + \frac{1}{1+t}dB_t$ *with* $X_0 = 0$.

Exercise 3.16 *Prove that* $X_t = (a^{1/3} + \frac{1}{3}B_t)^3$ *is a solution to* $dX_t = \frac{1}{3}X_t^{1/3} dt + X_t^{2/3} dB_t$, $X_0 = a > 0$.

Exercise 3.17 *Write the stochastic differential equation for* $Y_t = \frac{1}{1+B_t}$, *with* $Y_0 = 1$.

Exercise 3.18 *Write the stochastic differential equation for* $X_t = t + (1 - t)$ $\int_0^t \frac{1}{1-s} dB_s$.

Exercise 3.19 *Let* X_t *be a stochastic process solution to* $dX_t = b(X_t)dt + \sigma(X_t)dB_t$ *and let* Y_t *be the Lamperti transform of* X_t, *i.e.* $Y_t = f(X_t)$ *with* $f(x) = \int_{x_0}^x \frac{1}{\sigma(u)} du$ *and* x_0 *any real number in the support of* X_t. *Assume that* $\sigma(\cdot)$ *is twice differentiable and derive the stochastic differential equation for* Y_t.

3.16   Girsanov's theorem for diffusion processes

Girsanov's theorem is a change-of-measure theorem for stochastic processes. This is a necessary tool to perform inference for stochastic processes because it gives as a result the likelihood function of the process but in finance it is also a necessary tool to obtain equivalent martingale measures. As we will see in Chapter 6 the existence of an equivalent martingale measure is a necessary condition to avoid arbitrage in the market. Consider the three stochastic differential equations

$$dX_t = b_1(X_t)dt + \sigma(X_t)dW_t, \qquad X_0^{(1)}, \qquad 0 \le t \le T,$$

$$dX_t = b_2(X_t)dt + \sigma(X_t)dW_t, \qquad X_0^{(2)}, \qquad 0 \le t \le T,$$

$$dX_t = \sigma(X_t)dW_t, \qquad X_0, \qquad 0 \le t \le T,$$

and denote by P_1, P_2, and P the probability measures of X_t under the three models.

Theorem 3.16.1 (Lipster and Shiryayev (1977)) *Assume that Assumptions 3.50 and 3.51 are satisfied. Assume further that the initial values are either random variables with densities* $f_1(\cdot)$, $f_2(\cdot)$, *and* $f(\cdot)$ *with the same common support or nonrandom and equal to the same constant. Then the three measures* P_1, P_2, *and* P *are all equivalent and the corresponding Radon-Nikodým derivatives* (Z_T, Z_T') *are*

$$Z_T = \frac{dP_1}{dP}(X) = \frac{f_1(X_0)}{f(X_0)} \exp\left\{ \int_0^T \frac{b_1(X_s)}{\sigma^2(X_s)} dX_s - \frac{1}{2} \int_0^T \frac{b_1^2(X_s)}{\sigma^2(X_s)} ds \right\} \quad (3.52)$$

and

$$Z_T' = \frac{dP_2}{dP_1}(X)$$

$$= \frac{f_2(X_0)}{f_1(X_0)} \exp\left\{ \int_0^T \frac{b_2(X_s) - b_1(X_s)}{\sigma^2(X_s)} dX_s - \frac{1}{2} \int_0^T \frac{b_2^2(X_s) - b_1^2(X_s)}{\sigma^2(X_s)} ds \right\}.$$

$$(3.53)$$

3.17   Local martingales and semimartingales

Definition 3.17.1 *A process* $\{X_t, t \geq 0\}$ *adapted to a filtration* $\mathcal{F}$ *is called local martingale if there exists a sequence of stopping times* $\tau_k : \Omega \to [0, +\infty)$ *such that*

(i) the sequence τ_k *is almost surely increasing, i.e.* $P(\tau_k < \tau_{k+1}) = 1$;

(ii) the sequence τ_k *diverges almost surely, i.e.* $P(\tau_k \to \infty, k \to \infty) = 1$;

(iii) the stopped process

$$\mathbf{1}_{\{\tau_k > 0\}} X_t^{\tau_k} = \mathbf{1}_{\{\tau_k > 0\}} X_{t \wedge \tau_k}$$

is a martingale for every k.

The sequence of stopping times in the definition of local martingale is usually called *localizing sequence*.

Example 3.17.2 *Let* $T = \min\{t : B_t = -1\}$. *We know that* T *is a stopping time. Consider the stopped process* $B_t^T = B_{t \wedge T}$. *Then* $\mathbb{E}(B_t^T) = 0$ *for all* t. *It is also straightforward to prove that it is a martingale. However, the following rescaled process*

$$X_t = \begin{cases} B_{\frac{t}{1-t}}^T, & 0 \leq t < 1 \\ -1, & t \geq 1 \end{cases}$$

is such that

$$\mathbb{E}(X_t) = \begin{cases} 0, & 0 \leq t < 1 \\ -1, & t \geq 1 \end{cases}$$

so it is clearly not a martingale. Nevertheless, it is a local martingale with respect to the sequence $\tau_k = \min\{t : X_t = k\}$ *although we do not give the proof here.*

In general, local martingales are not martingales and local martingales arise often in the definition of the stochastic integral with respect to martingales. It turns out that such a stochastic integral is not a martingale but just a local martingale.

Definition 3.17.3 *A real-valued stochastic process* $\{X_t, t \geq 0\}$, *adapted to a filtration* $\{\mathcal{F}_t, t \geq 0\}$, *is called semimartingale if it can be decomposed as follows:*

$$X_t = M_t + A_t$$

where $\{M_t, t \geq 0\}$ *is a local martingale and* $\{A_t, t \geq 0\}$ *is a càdlàg adapted process of locally bounded variation.*

Most of the processes discussed in this book are semimartingales and semimartingales are the wider class of stochastic processes for which the stochastic integral can be defined. We will use the notion of local martingale and semimartingale whenever needed without going into deep details. A good reference on the topic is Protter (2004).

3.18   Lévy processes

Lévy processes were introduced as the sum of a compound and compensated Poisson process and a Brownian motion with drift. The original idea was to construct a family of processes wide enough to comprise a variety of well-known other stochastic processes and more. Most relevant references are Lévy (1954), Bertoin (1998) and Sato (1999) but many new publications appeared recently due to applications of these models in finance. We start with a gentle introduction before giving a formal definition. Assume we have a two processes X_t and M_t

$$X_t = \mu t + \sigma B_t, \quad \mu \in \mathbb{R}, \sigma \geq 0,$$

where B_t is a Brownian motion and

$$M_t = \sum_{i=0}^{N_t} Y_{\tau i} - \lambda t \mathbb{E}(Y_{\tau_i}), \quad \lambda \geq 0,$$

with N_t an homogenous Poisson process and Y_{τ_i} a sequence of i.i.d. random variables such that $\mathbb{E}(Y_{\tau_i}) < \infty$. We assume that N_t, W_t and the Y_{τ_i}'s are all independent. We introduce the process Z_t, which we can think of as an embryo of a Lévy process,

$$Z_t = X_t + M_t. \tag{3.54}$$

Remember that M_t is a martingale by construction and if $\mu = 0$ then also X_t is a martingale due to the fact that Brownian motion is itself a martingale. Therefore, we immediately see that also Z_t is a martingale. For Z_t in (3.54) we can also easily derive the characteristic function due to independence of X_t and M_t. Indeed,

$$\varphi_{X_t}(u) = \exp\left\{t\left(i\mu u - \frac{1}{2}\sigma^2 u^2\right)\right\}$$

and

$$\varphi_{M_t}(u) = \exp\left\{\lambda t \int_{-\infty}^{\infty} (e^{iux} - 1 - iux)\mathrm{d}F(x)\right\}$$

where $F(\cdot)$ is the common distribution function of the markers Y_{τ_i}. Then, we obtain the so-called *de Finetti characteristic function* for Z_t

$$\varphi_{Z_t}(u) = \exp\left\{t\left(i\mu u - \frac{1}{2}\sigma^2 u^2 + \lambda \int_{-\infty}^{\infty} (e^{iux} - 1 - iux)\mathrm{d}F(x)\right)\right\}.$$

This is nice but not enough to describe the general framework of Lévy processes. Assume that we add another independent Brownian motion to Z_t, then, the resulting characteristic function does not change too much because the sum of independent Gaussian random variables is a new Gaussian random variable, so $\varphi_{Z_t}(u)$ will remain in essentially the same form. But things do change if we add another jump component. Assume further that there are two compound and compensated Poisson processes, say M_t^1 and M_t^2. Let M_t^i, $i = 1, 2$ have intensity λ_i and the corresponding markers a distribution $F_i(x)$. Then, the characteristic function of $M_t^1 + M_t^2$ will have an exponent like the following

$$\lambda_1 t \int_{-\infty}^{\infty} (e^{iux} - 1 - iux)\mathrm{d}F_1(x) + \lambda_2 t \int_{-\infty}^{\infty} (e^{iux} - 1 - iux)\mathrm{d}F_2(x)$$

$$= t \int_{-\infty}^{\infty} (e^{iux} - 1 - iux)v(\mathrm{d}x)$$

with $v(\mathrm{d}x) = \lambda_1 \mathrm{d}F_1(x) + \lambda_2 \mathrm{d}F_2(x)$. The measure $v(\cdot)$ plays a central role in the construction of the Lévy process and it will be indeed called the *Lévy measure*. The idea of the Lévy measure is that it contains all the information about the jumps of the process Z_t and so in general one wants to model directly the measure $v(\cdot)$ instead of the various $F_i(\cdot)$. Due to the fact that jumps can be negative or positive but reasonably not null, the measure $v(\cdot)$ will have a discontinuity at 0. Moreover, while being surely positive, for any two numbers $\lambda_1 \geq 0$ and $\lambda_2 \geq 0$, the measure $v(\cdot)$ does not necessarily integrate to 1. So the general construction of Lévy processes requires proper restrictions to the general form of $v(\cdot)$. We now give the general definition of a Lévy process.

Definition 3.18.1 *A càdlàg, adapted, real valued stochastic process* $\{Z_t, 0 \leq t \leq T\}$, *with* $Z_0 = 0$ *almost surely, is called a Lévy process if the following properties are satisfied*

(i) Z_t *has independent increments, i.e.* $Z_t - Z_s$ *is independent of* $\mathcal{F}_s$, *for any* $0 \leq s < t \leq T$;

(ii) Z_t *has stationary increments, i.e. for any* $0 \leq s, t \leq T$ *the distribution of* $Z_{t+s} - Z_t$ *does not depend on* t;

(iii) Z_t *is stochastically continuous, i.e. for every* $0 \leq t \leq T$ *and* $\epsilon > 0$, *we have that*
$$\lim_{s \to t} P(|Z_t - Z_s| > \epsilon) = 0.$$

It is important to notice that property (iii) is not related to the continuity of the paths (indeed, it is a càdlàg process) but only to the property of the distributions of the increments. The simplest Lévy process is the deterministic linear drift process, i.e. for some constant μ, $Z_t = \mu t$, which has continuous paths and the only nondeterministic Lévy process with continuous paths is the Brownian motion. The Poisson and compound Poisson processes are also Lévy process and of course the process Z_t in (3.54) is also a Lévy process. The process in (3.54) is sometimes called *Lévy jump diffusion* to distinguish it from other jump diffusion processes which are not of Lévy type.

3.18.1 Lévy-Khintchine formula

If we look at the characteristic function of the random variable Z_t in (3.54) for Lévy jump diffusion

$$\varphi_{Z_t}(u) = \exp\left\{t\left(i\mu u - \frac{1}{2}\sigma^2 u^2 + \lambda \int_{-\infty}^{\infty}(e^{iux} - 1 - iux)\mathrm{d}F(x)\right)\right\}$$

and we compare it with the Lévy-Khintchine formula (2.16), we see that for this process the characteristic triplet is

$$(b = \mu t, c = \sigma^2 t, \nu = (\lambda F) \cdot t)$$

where b is called the *drift term*, c the Gaussian or *diffusion term* and ν is the Lévy measure. This means that the marginal law of the simple jump diffusion process Z_t is infinitely divisible. This fact is actually true for any Lévy process. We can indeed rewrite a general Lévy process Z_t of Definition 3.18.1 using increments over a regular grid of time points of size t/n for some integer n, i.e. we can write the following telescopic sum of increments

$$Z_t = Z_{\frac{t}{n}} + \left(Z_{\frac{2t}{n}} - Z_{\frac{t}{n}}\right) + \cdots + \left(Z_t - Z_{\frac{(n-1)t}{n}}\right).$$

By stationarity and independence of the increments $I_k = \left(Z_{\frac{kt}{n}} - Z_{\frac{(k-1)t}{n}}\right)$, $k = 1, \ldots, n$, is a sequence of i.i.d. random variables, so we can readily see that Z_t is, in the general case, also infinitely divisible.

Theorem 3.18.2 *Let Z_t be any Lévy process and let $\psi(u)$ be the characteristic exponent of the random variable Z_1. Then*

$$\mathbb{E}\left\{e^{iuZ_t}\right\} = e^{t\psi(u)} = \exp\left\{t\left(ibu - \frac{cu^2}{2} + \int_{\mathbb{R}}(e^{iux} - 1 - iux)\mathbf{1}_{\{|x|<1\}}\nu(\mathrm{d}x)\right)\right\}.$$

Proof. We present some hints of the proof, because it is instructive. The first fact that we notice is the following decomposition of the characteristic function of Z_{t+s}

$$\varphi_{Z_{t+s}}(u) = \mathbb{E}\left\{e^{iuZ_{t+s}}\right\} = \mathbb{E}\left\{e^{iu(Z_{t+s}-Z_s)}e^{iuZ_s}\right\} = \text{(by independence)}$$

$$= \mathbb{E}\left\{e^{iuZ_{t+s}}\right\} \cdot \mathbb{E}\left\{e^{iuZ_s}\right\} = \text{(by stationarity)}$$

$$= \mathbb{E}\left\{e^{iuZ_t}\right\} \cdot \mathbb{E}\left\{e^{iuZ_s}\right\}$$

$$= \varphi_{Z_t}(u)\varphi_{Z_s}(u).$$

Let us denote $\varphi_{Z_t}(u)$ by $\varphi_t(u)$. Then, we have obtained the so-called *Cauchy functional equation*

$$\varphi_{t+s}(u) = \varphi_t(u)\varphi_s(u)$$

and we know that $\varphi_t(0) = 1$ for all $t \geq 0$ and that, as a function of t, the characteristic function of the Lévy process is continuous due to the property of stochastic continuity. The only continuous solution to this Cauchy functional equation is the function of the form:

$$\varphi_t(u) = e^{tg(u)}, \qquad \text{with} \quad g(u) : \mathbb{R} \mapsto \mathbb{C}.$$

Since Z_1 is an infinitely divisible distribution, the statement of the theorem holds with $g(u) = \psi(u)$.

Theorem (3.18.2) is rather important because it says that the law of Z_t can be obtained from the law of Z_1, but because the law of Z_1 is infinitely divisible, so is the law of Z_t. But this also means that, for a given random variable X with an infinitely divisible distribution we can always build a Lévy process with that distribution simply setting $Z_1 \sim X$.

3.18.2   Lévy jumps and random measures

Let $\Delta Z = \{\Delta Z_t, 0 \leq t \leq T\}$ be the jump process associated with the Lévy process Z, defined for each t as

$$\Delta Z_t = Z_t - Z_{t-},$$

where $Z_{t-} = \lim_{s \to t} Z_s$, is the limit from the left. By the stochastic continuity of the Lévy process, if we consider a fixed t, then $\Delta Z_t = 0$ almost surely. This means that a Lévy process has no fixed times of discontinuity, but nevertheless it has jumps and usually the total sum of jumps may even diverge, i.e.

$$\sum_{s \leq t} |\Delta Z_s| = \infty \quad a.s.$$

but we always have that

$$\sum_{s \leq t} |\Delta Z_s|^2 < \infty \quad a.s.$$

which makes the Lévy process treatable as a martingale. Consider now a set $A \in \mathcal{B}(\mathbb{R}\backslash\{0\})$ and such that $0 \notin \bar{A}$ and let $0 \leq t \leq T$. We define the following measure of the jumps of the Lévy process Z as follows:

$$\mu^Z(\omega; t, A) = \#\{0 \leq s \leq t; \Delta L_s(\omega) \in A\} = \sum_{s \leq t} \mathbf{1}_A(\Delta L_s(\omega)).$$

The measure μ^Z is a *random measure* which counts the jumps of the process Z of size A up to time t. The random measure has several properties, e.g.

$$\mu^Z(t, A) - \mu^Z(s, A) \in \sigma(\{Z_u - Z_v; s \le v < u \le t\})$$

and hence $\mu^Z(t, A) - \mu^Z(s, A)$ is independent of $\mathcal{F}_s$ and has independent increments. Further, $\mu^Z(t, A) - \mu^Z(s, A)$ is the number of jumps of $Z_{s+u} - Z_s$ in A for $0 \le u \le t - s$. Due to the stationarity of the increments of Z_t we have stationarity of the increments of μ^Z, i.e.

$$\mu^Z(t, A) - \mu^Z(s, A) \sim \mu^Z(t - s, A).$$

From the above we obtain that $\mu^Z(\cdot, A)$ is a Poisson process and μ^Z is then called the *Poisson random measure*. The intensity of this Poisson process is $v(A) = \mathbb{E}\{\mu^Z(1, A)\}$.

Definition 3.18.3 *The measure v defined by*

$$v(A) = \mathbb{E}\{\mu^Z(1, A)\} = \mathbb{E}\left\{\sum_{s \le 1} \mathbf{1}_A(\Delta Z_s(\omega))\right\}$$

is the Lévy measure of the Lévy process Z.

Notice that, if $f : \mathbb{R} \to \mathbb{R}$ is a Borel measurable function and finite on A, it is possible to define the integral with respect to μ^Z as follows:

$$\int_A f(x)\mu^Z(\omega; t, \mathrm{d}x) = \sum_{s \le t} f(\Delta Z_s)\mathbf{1}_A(\Delta Z_s(\omega)).$$

Moreover, $\int_A f(x)\mu^Z(t, \mathrm{d}x)$ defines a real-valued random variable for each t and hence we can construct a càdlàg stochastic process as follows:

$$G_t = \int_0^t \int_A f(x)\mu^Z(\mathrm{d}s, \mathrm{d}x), \quad 0 \le t \le T. \tag{3.55}$$

The next theorem (for a proof see Applebaum (2004), Th. 2.3.8) presents the main properties of the process G_t.

Theorem 3.18.4 *Consider a set $A \in \mathcal{B}(\mathbb{R}\backslash\{0\})$ with $0 \notin \bar{A}$ and let $f : \mathbb{R} \to \mathbb{R}$, a Borel measurable function and finite on A. Then, the stocastic process G_t defined in (3.55) is a compound Poisson process such that*

(i) *its characteristic function is given by*

$$\varphi_{G_t}(u) = \mathbb{E}\left\{e^{iuG_t}\right\} = \exp\left\{t\int_A \left(e^{iuf(x)} - 1\right)v(\mathrm{d}x)\right\};$$

(ii) if $f \in L^1(A)$, then

$$\mathbb{E}\{G_t\} = t \int_A f(x)v(\mathrm{d}x);$$

(iii) is $f \in L^2(A)$, then

$$\mathrm{Var}\{|G_t|\} = t \int_A |f(x)|^2 v(\mathrm{d}x).$$

3.18.3   Itô-Lévy decomposition of a Lévy process

Consider the following decomposition of the characteristic exponent of a Lévy process:

$$\psi(u) = \left(ibu - \frac{cu^2}{2} + \int_{\mathbb{R}} (e^{iux} - 1 - iux)\mathbf{1}_{\{|x|<1\}}v(\mathrm{d}x) \right)$$
$$= \phi^{(1)}(u) + \phi^{(2)}(u) + \phi^{(3)}(u) + \phi^{(4)}(u),$$

with

$$\phi^{(1)}(u) = iub, \quad \phi^{(2)}(u) = \frac{u^2 c}{2},$$

$$\phi^{(3)}(u) = \int_{|x|\geq 1} \left(e^{iux} - 1 \right) v(\mathrm{d}x),$$

$$\phi^{(4)}(u) = \int_{|x|<1} \left(e^{iux} - 1 - iux \right) v(\mathrm{d}x).$$

The first term corresponds to a deterministic linear process, say $Z_t^{(1)}$; the second $Z_t^{(2)}$ to a Brownian motion rescaled by $\sqrt{c}$ and the third term $Z_t^{(3)}$ corresponds to a compound Poisson process with $\lambda = v(\mathbb{R}\backslash(-1, 1))$ and distribution of the markers (the jumps) $F(\mathrm{d}x) = \frac{v(\mathrm{d}x)}{\lambda}\mathbf{1}_{\{|x|\geq 1\}}$. The last term $Z_t^{(4)}$ is more difficult to describe but, intuitively, it is associated with a jump process, say $Z_t^{(4,\epsilon)}$, $\epsilon > 0$, defined as

$$Z_t^{(4,\epsilon)} = \int_0^t \int_{\epsilon < |x| < 1} x\mu^Z(\mathrm{d}x, \mathrm{d}s) - t \left(\int_{1 > |x| > \epsilon} xv(\mathrm{d}x) \right),$$

with characteristic function

$$\varphi^{(4,\epsilon)}(u) = \int_{\epsilon < |x| < 1} \left(e^{iux} - 1 - iux \right) v(\mathrm{d}x).$$

Then $\phi^{(4)}(u)$ is the characteristic function of the process $Z_t^{(4)} = \lim_{\epsilon \to 0^+} Z_t^{(4,\epsilon)}$. The next result states more precisely what we have shown in the above and it is known as the *Itô-Lévy decomposition*. The proof can be found, e.g., in Sato (1999).

Theorem 3.18.5 *Consider the triplet (b, c, v) where $b \in \mathbb{R}$, $c \geq 0$, and v a measure such that $v(\{0\}) = 0$ and $\int_{\mathbb{R}} (1 \wedge |x|^2) v(dx) < \infty$. Then, there exist, on some probability space, four independent stochastic processes $Z^{(1)}$, $Z^{(2)}$, $Z^{(3)}$ and $Z^{(4)}$, where $Z^{(1)}$ is a constant drift process; $Z^{(2)}$ is a Brownian motion; $Z^{(3)}$ is a compound Poisson process and $Z^{(4)}$ is a square integrable pure jump martingale with, almost surely, a countable number of jumps of magnitude less than 1 on each finite time interval. Moreover, the process defined as $Z = Z^{(1)} + Z^{(2)} + Z^{(3)} + Z^{(4)}$ is a Lévy process with characteristic exponent*

$$\varphi(u) = iub - \frac{u^2 c}{2} + \int_{\mathbb{R}} \left(e^{iux} - 1 - iux \mathbf{1}_{|x|<1} \right) v(dx), \quad u \in \mathbb{R}.$$

Thanks to the Itô-Lévy decomposition, we can always rewrite a Lévy process as follows:

$$
\begin{aligned}
Z_t &= bt + \sqrt{c} B_t + \int_0^t \int_{|x| \geq 1} x \mu^Z(ds, dx) + \int_0^t \int_{|x| < 1} x (\mu^Z - v^Z)(ds, dx) \\
&= Z^{(1)} + Z^{(2)} + Z^{(3)} + Z^{(4)},
\end{aligned}
$$

where $v^Z(ds, dx) = v(dx)ds$.

3.18.4 More on the Lévy measure

The basic requirements for the Lévy measure v in the triplet (b, c, v) are as follows:

$$v(\{0\}) = 0 \quad \text{and} \quad \int_{\mathbb{R}} (1 \wedge |x|^2) v(dx) < \infty.$$

Roughly speaking, the Lévy measure describes the expected number of jumps of a certain amplitude in a time interval of length 1. The Lévy measure has no mass at the origin, but many jumps may occur around the origin, i.e. there may be a big quantity of very small jumps. It is also true that jumps which are far away from the origin have bounded mass. If v is a finite measure, we have seen that, defining $\lambda = v(\mathbb{R}) < \infty$, $F(dx) = v(dx)/\lambda$ is a probability measure and λ is just the expected number of jumps with $F(\cdot)$ the distribution of the jumps. But if $v(\mathbb{R}) = \infty$, then an infinite number of small jumps is expected. In this second case, the Lévy process is said to have *infinite activity*. The next sets of results relate the Lévy measure with other aspects of the paths of the Lévy process. All the proofs are omitted but can be found in Sato (1999).

Theorem 3.18.6 *Let Z be a Lévy process with triplet (b, c, v).*

(i) *if $v(\mathbb{R}) < \infty$, then almost all paths of Z have a finite number of jumps on every compact interval. In this case, the Lévy process has finite activity;*

(ii) *if $v(\mathbb{R}) = \infty$, then almost all paths of Z have an infinite number of jumps on every compact interval. In this case, the Lévy process has infinite activity.*

Theorem 3.18.7 *Let Z be a Lévy process with triplet (b, c, v).*

(i) *if $c = 0$ and $\int_{|x| \leq 1} |x| v(\mathrm{d}x) < \infty$, then almost all paths of Z have finite variation;*

(ii) *if $c \neq 0$ and $\int_{|x| \leq 1} |x| v(\mathrm{d}x) = \infty$, then almost all paths of Z have infinite variation.*

The above results say that, when the Brownian part is missing and the sum of small jumps does not diverge, then the process Z has finite variation and viceversa. The final result concerns the finiteness of the moments of the Lévy process. The knowledge of this property will be particularly relevant in financial applications as we will see.

Theorem 3.18.8 *Let Z be a Lévy process with triplet (b, c, v).*

(i) *Z_t has finite p-th moment for $p \geq 0$, i.e. $\mathbb{E}|Z_t|^p < \infty$, if and only if $\int_{|x| \geq 1} |x|^p v(\mathrm{d}x) < \infty$;*

(ii) *Z_t has finite p-th exponential moment for $p \geq 0$, i.e. $\mathbb{E}\left(e^{pZ_t}\right) < \infty$, if and only if $\int_{|x| \geq 1} e^{px} v(\mathrm{d}x) < \infty$.*

To summarize the above results, the small jumps (and the Brownian part) determine the variation of the process; the big jumps affect the existence of the higher moments of the process; the activity depends on the whole set of jumps of the process.

3.18.5   The Itô formula for Lévy processes

It is possible to derive the Itô formula for Lévy processes. This formula is very useful when modeling financial data through Lévy processes and in particular for the exponential Lévy model. Next results is stated without proof but reader can, for example, check Proposition 8.15 in Cont and Tankov (2004).

Theorem 3.18.9 (Itô Formula) *Let Z_t be a Lévy process with Lévy triplet (b, c, v) and $f : \mathbb{R} \to \mathbb{R}$ a measurable and twice differentiable function. Then*

$$
\begin{aligned}
f(Z_t) = {}& f(Z_0) + \frac{c}{2} \int_0^t f''(Z_s)\mathrm{d}s + \int_0^t f'(X_{s-})\mathrm{d}Z_s \\
& + \sum_{0 \leq s < t, \Delta Z_s \neq 0} \left\{ f(Z_{s-} + \Delta Z_s) - f(Z_{s-}) - \Delta Z_s f'(Z_{s-}) \right\}.
\end{aligned}
\tag{3.56}
$$

Example 3.18.10 *Let Z_t be a Lévy process with Lévy triplet (b, c, v) with Itô-Lévy decomposition*

$$
Z_t = bt + \sqrt{c}B_t + \int_0^t \int_{\mathbb{R}} x(\mu^Z - v^Z)(\mathrm{d}s, \mathrm{d}x)
$$

and define $S_t = S_0 e^{Z_t}$, with S_0 some constant. Then S_t satisfies the stochastic differential equation

$$dS_t = S_{t-}\left\{dZ_t + \frac{c}{2}dt\int_{\mathbb{R}}\left(e^x - 1 - x\right)\mu^Z(dt, dx)\right\}.$$

Indeed,

$$S_t = S_0 + \frac{c}{2}\int_0^t e^{Z_s}ds + \int_0^t e^{Z_{s-}}dZ_s + \sum_{0\le s<t, \Delta Z_s\neq 0}\left\{e^{Z_{s-}+\Delta Z_s} - e^{Z_{s-}} - \Delta Z_s e^{Z_{s-}}\right\}$$

$$= S_0 + \frac{c}{2}\int_0^t e^{Z_s}ds + \int_0^t e^{Z_{s-}}dZ_s + \sum_{0\le s<t, \Delta Z_s\neq 0} e^{Z_{s-}}\left\{e^{\Delta Z_s} - 1 - \Delta Z_s\right\}$$

$$= S_0 + \frac{c}{2}\int_0^t S_s ds\int_0^t S_{s-}dZ_s + \sum_{0\le s<t, \Delta Z_s\neq 0} S_{s-}\left\{e^{\Delta Z_s} - 1 - \Delta Z_s\right\}$$

$$= S_0 + \frac{c}{2}\int_0^t S_s ds + \int_0^t S_{s-}dZ_s + \int_0^t\int_{\mathbb{R}} S_{s-}(e^x - 1 - x)\mu^Z(ds, dx)$$

and, in differential form,

$$dS_t = \frac{c}{2}S_t dt + S_{t-}dZ_t + \int_{\mathbb{R}} S_{t-}(e^x - 1 - x)\mu^Z(dt, dx)$$

$$= S_{t-}\left\{\frac{c}{2}dt + dZ_t + \int_{\mathbb{R}}(e^x - 1 - x)\mu^Z(dt, dx)\right\}.$$

3.18.6 Lévy processes and martingales

Next theorem applies to the wide class of one-dimensional stochastic processes with independent increments and will be used in option pricing to study martingale properties of, e.g., discounted price processes.

Theorem 3.18.11 *Let $\{X_t, t \ge 0\}$ be a real-values stochastic process with independent increments. Then*

(i) *the process*

$$M_t = \frac{e^{iuX_t}}{\mathbb{E}\left(e^{iuX_t}\right)}, \quad t \ge 0,$$

is a martingale for all $u \in \mathbb{R}$;

(ii) *if, for some $u \in \mathbb{R}$, $\mathbb{E}(e^{uX_t}) < \infty$ and for all $t \ge 0$, then*

$$M_t = \frac{e^{uX_t}}{\mathbb{E}\left(e^{uX_t}\right)}, \quad t \ge 0,$$

is a martingale (see e.g. Exercise 3.2);

(iii) *if* $\mathbb{E}(X_t) < \infty$ *for all* $t \geq 0$, *then*

$$M_t = X_t - \mathbb{E}(X_t), \quad t \geq 0,$$

is a martingale with independent increments;

(iv) *if* $\mathrm{Var}(X_t) < \infty$ *for all* $t \geq 0$, *then*

$$Q_t = M_t^2 - \mathbb{E}(M_t^2), \quad t \geq 0,$$

is a martingale, with $M_t = X_t - \mathbb{E}(X_t)$ *(see e.g. Exercise 3.5).*

Notice that, thanks to Theorem 3.18.8, for the Lévy processes all the above statements are true if the corresponding moments of the Lévy measure are finite for at least one t. A proof can be found in Sato (1999).

Theorem 3.18.12 *Consider again a Lévy process with Lévy triplet* (b, c, ν) *and Itô-Lévy decomposition*

$$Z_t = bt + \sqrt{c}B_t + \int_0^t \int_{\mathbb{R}} x(\mu^Z - \nu^Z)(\mathrm{d}s, \mathrm{d}x)$$

such that $\mathbb{E}|L_1| < \infty$. *Then* Z_t *is a martingale if and only if* $b = 0$.

Proof. The process Z_t, according to Definition 3.17.3 can be seen as a semimartingale of the form:

$$Z_t = M_t + A_t$$

where

$$A_t = \int_0^t \int_{\mathbb{R}} x(\mu^Z - \nu^Z)(\mathrm{d}s, \mathrm{d}x)$$

is the càdlàg process. Moreover, given that

$$\int_0^t \int_{\mathbb{R}} x\mu^Z(\mathrm{d}s, \mathrm{d}x) = \sum_{0 \leq s \leq t} \Delta Z_s,$$

we also have that

$$\mathbb{E}\left\{\int_0^t \int_{\mathbb{R}} x\mu^Z(\mathrm{d}s, \mathrm{d}x)\right\} = \int_0^t \int_{\mathbb{R}} x\nu^Z(\mathrm{d}s, \mathrm{d}x) = \int_0^t \int_{\mathbb{R}} x\nu(\mathrm{d}x)\mathrm{d}t = t\int_{\mathbb{R}} x\nu(\mathrm{d}x).$$

Thus, $\mathbb{E}\{Z_t\} = bt$ and it turns out that such a Lévy process is a martingale if and only if $b = 0$.

We now introduce an important tool which is needed to operate the change of measure in the case of Lévy processes. This theorem is again the Girsanov's theorem and it makes use of the Radon-Nikodým derivative introduced in Definition 2.3.5. We omit all the proofs which can be found in, e.g. Papapantoleon (2008).

Let P and $\tilde{P}$ be two measures on $(\Omega, \mathcal{A})$ and let $\mathcal{F} = \{\mathcal{F}_t, 0 \leq t \leq T\}$ be a filtration on this probability space. If P and $\tilde{P}$ are equivalent, then there exists a unique, positive martingale $\{M_t, 0 \leq t \leq T\}$ with respect to P such that $M_t = \mathbb{E}\left\{\frac{\mathrm{d}\tilde{P}}{\mathrm{d}P} \middle| \mathcal{F}_t\right\}$. Conversely, given P and a positive martingale M_t with respect to P, one can define an equivalent measure $\tilde{P}$ using the Radon-Nikodým derivative as follows $M_T = \mathbb{E}\left\{\frac{\mathrm{d}\tilde{P}}{\mathrm{d}P} \middle| \mathcal{F}_T\right\}$.

Theorem 3.18.13 (Jacod and Shiryaev (2003)) *Let $\{Z_t, 0 \leq t \leq T\}$ be a Lévy process with characteristic triplet (b, c, v) under the measure P, with finite first moment and canonical decomposition*

$$Z_t = bt + \sqrt{c}w_t + \int_0^t \int_{\mathbb{R}} x \left(\mu^Z - v^Z\right)(\mathrm{d}s, \mathrm{d}x).$$

(i) *assume that P and $\tilde{P}$ are two equivalent measures. Then, there exist a deterministic process β and a measurable non-negative deterministic process Y such that, almost surely under $\tilde{P}$, we have*

$$\int_0^t \int_{\mathbb{R}} |x(Y(s, x) - 1)|v(\mathrm{d}x) < \infty$$

and

$$\int_0^t (c\beta_s^2)\mathrm{d}s < \infty.$$

(ii) *conversely, if $\{M_t, 0 \leq t \leq T\}$ is a positive martingale of the form:*

$$M_t = \exp\Bigg\{ \int_0^t \beta_s \sqrt{c}\,\mathrm{d}W_s$$
$$- \frac{1}{2}\int_0^t \beta_s^2 \mathrm{d}s + \int_0^t \int_{\mathbb{R}} (Y(s, x) - 1) \left(\mu^Z - v^Z\right)(\mathrm{d}s, \mathrm{d}x)$$
$$- \int_0^t \int_{\mathbb{R}} (Y(s, x) - 1 - \log(Y(s, x)))\mu^Z(\mathrm{d}s, \mathrm{d}x)\Bigg\}$$

then it defines a probability measure $\tilde{P}$ equivalent to P.

(iii) *in both cases $\tilde{W}_t = W_t - \int_0^t \sqrt{c}\beta_s \mathrm{d}s$ is a Brownian motion with respect to $\tilde{P}$, $\tilde{v}^Z(\mathrm{d}s, \mathrm{d}x) = Y(s, x)v^Z(\mathrm{d}s, \mathrm{d}x)$ is the compensator of μ^Z under $\tilde{P}$ and Z has the following canonical representation under $\tilde{P}$:*

$$Z_t = \tilde{b}t + \sqrt{c}\tilde{W}_t + \int_0^t \int_{\mathbb{R}} x \left(\mu^Z - \tilde{v}^Z\right)(\mathrm{d}s, \mathrm{d}x),$$

with

$$\tilde{b}t = bt + \int_0^t c\beta_s \mathrm{d}s + \int_0^t \int_{\mathbb{R}} x \left(Y(s, x) - 1\right) v^Z(\mathrm{d}s, \mathrm{d}x).$$

3.18.7   Stochastic differential equations with jumps

Consider the *compensated Poisson random measure*

$$\tilde{\mu}(\mathrm{d}t, \mathrm{d}z) = \mu(\mathrm{d}t, \mathrm{d}z) - \nu(\mathrm{d}z)\mathrm{d}t$$

where $\nu(\cdot)$ is a Lévy measure

$$\nu(\{0\}) = 0 \quad \text{and} \quad \int_{\mathbb{R}} (1 \wedge |x|^2)\nu(\mathrm{d}x) < \infty.$$

Let $\mathcal{F}_t$ be the σ-algebra generated by the Brownian motion B_s and $\tilde{\mu}(\mathrm{d}s, \mathrm{d}z)$ for $z \in \mathbb{R}$ and $s \leq t$, and enlarged by all the sets of P-zero probability. Consider the filtration $\mathcal{F} = \{\mathcal{F}_t, t \geq 0\}$.

Definition 3.18.14 (Itô-Lévy process) *Let $a(t)$, $b(t)$ and $c(t, z)$ predictable process, for all $t > 0$, $z \in \mathbb{R}$, such that*

$$\int_0^t \left(|a(s)| + b^2(s) + \int_{\mathbb{R}} c^2(s, z)\nu(\mathrm{d}z) \right) \mathrm{d}s < \infty \quad a.s. \qquad (3.57)$$

The stochastic process $\{X(t), t \geq 0\}$, $X_0 = x$, admitting the stochastic integral representation

$$X_t = x + \int_0^t a(s)\mathrm{d}s + \int_0^t b(s)\mathrm{d}B_s + \int_0^t \int_{\mathbb{R}} c(s, z)\tilde{\mu}(\mathrm{d}s, \mathrm{d}z), \qquad (3.58)$$

is called Itô-Lévy process. The stochastic differential equation for X_t is written as follows:

$$\mathrm{d}X_t = a(t)\mathrm{d}t + b(t)\mathrm{d}B_t + \int_{\mathbb{R}} c(t, z)\tilde{\mu}(\mathrm{d}t, \mathrm{d}z), \quad X_0 = x.$$

Under condition (3.57) the stochastic integrals are well-defined and local martingales. If we replace condition (3.57) with the next one

$$\mathbb{E}\left\{ \int_0^t \left(|a(s)| + b^2(s) + \int_{\mathbb{R}} c^2(s, z)\nu(\mathrm{d}z) \right) \mathrm{d}s \right\} < \infty, \qquad (3.59)$$

the stochastic integrals become martingales.

Example 3.18.15 *In applications we often take $a(t) = a(X_t)$, $b(t) = b(X_t)$ and $c(t, z) = c(X_{t-}, z)$ and write the stochastic differential equation as follows:*

$$\mathrm{d}X_t = a(X_t)\mathrm{d}t + b(X_t)\mathrm{d}W_t + \int_{|z| > 1} c(X_{t-}, z)\mu(\mathrm{d}t, \mathrm{d}z)$$

$$+ \int_{0 < |z| \leq 1} c(X_{t-}, z)\{\mu(\mathrm{d}t, \mathrm{d}z) - \nu(\mathrm{d}z)\mathrm{d}t\},$$

where μ is the random measure associated with jumps of X,

$$\mu(dt, dz) = \sum_{s > 0} \mathbf{1}_{\{\Delta Z_s \neq 0\}} \delta_{(s, \Delta Z_s)}(dt, dz),$$

and δ denotes the Dirac measure. The process Z_t is the driving pure-jump Lévy process of the form:

$$Z_t = \int_0^t \int_{|z| \leq 1} z\{\mu(ds, dz) - \nu(dz)ds\} + \int_0^t \int_{|z| > 1} z\mu(ds, dz);$$

3.18.8   Itô formula for Lévy driven stochastic differential equations

Theorem 3.18.16 *Let $\{X_t, t \geq 0\}$ be an Itô-Lévy process as in (3.58) and let $f : (0, \infty) \times \mathbb{R} \to \mathbb{R}$ be a continuous measurable function with first derivative with respect to the first argument and twice differentiable with respect to the second argument. Set $Y_t = f(t, X_t)$, $t \geq 0$. Then Y_t is also an Itô-Lévy process which satisfies*

$$dY_t = \frac{\partial}{\partial t} f(t, X_t)dt + \frac{\partial}{\partial x} f(t, X_t)a(t)dt + \frac{\partial}{\partial x} f(t, X_t)b(t)dB_t$$

$$+ \frac{1}{2} \frac{\partial^2}{\partial x^2} f(t, X_t)b^2(t)dt$$

$$+ \int_{\mathbb{R}} \left(f(t, X_t + c(t, z)) - f(t, X_t) - \frac{\partial}{\partial x} f(t, X_t)c(t, z) \right) \nu(dz)dt$$

$$+ \int_{\mathbb{R}} (f(t, X_{t^-} + c(t, z)) - f(t, X_{t^-})) \tilde{\mu}(dt, dx).$$

Next example is taken from Di Nunno *et al.* (2009).

Example 3.18.17 (Generalized geometric Lévy process) *Consider the stochastic differential equation for the càdlàg process Z solution to*

$$dZ_t = Z_{t^-} \left\{ a(t)dt + b(t)dB_t + \int_{\mathbb{R}} c(t, z)\tilde{\mu}(dt, dz) \right\}, \quad Z_0 = z_0 > 0,$$

where $a(t)$, $b(t)$ and $c(t, z) > -1$ are given predictable processes satisfying (3.57). We now prove that the solution of the above stochastic differential equation is

$$Z_t = z_0 e^{X_t}, \quad t \geq 0,$$

where

$$X_t = \int_0^t \left\{ a(s) - \frac{1}{2}b^2(s) + \int_{\mathbb{R}} (\log(1 + c(s, z)) - c(s, z)) \, \nu(\mathrm{d}z) \right\} \mathrm{d}s$$

$$+ \int_0^t b(s)\mathrm{d}B_s + \int_0^t \int_{\mathbb{R}} \log(1 + c(s, z))\tilde{\mu}(\mathrm{d}s, \mathrm{d}z).$$

We use Itô formula for $Y_t = f(t, X_t)$ *with* $f(t, x) = z_0 e^x$.

$$\mathrm{d}Y_t = z_0 e^{X_t} \left\{ \left(a(t) - \frac{1}{2}b^2(t) + \int_{\mathbb{R}} (\log(1 + c(t, z)) - c(t, z))\nu(\mathrm{d}z) \right) \mathrm{d}t + b(t)\mathrm{d}B_t \right\}$$

$$+ z_0 e^{X_t} \frac{1}{2}b^2(t)\mathrm{d}t$$

$$+ \int_{\mathbb{R}} z_0 \left(e^{X_t + \log(1 + c(t, z))} - e^{X_t} - e^{X_t} \log(1 + c(t, z)) \right) \nu(\mathrm{d}z)\mathrm{d}t$$

$$+ \int_{\mathbb{R}} z_0 \left(e^{X_{t^-} + \log(1 + c(t, z))} - e^{X_{t^-}} \right) \tilde{\mu}(\mathrm{d}t, \mathrm{d}z)$$

$$= Y_{t^-} \left(a(t)\mathrm{d}t + b(t)\mathrm{d}B_t + \int_{\mathbb{R}} c(t, z)\tilde{\mu}(\mathrm{d}t, \mathrm{d}z) \right)$$

which is the required stochastic differential equation.

3.19   Stochastic differential equations in $\mathbb{R}^n$

We briefly describe the multidimensional version of stochastic differential equations and the Itô formula in the multidimensional case. We first start with the definition of the multidimensional Brownian motion.

Definition 3.19.1 (multidimensional Brownian motion) *Let* $\{B_1(t), t \geq 0\}$, $\{B_2(t), t \geq 0\}$, ..., $\{B_m(t), t \geq 0\}$ *be* m *independent Brownian motions. The vector*

$$\mathbf{B}_t = (B_1(t), B_2(t), \ldots, B_m(t)), \quad t \geq 0,$$

is called m-dimensional Brownian motion.

The multidimensional Brownian motion is such that

$$\mathbf{B}_t - \mathbf{B}_s \sim \mathrm{N}(\mathbf{0}, I(t - s))$$

where I is the $m \times m$ identity matrix, $\mathbf{0}$ is the zero vector of $\mathbb{R}^m$, and $N(\cdot, \cdot)$ is the multivariate normal distribution.

Definition 3.19.2 *A Itô diffusion is a stochastic process $\{X_t, 0 \leq t \leq T\}$ with $X_t(\omega) : [0, T] \times \Omega \to \mathbb{R}^n$ satisfying a stochastic differential equation of the form:*

$$\mathrm{d}X_t = b(t, X_t)\mathrm{d}t + \sigma(t, X_t)\mathrm{d}B_t, \quad t \geq 0, \quad X_0 = x, \tag{3.60}$$

where $\{B_t, t \geq 0\}$ is an m-dimensional Brownian motion, $b : [0, T] \times \mathbb{R}^n \to \mathbb{R}^n$, $\sigma : [0, T] \times d\,R^n \to \mathbb{R}^{n \times m}$ such that

$$|b(t, x) - b(t, y)| + |\sigma(x) - \sigma(y)| \leq L|x - y|, \quad x, y \in \mathbb{R}^n, t \in [0, T],$$

where $|\sigma|^2 = \sum |\sigma_{ij}|^2$ and

$$|b(t, x)| + |\sigma(t, x)| \leq L(1 + |x|), \quad x \in \mathbb{R}^n, t \in [0, T].$$

If there exists a random variable Z independent of $\{B_t, t \geq 0\}$ such that $\mathbb{E}|Z|^2 < \infty$, then the stochastic differential equation has also a unique solution and it is such that

$$\mathbb{E}\left\{\int_0^T |X_t|^2 \mathrm{d}t\right\} < \infty.$$

The stochastic differential equation (3.60) has to be understood in matrix form with

$$X_t = \begin{pmatrix} X_1(t) \\ \vdots \\ X_n(t) \end{pmatrix}, \quad b(t, X_t) = \begin{pmatrix} b_1(t, X_t) \\ \vdots \\ b_n(t, X_t) \end{pmatrix},$$

$$\sigma = \begin{bmatrix} \sigma_{11}(t, X_t) & \cdots & \sigma_{1m}(t, X_t) \\ \vdots & & \vdots \\ \sigma_{n1}(t, X_t) & \cdots & \sigma_{nm}(t, X_t) \end{bmatrix}.$$

More precisely,

$$\begin{cases} \mathrm{d}X_1(t) = b_1(t, X_t)\mathrm{d}t + \sum_{j=1}^m \sigma_{1j}(t, X_t)\mathrm{d}B_j(t) \\ \mathrm{d}X_2(t) = b_2(t, X_t)\mathrm{d}t + \sum_{j=1}^m \sigma_{2j}(t, X_t)\mathrm{d}B_j(t) \\ \vdots \\ \mathrm{d}X_n(t) = b_n(t, X_t)\mathrm{d}t + \sum_{j=1}^m \sigma_{nj}(t, X_t)\mathrm{d}B_j(t) \end{cases}$$

Theorem 3.19.3 (Itô formula) *Let B_t be a m-dimensional Brownian motion and $X(t)$ and n-dimensional Itô process solution to a stochastic differential equation* like (3.60). Let $f : [0, T] \times \mathbb{R}^n \to \mathbb{R}$ one time differentiable in t and twice with respect to the second argument. Then

$$\mathrm{d}f(t, X_t) = \frac{\partial}{\partial t} f(t, X_t)\mathrm{d}t$$

$$+ \sum_{i=1}^{n} \frac{\partial}{\partial x_i} f(t, X_t)\mathrm{d}X_i(t) + \frac{1}{2} \sum_{i=1, j}^{n} \frac{\partial^2}{\partial x_i x_j} f(t, X_t)\mathrm{d}X_i(t)\mathrm{d}X_j(t)$$

with the usual rules $\mathrm{d}B_i\mathrm{d}B_j = \delta_{ij}\mathrm{d}t$, $\mathrm{d}B_i\mathrm{d}t = \mathrm{d}t\mathrm{d}B_i = 0$, where $\delta_{ij} = 1$ only if $i = j$ and 0 otherwise.

3.20   Markov switching diffusions

Let α_t be a finite-state Markov chain in continuous time with state space $\mathcal{S}$ and generator $Q = [q_{ij}]$ (see Definition 3.4.17), i.e. the entries of Q are such that $q_{ij} \geq 0$ for $i \neq j$, $\sum_{j \in \mathcal{S}} q_{ij} = 0$ for each $i \in \mathcal{S}$. Suppose that $f(\cdot, \cdot) : \mathbb{R}^r \times \mathcal{S} \to \mathbb{R}^n$ and $g(\cdot, \cdot) : \mathbb{R}^n \times \mathcal{S} \to \mathbb{R}^{n \times n}$. Consider the so called n-dimensional hybrid diffusion system or stochastic differential equation with Markov switching

$$\mathrm{d}X_t = f(X_t, \alpha_t)\mathrm{d}t + g(X_t, \alpha_t)\mathrm{d}\mathbf{B}_t, \quad X_0 = x_0, \alpha(0) = \alpha, \quad (3.61)$$

where $\{\mathbf{B}_t, t \geq 0\}$ is a n-dimensional Brownian motion. The initial value x, the Brownian motion $\mathbf{B}$ and the Markov chain α_t are all mutually independent. The functions $f(\cdot, \cdot)$ and $g(\cdot, \cdot)$ satisfy certain regularity conditions which can be found, for example, in Mao and Yuan (2006) but usual Lipschitz and growth conditions are enough. In this section we assume that if, for each state $i \in \mathcal{S}$ the functions $f(\cdot, i)$ and $g(\cdot, i)$ satisfy the usual Lipschitz and growth conditions so that (3.61) has a unique solution in distribution for each initial condition (see, e.g. Yin et al. (2005)) as stated by the next result.

Theorem 3.20.1 (Mao 1999) Let f and g be globally Lipschitz continuous, i.e., there exists a constant $L > 0$ such that

$$\min \left(|f(u, j) - f(v, j)|^2, |g(u, j) - g(v, j)|^2 \right)$$
$$\leq L|u - v|^2, \quad \forall u, v, \in \mathbb{R}^n, j \in \mathcal{S}.$$

Then, (3.61) has a unique solution for any given initial value $X_0 = x_0 \in \mathbb{R}^n$ and $\alpha_0 = i \in \mathcal{S}$. Moreover, the joint process

$$Z_t = (X_t, \alpha_t), \quad t \geq 0,$$

is a time-homogeneous Markov process.

For the switching diffusion model, there exist an associated operator defined as

$$\mathcal{L}h(x, i) = \frac{1}{2}\mathrm{tr}[\nabla^2 h(x, i)g(x, i)g(x, i)'] + \nabla h(x, i)' f(x, i) + Qh(x, \cdot)(i)$$

where ∇h and $\nabla^2 h$ denote the gradient and the Hessian of h and $\text{tr} A$ is the trace of matrix A and

$$Qh(x, \cdot)(i) = \sum_{j \in \mathcal{S}} q_{ij} h(x, j) = \sum_{j \neq i} q_{ij} (h(x, j) - h(x, i)).$$

Associated with (3.61) there is a martingale problem formulation.

Definition 3.20.2 *A process* (X_t, α_t), $t \geq 0$, *is said to be a solution of the martingale problem operator* $\mathcal{L}$ *if*

$$f(X_t, \alpha_t) - \int_0^t \mathcal{L}h(X_s, \alpha_s) \mathrm{d}s$$

is a martingale for any function $h : \mathbb{R}^n \times \mathcal{S} \to \mathbb{R}$, *such that for each* $i \in \mathcal{S}$, *the function* $h(\cdot, i)$ *is twice continuously differentiable with respect to the first argument and with compact support.*

Under the conditions of Theorem 3.20.1, it is possible to prove that (3.61) has a unique solution associated with the martingale problem. For these Markov switching systems one interesting field of investigation is the asymptotic stability and the existence of an invariant measure.

Definition 3.20.3 *The process* $Z_t = (X_t, \alpha_t)$ *is said to be asymptotically stable in distribution if there exists a probability measure* $\pi(\cdot \times \cdot)$ *on* $\mathbb{R}^n \times \mathcal{S}$ *such that the transition probability* $p(t, y, i, \mathrm{d}x \times \{j\})$ *of* X_t *converges weakly to* $\pi(\mathrm{d}x \times \{j\})$ *as* $t \to \infty$ *for every* $(y, i) \in \mathbb{R}^n \times \mathcal{S}$. *The corresponding stochastic differential equation* (3.61) *is also said to be asymptotically stable in distribution.*

When a switching diffusion is asymptotically stable in distribution then X_t has a unique invariant probability measure. Under the conditions of Theorem 3.20.1 and additional tightness conditions, it is possible to prove that system (3.61) is asymptotically stable in distributions. In the following, we will always work under these conditions unless explicitly mentioned.

3.21   Solution to exercises

Solution 3.1 (to Exercise 3.1) *By using Jensen's inequality* (2.27) *for the convex function* $f(x) = |x|$ *we get*

$$\mathbb{E}\{Z_n | \mathcal{F}_{n-1}\} = \mathbb{E}\{|S_n| | \mathcal{F}_{n-1}\} \geq |\mathbb{E}\{S_n | \mathcal{F}_{n-1}\}| = |S_{n-1}| = Z_{n-1}.$$

where in the last passage we used the fact that the random walk S_n *is a martingale.*

Solution 3.2 (Exercise 3.2) *Clearly, Z_n is $\mathcal{F}_n$-measurable. Further,*

$$\mathbb{E}\{Z_n|\mathcal{F}_{n-1}\} = \mathbb{E}\left\{M(\alpha)^{-n}e^{\alpha S_n}\,\big|\,\mathcal{F}_{n-1}\right\} = M(\alpha)^{-n}\mathbb{E}\left\{e^{\alpha S_n}\,\big|\,\mathcal{F}_{n-1}\right\}$$

$$= M(\alpha)^{-n}\mathbb{E}\left\{e^{\alpha(S_{n-1}+X_n)}\,\big|\,\mathcal{F}_{n-1}\right\} = M(\alpha)^{-n}e^{\alpha S_{n-1}}\mathbb{E}\left\{e^{\alpha X_n}\,\big|\,\mathcal{F}_{n-1}\right\}$$

$$= M(\alpha)^{-n}e^{\alpha S_{n-1}}\mathbb{E}\left(e^{\alpha X_n}\right) = M(\alpha)^{-n+1}e^{\alpha S_{n-1}} = Z_{n-1}.$$

We need to prove that $\mathbb{E}|Z_n| < \infty$. Indeed, by using (2.9) in our context, we get that $\mathbb{E}(e^{\alpha S_n}) = M(\alpha)^n$. Therefore, noticing that Z_n is non-negative, we have $\mathbb{E}|Z_n| = M(\alpha)^{-n}\mathbb{E}(e^{\alpha S_n}) = 1 < \infty$.

Solution 3.3 (Exercise 3.3) *Clearly Z_n is $\mathcal{F}_n$-measurable. Remember that $|a - b| \le |a| + |b|$. Hence*

$$\mathbb{E}|Z_n| = \mathbb{E}\left\{\left|\left(\sum_{i=1}^{n}X_i\right)^2 - n\sigma^2\right|\right\} \le \mathbb{E}\left\{\left(\sum_{i=1}^{n}X_i\right)^2 + n\sigma^2\right\}$$

$$= \mathbb{E}\left\{\sum_{i=1}^{n}X_i^2 + \sum_{i\neq j}^{n}X_iX_j\right\} + n\sigma^2 = \sum_{i=1}^{n}\mathbb{E}\left\{X_i^2\right\} + \sum_{i\neq j}^{n}\mathbb{E}\{X_iX_j\} + n\sigma^2$$

$$= 2n\sigma^2 < \infty.$$

Finally,

$$\mathbb{E}\{Z_n|\mathcal{F}_{n-1}\}$$

$$= \mathbb{E}\left\{\left(\sum_{i=1}^{n}X_i\right)^2 - n\sigma^2\,\bigg|\,\mathcal{F}_{n-1}\right\} = \mathbb{E}\left\{\left(X_n + \sum_{i=1}^{n-1}X_i\right)^2 - n\sigma^2\,\bigg|\,\mathcal{F}_{n-1}\right\}$$

$$= \mathbb{E}\left\{X_n^2 + 2X_n\sum_{i=1}^{n-1}X_i + \left(\sum_{i=1}^{n-1}X_i\right)^2 - n\sigma^2\,\bigg|\,\mathcal{F}_{n-1}\right\}$$

$$= \mathbb{E}\left\{\left(\sum_{i=1}^{n-1}X_i\right)^2 - (n-1)\sigma^2 + X_n^2 + 2X_n\sum_{i=1}^{n-1}X_i - \sigma^2|\mathcal{F}_{n-1}\right\}$$

$$= \mathbb{E}\{Z_{n-1}|\mathcal{F}_{n-1}\} + \mathbb{E}\left\{X_n^2|\mathcal{F}_{n-1}\right\} + 2\sum_{i=1}^{n-1}X_i\mathbb{E}\{X_n|\mathcal{F}_{n-1}\} - \sigma^2$$

$$= Z_{n-1} + \mathbb{E}\left\{X_n^2\right\} + 2\sum_{i=1}^{n-1}X_i\mathbb{E}\{X_n\} - \sigma^2$$

$$= Z_{n-1} + \sigma^2 - \sigma^2 = 0.$$

Solution 3.4 (to Exercise 3.4) *We need to rewrite* inf *and* sup *appropriately, then the properties of the filtration ensure the result. Indeed,*

$$\left(\sup_{n \geq 1} T_n \leq t \right) = \bigcap_{n \geq 1} (T_n \leq t) \in \mathcal{F}_t$$

and

$$\left(\inf_{n \geq 1} T_n \leq t \right) = \bigcup_{n \geq 1} (T_n \leq t) \in \mathcal{F}_t.$$

Solution 3.5 (to Exercise 3.5) *Clearly X_t is $\mathcal{F}_t$-measurable and integrable. We can rewrite $X_t = B_t^2 - t = (B_t - B_s + B_s)^2 - t$ for $s < t$. Hence $X_t = (B_t - B_s)^2 + B_s^2 - 2(B_t - B_s)B_s - t$ and*

$$
\begin{aligned}
\mathbb{E}\{X_t | \mathcal{F}_s\} &= \mathbb{E}\{(B_t - B_s)^2 + B_s^2 - 2(B_t - B_s)B_s - t | \mathcal{F}_s\} \\
&= \mathbb{E}(B_t - B_s)^2 + B_s^2 - 2B_s \mathbb{E}(B_t - B_s) - t \\
&= (t - s) + B_s^2 - t = B_s^2 - s.
\end{aligned}
$$

Solution 3.6 (to Exercise 3.6) *The process X_t is clearly $\mathcal{F}_t$-measurable. Using calculations similar to that of Exercise 2.11, one can easily show that if $X \sim N(0, \sigma^2)$ then $\mathbb{E}(e^X) = e^{\frac{1}{2}\sigma^2}$ and also that $\mathbb{E}|X_t| < \infty$. We first assume that X_t is a martingale. Hence*

$$\mathbb{E}(X_t) = \mathbb{E}\left(e^{\mu t + \sigma B_t} \right) = e^{\mu t} \mathbb{E}\left(e^{\sigma B_t} \right) = e^{\mu t} e^{\frac{1}{2}\sigma^2 t} = e^{(\mu + \frac{1}{2}\sigma^2)t}.$$

But, X_t is a martingale, hence $\mathbb{E}(X_t) = \mathbb{E}(X_s) = k$, for all t and s, where k is some constant. Therefore, $\mathbb{E}(X_t)$ should not depend on t and this is true only if $\mu = -\frac{1}{2}\sigma^2$. We now assume that $\mu = -\frac{1}{2}\sigma^2$ and show that X_t is a martingale.

$$
\begin{aligned}
\mathbb{E}\{X_t | \mathcal{F}_s\} &= e^{\mu t} \mathbb{E}\left\{ e^{\sigma B_t} \big| \mathcal{F}_s \right\} = e^{\mu t} e^{\sigma B_s} \mathbb{E}\left\{ e^{\sigma(B_t - B_s)} \big| \mathcal{F}_s \right\} = e^{\mu t + \sigma B_s} \mathbb{E}\left(e^{\sigma(B_t - B_s)} \right) \\
&= e^{\mu t + \sigma B_s} e^{\frac{1}{2}\sigma^2(t-s)} = e^{\mu t + \sigma B_s + \mu(t-s)} = e^{\mu s + \sigma B_s} = X_s.
\end{aligned}
$$

Solution 3.7 (to Exercise 3.7) *We need to identify an $f(\cdot)$ such that $f'(x) = x^2$. This is clearly $f(x) = \frac{1}{3}x^3$, hence we apply (3.44) for this $f(\cdot)$ and obtain*

$$\frac{B_t^3}{3} = \int_0^t B_s^2 dB_s + \frac{1}{2} \int_0^t 2B_s ds$$

because $f''(x) = 2x$. Hence

$$\int_0^t B_s^2 dB_s = \frac{B_t^3}{3} - \int_0^t B_s ds.$$

Solution 3.8 (to Exercise 3.8) *In this case* $f_x(t, x) = t$ *hence*

$$f(t, x) = tx, \quad f_t(t, x) = x, \quad f_x(t, x) = t, \quad f_{xx}(t, x) = 0,$$

then, using (3.45), we obtain

$$t B_t = 0 + \int_0^t B_s \mathrm{d}s + \int_0^t s \mathrm{d}B_s + \frac{1}{2} \int_0^t 0 \mathrm{d}B_s,$$

therefore

$$\int_0^t s \mathrm{d}B_s = t B_t - \int_0^t B_s \mathrm{d}s.$$

Solution 3.9 (to Exercise 3.9) *Consider the function* $f(t, x) = \frac{2x^{n+1}}{n+1} - t^2 x$.
Then, $f_x(t, x) = 2x^n - t^2$, $f_{xx}(t, x) = 2nx^{n-1}$ *and* $f_t(t, x) = -2tx$. *Apply Itô*
formula to $f(t, B_t) = \frac{2B_t^{n+1}}{n+1} - t^2 B_t$

$$
\begin{aligned}
\frac{2B_t^{n+1}}{n+1} - t^2 B_t &= \int_0^t f_x(s, B_s) \mathrm{d}B_s + \frac{1}{2} \int_0^t f_{xx}(s, B_s) \mathrm{d}s + \int_0^t f_t(s, B_s) \mathrm{d}s \\
&= \int_0^t (2B_s^n - s^2) \mathrm{d}B_s + \int_0^t n B_s^{n-1} \mathrm{d}s - \int_0^t 2s B_s \mathrm{d}s \\
&= \int_0^t (2B_s^n - s^2) \mathrm{d}B_s + \int_0^t (n B_s^{n-1} - 2s B_s) \mathrm{d}s.
\end{aligned}
$$

Therefore,

$$\int_0^t (2B_s^n - s^2) \mathrm{d}B_s = \frac{2B_t^{n+1}}{n+1} - t^2 B_t - \int_0^t (n B_s^{n-1} - 2s B_s) \mathrm{d}s.$$

Solution 3.10 (to Exercise 3.10) *We apply (3.47) with* $f(t, x) = x^2$ *and*
$X_t = B_t$. *Hence* $f_t(t, x) = 0$, $f_x(t, x) = 2x$ *and* $f_{xx}(t, x) = 2$.

$$\mathrm{d}Y_t = 2B_t \mathrm{d}B_t + \frac{1}{2} 2(\mathrm{d}B_t)^2 = 2B_t \mathrm{d}B_t + \mathrm{d}t.$$

Solution 3.11 (to Exercise 3.11) *We apply (3.47) with* $f(t, x) = 2 + t + e^x$ *and*
$X_t = B_t$. *Hence* $f_t(t, x) = 1$, $f_x(t, x) = e^x$ *and* $f_{xx}(t, x) = e^x$.

$$\mathrm{d}Y_t = \mathrm{d}t + e^{B_t} \mathrm{d}B_t + \frac{1}{2} e^{B_t} \mathrm{d}t = \left(1 + \frac{1}{2} e^{B_t}\right) \mathrm{d}t + e^{B_t} \mathrm{d}B_t.$$

Solution 3.12 (to Exercise 3.12) *Denote by* $Z_t = \int_0^t e^{\lambda s} \mathrm{d}B_s$ *and define*
$f(t, x) = xe^{-\lambda t}$, *therefore* $U_t = f(t, Z_t)$. *We now apply Itô formula to this*
$f(t, x)$.

$$f_x(t, x) = e^{-\lambda t}, \quad f_{xx}(t, x) = 0, \quad f_t(t, x) = -\lambda x e^{-\lambda t} = -\lambda f(t, x).$$

We now apply (3.47) to $df(t, Z_t) = dU_t$:

$$dU_t = -\lambda Z_t e^{-\lambda t} dt + e^{-\lambda t} dZ_t.$$

Now notice that $Z_t e^{-\lambda t} = U_t$ *and* $dZ_t = e^{\lambda t} B_t$. *Thus*

$$dU_t = -\lambda U_t dt + dB_t.$$

The process U_t *is called the Ornstein-Uhlenbeck process.*

Solution 3.13 (to Exercise 3.13) *Note that* $f_x(x) = \cos(x)$, $f_{xx}(x) = -\sin(x)$. *An application of the Itô formula gives*

$$Y_t = Y_0 + \int_0^t f_x(X_s)dX_s + \frac{1}{2}\int_0^t f_{xx}(X_s)(dX_s)^2.$$

But

$$\int_0^t f_x(X_s)dX_s = \int_0^t \cos(X_s)X_s ds + \sigma \int_0^t \cos(X_s)dB_s$$

and

$$\frac{1}{2}\int_0^t f_{xx}(X_s)dX_s = -\frac{1}{2}\int_0^t \sin(X_s)(dX_s)^2$$

$$= -\frac{1}{2}\int_0^t \sin(X_s)(X_s^2(ds)^2 + \sigma^2(dB_s)^2 + 2\sigma X_s dt dB_s)$$

$$= -\frac{\sigma^2}{2}\int_0^t \sin(X_s)ds.$$

Therefore

$$Y_t = Y_0 + \int_0^t \left(\cos(X_s)X_s - \frac{\sigma^2}{2}\sin(X_s)\right)ds + \sigma \int_0^t \cos(X_s)dB_s$$

or

$$dY_t = \left(\cos(X_t)X_t - \frac{\sigma^2}{2}\sin(X_t)\right)dt + \sigma \cos(X_t)dB_t.$$

Solution 3.14 (to Exercise 3.14) *We try to apply Itô formula to derive the result. Consider the function* $f(t, x) = \log(x)$. *Then,* $f_x(t, x) = \frac{1}{x}$, $f_{xx}(t, x) = -\frac{1}{x^2}$ *and* $f_t(t, x) = 0$. *Therefore,*

$$\log(X_t) = \log(X_0) + \int_0^t f_x(s, X_s)dX_s + \frac{1}{2}\int_0^t f_{xx}(s, X_s)(dX_s)^2$$

$$= \log(X_0) + \int_0^t \frac{dX_s}{X_s} - \frac{1}{2}\int_0^t \frac{(dX_s)^2}{X_s^2}$$

$$= \log(X_0) + \int_0^t \frac{1}{X_s}(b(s)X_s ds + \sigma(s)X_s dB_s)$$

$$- \frac{1}{2} \int_0^t \frac{1}{X_s^2} (b^2(s) X_s^2 (ds)^2 + \sigma^2(s) X_s^2 (dB_s)^2 + b(s)\sigma(s) X_s^2 ds dB_s)$$

$$= \log(X_0) + \int_0^t \left(b(s) - \frac{1}{2}\sigma^2(s) \right) ds + \int_0^t \sigma(s) dB_s.$$

Remember that $\exp\{a + b\} = \exp\{a\} \exp\{b\}$, *then*

$$X_t = X_0 \exp \left\{ \int_0^t \left(b(s) - \frac{1}{2}\sigma^2(s) \right) ds + \int_0^t \sigma(s) dB_s \right\}.$$

To end the proof, we need to prove that X_t is a solution of the proposed stochastic differential equation. This is trivially true and left to the reader.

Solution 3.15 (to Exercise 3.15) *Let* $f(t, x) = \frac{x}{1+t}$, *then* $f_x(t, x) = \frac{1}{1+t}$, $f_{xx}(t, x) = 0$ *and* $f_t(t, x) = -\frac{x}{(1+t)^2}$. *In order to show that X_t is a solution of the stochastic differential equation considered, we apply Itô formula to* $f(t, B_t) = X_t = \frac{B_t}{1+t}$. *Indeed,*

$$X_t = f(t, B_t) = \int_0^t f_x(s, B_s) dB_s + \frac{1}{2} \int_0^t f_{xx}(s, B_s) ds + \int_0^t f_t(s, B_s) ds$$

$$= \int_0^t \frac{dB_s}{1+s} - \int_0^t \frac{B_s}{(1+s)^2} ds$$

$$= \int_0^t \frac{dB_s}{1+s} - \int_0^t \frac{X_s}{(1+s)} ds$$

and hence, in differential form, we have

$$dX_t = -\frac{1}{1+t} X_t dt + \frac{1}{1+t} dB_t.$$

Solution 3.16 (to Exercise 3.16) *Let* $f(t, x) = (a^{1/3} + \frac{1}{3}x)^3$ *and notice that* $f_x(t, x) = (a^{1/3} + \frac{1}{3}x)^2$, $f_{xx}(t, x) = \frac{2}{3}(a^{1/3} + \frac{1}{3}x)$ *with* $f_t(t, x) = 0$. *As in Exercise 3.15, we apply Itô formula to* $f(t, B_t) = X_t = (a^{1/3} + \frac{1}{3}B_t)^3$. *Hence,*

$$X_t = f(t, B_t) = \int_0^t f_x(s, B_s) dB_s + \frac{1}{2} \int_0^t f_{xx}(s, B_s) ds + \int_0^t f_t(s, B_s) ds$$

$$= a + \int_0^t (a^{1/3} + \frac{1}{3}B_s)^2 dB_s + \frac{1}{3} \int_0^t (a^{1/3} + \frac{1}{3}B_s) ds$$

$$= X_0 + \int_0^t X_s^{2/3} dB_s + \frac{1}{3} \int_0^t X_s^{1/3} ds$$

or $dX_t = \frac{1}{3} X_t^{1/3} dt + X_t^{2/3} dB_t$.

Solution 3.17 (to Exercise 3.17) *Consider* $f(t, x) = \frac{1}{1+x}$. *Then,* $f_x(t, x) =$
$-\frac{1}{(1+x)^2}$, $f_{xx}(t, x) = \frac{2}{(1+x)^3}$ *and* $f_t(t, x) = 0$. *We apply Itô formula to* $Y_t =$
$f(t, B_t) = \frac{1}{1+B_t}$ *and obtain*

$$Y_t = f(t, B_t) = \int_0^t f_x(s, B_s)dB_s + \frac{1}{2}\int_0^t f_{xx}(s, B_s)ds + \int_0^t f_t(s, B_s)ds$$

$$= 1 - \int_0^t \frac{1}{(1 + B_s)^2}dB_s + \int_0^t \frac{1}{(1 + B_t)^3}ds$$

$$= 1 - \int_0^t Y_s^2 dB_s + \int_0^t Y_s^3 ds$$

and so $dY_t = Y_t^3 dt - Y_t^2 dB_t$.

Solution 3.18 (to Exercise 3.18) *Let* $Y_t = \int_0^t \frac{1}{1-s}dB_s$. *Then, we should uso Itô
formula in differential form for* $f(t, y) = t + (1 - t)y$ *so that* $X_t = f(t, Y_t)$.
Hence, $f_t(t, y) = 1 - y$, $f_y(t, y) = 1 - t$ *and* $f_{yy}(t, y) = 0$. *Notice that* $dY_t =$
$\frac{dB_t}{1-t}$. *There, by Itô formula we get*

$$dX_t = df(t, Y_t) = (1 - Y_t)dt + (1 - t)dY_t = (1 - Y_t)dt + dB_t$$

$$= \left(1 - \int_0^t \frac{1}{1 - s}dB_s\right)dt + dB_t = \frac{(1 - t) - (1 - t)\int_0^t \frac{1}{1-s}dB_s}{1 - t}dt + dB_t$$

$$= \frac{1 - \left(t + (1 - t)\int_0^t \frac{1}{1-s}dB_s\right)}{1 - t}dt + dB_t = \frac{1 - X_t}{1 - t}dt + dB_t.$$

Solution 3.19 (to Exercise 3.19) *We need the first two derivatives of* $f(\cdot)$

$$f_x(u) = \frac{1}{\sigma(u)} \qquad f_{xx}(u) = -\frac{\sigma_x(u)}{\sigma^2(u)}$$

and we can apply Itô formula

$$df(X_t) = f_x(X_t)dX_t + \frac{1}{2}f_{xx}(X_t)(dX_t)^2$$

$$= \frac{\mu(X_t)dt + \sigma(X_t)dB_t}{\sigma(X_t)} - \frac{1}{2}\frac{\sigma_x(X_t)(\mu(X_t)dt + \sigma(X_t)dB_t)^2}{\sigma^2(X_t)}$$

$$= \frac{\mu(X_t)}{\sigma(X_t)}dt + dB_t$$

$$- \frac{1}{2}\frac{\sigma_x(X_t)(\mu^2(X_t)(dt)^2 + \sigma^2(X_t)(dB_t)^2 + 2\mu(X_t)\sigma(X_t)(dtdB_t)}{\sigma^2(X_t)}$$

$$= \frac{\mu(X_t)}{\sigma(X_t)}dt + dB_t - \frac{1}{2}\sigma_x(X_t)dt$$

therefore

$$dY_t = df(X_t) = \left(\frac{\mu(X_t)}{\sigma(X_t)} - \frac{1}{2}\sigma_x(X_t) \right) dt + dB_t.$$

Notice that the Lamperti transform has the effect of making the diffusion coefficient unitary.

3.22 Bibliographical notes

Classical references on stochastic processes, Brownian motion, stochastic differential equations and Itô formula are Karlin and Taylor (1981), Karatzas and Shrevre (1988) and Rogers and Williams (1987). For additional examples and exercises on Itô calculus, one can refer to Øksendal (1998). The books of Sato (1999), Bertoin (1998) and Applebaum (2004) contain rigorous treatments of the theory of Lévy processes and stochastic calculus with jump processes. For abstract stochastic analysis and calculus for several classes of processes we refer to Protter (2004). We also suggest the textbook of Di Nunno *et al.* (2009) which is a gentle introduction to Malliavin calculus in view of applications in finance. Finally, the books of Cont and Tankov (2004) and Schoutens (2003) review several types of Lévy processes along with their characteristic triplets, simulation schemes, etc.

References

Abramowitz, M. and Stegun, I. (1964). *Handbook of Mathematical Functions*. Dover Publications, New York.

Aldous, D. and Eagleson, G. (1978). On mixing and stability of limit theorems. *Ann. Probab.* **6**, 325–331.

Applebaum, D. (2004). *Lévy Processes and Stochastic Calculus*. Cambridge University Press, Cambridge.

Bertoin, J. (1998). *Lévy Processes*. Cambridge University Press, Cambridge.

Billingsley, P. (1986). *Probability and Measure, Second Edition*. John Wiley & Sons, Inc., New York.

Cont, R. and Tankov, P. (2004). *Financial Modelling With Jump Processes*. Chapman & Hall/CRC, Boca Raton.

Cox, D. and Miller, H. (1965). *The theory of stochastic processes*. Chapman & Hall, Bristol.

Di Crescenzo, A. and Martinucci, B. (2006). On the effect of random alternating perturbations on hazard rates. *Scientiae Mathematicae Japonicae* **64**, 2, 381–394.

Di Nunno, G., Øksendal, B., and Proske, F. (2009). *Malliavin Calculus for Lévy Processes with Applications to Finance*. Springer, New York.

Feller, W. (1968). *An Introduction to Probability Theory and Its Applications, Vol I. (3rd ed.)*. John Wiley & Sons, Inc., New York.

Feller, W. (1971). *An Introduction to Probability Theory and Its Applications, Vol II. (2nd ed.)*. John Wiley & Sons, Inc., New York.

Fong, S. and Kanno, S. (1994). Properties of the telegrapher's random process with or without a trap. *Stochastic Processes and their Applications* **53**, 147–173.

Goldstein, S. (1951). On diffusion by discontinuous movements and the telegraph equation. *The Quarterly Journal of Mechanics and Applied Mathematics* **4**, 129–156.

Iacus, S. (2001). Statistic analysis of the inhomogeneous telegrapher's process. *Statistics and Probability Letters* **55**, 1, 83–88.

Iacus, S. and Yoshida, N. (2008). Estimation for the discretely observed telegraph process. *Theory of Probability and Mathematical Statistics* **78**, 33–43.

Itô, K. and McKean, P. (1996). *Diffusion Processes and their Sample Paths*. Springer, New York.

Jacod, J. (1997). On continuous conditional gaussian martingales and stable convergence in law. *Séminaire de probabilités (Strasbourg)* **31**, 232–246.

Jacod, J. (2002). On processes with conditional independent increments and stable convergence in law. *Séminaire de probabilités (Strasbourg)* **36**, 383–401.

Jacod, J. and Shiryaev, A. (2003). *Limit Theorems for Stochastic Processes (2nd ed.)*. Springer, New York.

Kac, M. (1974). A stochastic model related to the telegrapher's equation. *Rocky Mountain Journal of Mathematics* **4**, 497–509.

Karatzas, I. and Shrevre, S. (1988). *Brownian motion and stochastic calculus*. Springer-Verlag, New York.

Karlin, S. and Taylor, H. (1981). *A Second Course in Stochastic Processes*. Academic Press, New York.

Klebaner, F. (2005). *Introduction to stochastic calculus with applications*. Imperial College Press (2nd ed.), London.

Kutoyants, Y. (1998). *Statistical inference for spatial Poisson processes*. Lecture Notes in Statistics, Springer-Verlag, New York.

Lévy, P. (1954). *Théorie de l'addition des variables aléatoires*. Gauthier-Villars, Paris.

Lipster, R. and Shiryayev, A. (1977). *Statistics of Random Processes, Volume 1*. Springer-Verlag, New York.

Mao, X. (1999). Stability of stochastic differential equations with markovian switching. *Stochastic Process. Appl.* **79**, 45–67.

Mao, X. and Yuan, C. (2006). *Stochastic Differential Equations With Markov Switching*. Imperial College Press, London.

Moler, C. and van Loan, C. (2003). Nineteen dubious ways to compute the exponential of a matrix, twenty-five years later. *SIAM Review* **45**, 3–49.

Øksendal, B. (1998). *Stochastic Differential Equations. An Introduction with Applications, 5th ed.* Springer-Verlag, Berlin.

Orsingher, E. (1985). Hyperbolic equations arising in random models. *Stochastic Processes and their Applications* **21**, 93–106.

Orsingher, E. (1990). Probability law, flow function, maximun distribution of wave-governed random motions and their connections with kirchoff's laws. *Stochastic Processes and their Applications* **34**, 49–66.

Papapantoleon, A. (2008). *An Introduction to Lévy Processes with Applications in Finance*. Lecture notes, TU Vienna, http://arxiv.org/abs/0804.0482, Vienna.

Pinsky, M. (1991). *Lectures on Random Evolution*. World Scientific, River Edge, New York.

Protter, P. (2004). *Stochastic Integration and Differential Equations (2nd ed.)*. Springer, New York.

Rényi, A. (1963). On stable sequences of events. *Sankyā Ser. A* **25**, 293–202.

Revuz, D. and Yor, M. (2004). *Continuous Martingales and Brownian Motion (3rd ed.)*. Springer, New York.

Rogers, L. and Williams, D. (1987). *Diffusions, Markov Processes, and Martingales, volume 2: Itô Calculus*. John Wiley & Sons, Inc., New York.

Sato, K. (1999). *Lévy Processes and Infinitely Divisible Distributions*. Cambridge University Press, Cambridge.

Schoutens, W. (2003). *Lévy Processes in Finance*. John Wiley & Sons, Ltd, Chichester.

Stadje, W. and Zacks, S. (2004). Telegraph processes with random velocities. *Journal of Applied Probability* **41**, 665–678.

Yin, G., Mao, X., and Yin, K. (2005). Numerical approximation of invariant measures for hybrid diffusion systems. *IEEE Transactions on Automatic Control* **50**, 7, 934–946.

4

Numerical methods

4.1 Monte Carlo method

Suppose we are given a random variable X and are interested in the evaluation of $\mathbb{E}(g(X))$ where $g(\cdot)$ is some known function. If we are able to draw n pseudo random numbers $x_1, \ldots, x_n$ from the distribution of X, then we can think about approximating $\mathbb{E}(g(X))$ with the sample mean of the $g(x_i)$,

$$\mathbb{E}(g(X)) \simeq \frac{1}{n} \sum_{i=1}^{n} g(x_i) = \bar{g}_n \,. \tag{4.1}$$

The expression (4.1) is not just symbolic but holds true in the sense of the law of large numbers whenever $\mathbb{E}|g(X)| < \infty$. Moreover, the central limit theorem guarantees that

$$\bar{g}_n \xrightarrow{d} N\left(\mathbb{E}(g(X)), \frac{1}{n}\mathrm{Var}(g(X))\right),$$

where $N(m, s^2)$ denotes the distribution of the Gaussian random variable with expected value m and variance s^2. In the end, the number we estimate with simulations will have a deviation from the true expected value $\mathbb{E}g(X)$ of order $1/\sqrt{n}$. Given that $P(|Z| < 1.96) \simeq 0.95$, $Z \sim N(0, 1)$, one can construct an interval for the estimate $\bar{g}_n$ of the form:

$$\left(\mathbb{E}(g(X)) - 1.96\frac{\sigma}{\sqrt{n}}, \mathbb{E}(g(X)) + 1.96\frac{\sigma}{\sqrt{n}}\right),$$

with $\sigma = \sqrt{\mathrm{Var}(g(X))}$, which is interpreted such that the Monte Carlo estimate of $\mathbb{E}(g(X))$ above is included in the interval above 95% of the time. The confidence

Option Pricing and Estimation of Financial Models with R, First Edition. Stefano M. Iacus.
© 2011 John Wiley & Sons, Ltd. Published 2011 by John Wiley & Sons, Ltd.

interval depends on $\text{Var}(g(X))$, and usually this quantity has to be estimated through the sample as well. Indeed, one can estimate it as the sample variance of Monte Carlo replications as

$$\hat{\sigma}^2 = \frac{1}{n-1} \sum_{i=1}^{n} (g(x_i) - \bar{g}_n)^2$$

and use the following 95% level Monte Carlo confidence interval[1] for $\mathbb{E}(g(X))$:

$$\left(\bar{g}_n - 1.96 \frac{\hat{\sigma}}{\sqrt{n}}, \bar{g}_n + 1.96 \frac{\hat{\sigma}}{\sqrt{n}} \right).$$

The quantity $\hat{\sigma}/\sqrt{n}$ is called the *standard error*. The standard error is itself a random quantity and thus subject to variability; hence one should interpret this value as a measure of accuracy.

One more remark is that the rate of convergence $\sqrt{n}$ is not particularly fast but at least is independent of the smoothness of $g(\cdot)$. Moreover, if we need to increase the quality of our approximation, we just need to draw additional new samples instead of rerunning the whole simulation.

4.1.1 An application

Suppose we have a function g defined as $[a, b]$ taking values in $[c, d]$, $c, d \geq 0$. Assume that we want to calculate the integral

$$\int_a^b g(x)\mathrm{d}x$$

Expected values are of course integrals, so we can apply the Monte Carlo method. The integral $\int_a^b g(x)\mathrm{d}x$ is the area under the curve, which we can then calculate as the proportion of points in the rectangle $[a, b] \times [c, d]$ under the curve g; of course we need to rescale by the total area of $[a, b] \times [c, d]$, i.e. by $A = (b - a) * (d - c)$. To transform this the Monte Carlo way, we need to use proper random variables. Which is the random variable involved? We can consider a 2-dimensional uniform random variable (X, Y) on the rectangle $[a, b] \times [c, d]$, consider the indicator function of the event '$g(X) < Y$' and take the expected value of this indicator function, i.e.

$$\mathbb{E}(\mathbf{1}_{\{Y<g(X)\}}) = P(Y < g(X))$$

so $1/A \times \mathbb{E}(\mathbf{1}_{\{g(X)<Y\}})$ is just the integral of g. The algorithm is as follows:

(1) $i = 1$, set $S_0 = 0$, $A = (b - a) * (d - c)$

(2) extract a uniform random number $x \sim U([a, b])$

---

[1] Again, this means that the interval covers the true value 95% of the time.

(3) extract a uniform random number $y \sim U([c, d])$

(4) if $y < g(x)$ then $S_i = S_{i-1} + 1$

(5) $i = i + 1$

(6) if $i = n$ exit, otherwise go to 2)

(7) $\int_a^b g(x)\mathrm{d}x \simeq A * S_n/n$

Next code is an implementation of the above. Just for checking the approximation, we use a polynomial function $g(x) = x^2$ and consider the interval $[a, b] = [0, 2]$, therefore $[c, d] = [0, 4]$. In this case,

$$\int_0^2 x^2\mathrm{d}x = \left[\frac{x^3}{3}\right]_0^2 = \frac{8}{3} = 2.\bar{6}$$

and the code is as follows:

```
R> set.seed(123)
R> g <- function(x) x^2
R> a <- 0
R> b <- 2
R> c <- 0
R> d <- 4
R> A <- (b - a) * (d - c)
R> n <- 1e+05
R> x <- runif(n, a, b)
R> y <- runif(n, c, d)
R> A * sum(y < g(x))/n

[1] 2.66792

R> integrate(g, a, b)

2.666667 with absolute error < 3.0e-14
```

Notice that instead of writing a loop we used vectorization capabilities of R. Instead of extracting a single number we generate two vectors of random numbers x and y of length n. Thus, y and g(x) are two vectors and y < g(x) returns a vector of comparisons of each element of y with the corresponding element of g(x), hence a vector of TRUE/FALSE. These are the indicator functions. The function sum applied to this vector first converts TRUE/FALSE into 1/0 and then sums this vector returning the number of ones. Divided by n is the Monte Carlo estimate of $\mathbb{E}(1_{\{Y<g(X)\}})$. Figure 4.1 shows how the random points fill the area under the curve $g(\cdot)$ as the number n of Monte Carlo replication increases.

Exercise 4.1 *Find a way to calculate* π *using the Monte Carlo method and write the corresponding* R *code. [Hint: use the formula of the area of the circle.]*

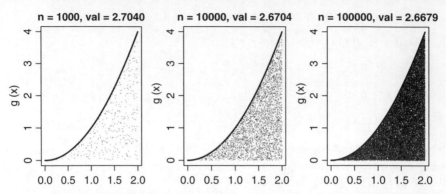

Figure 4.1   Approximation of the area as a function of the number of Monte Carlo replications $n = 1000, 10000, 10000$.

4.2   Numerical differentiation

Consider a function $f(\cdot)$ such that first order Taylor expansion up to second order is admissible around some point x_0, i.e. in the interval $[x_0 - \epsilon, x_0 + \epsilon]$. Then, the derivative of $f'(x)$ of f at point $x \in [x_0 - \epsilon, x_0 + \epsilon]$ can be written as follows:

$$f(x) = f(x_0) + f'(x_0)(x - x_0) + \frac{1}{2} f''(x_0)(x - x_0)^2 + O((x - x_0)^2).$$

For simplicity, let us denote by $h = x - x_0$, $|h| < \epsilon$, then

$$f'(x_0) = \frac{f(x_0 + h) - f(x_0)}{h} - \frac{1}{2} f''(x_0)h + O(h).$$

By definition of first derivative we know that

$$f'(x_0) = \lim_{h \to 0} \frac{f(x_0 + h) - f(x_0)}{h}.$$

Thus, one can think of approximating the first derivate of a function with the incremental ratio

$$f'(x_0) \simeq \frac{f(x_0 + h) - f(x_0)}{h}$$

for very small positive h and the residual error

$$f'(x_0) - \frac{f(x_0 + h) - f(x_0)}{h} = -\frac{1}{2} f''(x_0)h$$

is proportional to h. For example, let $x > 0$ and $f(x) = x^x$ and suppose that we want to calculate the first derivative of f at point $x = 1$, i.e. $f'(1) = 1$. While the numerical derivative if $h = 0.01$ is

```
R> h <- 0.01
R> x0 <- 1
R> err <- 1 - ((x0 + h)^(x0 + h) - x0^x0)/h
R> err
```

```
[1] -0.01005033
```

which is of order h. If we take $h = 0.001$ we get

```
R> h <- 0.001
R> x0 <- 1
R> err <- 1 - ((x0 + h)^(x0 + h) - x0^x0)/h
R> err
```

```
[1] -0.001000500
```

so it is clearly proportional to h. Unfortunately, in the above approximations there are two kinds of errors. The first is the *truncation error*, which comes from the higher order terms in the Taylor expansion. The second is the *roundoff error* which is strictly related to the internal binary representation of numbers used by computers. Every number a in a computer is internally represented as $a(1 + \varepsilon)$ where, in single floating point precision, is of order $\sim 10^{-7}$. Suppose we choose $x_0 = 10.3$ and $h = 0.0001 = 10^{-5}$. Then, inside the computer both $x_0 = 10.3$ and $x_0 + h = 10.3001$ admit a representation which is exact only up to the ε error, i.e. each number x is represented internally as $x(1 + \varepsilon)$, thus $x(1 + \varepsilon)/h$ in our example is of order 10^{-2}. It is clear that this roundoff error propagates into the calculation of the numerical derivative as well. It is possible to reduce the effect of the roundoff error by appropriate optimal choices of h Press *et al.* (2007) and at the same time one can use it to refine the approximation of the derivative in order to reduce the truncation error. Here we explain how to reduce the truncation error. For example, if we consider the third order Taylor expansion of both $f(x_0 + h)$ and $f(x_0 - h)$, i.e.

$$f(x_0 + h) = f(x_0) + f'(x_0)h + \frac{1}{2}f''(x_0)h^2 + \frac{1}{3!}f'''(x_0)h^3 + O(h^3)$$

$$f(x_0 - h) = f(x_0) - f'(x_0)h + \frac{1}{2}f''(x_0)h^2 - \frac{1}{3!}f'''(x_0)h^3 + O(h^3)$$

we get that

$$f(x_0 + h) - f(x_0 - h) = 2f'(x_0)h + \frac{1}{3}f'''(x_0)h^3 + O(h^3)$$

and hence, we can introduce the symmetrized numerical derivative

$$f'(x_0) \simeq \frac{f(x_0 + h) - f(x_0 - h)}{2h} - \frac{1}{3}f'''(x_0)h^2$$

and the reminder is of order h^2. We can see it empirically as well.

```
R> h <- 0.01
R> x0 <- 1
R> err <- 1 - ((x0 + h)^(x0 + h) - (x0 - h)^(x0 - h))/(2 * h)
R> err

[1] -5.000083e-05

R> h^2

[1] 1e-04

R> h <- 0.001
R> x0 <- 1
R> err <- 1 - ((x0 + h)^(x0 + h) - (x0 - h)^(x0 - h))/(2 * h)
R> err

[1] -5e-07

R> h^2

[1] 1e-06
```

Still, the roundoff error is present, but the overall precision increases to h^2 without the need for explicit calculation of higher order derivatives, i.e. without the need to use explicit second order Taylor expansion. Another approach is the Richardson's extrapolation method (Richardson 1911, 1927), which is an acceleration method such that, for an approximation method of order h^p transforms it into one of order h^{p+1}. The symmetrized numerical derivative is not unrelated to this. The idea is the following: suppose we have an approximation method such that a certain quantity Q can be approximated as follows:

$$Q = Q(h) + ah^p + O\left(h^{p+1}\right)$$

where the term a is independent of h (Taylor expansion is one such method). The idea is to eliminate a taking two different values of h, say h_1 and h_2 so that the term of order h^p. Thus, for example

$$Q = Q(h_1) + ah_1^p + O\left(h_1^{p+1}\right)$$
$$Q = Q(h_2) + ah_2^p + O\left(h_2^{p+1}\right)$$

and, given that $ah_1^p h_2 = O\left(h^{p+1}\right)$ and $ah_2^p h_1 = O\left(h^{p+1}\right)$, we can write

$$(h_2^p - h_1^p)Q = h_2^p Q(h_1) - h_1^p Q(h_2) + O\left(h^{p+1}\right)$$
$$Q = \frac{h_2^p Q(h_1) - h_1^p Q(h_2)}{h_2^p - h_1^p} + O\left(h_2^{p+1}\right)$$

Thus, the Richardson's extrapolation is

$$Q_R = \frac{h_2^p Q(h_1) - h_1^p Q(h_2)}{h_2^p - h_1^p}.$$

The usual choice is to take $h_1 = h$ and $h_2 = \frac{1}{2}h$ so that

$$Q_R = \frac{\left(\frac{h}{2}\right)^p Q(h) - h^p Q\left(\frac{h}{2}\right)}{\left(\frac{h}{2}\right)^p - h^p} = \frac{2^p Q\left(\frac{h}{2}\right) - Q(h)}{2^p - 1}.$$

If we apply this to the previous example with $p = 1$ we obtain

```
R> h <- 0.001
R> x0 <- 1
R> err <- 1 - (2 * (x0 + h/2)^(x0 + h/2) - (x0 + h)^(x0 + h))
R> err
```

```
[1] 5.003753e-07
```

The symmetrized derivative is clearly an application of this method. The same idea can be iterated to increase the calculation of each term on the right-hand side of the naive Richardson's extrapolation to further increase the quality of the approximation. For our purposes the symmetrized derivative is enough, but in general one can obtain numerical derivatives in R using different packages. One option is the **numDeriv** package. The package has two main functions to calculate the numerical gradients grad and hessians hessian of multidimensional functions using the Richardson's extrapolation method. The use is quite simple and we present here an example, such as:

```
R> require(numDeriv)
R> f <- function(x) x^x
R> grad(f, x = 1)
```

```
[1] 1
```

```
R> grad(f, x = 1, method = "simple")
```

```
[1] 1.0001
```

Notice that the package **numDeriv** implements a refined version of the Richardson's extrapolation method. When we specify 'simple' in the argument method we get the naive result.

4.3   Root finding

Calibration of financial models or the methods of moments in statistics more or less correspond to the problem of finding the roots of a function. For one-dimensional function there are many specific algorithms while for the multidimensional case the problem of root finding is sometimes translated into an optimization (minimization/maximization) problem. We consider optimization in the next section; here we focus on the problem of root finding. If the function is a polynomial function, like

$$p(x) = a_0 + a_1 \cdot x + \ldots + a_n \cdot x^n$$

then $p(x) = 0$ always has n solutions (at least in the complex plane). For example, the second order polynomial $p(x) = a \cdot x^2 + b \cdot x + c$ has two well-known solutions:

$$x_i = \frac{-b \pm \sqrt{b^2 - 4ac}}{2a}, \quad i = 1, 2$$

which can be either real or complex, depending on the sign of $\Delta = b^2 - 4ac$. In R these solutions can be found using the `polyroot` function, specifying only the vector of coefficients $a_0, a_1, \ldots, a_n$. For example, suppose we want to find the solutions of

$$1 + 2x + x^2 = 0$$

which we already know are equal to -1 with multiplicity 2. Then, we use

```
R> polyroot(c(1, 2, 1))

[1] -1-0i -1+0i
```

For general functions, like the calculations of implied volatilities (see Section 6.6) we can use the `uniroot` function. This method requires the specification of the range of values in which to find the solutions and, optionally, the precision required and maximal number of iterations of the numerical method used. The method requires that the function for which we want to find the roots has opposite signs in the extremes of the specified interval. This allow the algorithm to improve the speed of convergence of the algorithm. For example, in the above case if we write something like

```
R> f <- function(x) 1 + 2 * x + x^2
R> uniroot(f, c(-2, 2))
```

R will raise an error, because that function `f` is always non-negative. But something like this:

```
R> f <- function(x) 2 * x + x^2 - 2
R> uniroot(f, c(-2, 2))

$root
[1] 0.732051

$f.root
[1] 4.266697e-07
```

```
$iter
[1] 7

$estim.prec
[1] 6.103516e-05
```

produces the correct result.

4.4 Numerical optimization

If the problem is the minimization or maximization of a function, as is required in maximum likelihood estimation, then the problem is relatively easy in the one-dimensional case, but quite defeating in the high-dimensional case. Even in the one-dimensional case, depending on the algorithm that is used and due to the iterative nature of most methods, it is not always clear if the maximum/minimum obtained is a global or a local stationarity point. The function `nlm` applies to the Newton-Raphson method, which is an iterative method based on the fact that, if x_{n+1} is the point such that $f(x_{n+1}) = 0$, then

$$f'(x_n) = \frac{f(x_n) - 0}{x_n - x_{n+1}}$$

and hence

$$x_{n+1} = x_n - \frac{f(x_n)}{f'(x_n)},$$

starting from an initial value and assuming that the gradient of the function f is known, is a relatively efficient method to look for zeros of a function. If the same idea is applied to the first derivative of f (and the second derivative of f is also known) then the Newton-Raphson method is fairly good at finding the points in which the first derivative is zero, i.e. the points of maximum and minimum of a f. But the function f should be fairly regular as well. For example,

```
R> f <- function(x) 2 * x + x^2 - 2
R> nlm(f, 0)

$minimum
[1] -3

$estimate
[1] -1

$gradient
[1] 1.000089e-06

$code
[1] 1

$iterations
[1] 1
```

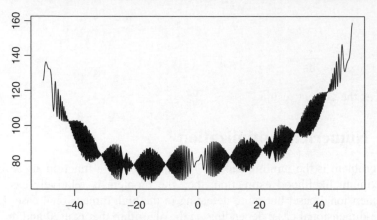

Figure 4.2   The 'wild' function is quite difficult to minimize.

finds the correct minimum. In the `nlm` command, the second argument is the initial value of the Newton-Raphson sequence. Consider now the so-called *wild* function in Figure 4.2:

```
R> f <- function(x) 10 * sin(0.3 * x) * sin(1.3 * x^2) + 1e-05 *
+    x^4+0.2 * x+80
```

Then, `nlm` finds only a local minimum

```
R> nlm(f, 0)$estimate
```

```
[1] -6.314295
```

```
R> nlm(f, 20)$estimate
```

```
[1] 19.93778
```

So, this is fast but unsatisfactory. An alternative solution in R is the function `optim` which is an interface to different optimization methods (for more information on each method, the reader should check the documentation page of the function `optim`). For example, the following code is able to find the real minimum of the wild function using the simulated annealing method

```
R> res <- optim(50, f, method = "SANN", control =
   list(maxit = 20000,
+     temp = 20, parscale = 20))
R> res
```

```
$par
[1] -15.81506
```

```
$value
[1] 67.4678
```

```
$counts
function gradient
   20000        NA

$convergence
[1] 0

$message
NULL

R> res <- optim(0, f, method = "SANN", control =
   list(maxit = 20000,
+      temp = 20, parscale = 20))
R> res$par

[1] -15.66174
```

which, as we see, produces a quite stable result compared to nlm. Both nlm
and optim accept functions f with vector arguments, so they also work in the
multidimensional case. The optim function also accepts constraints, which is
very useful in quasi-maximum likelihood estimation for diffusion processes (for
example, in the estimation of the parameters of the volatility in interest rates
models). The mle function indeed uses the optim function internally.

4.5   Simulation of stochastic processes

4.5.1   Poisson processes

The simulation of a homogenous Poisson process is quite easy, in that interval
between two Poissonian event are exponentially distributed. Hence the simulation
of exponential random variables with rate λ is enough. The only problem is that,
for a given time length, say $[0, T]$ it is not known a priori the number of events
that will occur but only the average number which is λT. Hence, a big number
of exponential events should be considered in order to have a full trajectory or,
otherwise use an iterative scheme. We present both schemes:

```
R> set.seed(123)
R> lambda <- 0.8
R> T <- 10
R> avg <- lambda * T
R> avg

[1] 8

R> t <- 0
R> N <- 0
R> k <- 0
R> continue <- TRUE
R> while (continue){
```

```
+        event <- rexp(1, lambda)
+        if (sum(t) + event < T){
+            k <- k + 1
+            N <- c(N, k)
+            t <- c(t, event)
+        }
+        else {
+            continue <- FALSE
+            t <- cumsum(t)
+            N <- c(N, k)
+            t <- c(t, T)
+        }
+ }
R> N
```

```
[1]   0  1  2  3  4  5  6  7  8  9 10 11 12 12
```

```
R> t
```

```
[1]    0.000000  1.054322  1.775084  3.436403  3.475875  3.546138
       3.941765
[8]    4.334549  4.516133  7.923928  7.960370  9.216408  9.816676
       10.000000
```

and the noninterative version

```
R> set.seed(123)
R> t <- cumsum(c(0, rexp(10 * avg, lambda)))
R> last <- which(t > T)[1]
R> t <- t[1:last]
R> t[last] <- T
R> N <- c(0, 1:(length(t) - 2), length(t) - 2)
R> N
```

```
[1]   0  1  2  3  4  5  6  7  8  9 10 11 12 12
```

```
R> t
```

```
[1]    0.000000  1.054322  1.775084  3.436403  3.475875  3.546138
       3.941765
[8]    4.334549  4.516133  7.923928  7.960370  9.216408  9.816676
       10.000000
```

```
R> plot(t, N, type = "s", main = "Poisson process",
       ylab = expression(N(t)),
+      xlim = c(0, T))
```

Figure 4.3 contains a plot of the simulated path. The simulation of a nonhomogeneous Poisson is a bit more complicated and is based on the acceptance/rejection or *thinning* method introduced by Lewis and Shedler (1979). We will refer to it as *Lewis method*. Optimized versions of it can be found in Ross (2006) or Ogata

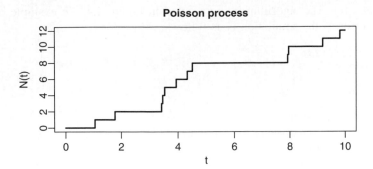

Figure 4.3   Simulation of homogenoeus Poisson process.

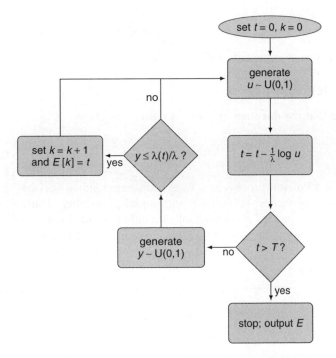

Figure 4.4   Algorithm to simulate inhomogeneous Poisson process.

(1981). The method is as follows: let $\lambda(t)$ be the intensity function and assume there exists a constant λ such that

$$\lambda(t) \leq \lambda, \quad 0 \leq t \leq T .$$

A homogeneous Poisson process with constant rate λ. When the event occurs at time t this event is considered as an event for the nonhomogeneous Poisson process with probability $\lambda(t)/\lambda$. The selected events are indeed Poissonian events with rate $\lambda(t)$. The algorithm is represented in Figure 4.4. An example of code which performs the simulation is as follows:

```
R> set.seed(123)
R> lambda <- 1.1
R> T <- 20
R> E <- 0
R> t <- 0
R> while (t < T) {
+       t <- t - 1/lambda * log(runif(1))
+       if (runif(1) < sin(t)/lambda)
+           E <- c(E, t)
+ }
R> plot(E, 0:(length(E) - 1), type = "s", ylim =
    c(-4, length(E)),
+       ylab = expression(N(t)), xlab = "t")
R> curve(-3 + sin(x), 0, 20, add = TRUE, lty = 2, lwd = 2)
```

The trajectory is represented in Figure 4.5.

4.5.2   Telegraph process

We remind that the telegraph process is defined as

$$X_t = x_0 + \int_0^t V_s ds, \quad V_t = V_0(-1)^{N_t}, \quad t > 0,$$

where N_t is a Poisson process and V_0 is a discrete random variable independent of N_t and taking values $+c$ and $-c$ with equal probability. Thus, simulation is very easy when a path of N_t is available we only need to simulate an initial value of the velocity V_0 and then discretize the above integral. More precisely, let τ_i be the length of the random interval between two subsequent Poisson events and for simplicity we set $\tau_{N_T} = T - \tau_{N_T-1}$. Then we can rewrite X_t as a random sum

$$X_t = x_0 + V_0 \sum_{i=1}^{N_T} \tau_i(-1)^i$$

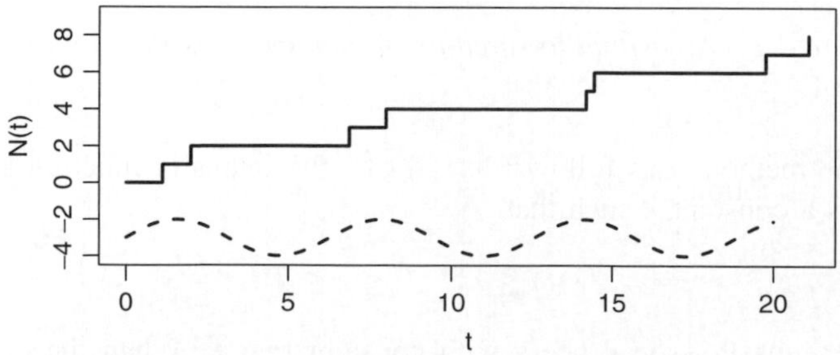

Figure 4.5 Simulation of nonhomogeneous Poisson process with rate $\lambda(t) = \sin(t)$.

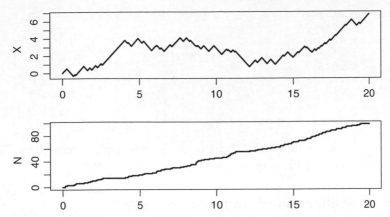

Figure 4.6   Simulation of the Telegraph process (up) based on the underline Poisson process (bottom).

```
R> T <- 20
R> lambda <- 5
R> avg <- lambda * T
R> t <- cumsum(c(0, rexp(10 * avg, lambda)))
R> last <- which(t > T)[1]
R> t <- t[1:last]
R> t[last] <- T
R> N <- c(0, 1:(length(t) - 2), length(t) - 2)
R> c <- 2
R> V0 <- sample(c(-c, +c), 1)
R> ds <- diff(t)
R> nds <- length(ds)
R> x0 <- 0
R> X <- c(x0, x0 + cumsum(V0 * ds * (-1)^(1:nds)))
```

We can now plot both trajectories together on the same graph

```
R> par(mfrow = c(2, 1))
R> par(mar = c(3, 4, 0.5, 0.1))
R> plot(t, X, type = "l")
R> plot(t, N, type = "s")
```

and the result is shown in Figure 4.6. For pure fun, we can see empirically what was explained in Section 3.12.3 on the convergence of the telegraph process to the Brownian motion. For this we need to set $\lambda = c^2$ and let $c \to \infty$, which in our case means a large value of c.

```
R> T <- 1
R> c <- 100
R> lambda <- c^2
R> avg <- lambda * T
R> t <- cumsum(c(0, rexp(10 * avg, lambda)))
```

```
R> last <- which(t > T)[1]
R> t <- t[1:last]
R> t[last] <- T
R> N <- c(0, 1:(length(t) - 2), length(t) - 2)
R> V0 <- sample(c(-c, +c), 1)
R> ds <- diff(t)
R> nds <- length(ds)
R> x0 <- 0
R> X <- c(x0, x0 + cumsum(V0 * ds * (-1)^(1:nds)))
```

Figure 4.7 shows a limiting trajectory for the telegraph process which looks qualitatively close to the path of a genuine Brownian motion.

4.5.3   One-dimensional diffusion processes

Let $X = \{X_t, t \geq 0\}$ a diffusion process solution to the stochastic differential equation

$$dX_t = b(t, X_t)dt + \sigma(t, X_t)dB_t \qquad (4.2)$$

with some initial condition X_0. We assume that the drift and diffusion coefficients satisfy the usual regularity assumptions as in Section 3.15.1 to ensure the existence of a solution of (4.2). The most used scheme to simulate stochastic differential equations is the Euler-Maruyama scheme. This method is far from being optimal but it is one of the few which is available also for the multidimensional case and for Lévy processes. We assume to discretize the interval $[0, T]$ into a grid of points $0 = t_0 < t_1 < \cdots < t_N = T$. This grid does not need to be regular. To simplify the exposition we use the notation $X(t_i) = X_i$. The approximated

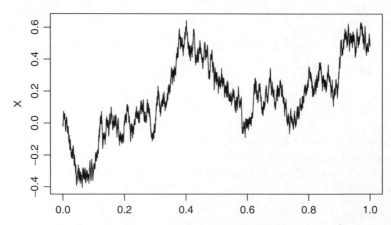

Figure 4.7   Limiting path of the telegraph process when $\lambda = c^2$ and $c \to \infty$. This trajectory looks qualitatively close to a Brownian motion path.

Euler-Maruyama solution is a new continuous stochastic process Y satisfying the iterative scheme

$$Y_{i+1} = Y_i + b(t_i, Y_i)(t_{i+1} - t_i) + \sigma(t_i, Y_i)(B_{i+1} - B_i), \quad i = 0, 1, \ldots, N - 1,$$
(4.3)

with $Y_0 = X_0$. Outside the points of the grid, the process is defined via linear interpolation. So, the implementation of this method only requires the simulation of the increments of the Brownian motion $B_{i+1} - B_i$ which are Gaussian distributed with zero mean and variance equal to the time interval $t_{i+1} - t_i$, so $B_{i+1} - B_i \sim \sqrt{t_{i+1} - t_i} \cdot N(0, 1)$. The algorithm is represented in Figure 4.9. The stability of this method depends on the regularity of the diffusion coefficient and on the distance between time points on the grid. Several modifications of this method exists in the one-dimensional case and mostly used is the Milstein (1978) scheme. The idea behind the method is to apply Itô's Lemma to obtain a second-order expansion and increase accuracy. The resulting approximating process Y satisfies the new iterative scheme

$$\begin{aligned} Y_{i+1} = \ & Y_i + b(t_i, Y_i)(t_{i+1} - t_i) + \sigma(t_i, Y_i)(B_{i+1} - B_i) \\ & + \frac{1}{2}\sigma(t_i, Y_i)\sigma_x(t_i, Y_i)\left\{(B_{i+1} - B_i)^2 - (t_{i+1} - t_i)\right\} \end{aligned}$$

where $\sigma_x(t, x)$ is the partial derivative of $\sigma(t, x)$ with respect to variable x. When the conditional distribution of $X_t | X_{t-\Delta_t}$ is available, it is possible to exploit the Markov property of the diffusion process and obtain an exact simulation of the path. Unfortunately, this is a rare event, so other approximation schemes exist. For an updated review, see Iacus (2008). The package **sde** implements most of the available methods in the literature via the `sde.sim` function. The package only simulates one-dimensional paths or M independent paths but not a real multidimensional process. The use is quite simple. Assume we want to simulate the stochastic process X solution to

$$dX_t = (1 - 2X_t)dt + \sqrt{(1 + X_t^2)}dB_t, \quad X_0 = 2, 0 \le T \le 1.$$

We first need to define R expressions to describe the drift and diffusion coefficients as functions and then pass them to the `sde.sim` function

```
R> require(sde)
R> set.seed(123)
R> b <- expression(1 - 2 * x)
R> s <- expression(sqrt(1 + x^2))
R> X <- sde.sim(X0 = 2, T = 1, drift = b, sigma = s)
```

The `sde.sim` function outputs an object of type `ts` which can be handled in R as such, e.g. it can plotted via `plot(X)`. The interface of `sde.sim` is quite flexible and by default, it implements the Euler-Maruyama scheme with predictor-corrector (see e.g. (Kloden *et al.* 2000)). If, for example, we want to use Milstein scheme, we can specify it in the following way:

```
R> X <- sde.sim(X0 = 2, T = 1, drift = b, sigma = s,
    method = "milstein")
```

For the few models for which the conditional distribution is known, one can specify the argument `model` and the additional vector of parameters `theta` which characterize the model. For example, the geometric Brownian motion or Black & Scholes model, which we analyze in detail in Chapter 5, satisfies the following stochastic differential equation:

$$dX_t = \theta_1 X_t dt + \theta_2 X_t dB_t, \quad X_0 = x_0.$$

Assume we want to simulate it for $\theta_1 = 1$, $\theta_2 = 0.5$ and $X_0 = 2$. Then we use the following code:

```
R> X <- sde.sim(X0 = 2, T = 1, model = "BS", theta = c(1, 0.5))
```

For more accurate simulation, one can increase the number of points in the grid by choosing the appropriate N, by default N = 100, or specifying the δ increment using argument `delta`. The next code makes use of the argument N and the plot can be seen in Figure 4.8.

```
R> set.seed(123)
R> X <- sde.sim(X0 = 2, T = 1, model = "BS", theta = c(1, 0.5),
+      N = 5000)

R> plot(X)
```

The `sde.sim` function implements Euler-Maruyama, both Milstein schemes, the KPSS method (Kloden *et al.* 1996), the Local Linearization Method for homogeneous (Ozaki 1985, 1992) and inhomogenous stochastic differential equations (Shoji 1995, 1998) and the Exact Algorithm (Beskos and Roberts 2005). For a detailed description of all arguments and options of the `sde.sim` function, please refer to the `man` page of the function in the package `pkg`.

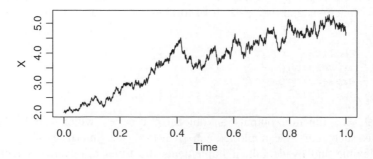

Figure 4.8   Simulation of the Black and Scholes model using `sde.sim`.

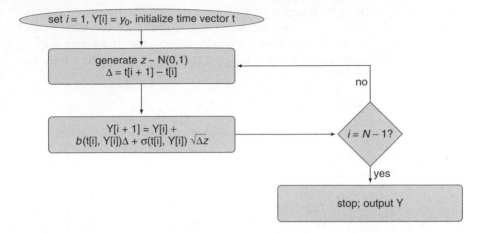

Figure 4.9   Algorithm to simulate stochastic differential equations.

4.5.4   Multidimensional diffusion processes

The package **yuima**[2] allows us to simulate multidimensional diffusion processes via Euler-Maruyama scheme. The choice between the **sde** and **yuima** packages depends on the specific application. The latter assumes the description of more structure and it is not limited to diffusion processes. With the **yuima** package the user first needs to construct the model before simulation. Let X be an m-dimensional diffusion process solution to

$$dX_t = b(t, X_t)dt + \sigma(t, X_t)dB_t, \quad t \in [0, T], \tag{4.4}$$

where B_t is an r-dimensional standard Brownian motion, i.e. a vector process $B_t = (B_t^1, B_t^2, \ldots, B_t^r)^\top$ where all components are independent standard Brownian motions. In the above $b(x, \theta)$ is an m-valued function and $\sigma(x, \theta)$ is an $m \times r$ matrix valued function. Let $\{t_i, i = 0, 1, \ldots, n\}$ be a grid of times such that $t_0 = 0$ and $t_n = T$ and let us denote X_{t_i} by X_i, $i = 0, 1, \ldots, n$. The Euler-Maruyama scheme approximates the solution of X_t with the discretization of the stochastic differential equation (4.4) in the following manner

$$X_i = X_{i-1} + b(t_{i-1}, X_{i-1})\Delta_i + \sigma(t_{i-1}, X_{i-1})\sqrt{\Delta}B_i \tag{4.5}$$

with $\Delta_i = t_i - t_{i-1}$ and $\Delta B_i = B_{t_i} - B_{t_{i-1}}$. Then, conditionally on the value of X_{i-1} the random variable X_i can be obtained by simulating the increments of the Brownian motion ΔB_i which is a multivariate Gaussian random variable with independent components. Theoretical properties of the Euler-Maruyama scheme can be found in Kloden and Platen (1999). Other approximation schemes for the one-dimensional case can be found in Iacus (2008). We now see some examples

---

[2] The package is available at http://R-Forge.R-Project.org/projects/yuima

on how to simulate multidimensional stochastic differential equations with the **yuima** package. Let us consider the following example of two-dimensional diffusion process $X = (X_t^1, X_t^2)$ driven by three independent Brownian motions $B_t = (B_t^1, B_t^2, B_t^3)$ and solution of the following system of stochastic differential equations

$$dX_t^1 = -3X_t^1 dt + dB_t^1 + X_t^2 dB_t^3$$
$$dX_t^2 = -(X_t^1 + 2X_t^2)dt + X_t^1 dB_t^1 + 3 dB_t^2.$$

In order to be described in a form which is suited for the **yuima** package, we rewrite it in matrix form as follows:

$$\begin{pmatrix} dX_t^1 \\ dX_t^2 \end{pmatrix} = \begin{pmatrix} -3X_t^1 + 0X_t^2 \\ -X_t^1 - 2X_t^2 \end{pmatrix} dt + \begin{bmatrix} 1 & 0 & X_t^2 \\ X_t^1 & 3 & 0 \end{bmatrix} \begin{pmatrix} dB_t^1 \\ dB_t^2 \\ dB_t^3 \end{pmatrix}.$$

Now we prepare the model using the `setModel` constructor function:

```
R> require(yuima)
R> sol <- c("x1", "x2")
R> b <- c("-3*x1", "-x1-2*x2")
R> s <- matrix(c("1", "x1", "0", "3", "x2", "0"), 2, 3)
R> model <- setModel(drift = b, diffusion = s,
    solve.variable = sol)
```

The vector `sol` defines the variables which are used to solve numerically the stochastic differential equations. If not specified, it is assumed to be variable `x` in each equation. Similarly for the time variable which, by default, is assumed to be `t`. Now we are ready to simulate the process using the generic R function `simulate` in the following way:

```
R> set.seed(123)
R> X <- simulate(model, n = 1000)

YUIMA: 'delta' (re)defined.
```

now `x` contains the simulated path. Notice that `x` is not a simple `ts` object but a `s4` object of class `yuima-data`. The internal storage is of class `zoo` and can be extracted from `x` using the `get.zoo.data` method. Clearly, other methods exist, such as, for instance the `plot` method. The output of `plot` is given in Figure 4.10.

```
R> plot(X, plot.type = "single", lty = c(1, 3), ylab = "X")
```

4.5.5 Lévy processes

As we have seen in Section 3.18, Lévy processes are characterized by the fact that they have independent increments. This makes the simulation of such processes

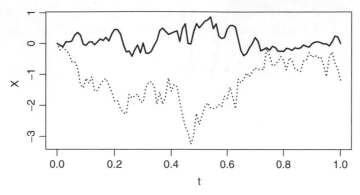

Figure 4.10   Simulation of two-dimensional diffusion process using `yuima`
package.

quite easy in some cases when it is possible to simulate random numbers from
the distribution of the increments. One particular process which will be used in
Chapter 8 for option pricing is the exponential Lévy model

$$S_t = S_0 e^{Z_t}$$

so that the log-returns of the price process S_t, i.e. $\log(S_{t+s}/S_t) = Z_{t+s} - Z_t$ are
distributed as the increments of the Lévy process Z_t. Thus, in this section we
show how to simulate several types of Lévy processes and in Chapter 5.4 we
describe some statistical techniques to estimate the parameters of these processes
from financial data. The **yuima** package is one of the preferred way to simulate
one or multidimensional Lévy paths but also stochastic differential equations
with jumps, which we consider separately in the next section. The package can
simulate Lévy processes with the following classes of increments:

- gamma, bilateral gamma and multivariate normal gamma studied in the
  context of option pricing in Küchler and Tappe (2008a,b, 2009);

- inverse Gaussian and multivariate normal inverse Gaussian, originally
  introduced in Tweedie (1947) and studied in the context of finance by,
  e.g. Barndorff-Nielsen and Shepard (2001);

- univariate stable (possibly skewed) and univariate exponentially tempered
  stable distributions.

Notice that some processes, like the more famous Variance Gamma process, are
special cases of the above. For more information on each of the distributions
it is possible to refer to the manual of the **yuima** package; here we present
some example of use. All the above Lévy processes can be simulated exactly
using the Euler approximation. In the **yuima** the approach used is to transform a
Wiener process into a pure jump Lévy process using the following approach: let

$Z_\Delta = Z_{t+\Delta} - Z_t$, the increment of the Lévy process. Then, it is possible to express Z_Δ as a normal variance mean mixture with some subordinator τ_Δ

$$Z_\Delta \sim \mu\Delta + \tau_\Delta \Lambda\beta + (\tau_\Delta \Lambda)^{\frac{1}{2}}\zeta$$

where

- ζ is a d-dimensional Gaussian random variable independent of τ_Δ;

- β is a d-vector related to the skewness of the distribution;

- μ is a d-dimensional vector of location parameters;

- Λ is a $d \times d$ symmetric and positive definite matrix which represents the matrix of scale parameters of the distribution;

- $(\tau_\Delta \Lambda)^{\frac{1}{2}}$ is the positive definite square root of $\tau_\Delta \Lambda$.

To avoid redundancy, it is necessary to set $\det(\Lambda) = 1$ and, in the one-dimensional case of $d = 1$, then $\Lambda = 1$ so that the mixture takes the form:

$$Z_\Delta \sim \mu\Delta + \tau_\Delta \Lambda\beta + \sqrt{\tau_\Delta}\zeta.$$

For example, consider the multivariate normal gamma distributed increments, i.e.

$$Z_\Delta \sim N\Gamma_d(\lambda\Delta, \alpha, \beta, \mu\Delta, \Lambda)$$

with $\lambda > 0$, $\alpha > 0$, $\alpha^2 > \beta^\top \Lambda\beta$, $\det(\Lambda) = 1$ and d-dimensional density

$p(z; \lambda\Delta, \alpha, \beta, \mu\Delta, \Lambda)$

$$= \frac{e^{\beta^\top(z-\mu\Delta)}(\alpha^2 - \beta^\top \Lambda\beta)^{\lambda\Delta} K_{\lambda\Delta-d/2}\left(\alpha Q(z; \mu\Delta, \Lambda)\right)\{Q(z; \mu\Delta, \Lambda)\}^{\lambda\Delta-d/2}}{\Gamma(\lambda\Delta)\pi^{d/2}2^{d/2+\lambda\Delta-1}\alpha^{\lambda\Delta-d/2}},$$

where $Q(z; \mu\Delta, \Lambda) = \sqrt{(z - \mu\Delta)^\top \Lambda^{-1}(z - \mu\Delta)}$ and K_λ denotes the modified Bessel function of the third kind with index λ which in R can be obtained with `besselK`. In this case, the algorithm is as follows:

- generate τ_Δ with `rgamma(1, λΔ, (α² − βᵀΛβ)/2)`;

- generate ζ with `rmvnorm`;

- set $Z_\Delta = \mu\Delta + \beta\tau_\Delta\Lambda\sqrt{\tau_\Delta\Lambda}\zeta$;

then Z_Δ is distributed as $N\Gamma_d(\lambda\Delta, \alpha, \beta, \mu\Delta, \Lambda)$. In the next section we see that these processes can be simulated as special cases of stochastic differential equations with jumps.

4.5.6   Simulation of stochastic differential equations with jumps

We have introduced stochastic differential equations with jumps in Section 3.18.7 of the form:

$$dX_t = a(X_t)dt + b(X_t)dW_t + \int_{|z|>1} c(X_{t-}, z)\mu(dt, dz)$$

$$+ \int_{0<|z|\le 1} c(X_{t-}, z)\{\mu(dt, dz) - \nu(dz)dt\},$$

where μ is the random measure associated with jumps of X,

$$\mu(dt, dz) = \sum_{s>0} \mathbf{1}_{\{\Delta Z_s \neq 0\}} \delta_{(s, \Delta Z_s)}(dt, dz),$$

and δ denotes the Dirac measure. The process Z_t is the driving pure-jump Lévy process of the form:

$$Z_t = \int_0^t \int_{|z|\le 1} z\{\mu(ds, dz) - \nu(dz)ds\} + \int_0^t \int_{|z|>1} z\mu(ds, dz).$$

To some extent, the **yuima** package covers most of the cases described by the above formulas. There are too many aspects which is not possible to discuss here and these includes the case of processes with finite or infinite activity, pure jump (as seen in the previous section) or compound type specification. The reader is invited to follow the developments of the Yuima Project and check the latest documentation available. Here we just show how simple it is to simulate stochastic differential equations with jumps in two simple situations. For example, suppose one wants to simulate Z_t which is a Compound Poisson Process (i.e. jumps follow some distribution, e.g. Gaussian). Then it is possible to consider the following SDE with jumps:

$$dX_t = a(X_t)dt + b(X_t)dB_t + dZ_t.$$

Let us set an intensity of $\lambda = 10$ in the Compound Poisson Process and choose standard Gaussian jumps. In this case, it is possible to extend the basic `yuima` model to allow for jumps with compound Poisson specification using the flag `CP` in the argument `measure.type`. For example, if we want to simulate the process

$$dX_t = -\theta X_t dt + \sigma dB_t + Z_t$$

where θ and σ are some parameters, we proceed as follows:

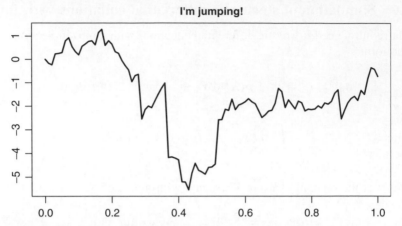

Figure 4.11   Simulation of stochastic differential equations with jumps using the compound Poisson specification.

```
R> require(yuima)
R> modCP <- setModel(drift = c("-theta*x"), diffusion = "sigma",
+      jump.coeff = "1", measure = list(intensity = "10",
       df = list("dnorm(z, 0, 1)")),
+      measure.type = "CP", solve.variable = "x")
```

and if we want to simulate it we simply proceed as in the diffusion case

```
R> set.seed(123)
R> X <- simulate(modCP, true.p = list(theta = 1, sigma = 3),
    n = 1000)

YUIMA: 'delta' (re)defined.
```

which we can plot in the usual way with

```
R> plot(X, main = "I'm jumping!")
```

and obtain the plot in Figure 4.11. If, instead, we want to simulate specifying the Lévy increments we use the switch code in argument measure.type. For example, suppose we want to simulate a pure jump process with drift but without the Gaussian component, e.g. the following Ornstein-Uhlenbeck process of Lévy type solution to the stochastic differential equation

$$dX_t = -x dt + dZ_t.$$

We proceed as follows:

```
R> modPJ <- setModel(drift = "-x", xinit = 1, jump.coeff = "1",
+      measure.type = "code", measure = list(df =
    "rIG(z, 1, 0.1)"))
```

```
YUIMA: Solution variable (lhs) not specified. Trying to
    use state variables.

R> set.seed(123)
R> Y <- simulate(modPJ, Terminal = 10, n = 10000)

YUIMA: 'delta' (re)defined.
```

and we can see its simulated path in Figure 4.12. It is also possible to specify the jump coefficient $c(\cdot)$ using the argument `jump.coeff` in the `setModel` function so the complete model of the form $dX_t = a(X_t)dt + b(X_t)dW_t + c(X_t)dZ_t$ can be fully specified. As mentioned, the number of Lévy processes which can be simulated with the **yuima** package is growing constantly and it is not worth listing the different options here. A nice review of simulation schemes for Lévy processes can be found in Schoutens (2003) and Cont and Tankov (2004).

4.5.7   Simulation of Markov switching diffusion processes

For simplicity in this section we consider the one-dimensional case only. Consider the Markov switching diffusion process X solution to the stochastic differential equation

$$dX_t = f(X_t, \alpha_t)dt + g(X_t, \alpha_t)dB_t, \quad X_0 = x_0, \alpha(0) = \alpha, \qquad (4.6)$$

where $\{B_t, t \geq 0\}$ is the standard Brownian motion and α_t is a finite-state Markov chain in continuous time with state space S and infinitesimal generator $Q = [q_{ij}]$ (see Definition 3.4.17), i.e. the entries of Q are such that $q_{ij} \geq 0$ for $i \neq j$, $\sum_{j \in S} q_{ij} = 0$ for each $i \in S$. Remember that the transition probability matrix, at

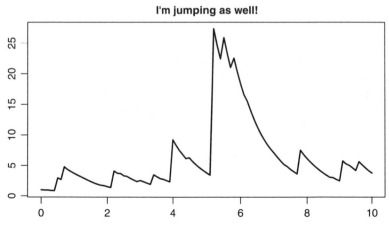

Figure 4.12   Simulation of an Ornstein-Uhlenbeck process of Lévy type without Gaussian component.

any given time instant and lag t is given by the matrix exponential

$$P(t) = \exp(Qt).$$

The drift and diffusion coefficient are such that Equation (4.6) has a unique solution in distribution for each initial condition. For any $\epsilon > 0$, define $\alpha_n = \alpha_{\epsilon n}$, this sequence $\{\alpha_n\}$ is called the ϵ-skeleton of the continuous-time Markov chain α_t (see Chung 1967). It can be shown that α_n is a discrete-time Markov chain with one-step transition probability matrix

$$P = [p_{ij}] = \exp(\epsilon Q).$$

Notice that while $P(t)$ above is time-dependent for α_t, the matrix P is not and the Markov chain α_n has stationary transition probabilities. We use the notation $\Delta t = \epsilon$ according to previous sections. As in the diffusion case, we can apply the Euler-Maruyama scheme to Equation (4.6) to obtain the following discretization

$$X_{n+1} = X_n + f(x_n, \alpha_n)\Delta t + g(x_n, \alpha_n)\sqrt{\Delta t}\, Z_i$$

with $Z_i \sim N(0, 1)$. This scheme converges weakly to the solution of (4.6) as $\Delta t \to 0$ as proved in Yin et al. (2005). Although we did not dedicate a specific section for the simulation of Markov chains, we now show a generic algorithm which we will use to simulate α_n. We keep this function simMarkov separate so one can use it individually.

```
R> simMarkov <- function(x0, n, x, P) {
+       mk <- numeric(n + 1)
+       mk[1] <- x0
+       state <- which(x == x0)
+       for (i in 1:n) {
+           mk[i + 1] <- sample(x, 1, prob = P[state, ])
+           state <- which(x == mk[i + 1])
+       }
+       return(ts(mk))
+ }
```

This function needs an initial state x0, the number of new observations to simulate n, a vector of states x representing the state space and a transition matrix P. The rows of the transition matrix are associated with the elements of the state space vector x. The result is a time series of class ts. Suppose we want to simulate a Markov chain with state space $S = \{1, 2, 3\}$ and transition matrix P

$$P = \begin{bmatrix} 0.1 & 0.1 & 0.8 \\ 0.5 & 0.2 & 0.3 \\ 0.3 & 0.3 & 0.4 \end{bmatrix}$$

then we will use the following R code

```
R> P <- matrix(c(0.1, 0.5, 0.3, 0.1, 0.2, 0.3, 0.8, 0.3, 0.4), 3,
+       3)
```

```
R> set.seed(123)
R> X <- simMarkov(1, 10, 1:3, P)
R> plot(X, type = "s")
```

the plot of the trajectory is given in Figure 4.13. For continuous time Markov
chains one can use the package **msm**, but we will only use this package to
calculate the exponential matrix later. Next follows the code to simulate the
diffusion process with Markov switching in Equation (4.6). This code is not
very sophisticated, but contains the elementary blocks to construct an efficient
simulator.

```
R> simMSdiff <- function(x0, a0, S, delta, n, f, g, Q){
+       require(msm)
+       P <- MatrixExp(delta * Q)
+       alpha <- simMarkov(a0, n, S, P)
+       x <- numeric(n + 1)
+       x[1] <- x0
+       for (i in 1:n) {
+           A <- f(x[i], alpha[i]) * delta
+           B <- g(x[i], alpha[i]) * sqrt(delta) * rnorm(1)
+           x[i + 1] <- x[i] + A + B
+       }
+       ts(x, deltat = delta, start = 0)
+ }
```

The function `simMSdiff` has arguments `x0` the initial value of the process, `a0`
the starting value of the Markov chain α_t, `S` the state space of α_t, `delta` the
discretization step, `n` the number of new observations to generate, `f` and `g` two
R functions of two arguments of the form $f(x, a)$ and $g(x, a)$ where x is the
state of X_t and a the value of the Markov chain and `Q` the generator of the
continuous time Markov chain. Notice that the code is very elementary and it is
up to the user to correctly specify the functions and the matrixes involved. Just

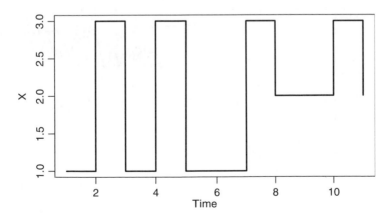

Figure 4.13   Example of plot of a Markov chain with three states simulated with
`simMarkov`.

as an example, we try to simulate a nonlinear Markov switching diffusion (4.6) where $f(\cdot, \cdot)$ and $g(\cdot, \cdot)$ are as follows:

$$f(x, a) = \begin{cases} (1 + \sin(x))x, & \text{if } a = 0, \\ (1 - \cos(x))x, & \text{if } a = 1, \end{cases}$$

$$g(x, a) = \begin{cases} \sigma_0 \cdot x = 0.5 \cdot x, & \text{if } a = 0, \\ \sigma_1 \cdot x = 0.2 \cdot x, & \text{if } a = 1. \end{cases}$$

and the generator Q is the following matrix:

$$Q = \begin{bmatrix} -6 & 6 \\ 12 & -12 \end{bmatrix}$$

with an initial values $(X_0, \alpha_0) = (5, 0)$. We set the discretization step $\Delta = 1/n$ with $n = 1000$. So, we proceed as follows:

```
R> Q <- matrix(c(-6, 12, 6, -12), 2, 2)
R> n <- 1000
R> s0 <- 0.5
R> s1 <- 0.2
R> f <- function(x, a) ifelse(a == 0, (1 + sin(x)) * x,
   (1 - cos(x)) *
+      x)
R> g <- function(x, a) ifelse(a == 0, s0 * x, s1 * x)
R> set.seed(123)
R> X <- simMSdiff(x0 = 5, a0 = 0, S = 0:1, delta = 1/n, n = 1000,
+      f, g, Q)
R> plot(X)
```

The simulated path is shown in Figure 4.14.

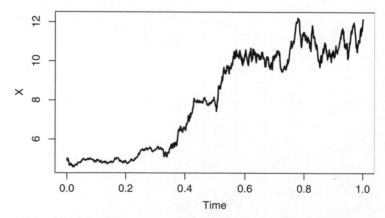

Figure 4.14   Example of trajectory of a Markov switching diffusion process using simMSdiff.

4.6   Solution to exercises

Solution 4.1 (to Exercise 4.1) *Remind that the equation of the semi-circle is* $g(x) = \sqrt{1 - x^2}$, $x \in [-1, 1]$. *Given that the area of the circle is* $\pi * r^2$ *where* r *is the radius* ($r = 1$ *in this case*), *we can calculate* π *as* $\pi = 2 * A * S_n/n$ *where* S_n *and* A *are as in Section 4.1.1. Then*

```
R> set.seed(123)
R> g <- function(x) sqrt(1 - x^2)
R> a <- -1
R> b <- +1
R> c <- 0
R> d <- 2
R> A <- (b - a) * (d - c)
R> n <- 1e+05
R> x <- runif(n, a, b)
R> y <- runif(n, c, d)
R> 2 * A * sum(y<g(x))/n

[1] 3.14272

R> 2 * integrate(g, a, b)$value

[1] 3.141593

R> pi

[1] 3.141593
```

4.7   Bibliographical notes

Simulation for stochastic differential equations driven by Brownian motion can be found in the classical book of Kloden and Platen (1999), Kloden *et al.* (2000) and, in view of inference for these models, in Iacus (2008). For jump processes the literature is sparse, but we can mention at least Platen and Bruti-Liberati (2010) which consider a general approach to simulation of jump processes with a view on finance applications; the monographs of Cont and Tankov (2004), Schoutens (2003), among the others, also contain section on simulation for these processes. For point processes one can found elements in Ripley (2006) and Ross (2006). Books on numerical methods with finance in view are, for example, Jäckel (2002) and Glasserman (2004).

References

Barndorff-Nielsen, O. and Shepard, N. (2001). Non-Gaussian Ornstein-Uhlenbeck-based models and some of their uses in financial economics. *Journal of the Royal Statistical Socieciety, Ser. B* **63**, 167–241.

Beskos, A. and Roberts, G. (2005). Exact simulation of diffusions. *Ann. Appl. Probab.* **4**, 2422–2444.

Chung, K. (1967). *Markov Chains with Stationary Transition Probabilities, 2nd ed.* Springer-Verlag, New York.

Cont, R. and Tankov, P. (2004). *Financial Modelling with Jump Processes.* Chapman & Hall/CRC, Boca Raton.

Glasserman, P. (2004). *Monte Carlo Methods in Financial Engineering.* Springer, New York.

Iacus, S. (2008). *Simulation and Inference for Stochastic Differential Equations. With R Examples.* Springer Series in Statistics, Springer, New York.

Jäckel, P. (2002). *Monte Carlo Methods in Finance.* John Wiley & Sons, Ltd, Chichester, England.

Kloden, P. and Platen, E. (1999). *Numerical Solution of Stochastic Differential Equations.* Springer, New York.

Kloden, P., Platen, E., and Schurz, H. (2000). *Numerical Solution of SDE through Computer Experiments.* Springer, Berlin.

Kloden, P., Platen, E., Schurz, H., and Sørensen, M. (1996). On effect of discretization on estimation of the drift parameters for diffusion processes. *J. Appl. Probab.* **33**, 4, 1061–1073.

Küchler, U. and Tappe, S. (2008a). Bilateral gamma distributions and processes in financial mathematics. *Stochastic Processes and their Applications* **118**, 2, 261–283.

Küchler, U. and Tappe, S. (2008b). On the shapes of bilateral gamma densities. *Statistics & Probability Letters* **78**, 15, 2478–2484.

Küchler, U. and Tappe, S. (2009). Option pricing in bilateral gamma stock models. http://www.math.ethz.ch/tappes/Bilateral_Pricing.pdf 1–13.

Lewis, A. and Shedler, G. (1979). Simulation of nonhomogeneous poisson processes by thinning. *Naval Res. Logistics Quart.* **26**, 3, 403–413.

Milstein, G. (1978). A method of second-order accuracy integration of stochastic differential equations. *Theory Probab. Appl.* **23**, 396–401.

Ogata, Y. (1981). On Lewis' simulation method for point processes. *IEEE Trans. Inf. Theory* **27**, 1, 23–31.

Ozaki, T. (1985). Non-linear time series models and dynamical systems. *Handbook of Statistics, Hannan, E.J. ed.* **5**, 25–83.

Ozaki, T. (1992). A bridge between nonlinear time series models and nonlinear stochastic dynamical systems: a local linearization approach. *Stat. Sinica* **2**, 25–83.

Platen, E. and Bruti-Liberati, N. (2010). *Numerical Solution of Stochastic Differential Equations with Jumps in Finance.* Springer, New York.

Press, W., Teukolsky, S., Vetterling, W., and Flannery, B. (2007). *Numerical Recipes 3rd Edition: The Art of Scientific Computing.* Cambridge University Press, Hong Kong.

Richardson, L. F. (1911). The approximate arithmetical solution by finite differences of physical problems including differential equations, with an application to the stresses in a masonry dam. *Philosophical Transactions of the Royal Society of London, Series A* **210**, 307–357.

Richardson, L. F. (1927). The deferred approach to the limit. *Philosophical Transactions of the Royal Society of London, Series A* **226**, 299–349.

Ripley, B. (2006). *Stochastic Simulation*. Wiley-Interscience.

Ross, S. (2006). *Simulation. Fourth Edition*. Academic Press, Boston.

Schoutens, W. (2003). *Lévy Processes in Finance*. John Wiley & Sons, Ltd, Chichester.

Shoji, L. (1995). *Estimation and inference for continuous time stochastic models*. PhD thesis, Institute of Statistical Mathematics, Tokyo.

Shoji, L. (1998). A comparative study of maximum likelihood estimators for nonlinear dynamical system models. *Int. J. Control* **71**, 3, 391–404.

Tweedie, M. (1947). Functions of a statistical variate with given means, with special reference to laplacian distributions. *Proceedings of the Cambridge Philosophical Society* **43**, 41–49.

Yin, G., Mao, X., and Yin, K. (2005). Numerical approximation of invariant measures for hybrid diffusion systems. *IEEE Transactions on Automatic Control* **50**, 7, 934–946.

5

Estimation of stochastic models for finance

In this chapter we review some of the basics models used to model asset prices, or market indexes, interest rates, etc. These processes may have different structures and we start from the very fundamental model of asset dynamics of the Black and Scholes (1973) and Merton (1973) model.

5.1 Geometric Brownian motion

Let us denote by $\{S_t = S(t), t \geq 0\}$ a stochastic process which represents the value of an asset (the asset price) at time $t \geq 0$. The process S is called *geometric Brownian motion* process if it is the solution to the following stochastic differential equation:

$$\mathrm{d}S_t = \mu S_t \mathrm{d}t + \sigma S_t \mathrm{d}W_t \tag{5.1}$$

with some initial value S_0, and some constants μ, $\sigma > 0$. As seen in Chapter 1, the fundamental idea was to describe the returns of S_t in the interval $[t, t + \mathrm{d}t)$ in terms of two components:

$$\frac{S_{t+\mathrm{d}t} - S_t}{S_t} = \frac{\mathrm{d}S_t}{S_t} = \text{deterministic contribution} + \text{stochastic contribution}$$

where the deterministic contribution is assumed to be proportional to time, i.e. $\mu \mathrm{d}t$, and the stochastic part is assumed to be of Gaussian type, i.e. $\sigma \mathrm{d}W_t$. A simple rewriting of

$$\frac{\mathrm{d}S_t}{S_t} = \mu \mathrm{d}t + \sigma \mathrm{d}W_t,$$

Option Pricing and Estimation of Financial Models with R, First Edition. Stefano M. Iacus.
© 2011 John Wiley & Sons, Ltd. Published 2011 by John Wiley & Sons, Ltd.

gives the stochastic differential Equation (5.1). The constant μ represent the *drift* of the process and σ^2 is called *volatility*. We will see that, in more advanced models, but μ and σ can be deterministic functions depending on the time t or the state of the process S_t or both or some other kind of stochastic processes such that the solution of the corresponding stochastic differential equation exists. Exercise 3.14 proves that

$$X_t = X_0 \exp\left\{\int_0^t \left(b(s) - \frac{1}{2}\sigma^2(s)\right)ds + \int_0^t \sigma(s)dB_s\right\}$$

is the solution of the stochastic differential equation

$$dX_t = b(t)X_t dt + \sigma(t)X_t dB_t$$

of which Equation (5.1) is just a particular case. So, the explicit solution of the geometric Brownian motion is

$$S_t = S_0 \exp\{\alpha t + \sigma B_t\}$$

where $\alpha = \mu - \frac{1}{2}\sigma^2$. Notice that, if $\sigma = 0$, i.e. absence of the stochastic noise, Equation (5.1) becomes the simple ordinary differential equation

$$ds_t = \mu s_t dt$$

which we can easily solve by first rewriting it as follows:

$$\frac{ds_t}{dt} = \mu s_t$$

which, in the limit as $dt \to 0$, is equivalent to writing:

$$\frac{d}{dt}s_t = \mu s_t$$

or

$$\frac{s_t'}{s_t} = \frac{d}{dt}\log s_t = \mu$$

whose solution is

$$\log s_t - \log s_0 = \int_0^t \mu ds = \mu(t - 0)$$

and, after integration, we get

$$s_t = s_0 \exp\{\mu t\}.$$

The main difference between the stochastic

$$S_t = S_0 \exp\left\{\mu t - \frac{1}{2}\sigma^2 t + \sigma B_t\right\}$$

and deterministic version is not only the stochastic term σB_t but also the compensating factor $\frac{1}{2}\sigma^2 t$ which is the contribution of the Itô formula and makes the stochastic version S_t a well-behaved stochastic process.

Two hypotheses which are always implicit in the Black and Scholes (1973) and Merton (1973) model are the following:

(i) the past history is entirely reflected in the present value of the asset and does not contain information from the future;

(ii) the market reacts immediately to each new information on the asset.

Hence, what can be modeled is the effect of new information on the asset price, i.e. the increments of the price process or the innovations. Indeed, in the previous derivation of the geometric Brownian motion we impose Gaussian returns. Processes are usually supposed to be Markovian as diffusion processes are.

5.1.1 Properties of the increments

From the explicit solution of S_t it is clear that the geometric Brownian motion is nothing but the exponential of Brownian motion which in turn, for each t, is a Gaussian random variable. In general, if $X \sim N$ and c is some constant, then ce^X has *log-normal* distribution, which means that $\log X \sim N$. For this reason, the geometric Brownian motion is well suited for financial applications where log-returns are more common that simple returns. To see this, let us denote by $S_i = S(t_i)$ the asset price at time $t_i \in [0, T]$, with T some fixed time horizon, $t_0 = 0 < t_1 < t_2 < \cdots < t_{n-1} < t_n = T$. The returns Y_i can be simply written as

$$Y_i = \frac{S_i - S_{i-1}}{S_{i-1}} = \frac{S_i}{S_{i-1}} - 1, \quad i = 1, 2, \ldots, n.$$

Now, consider the Taylor expansion of $\log(1 + z)$

$$\log(1 + z) = z - \frac{1}{2}z^2 + o(z^2) \simeq z$$

if z is very small. Now, let $X_i = 1 + Z_i = S_i/S_{i-1} = 1 + Y_i$, i.e. $Z_i = Y_i$. Then

$$\log X_i = \log(1 + Z_i) = \log(1 + Y_i) \simeq Y_i$$

which means that the true returns Y_i are almost identical to the *log-returns*

$$X_i = \log S_i - \log S_{i-1}, \quad i = 1, 2, \ldots, n.$$

These new values have very appealing properties. Let us denote with $\Delta t = t_i - t_{i-1}$, $i = 1, \ldots, n$, the time increment between two observations (usually $\Delta t = 1/252$ if the time reference is the year and S_i represent daily data). The log-returns X_i are *additive*, indeed

$$X_i + \cdots + X_{i+n-1} = \log\left(\frac{S_{i+n-1}}{S_{i-1}}\right) = \log S_{i+n-1} - \log S_{i-1}$$

while the Y_i are not additive. Now, given that S_t is a geometric Brownian motion, we have that

$$X_i = X(t_i) = \log\left(\frac{S(t_i)}{S(t_{i-1})}\right)$$

$$= \alpha \Delta t + \sigma(B(t_i) - B(t_{i-1})) \sim \alpha \Delta t + \sigma \sqrt{\Delta t} N(0, 1)$$

therefore

$$X_i \sim N(\alpha \Delta t, \sigma^2 \Delta t).$$

Moreover, $X(t_i)$ is independent of $X(t_j)$ for $i \neq j$ because they depend on the increments of the Brownian motion on disjoint intervals.

5.1.2   Estimation of the parameters

So if we consider the values X_i, $i = 1, \ldots, n$, they constitute a sample of i.i.d. random variables with common distribution $N(\alpha \Delta t, \sigma^2 \Delta t)$. So, from Chapter 2 we know how to obtain unbiased and efficient estimators of the mean and the variance of a sample of i.i.d. random variables. For example, $\bar{X}_n = \frac{1}{n} \sum_{i=1}^{n} X_i$ is an estimator of $\alpha \Delta t$ and $\frac{1}{n-1} \sum_{i=1}^{n} \left(X_i - \bar{X}_n\right)^2$ is an estimator of $\sigma^2 \Delta t$, thus

$$\hat{\alpha}_n = \frac{1}{\Delta t} \frac{1}{n} \sum_{i=1}^{n} X_i = \frac{\bar{X}_n}{\Delta t}$$

$$\hat{\sigma}_n^2 = \frac{1}{\Delta t} \frac{1}{n-1} \sum_{i=1}^{n} \left(X_i - \bar{X}_n\right)^2 = \frac{\bar{S}_n^2}{\Delta t}$$

from which we obtain immediately the following estimator of μ

$$\hat{\mu}_n = \hat{\alpha}_n + \frac{1}{2}\hat{\sigma}_n^2.$$

In the next code we simulate a trajectory of the geometric Brownian motion and we fit the parameters in an elementary way. To this end, we use the sde.sim function specifying the option BS in the argument model, with a vector of parameters theta = ($\mu = 1, \sigma = 0.5$), over a time interval $[0, T = 100]$, with 10000 equally spaced observations.

```
R> require(sde)
R> set.seed(123)
R> sigma <- 0.5
R> mu <- 1
R> S <- sde.sim(X0 = 100, model = "BS", theta = c(mu, sigma),
    T = 100,
+      N = 10000)
```

We then create the log-returns using `diff(log(S))` and estimate the parameters

```
R> X <- diff(log(S))
R> sigma.hat <- sqrt(var(X)/deltat(S))
R> alpha.hat <- mean(X)/deltat(S)
R> mu.hat <- alpha.hat + 0.5 * sigma.hat^2
R> sigma.hat
```

```
[1] 0.4993183
```

```
R> mu.hat
```

```
[1] 0.9878009
```

5.2   Quasi-maximum likelihood estimation

For diffusion process solutions of stochastic differential equations and observed at discrete times, it is possible to define the likelihood function making use of the Markov property. We consider multidimensional diffusion process and then particularize it to the one-dimensional case. Let X be an m-dimensional diffusion process solution to

$$dX_t = b(X_t, \theta)dt + \sigma(X_t, \theta)dB_t, \quad t \in [0, T], \tag{5.2}$$

where B_t is an r-dimensional standard Brownian motion, $b(x, \theta)$ is an m-valued function, and $\sigma(x, \theta)$ is an $m \times r$ matrix valued function. The parameter θ belongs to Θ which is a bounded domain in $\mathbb{R}^d$, $d \geq 1$. Let $\{t_i, i = 0, 1, \ldots, n\}$ be a grid of times such that $t_0 = 0$ and $t_n = T$ and let us denote X_{t_i} by X_i, $i = 0, 1, \ldots, n$. We can construct the likelihood function $L_n(\theta)$ of the Markovian process X using the conditional distributions, i.e. $p_\theta(\Delta, x|y)$ is the conditional density of X_i given $X_{i-1} = y$ with $\Delta_i = t_i - t_{i-1}$. Indeed, we can write

$$L_n(\theta) = \prod_{i=1}^{n} p_\theta(\Delta_i, X_i|X_{i-1}) \, p_\theta(X_0), \tag{5.3}$$

where $p_\theta(X_0)$ is the distribution of the initial value X_0. For example, if X is the one-dimensional geometric Brownian motion, then $\theta = (\mu, \sigma)$ and the conditional density is the log-Normal distribution, while the initial condition can be taken as nonrandom so $p_\theta(X_0) = 1$. In general, the initial distribution $p_\theta(X_0)$

is either discarded or taken as equal to one in the asymptotic theory. This is due to the fact that it is usually impossible to estimate it from the data and, given the high number of observations in the asymptotic framework, it is assumed that the initial distribution gives a marginal and small contribution to the full likelihood $L_n(\theta)$. As usual, we denote by $\ell_n(\theta) = \log L_n(\theta)$ the log-likelihood function

$$\ell_n(\theta) = \log L_n(\theta) = \sum_{i=1}^{n} \ell_i(\theta) + \log(p_\theta(X_0))$$

$$= \sum_{i=1}^{n} \log p_\theta(\Delta_i, X_i | X_{i-1}) + \log(p_\theta(X_0)).$$

and, from now on, we take, without any further comment, $p_\theta(X_0) = 1$, thus

$$\ell_n(\theta) = \sum_{i=1}^{n} \log p_\theta(\Delta_i, X_i | X_{i-1}). \tag{5.4}$$

When the transition densities $p_\theta(s, x|y)$ are known in explicit form, it is possible to proceed with exact likelihood inference although the results depend, clearly, on the regularity of the drift and diffusion coefficients, but also on the asymptotic scheme. Unfortunately, the likelihood function is rarely known in explicit form and, when known it is usually for the one-dimensional case. In any case, when the model is regular, the MLE based on the exact likelihood function possesses the usual good properties but we need to distinguish the case of parameters in the drift and the diffusion coefficient. For the parameters in the diffusion coefficient, the rate of convergence is $\sqrt{n}$ in all asymptotic schemes. In this case, the MLE is also asymptotically efficient as in the i.i.d. case. In case of the drift, the rate of convergence of (any) estimator is of order $\sqrt{T}$ and, if T is not large enough or is fixed, it is impossible to obtain consistent estimates of the drift parameters. Luckily enough, the most important quantity in finance is volatility rather than the trend. Usually the grid of time points is such that $n\Delta_n = T$, where Δ_n is taken equal for all time points, i.e. $t_i - t_{i-1} = \Delta_n$ and $t_i = i\Delta_n, i = 0, 1, \ldots, n$. In this case, the rate of convergence of the drift is written as $\sqrt{n\Delta_n}$ and hence, it is required that $n\Delta_n = T \to \infty$ in order to get consistent estimators of the parameters along with some conditions on the stationarity or ergodicity of the process X. Here we present only quasi-maximum likelihood estimation properties for two different asymptotic schemes. For more details on other approaches like estimating functions, method of the moments, etc., the reader can consult Iacus (2008), Phillips and Yu (2009) and Sørensen (2009). Let us consider $\Delta_n \to 0$, $n\Delta = T \to \infty$ as $n \to \infty$. Consider the Euler-Maruyama discretization (4.5) for the time-homogenous stochastic differential equation (5.2). Then, the conditional distribution $p_\theta(\Delta_n, X_i | X_{i-1})$ is a multivariate Gaussian random variable with mean $X_{i-1} + b(t_{i-1}, X_{i-1})\Delta_n$ and variance-covariance matrix $\Delta_n S(x, \theta)$ with $S(x, \theta) = \sigma(x, \theta)^{\otimes 2}$. Here, for a matrix A, we denote by $A^\top$ the transpose of A, $\mathrm{Tr}(A)$ the trace of A, $A^{\otimes 2} = A \cdot A^\top$ and by A^{-1} the inverse of A. Then, the corresponding approximated log-likelihood function for the Euler-Maruyama scheme takes the form:

$$\tilde{\ell}_n(\theta) = \sum_{i=1}^{n} \log \det S(X_{i-1}, \theta) + \Delta_n^{-1} S(X_{i-1}, \theta)^{-1}[(\Delta X_i - b(t_{i-1}, X_{i-1}))^{\otimes 2}]$$

where $\Delta X_i = X_i - X_{i-1}$ and we have dropped the terms with the constants $\frac{1}{2}$ and $\sqrt{2\pi}$. The Quasi-maximum likelihood estimator is the solution of

$$\hat{\theta}_n = \arg\min_{\theta \in \Theta} \tilde{\ell}_n(\theta).$$

In order to get good asymptotic properties for $\hat{\theta}_n$ it is also required that $n\Delta_n^2 \to 0$ and several other regularity properties which we do not mention here but can be found in the above mentioned references. The only necessary condition to explicit is that it is needed to require the separation of the vector θ into two subvectors, i.e. $\theta = (\alpha, \beta)$, with $\alpha \in \Theta_p$ and $\beta \in \Theta_q$ such that $\Theta_p \times \Theta_q = \Theta$. Let us denote by $\mu_\theta(\cdot)$ the invariant measure of the process and define the Fisher information matrix for this experiment as

$$\mathcal{I}(\theta) = \begin{pmatrix} [\mathcal{I}_b^{kj}(\alpha)]_{k,j=1,\ldots,p} & 0 \\ 0 & [\mathcal{I}_\sigma^{kj}(\beta)]_{k,j=1,\ldots,q} \end{pmatrix}$$

where

$$\mathcal{I}_b^{kj}(\alpha) = \int \frac{1}{\sigma^2(\beta, x)} \frac{\partial b(\alpha, x)}{\partial \alpha_k} \frac{\partial b(\alpha, x)}{\partial \alpha_j} \mu_\theta(dx),$$

$$\mathcal{I}_\sigma^{kj}(\beta) = 2 \int \frac{1}{\sigma^2(\beta, x)} \frac{\partial \sigma(\beta, x)}{\partial \beta_k} \frac{\partial \sigma(\beta, x)}{\partial \beta_j} \mu_\theta(dx).$$

Consider the matrix

$$\varphi(n) = \begin{pmatrix} \frac{1}{n\Delta_n}\mathbf{I}_p & 0 \\ 0 & \frac{1}{n}\mathbf{I}_q \end{pmatrix}$$

where $\mathbf{I}_p$ and $\mathbf{I}_q$ are respectively the identity matrix of order p and q. Then, is possible to show that

Theorem 5.2.1 (Ergodic case)

$$\varphi(n)^{-1/2}(\hat{\theta}_n - \theta) \xrightarrow{d} N(0, \mathcal{I}(\theta)^{-1}).$$

The proof of this theorem can be found, e.g., in Yoshida (1992), where the author also presented milder hypotheses and showed that, given that the estimator of β has a faster rate of convergence, one can first estimate β with $\hat{\beta}_n$ with a partial quasi-likelihood function and the plug $\hat{\beta}_n$ in the above quasi-likelihood function and estimate α. In the case when T is fixed and hence $\Delta_n \to 0$ but $n\Delta_n$ does not diverge, then the asymptotic theory changes and, in particular, only the parameter β in the volatility can be estimated efficiently. In this case, there is no need to assume ergodicity of the process X but the problem is now that the Fisher information matrix is a random matrix and the limit theorems are of mixed-normal type. In this case, we have this limit theorem for the quasi-MLE, with all parameters θ in the diffusion coefficient

Theorem 5.2.2 (Pure high frequency)

$$\sqrt{n}(\hat{\theta}_n - \theta) \overset{d_s}{\to} \Gamma_d N(0, I_d)$$

with I_d a $d \times d$ identity matrix and Γ_d a random matrix. The asymptotic result is based on the stable convergence Jacod (1997, 2002). There are several versions of this results proved under different regularity conditions and one milestone reference is surely Genon-Catalot and Jacod (1993).

From the numerical point of view, the two cases are equivalent in that there is only the need to maximize the quasi-likelihood. The difference is in the estimation of the standard errors of the estimates. But these are usually obtained by numerical approximation of the hessian matrix. The package **yuima** implements the quasi-likelihood method for multidimensional diffusion processes and the package sde implements exact, quasi-likelihood and other approximation schemes for the one-dimensional case. In some cases, when e.g. the Δ between observations is not so small and the diffusion process is one-dimensional, the use is preferred of exact methods and other approximation schemes based on asymptotic expansions which are available through the sde package. In the multidimensional case, there is no other option than the **yuima** package at present. Here we present an example based on the function qmle for quasi maximum likelihood estimation of multidimensional diffusion processes.

```
R> require(yuima)
R> diff.matrix <- matrix(c("alpha1", "alpha2", "1", "1"), 2, 2)
R> drift.c <- c("-1*beta1*x1", "-1*beta2*x2", "-1*beta2",
   "-1*beta1")
R> drift.matrix <- matrix(drift.c, 2, 2)
R> ymodel <- setModel(drift = drift.matrix, diffusion =
   diff.matrix,
+      time.variable = "t", state.variable = c("x1", "x2"),
       solve.variable = c("x1",
+         "x2"))
R> n <- 100
R> ysamp <- setSampling(Terminal = (n)^(1/3), n = n)
R> yuima <- setYuima(model = ymodel, sampling = ysamp)
R> set.seed(123)
R> truep <- list(alpha1 = 0.5, alpha2 = 0.3, beta1 = 0.6,
   beta2 = 0.2)
R> yuima <- simulate(yuima, xinit = c(1, 1),
   true.parameter = truep)
R> opt <- qmle(yuima, start = list(alpha1 = 0.7,
   alpha2 = 0.2, beta1 = 0.8,
+      beta2 = 0.2))
```

and now `opt` is an object of class `mle` and we can extract the estimates of our parameters

```
R> opt@coef

   alpha1    alpha2     beta1     beta2
0.3842415 0.1950523 0.5899493 0.1670455

R> unlist(truep)

alpha1 alpha2  beta1  beta2
   0.5    0.3    0.6    0.2
```

5.3   Short-term interest rates models

In the option pricing formulas of the standard Black and Scholes (1973) and Merton (1973) model, which we discuss in Chapter 6, the short-term interest rate is assumed to be constant. But, for example, if we look at Figure 5.1 we see that the U.S. interest rates are clearly changing over time. Thus, it makes sense to try to model interest rates using stochastic processes like diffusion processes solutions to differential equations. Figure 5.1 is obtained using the R package **Ecdat** and the following code

```
R> library(Ecdat)
R> library(sde)
R> data(Irates)
R> X <- Irates[, "r1"]
R> plot(X)
```

Here, we briefly introduce the most common known models and their properties. We start with the general Chan *et al.* (1992) model (CKLS) which encompasses

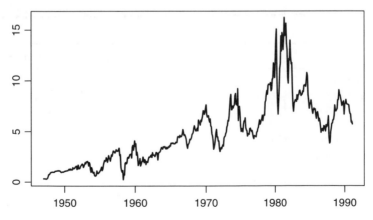

Figure 5.1   The U.S. Interest Rates monthly data from 06/1964 to 12/1989.

many other sub models presented in the literature (see Table 5.1). The CKLS process is the solution to the following stochastic differential equation:

$$dX_t = (\alpha + \beta X_t)dt + \sigma X_t^\gamma dB_t.$$

The parameters must satisfy the constraints $\sigma > 0$, $\gamma \geq 0$ and α, β may, in principle, take any real value, but for meaningful applications, the process must be non-negative. This process is non-negative when $\alpha > 0$ and $\beta \leq 0$ and $\gamma > \frac{1}{2}$ and the initial condition $X_0 > 0$. Processes of this type may possess the *mean reverting* property when they tend to oscillate around some mean level represented by the parameter α. The speed at which the process tends toward the mean level is the speed of mean reversion, and this is captured by the parameter β. Table 5.2 shows conditions on the parameters and the property of mean reversion for the different sub-models.

Table 5.1   The family of one-factor short-term interest rates models seen as special cases of the general CKLS model.

Reference	Model	α	β	γ
Merton (1973)	$dX_t = \alpha dt + \sigma dB_t$		0	0
Vasicek (1977)	$dX_t = (\alpha + \beta X_t)dt + \sigma dB_t$			0
Cox *et al.* (1985)	$dX_t = (\alpha + \beta X_t)dt + \sigma \sqrt{X_t} dB_t$			1/2
Dothan (1978)	$dX_t = \sigma X_t dB_t$	0	0	1
Geometric Brownian Motion	$dX_t = \beta X_t dt + \sigma X_t dB_t$	0		1
Brennan and Schwartz (1980)	$dX_t = (\alpha + \beta X_t)dt + \sigma X_t dB_t$			1
Cox *et al.* (1980)	$dX_t = \sigma X_t^{3/2} dB_t$	0	0	3/2
Constant Elasticity Variance	$dX_t = \beta X_t dt + \sigma X_t^\gamma dB_t$	0		
Chan *et al.* (1992)	$dX_t = (\alpha + \beta X_t)dt + \sigma X_t^\gamma dB_t$			

Table 5.2   Mean reversion property of the members in the family of CKLS model.

Model	α	β	γ	mean reverting?
Merton (1973)	$\mathbb{R}$	0	0	no
Vasicek (1977)	$\mathbb{R}$	$\mathbb{R}$	0	yes
Cox *et al.* (1985)	$\mathbb{R}$	$\mathbb{R}$	1/2	yes
Dothan (1978)	0	0	1	no
Geometric Brownian Motion	0	$\mathbb{R}$	1	yes
Brennan and Schwartz (1980)	$\mathbb{R}$	$\mathbb{R}$	1	yes
Cox *et al.* (1980)	0	0	3/2	no
Constant Elasticity Variance	0	$\mathbb{R}$	$\mathbb{R}$	yes

5.3.1 The special case of the CIR model

Introduced in Cox *et al.* (1985), this is one of the few examples in the class CKLS for which the transition density is known and hence exact maximum likelihood estimation is possible. Indeed, the conditional density of $X_{t+\Delta}|X_t = x$ for the Cox-Ingersoll-Ross (CIR) process solution of

$$dX_t = (\alpha + \beta X_t)dt + \sigma\sqrt{X_t}dB_t$$

is a noncentral χ^2 distribution, $\beta < 0$, $\sigma > 0$. In particular, the transition density $p_\theta(\Delta, y|x)$ can be rewritten in terms of the transition density of $Y_t = 2cX_t$, which has a χ^2 distribution with $\nu = 4\alpha/\sigma^2$ degrees of freedom and noncentrality parameter $Y_s e^{\beta t}$, where $c = -2\beta/(\sigma^2(1 - e^{\beta t}))$. The process also has an explicit solution of the form:

$$X_t = \left(X_0 + \frac{\alpha}{\beta}\right)e^{\beta t} + \theta_3 e^{\beta t}\int_0^t e^{-\beta u}\sqrt{X_u}dB_u.$$

and, when it exists, the stationary distribution of the CIR process is a Gamma law with shape parameter $2\alpha/\sigma^2$ and scale parameter $-\sigma^2/2\beta$. Hence the stationary law has mean equal to $-\frac{\alpha}{\beta}$ and variance $\alpha\sigma^2/(2\beta^2)$. The **sde** package implements [rpdq]cCIR and [rpdq]sCIR for random number generation, cumulative distribution function, density function, and quantile calculations, respectively, for the conditional and stationary laws. It also provides an explicit simulation scheme in the sde.sim function. One should know that the parametrization in the **sde** package is the following:

$$dX_t = (\theta_1 - \theta_2 X_t)dt + \theta_3\sqrt{X_t}dB_t.$$

Thus, for example, if we want to simulate the process with $\theta = (1, 0.3, .1)$ we should use

```
R> sde.sim(model = "CIR", theta = c(1, 0.3, 0.1))
```

If we want to fit this model for the U.S. Interest Rates monthly data above, we can use exact maximum likelihood estimation. Thus, we can construct the true likelihood with

```
R> CIR.loglik <- function(theta1, theta2, theta3) {
+       n <- length(X)
+       dt <- deltat(X)
+       -sum(dcCIR(x = X[-1], Dt = dt, x0 = X[-n], theta = c(theta1,
+           theta2, theta3), log = TRUE))
+ }
```

and then use the standard `mle` function to estimate the parameters

```
R> fit <- mle(CIR.loglik, start = list(theta1 = 0.1, theta2 = 0.1,
+      theta3 = 0.3), method = "L-BFGS-B", lower = rep(0.001, 3),
+      upper = rep(1, 3))
R> coef(fit)

   theta1     theta2     theta3
0.9194592 0.1654958 0.8255179
```

It is also possible to estimate the full CKLS model using the quasi-maximum likelihood estimator. In Section 9.3 we discuss a model selection strategy for the parameters of the CKLS model making use of the **yuima** package.

5.3.2   Ahn-Gao model

The CKLS model introduces the mean reverting property but still assumes a linear drift. The literature have proposed several alternative models which admit also nonlinear drift. One example is given by the process solution to the stochastic differential equation

$$dX_t = X_t(\theta_1 - (\theta_3^3 - \theta_1\theta_2)X_t)dt + \theta_3 X_t^{\frac{3}{2}}dB_t.$$

The conditional distribution of this process is also known and it reads as

$$p_\theta(t, y|x_0) = \frac{1}{y^2} p_\theta^{CIR}\left(t, \frac{1}{y}\,\middle|\,\frac{1}{x_0}\right),$$

where p_θ^{CIR} is the conditional density of the CIR model of previous section. It is left as an exercise to construct the maximum likelihood estimator for this model.

Exercise 5.1 *Write the R code to obtain maximum likelihood estimator of the Ahn-Gao model.*

5.3.3   Aït-Sahalia model

Later Aït-Sahalia (1996a, b) proposed a more sophisticated model to include other polynomial terms. The simplest version of Aït-Sahalia Model is a diffusion process solution to the stochastic differential equation

$$dX_t = (\alpha_{-1}X_t^{-1} + \alpha_0 + \alpha_1 X_t + \alpha_2 X_t^2)dt + \beta_1 X_t^\rho dB_t$$

which is further generalized into

$$dX_t = (\alpha_{-1}X_t^{-1} + \alpha_0 + \alpha_1 X_t + \alpha_2 X_t^2)dt + \sqrt{\beta_0 + \beta_1 X_t + \beta_2 X_t^{\beta_3}}\,dB_t.$$

Unfortunately, there are natural but complex constraints that the set of coefficients in the drift and diffusion coefficient must satisfy in order to have a well-defined stochastic differential equation. We refer the reader to the original paper. Suppose these constraints are in place. Then the problem is how to estimate this highly parametrized model. One approach is to use quasi-maximum likelihood estimation with the LASSO method as in Section 9.3. Another approach is to use the two stage least squares approach in the following manner. First estimate the drift coefficients with a simple linear regression like

$$\mathbb{E}(X_{t+1} - X_t | X_t) = (\alpha_{-1} X_t^{-1} + \alpha_0 + (\alpha_1 - 1) X_t + \alpha_2 X_t^2$$

then, regress the residuals ϵ_{t+1}^2 from the first regression to obtain the estimates of the coefficients in the diffusion term with

$$\mathbb{E}(\epsilon_{t+1}^2 | X_t) = \beta_0 + \beta_1 X_t + \beta_2 X_t^{\beta_3}$$

and finally, use the fitted values from the last regression to set the weights in the second stage regression for the drift. Here is an example of R code to obtain the two stage least squares regression as proposed in Aït-Sahalia (1996b).

```
R> library(Ecdat)
R> data(Irates)
R> X <- Irates[, "r1"]
R> Y <- X[-1]
R> stage1 <- lm(diff(X) ~ I(1/Y) + Y + I(Y^2))
R> coef(stage1)

 (Intercept)          I(1/Y)               Y           I(Y^2)
 0.253478884 -0.135520883 -0.094028341   0.007892979

R> eps2 <- residuals(stage1)^2
R> mod <- nls(eps2 ~ b0 + b1 * Y + b2 * Y^b3, start = list(b0 = 1,
+       b1 = 1, b2 = 1, b3 = 0.5), lower = rep(1e-05, 4),
        upper = rep(2,
+       4), algorithm = "port")
R> w <- predict(mod)
R> stage2 <- lm(diff(X) ~ I(1/Y) + Y + I(Y^2), weights = 1/w)
R> coef(stage2)

 (Intercept)          I(1/Y)               Y           I(Y^2)
 0.189414678 -0.104711262 -0.073227254   0.006442481

R> coef(mod)

          b0            b1            b2            b3
 0.00001000  0.09347354  0.00001000  0.00001000
```

These estimates are biased because the empirical conditional moments of the discretized stochastic differential equations do not coincide with the real

conditional moments, but they can still provide a set of initial values to pass
to the quasi-maximum likelihood optimizer although it is very unlikely to get
satisfying solutions in such high dimensions by most algorithms.

```
R> summary(stage2)

Call:
lm(formula = diff(X) ~ I(1/Y) + Y + I(Y^2), weights = 1/w)

Residuals:
     Min       1Q    Median       3Q      Max
-4.86672 -0.25629   0.04789  0.40901  2.74668

Coefficients:
              Estimate Std. Error t value Pr(>|t|)
(Intercept)   0.189415   0.063891   2.965  0.00317 **
I(1/Y)       -0.104711   0.026529  -3.947 8.99e-05 ***
Y            -0.073227   0.025867  -2.831  0.00482 **
I(Y^2)        0.006442   0.002249   2.865  0.00434 **
---
Signif. codes:  0 '***' 0.001 '**' 0.01 '*' 0.05 '.' 0.1 ' ' 1

Residual standard error: 0.8019 on 526 degrees of freedom
Multiple R-squared: 0.03467,     Adjusted R-squared: 0.02917
F-statistic: 6.298 on 3 and 526 DF,  p-value: 0.0003339

R> summary(mod)

Formula: eps2 ~ b0 + b1 * Y + b2 * Y^b3

Parameters:
     Estimate Std. Error  t value Pr(>|t|)
b0    0.00001 151.78899 6.59e-08        1
b1    0.09347   0.01940    4.818 1.9e-06 ***
b2    0.00001 151.77379 6.59e-08        1
b3    0.00001 588.06245 1.70e-08        1
---
Signif. codes:  0'***' 0.001 '**' 0.01 '*' 0.05 '.' 0.1 ' ' 1

Residual standard error: 1.305 on 526 degrees of freedom

Algorithm "port", convergence message: relative convergence (4)
```

From the above we see that the two stage approach selects the model

$$dX_t = (\alpha_{-1}X_t^{-1} + \alpha_0 + \alpha_1 X_t + \alpha_2 X_t^2)dt + \sqrt{\beta_1 X_t}dB_t$$

and we can consider estimating the parameters further by quasi-maximum likelihood estimation.

5.4   Exponential Lévy model

As seen in Section 4.5.5 the common form of Lévy processes studied in empirical finance is the exponential Lévy process

$$S_t = S_0 e^{Z_t} \qquad (5.5)$$

so that the log-returns of the price process S_t, i.e. $\log(S_{t+s}/S_t) = Z_{t+s} - Z_t$ are distributed as the increments of the Lévy process Z_t. The package **fBasics** offers several routines to fit the parameters of several distributions commonly adopted in finance. For example, the `nigFit` for the Normal Inverse Gaussian distribution; `hypFit` for the hyperbolic distribution; `ghFit` for the generalized hyperbolic distribution; `stableFit` for the stable distribution and others. The idea is to take the log-returns of the stochastic process S_t and fit one of these distributions. As an example, we try to fit the Gaussian, the Normal Inverse Gaussian, the hyperbolic and the generalized hyperbolic distribution to the ENI.MI prices for the year 2009.

```
R> require(fImport)
R> data <- yahooSeries("ENI.MI", from = "2009-01-01",
    to = "2009-12-31")
R> S <- data[, "ENI.MI.Close"]
R> X <- returns(S)
```

The time series and the returns are depicted in Figure 5.2 via the **quantmod** package using these commands

```
R> require(quantmod)
R> lineChart(S, layout = NULL, theme = "white")
R> lineChart(X, layout = NULL, theme = "white")
```

Figure 5.3 shows the plot of the estimated densities on the logarithmic scale. We can see that both the Gaussian model (which means the geometric Brownian motion process) and the generalized hyperbolic distribution provide a very bad fit, while the Normal inverse Gaussian and the hyperbolic distribution provide a similar fit.

```
R> library(fBasics)
R> nFit(X)
R> nigFit(X, trace = FALSE)
R> hypFit(X, trace = FALSE)
R> ghFit(X, trace = FALSE)
```

5.4.1   Examples of Lévy models in finance

As seen, log-returns of the exponential Lévy process allow us to estimate the parameters of the underlying distribution of Z_1. Here we present a small

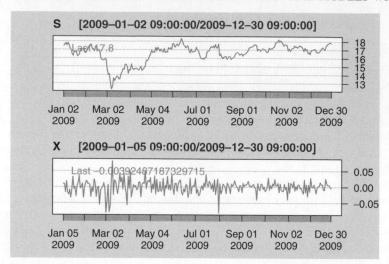

Figure 5.2   Asset prices and log-returns of the time series ENI.MI for the year 2009.

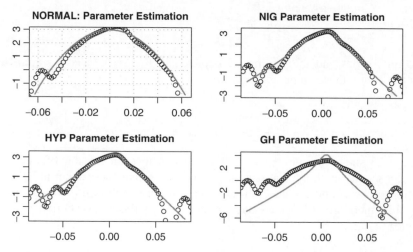

Figure 5.3   Several fitted densities on the logarithm scale for the series of returns of the ENI.MI title of Figure 5.2.

collection of Lévy processes Z_t which are often used in finance to construct the exponential Lévy process in (5.5), although we have anticipated, in the previous section, ways to estimate their parameters. We present their canonical decomposition as well as their characteristic functions (when not enumerated in Section 2.3.3).

5.4.1.1 Generalized hyperbolic process

Making use of the GH distribution (2.13), we can construct a Lévy process Z_t such that $Z_1 \sim$ GH with Itô-Lévy decomposition

$$Z_t = t\mathbb{E}Z_1 + \int_0^t \int_{\mathbb{R}} x \left(\mu^Z - \nu^{GH}\right) (\mathrm{d}s, \mathrm{d}x)$$

where $\nu^{GH}(\cdot)$ is the following Lévy measure

$$\nu^{GH}(\mathrm{d}x) = \frac{e^{\beta x}}{|x|} \left(\int_0^\infty \frac{\exp\left\{-\sqrt{2y + \alpha^2}|x|\right\}}{\pi^2 y \left\{J_{|\lambda|}^2(\delta\sqrt{2y}) + Y_{|\lambda|}^2(\delta\sqrt{2y})\right\}} \mathrm{d}y + \lambda e^{-\alpha|x|} \mathbf{1}_{\{\lambda \geq 0\}} \right) \mathrm{d}x$$

with J_λ and Y_λ denote the Bessel functions of the first and the second kind with index λ (see, (Abramowitz and Stegun 1964)). Its Lévy triplet is $\left(\mathbb{E}Z_1, 0, \nu^{GH}\right)$. For more details on this process see also Raible (2000).

5.4.1.2 Tempered stable process

This process is such that $Z_1 \sim$ TS(k, α, β), the *tempered stable distribution* with characteristic function

$$\varphi_{Z_1}(t) = \mathbb{E}\, e^{itZ_1} = \exp\left\{\alpha\beta - \alpha\left(\beta^{\frac{1}{k}} - 2it\right)^k\right\}$$

with density not available in explicit form. This law is such that

$$\mathbb{E}Z_1 = 2\alpha k\beta^{\frac{k-1}{k}}, \quad \mathrm{Var}Z_1 = 4\alpha k(1-k)\beta^{\frac{k-2}{k}}.$$

The process has Itô-Lévy decomposition

$$Z_t = t\mathbb{E}Z_1 + \int_0^t \int_{\mathbb{R}} x \left(\mu^Z - \nu^{TS}\right) (\mathrm{d}s, \mathrm{d}x)$$

where $\nu^{TS}(\cdot)$ is the following Lévy measure

$$\nu^{TS}(\mathrm{d}x) = \alpha 2^k \frac{k}{\Gamma(1-k)} x^{-k-1} \exp^{-\frac{1}{2}\beta^{\frac{1}{k}}x} \mathbf{1}_{\{x>0\}}\mathrm{d}x$$

Its Lévy triplet is given by $\left(\mathbb{E}Z_1, 0, \nu^{TS}\right)$. The stable distribution was introduced in Tweedie (1984) and extended later in many directions in Barndorff-Nielsen and Shepard (2001).

5.4.1.3 Normal inverse Gaussian process

In this case the Lévy process Z_t is such that $Z_1 \sim$ NIG from (2.12), with Itô-Lévy decomposition

$$Z_t = t\mathbb{E}Z_1 + \int_0^t \int_{\mathbb{R}} x\left(\mu^Z - \nu^{NIG}\right)(\mathrm{d}s, \mathrm{d}x)$$

where $\nu^{GH}(\cdot)$ is the following Lévy measure

$$\nu^{NIG}(\mathrm{d}x) = e^{\beta x} \frac{\delta\alpha}{\pi|x|} K_1(\alpha|x|)\mathrm{d}x.$$

Its Lévy triplet is given by $\left(\mathbb{E}Z_1, 0, \nu^{NIG}\right)$. See Barndorff-Nielsen (1997).

5.4.1.4 Meixner process

The Meixner process with $Z_1 \sim$ Meixner as in (2.14), has the following Lévy measure

$$\nu^{Meixner}(\mathrm{d}x) = \delta \frac{e^{\frac{\beta x}{\alpha}}}{x \sinh\left(\frac{\pi x}{\alpha}\right)}\mathrm{d}x$$

with Itô-Lévy decomposition

$$Z_t = t\mathbb{E}Z_1 + \int_0^t \int_{\mathbb{R}} x\left(\mu^Z - \nu^{Meixner}\right)(\mathrm{d}s, \mathrm{d}x)$$

and triplet $\left(\mathbb{E}Z_1, 0, \nu^{Meixner}\right)$. See Schoutens (2003) and Schoutens and Teugels (1998).

5.4.1.5 Variance Gamma process

The Variance Gamma is a pure jump process with characteristic function of Z_1 in the form:

$$\varphi_{Z_1}(t) = \mathbb{E}e^{itZ_1} = \left(\frac{1}{1 - i\theta kt + \frac{t^2\sigma^2 k}{2}}\right)^{\frac{1}{k}}.$$

The parameters θ and k control respectively the skewness and the kurtosis of the increments of Z_t. This process has Itô-Lévy decomposition

$$Z_t = t\mathbb{E}Z_1 + \int_0^t \int_{\mathbb{R}} x\left(\mu^Z - \nu^{VG}\right)(\mathrm{d}s, \mathrm{d}x)$$

and triplet $\left(\mathbb{E}Z_1, 0, \nu^{VG}\right)$ with Lévy measure

$$\nu^{VG}(\mathrm{d}x) = \frac{1}{k|x|} e^{\frac{\theta x}{\sigma^2} - \frac{\sqrt{\theta^2 + \frac{2\sigma^2}{k}}}{\sigma^2}|x|}\mathrm{d}x.$$

The first two moments of the Variance Gamma process are

$$\mathbb{E}Z_t = \theta t, \quad \mathrm{Var}Z_t = (2\theta^3 k^2 + 3\sigma^2 \theta k)t.$$

This process was introduced in Madan and Seneta (1990) and further extended in Madan *et al.* (1998).

5.4.1.6   CGMY process

This process, called CGMY from the authors Carr *et al.* (2002) is also called *generalized tempered stable process*. The characteristic function of Z_1 has the form:

$$\varphi_{Z_1}(t) = \mathbb{E}e^{itZ_1} = \exp\left\{C\Gamma(-Y)\left((M - it)^Y + (G + it)^Y - M^Y - G^Y\right)\right\}$$

where $C > 0$, $G > 0$, $M > 0$ and $Y < 2$, but its density is not known in closed form. Its Lévy measure is given by

$$\nu^{CGMY}(\mathrm{d}x) = C\frac{e^{-Mx}}{x^{1+Y}}\mathbf{1}_{\{x > 0\}}\mathrm{d}x + C\frac{e^{Gx}}{|x|^{1+Y}}\mathbf{1}_{\{x < 0\}}\mathrm{d}x.$$

This process has Itô-Lévy decomposition

$$Z_t = t\mathbb{E}Z_1 + \int_0^t \int_{\mathbb{R}} x\left(\mu^Z - \nu^{CGMY}\right)(\mathrm{d}s, \mathrm{d}x)$$

and triplet $\left(\mathbb{E}Z_1, 0, \nu^{CGMY}\right)$. See Cont and Tankov (2004) for additional details. This model contains, as special cases, the variance gamma and the bilateral gamma processes. The distribution of Z_1 has finite moments of all orders.

5.4.1.7   Merton process

Merton (1976) introduced one of the first jump models in finance. We start from its Itô-Lévy decomposition which reads

$$Z_t = \mu t + \sigma W_t + \sum_{k=1}^{N_t} J_k$$

with $J_k \sim \mathrm{N}(\mu_J, \sigma_J^2)$, $k = 1, \ldots$ The characteristic function of Z_1 is then

$$\varphi_{Z_1}(t) = \mathbb{E}E^{itZ_1} = \exp\left\{i\mu t - \frac{\sigma^2 t^2}{2} + \lambda\left(e^{i\mu_J t - \frac{\sigma_J^2 t^2}{2}} - 1\right)\right\} \tag{5.6}$$

with Lévy triplet $(\mu, \sigma^2, \lambda \times f_J)$, with $f_J(\cdot)$ the density of J_k and λ the intensity of the Poisson process N_t, $t \geq 0$. The density of Z_1 is not known in closed form but it is possible to obtain

$$\mathbb{E}Z_1 = \mu + \lambda\mu_J, \quad \mathrm{Var}Z_1 = \sigma^2 + \lambda\mu_J^2 + \lambda\sigma_J^2$$

from the characteristic function (5.6).

5.4.1.8   Kou process

The Kou (2002) process replaces the symmetric distribution of the jumps J_K in the Merton model with an asymmetric double exponential distribution. This distribution has density

$$f(x) = p\theta_1 e^{-\theta_1 x} \mathbf{1}_{\{x<0\}} + (1-p)\theta_2 e^{\theta_2 x} \mathbf{1}_{\{x>0\}} \tag{5.7}$$

with $p \in (0,1)$, θ_1 and θ_2 positive parameters. This is a mixture of the densities of two exponential random variables weighted by p and $1-p$. When a random variable X has a distribution with density (5.7), then we write $X \sim \mathrm{DExp}(p, \theta_1, \theta_2)$. The Kou process has then Itô-Lévy decomposition

$$Z_t = \mu t + \sigma W_t + \sum_{k=1}^{N_t} J_k$$

with $J_k \sim \mathrm{DExp}(p, \theta_1, \theta_2)$. The characteristic function of Z_1 is given by

$$\varphi_{Z_1}(t) = \mathbb{E}E^{it Z_1} = \exp\left\{ it\mu - \frac{\sigma^2 t^2}{2} + \lambda\left(\frac{p\theta_1}{\theta_1 - it} - \frac{(1-p)\theta_2}{\theta_2 + it} - 1 \right) \right\}$$

an the Lévy triplet of Z_t is $(\mu, \sigma^2, \lambda \times f)$, with $f(\cdot)$ from (5.7).

5.4.1.9   Geometric Brownian motion as exponential Lévy process

Clearly, the geometric Brownian motion is a particular case of the exponential Lévy model, with $Z_1 \sim N(\mu, \sigma^2)$, Itô-Lévy decomposition $Z_t = \mu t + \sigma W_t$ and Lévy triplet $(\mu, \sigma^2, 0)$.

The GH, NIG and Meixner process do not contain the Gaussian part (indeed the second term of the Lévy triplet is always zero) and the jumps essentially drive their excursions. The Merton and Kou process contain both the Gaussian part and compound Poisson-type jumps. Finally, the Black and Scholes has no jump. From this perspective it is clear that such a way of modelling financial asset is particularly rich.

5.5   Telegraph and geometric telegraph process

Let X_t be a telegraph process, i.e.

$$X_t = x_0 + \int_0^t V_s \mathrm{d}s, \quad V_t = V_0(-1)^{N_t}, \quad t > 0.$$

where N_t is a Poisson process and V_0 is a discrete random variable independent of N_t and taking values $+c$ and $-c$ with equal probability. The so-called geometric telegraph process was proposed in finance by Di Crescenzo and Pellerey (2002) and has the following form:

$$S_t = S_0 \exp\{\alpha t + \sigma X_t\}$$

with $\alpha = \mu - \frac{1}{2}\sigma^2$, $\mu \in \mathbb{R}$ and $\sigma > 0$ in analogy to the geometric Brownian motion case, and S_0 a constant. Due to the fact that the telegraph process has finite variation, the same occurs for its geometric version so it is not good because pricing under this model admits arbitrage opportunities. But these (standard and geometric) telegraph processes are the building block of other stochastic models for finance not affected by this problem. Thus inference for these models is an important task to solve. Estimation of telegraph process from discrete observations was considered in De Gregorio and Iacus (2008) using approximate likelihood method and in Iacus and Yoshida (2008) via the method of the moments. We consider first the pseudo-likelihood approach. By taking into account the transition density (3.26), the quasi-likelihood function for the discretely observed telegraph process is given in the following form:

$$L_n(\lambda) = L_n(\lambda | X_0, X_1, \ldots, X_n) = \prod_{i=1}^{n} p(X_i, \Delta_n; X_{i-1}, t_{i-1}) \tag{5.8}$$

$$= \prod_{i=1}^{n} \left\{ \frac{e^{-\lambda \Delta_n}}{2c} \left\{ \lambda I_0 \left(\frac{\lambda}{c} \sqrt{u_{n,i}} \right) + \frac{c \lambda \Delta_n I_1 \left(\frac{\lambda}{c} \sqrt{u_{n,i}} \right)}{\sqrt{u_{n,i}}} \right\} \chi_{\{u_{n,i} > 0\}} \right.$$
$$\left. + \frac{e^{-\lambda \Delta_n}}{2} \delta(u_{n,i} = 0) \right\}$$

where $u_{n,i} = u_n(X_i, X_{i-1}) = c^2 \Delta_n^2 - (X_i - X_{i-1})^2$. The density $p(X_i, \Delta_n; X_{i-1}, t_{i-1})$ appearing in (5.8) is to be interpreted as the probability law of a telegraph process initially located in X_{i-1}, that reaches the position X_i at time t_i. The construction of $L_n(\lambda)$ is based on the following assumption: the observed increments $X_i - X_{i-1}$ are n copies of the process $X(\Delta_n)$ (i.e. the process $X(t)$ up to time Δ_n) and treated as if they were independent. This is of course untrue, but the estimators based on $L_n(\lambda)$ possess reasonable properties. Notice that at this stage, the only parameter of interest is the intensity of the underlying homogeneous Poisson process λ. Given the following contrast function

$$F(\lambda; X_1, \ldots, X_n) = \frac{\partial}{\partial \lambda} \log L_n(\lambda) \tag{5.9}$$

we consider the estimator $\hat{\lambda}_n$ solution to $F(\lambda) = 0$, i.e.

$$\hat{\lambda}_n : F(\lambda = \hat{\lambda}_n; X_1, \ldots, X_n) = 0. \tag{5.10}$$

It is is possible to prove uniqueness of this estimator as solution of (5.10) and, under the additional conditions $n\Delta_n = T$, $\Delta_n \to 0$ as $n \to \infty$ and T fixed, the estimator $\hat{\lambda}_n$ converges to N_T/T which is the optimal estimator of λ for the continuous time observations. For distributional theory and consistency we need to let $T \to \infty$ as in the continuous case. This is exploited using the method of the moments. Let us consider the increments η_i of the telegraph process defined as

$$\eta_i = X_i - X_{i-1} = V_0 \int_{t_{i-1}}^{t_i} (-1)^{N(s)} ds = V_0 (-1)^{N(t_{i-1})} \int_{t_{i-1}}^{t_i} (-1)^{N(s)-N(t_{i-1})} ds.$$

These increments are stationary but not independent. Conversely, the squared increments

$$\eta_i^2 = c^2 \left(\int_{t_{i-1}}^{t_i} (-1)^{N(s)-N(t_{i-1})} ds \right)^2$$

(or the absolute increments $|\eta_i|$) are independent. This allow to introduce an estimator of the method of the moments for λ. Indeed, the statistics

$$V_n = \frac{1}{n} \sum_{i=1}^{n} \frac{\eta_i^2}{\Delta_n^2}$$

is an unbiased estimator of

$$g_n(\lambda) = \frac{c^2}{\lambda \Delta_n^2} \left(\Delta_n - \frac{1 - e^{-2\lambda \Delta_n}}{2\lambda} \right).$$

We define the moment type estimator $\check{\lambda}_n$ as the unique solution to the equation

$$\frac{1}{n} \sum_{i=1}^{n} \eta_i^2 - \frac{c^2}{\lambda} \left(\Delta_n - \frac{1 - e^{-2\lambda \Delta_n}}{2\lambda} \right) = 0. \tag{5.11}$$

This estimator is given in implicit form, but using Taylor expansion we can derive and approximate but explicit moment type estimator. Taking into account the moment formula (3.30) we obtain:

$$\mathbb{E}\{X_i - X_{i-1}\}^2 = \frac{c^2}{\lambda} \left(\Delta_n - \frac{1 - e^{-2\lambda \Delta_n}}{2\lambda} \right)$$

$$= \frac{c^2}{\lambda} \left(\Delta_n - \frac{2\lambda \Delta_n - \frac{1}{2}(-2\lambda \Delta_n)^2) - \frac{1}{6}(-2\lambda \Delta_n)^3) + o(\Delta_n^3)}{2\lambda} \right)$$

$$= c^2 \Delta_n^2 - \frac{2}{3} c^2 \lambda \Delta_n^3 + o(\Delta_n^3).$$

Therefore, an approximate moment type estimator is the following:

$$\lambda_n^* = \frac{3}{2} \frac{1}{nc^2 \Delta_n^3} \sum_{i=1}^{n} \{c^2 \Delta_n^2 - (X_i - X_{i-1})^2\} = \frac{3}{2} \frac{1}{n \Delta_n} \sum_{i=1}^{n} \left\{ 1 - \frac{\eta_i^2}{c^2 \Delta_n^2} \right\} \tag{5.12}$$

and λ_n^* is a weighted sum of the independent random variables η_i^2. Let λ_0 denote the true value of the parameter λ, then

Theorem 5.5.1 *Let $\hat{\lambda}_n$ any estimator among $\check{\lambda}_n$ and λ_n^* and suppose that $n\Delta_n \to \infty$, $\Delta_n \to 0$ as $n \to \infty$. Then, $\hat{\lambda}_n$ is a consistent estimator of λ_0. Moreover,*

$$\sqrt{n\Delta_n}(\check{\lambda}_n - \lambda_0) \overset{d}{\to} N\left(0, \frac{6}{5}\lambda_0\right)$$

as $n \to \infty$, where $\overset{d}{\to}$ denotes the convergence in distribution. The estimator λ_n^* has the same asymptotic distribution of $\check{\lambda}_n$ under the additional condition $n\Delta_n^3 \to 0$.

Thus moment type estimator for the discretely observed telegraph process is in general not asymptotically efficient because the asymptotic variance of these estimator is $\frac{6}{5}\lambda_0$ while the optimal variance in this setting is λ_0. We finally show how to obtain an asymptotically efficient estimator and we also show the proof of the result because it makes use of the material in Chapter 2 and the properties of the Poisson process. Let us use the notation $\mathbb{E}_0$ and Var_0 to indicate respectively the expected value and the variance operator under P_0, the law corresponding to true value of the parameter $\lambda = \lambda_0$. Consider the following statistic

$$\tilde{\lambda}_n = \frac{1}{n\Delta_n} \sum_{i=1}^n \mathbf{1}_{\{|\eta_i| < c\Delta_n\}} = \frac{1}{n\Delta_n} \sum_{i=1}^n \mathbf{1}_{\{N([t_{i-1},t_i)) \geq 1\}}. \tag{5.13}$$

The statistic $\tilde{\lambda}_n$ is not a good estimator of λ for fixed Δ_n. Indeed,

$$\mathbb{E}_0\{\tilde{\lambda}_n\} = \frac{1}{n\Delta_n} \sum_{i=1}^n \mathbb{E}_0\{\mathbf{1}_{\{|\eta_i| < c\Delta_n\}}\} = \frac{1 - e^{-\lambda_0\Delta_n}}{\Delta_n}.$$

Therefore, we consider the following estimator

$$\hat{\lambda}_n = -\frac{1}{\Delta_n} \log\left(1 - \Delta_n\tilde{\lambda}_n\right) \tag{5.14}$$

and Theorem 5.5.2 proves that it is the efficient estimator in this context.

Theorem 5.5.2 (Iacus and Yoshida 2008) *Let $\Delta_n \to 0$, $n\Delta_n \to \infty$ as $n \to \infty$. Then the estimator $\hat{\lambda}_n$ in (5.14) is consistent, asymptotically normal and attains the minimal variance, i.e. it is asymptotically efficient:*

$$\sqrt{n\Delta_n}(\hat{\lambda}_n - \lambda_0) \overset{d}{\longrightarrow} N(0, \lambda_0)$$

Proof. In order to prove consistency and asymptotic normality of $\hat{\lambda}_n$ we first prove the same properties for $\tilde{\lambda}_n$. Let us consider the following quantity

$$
\begin{aligned}
U_n &= \sqrt{n\Delta_n}\left(\tilde{\lambda}_n - \mathbb{E}_0\{\tilde{\lambda}_n\}\right) \\
&= \frac{1}{\sqrt{n\Delta_n}}\sum_{i=1}^{n}\left\{\mathbf{1}_{\{|\eta_i|<c\Delta_n\}} - \mathbb{E}_0\left\{\mathbf{1}_{\{|\eta_i|<c\Delta_n\}}\right\}\right\} \\
&= \frac{1}{\sqrt{n\Delta_n}}\sum_{i=1}^{n}\left\{\mathbf{1}_{\{|\eta_i|<c\Delta_n\}} - \left(1 - e^{-\lambda_0\Delta_n}\right)\right\} \\
&= \sum_{i=1}^{n}\xi_i
\end{aligned}
$$

with

$$
\xi_i = \frac{1}{\sqrt{n\Delta_n}}\left\{\mathbf{1}_{\{|\eta_i|<c\Delta_n\}} - \left(1 - e^{-\lambda_0\Delta_n}\right)\right\}.
$$

We have that $\mathbb{E}_0\{\xi_i\} = 0$ thus $\mathbb{E}_0\{U_n\} = 0$. Moreover,

$$
\begin{aligned}
n\Delta_n \mathrm{Var}_0\{\xi_i\} &= \mathrm{Var}_0\{\mathbf{1}_{\{|\eta_i|<c\Delta_n\}}\} = \mathbb{E}_0\{\mathbf{1}_{\{|\eta_i|<c\Delta_n\}}\}(1 - \mathbb{E}_0\{\mathbf{1}_{\{|\eta_i|<c\Delta_n\}}\}) \\
&= (1 - e^{-\lambda_0\Delta_n})e^{-\lambda_0\Delta_n} = \lambda_0\Delta_n + o(\lambda_0\Delta_n)
\end{aligned}
$$

hence

$$
\mathrm{Var}_0\{U_n\} = \frac{1}{n\Delta_n}n(\lambda_0\Delta_n + o(\lambda_0\Delta_n)) = \lambda_0 + o(1).
$$

Finally, the ξ_i's are independent because they only involve the absolute value of the increments η_i, Since $|\xi_i| \leq 1/\sqrt{n\Delta_n}$, then the Lindeberg's condition in Theorem 2.4.17 trivially holds true:

$$
\sum_{i=1}^{n}\mathbb{E}_0\left\{\mathbf{1}_{\{|\xi_i|\geq\epsilon\}}\xi_i^2\right\} \to 0
$$

therefore $U_n \xrightarrow{d} N(0, \lambda_0)$. Now we need to prove asymptotic normality of $\hat{\lambda}_n$ in (5.14). Let $f_n(u) = -\frac{1}{\Delta_n}\log(1 - u\Delta_n)$, then $f_n'(u) = \frac{d}{du}f_n(u) = \frac{1}{1-u\Delta_n}$, and $\hat{\lambda}_n = f_n(\tilde{\lambda}_n)$. Further,

$$
\lambda_0 = f_n(\tilde{\lambda}_0) \quad \text{where} \quad \tilde{\lambda}_0 = \frac{1 - e^{-\lambda_0\Delta_n}}{\Delta_n} \equiv \mathbb{E}_0\{\tilde{\lambda}_n\}.
$$

By using the δ-method (2.29), we obtain

$$\sqrt{n\Delta_n}(\hat{\lambda}_n - \lambda_0) = \sqrt{n\Delta_n}(f_n(\tilde{\lambda}_n) - f_n(\tilde{\lambda}_0))$$

$$= \sqrt{n\Delta_n}(\tilde{\lambda}_n - \tilde{\lambda}_0)f_n'(\tilde{\lambda}_0) + o_p(\sqrt{n\Delta_n}|\tilde{\lambda}_n - \tilde{\lambda}_0|)$$

$$= U_n \frac{1}{1 - \lambda_0\Delta_n} + o_p(1)$$

hence $\sqrt{n\Delta_n}(\hat{\lambda}_n - \lambda_0) \xrightarrow{d} N(0, \lambda_0)$.

Consider now the geometric telegraph process

$$S_t = S_0 \exp\{\alpha t + \sigma X_t\}$$

and assume we observe the log-returns

$$Y_i = \log \frac{S_i}{S_{i-1}} = \alpha\Delta_n + \sigma(X_i - X_{i-1})$$

where $S_i = S(t_i)$ are discrete observations from the geometric telegraph process. We assume μ to be known, which is usually the case in finance where μ is related to the expected return of nonrisky assets like bonds, etc. The parameters σ and λ are to be estimated. We assume Y_i to be n copies of the process

$$Y(\Delta_n) = \alpha\Delta_n + \sigma X(\Delta_n)$$

with $X(\Delta_n) = X_i - X_{i-1}$ and $X(0) = x_0 = 0$. Therefore, by (3.29), we have

$$\mathbb{E}\{Y(\Delta_n)\} = \alpha\Delta_n$$

and by (3.30) we obtain

$$\text{Var}\{Y(\Delta_n)\} = \sigma^2 \text{Var} X(\Delta_n) = \sigma^2 \frac{c^2}{\lambda}\left(\Delta_n - \frac{1 - e^{-2\lambda\Delta_n}}{2\lambda}\right). \tag{5.15}$$

A good estimator of the volatility σ can be derived from the sample mean of the log returns. Indeed,

$$\bar{Y}_n = \frac{1}{n}\sum_{i=1}^{n} Y_i = \alpha\Delta_n + \frac{\sigma}{n}X_n \tag{5.16}$$

and

$$\mathbb{E}\bar{Y}_n = \alpha\Delta_n + \frac{\sigma}{n}\mathbb{E}X_n = \alpha\Delta_n = \left(\mu - \frac{1}{2}\sigma^2\right)\Delta_n \tag{5.17}$$

again by (3.29) and for the properties of the log-returns. From (5.17) we have that

$$\sigma^2 = 2 \left(\mu - \frac{\mathbb{E}\bar{Y}_n}{\Delta_n} \right)$$

from which the following unbiased estimator of σ^2 can be derived

$$\hat{\sigma}_n^2 = 2 \left(\mu - \frac{\bar{Y}_n}{\Delta_n} \right).$$

Therefore, a reasonable moment type estimator of σ is

$$\hat{\sigma}_n = \sqrt{2 \left(\mu - \frac{\bar{Y}_n}{\Delta_n} \right)} \tag{5.18}$$

which not always exists because there is no guarantee that $\mu > \bar{Y}_n/\Delta_n$. We then use $\hat{\sigma}_n$ to estimate λ making use of (5.15). Let

$$\bar{s}_Y^2 = \frac{1}{n} \sum_{i=1}^{n} (Y_i - \bar{Y}_n)^2$$

then the proposed estimator of λ is

$$\hat{\lambda}_n = \arg\min_{\lambda > 0} \left(\bar{s}_Y^2 - \hat{\sigma}_n^2 \frac{c^2}{\lambda} \left(\Delta_n - \frac{1 - e^{-2\lambda\Delta_n}}{2\lambda} \right) \right)^2. \tag{5.19}$$

5.5.1 Filtering of the geometric telegraph process

If the velocity c is not known one can proceed as follows: set

$$Z_i = \frac{Y_i - \mathbb{E}\bar{Y}_n}{\sigma} = X_i - X_{i-1} = X(\Delta_n),$$

an estimator of the increments of the telegraph process is

$$\hat{Z}_i = \frac{Y_i - \bar{Y}_n}{\hat{\sigma}_n} = \hat{X}(\Delta_n), \quad i = 1, \ldots, n.$$

Then
$$\hat{Z}_1 = \hat{X}_1, \qquad \hat{Z}_2 + \hat{Z}_1 = \hat{X}_2, \qquad \hat{Z}_3 + \hat{Z}_2 + \hat{Z}_1 = \hat{X}_3, \ldots$$

where $\hat{X}_i$ are the estimated states of the underlying telegrapher's process. From these estimates, one can proceed as in previous sections and estimate both λ and c. From both estimated and true increments of the underlying telegraph process it is possible to obtain the following consistent estimator of c

$$\hat{c}_n = \frac{1}{n} \sum_{i=1}^{n} \frac{|\eta_i|}{\Delta_n}.$$

This is a consistent estimator of c, because

$$P\left\{\frac{|\eta_i|}{\Delta_n} = c\right\} = e^{-\lambda\Delta_n} \to 1,$$

then $\hat{c}_n \xrightarrow{p} c$ as $\Delta_n \to 0$. With this estimator one can subsequently use it to estimate λ, σ and α by replacing $\hat{c}_n$ in place of c in all expressions.

5.6 Solution to exercises

Solution 5.1 (to Exercise 5.1) *We can write the log-likelihood as*

$$-2\sum_{i=2}^{n}\log X_i + \sum_{i=2}^{n}\log\left\{p_{\theta}^{CIR}\left(\Delta, \frac{1}{X_i}\middle|\frac{1}{X_{i-1}}\right)\right\}$$

and hence the corresponding negative log-likelihood can be coded as follows:

```
R> library(sde)
R> AhnGao.loglik <- function(theta1, theta2, theta3){
+      n <- length(X)
+      dt <- deltat(X)
+      -sum(dcCIR(x = 1/X[-1], Dt = dt, x0 = 1/X[-n],
         theta = c(theta1,
+          theta2, theta3), log = TRUE)) + 2 * sum(log(X[-1]))
+ }
```

To test it we use the U.S. interest rates again

```
R> library(Ecdat)
R> data(Irates)
R> X <- Irates[, "r1"]
R> fit <- mle(AhnGao.loglik, start = list(theta1 = 0.1,
      theta2 = 0.1,
+      theta3 = 0.3), method = "L-BFGS-B", lower = rep(0.001, 3),
+      upper = rep(1, 3))
```

and the final estimates are as follows:

```
R> coef(fit)

    theta1     theta2     theta3
0.3200484  0.9691536  0.5215643
```

5.7 Bibliographical notes

Inference for stochastic processes is a wide field still in development. Most continuous time processes in finance are diffusion processes. To deeply understand the theory of estimation starting from continuous time, one should refer at least to

Kutoyants (2004). Books which consider discrete time observations are Prakasa Rao (1999) and Iacus (2008). For Poisson processes one good starting reference is Kutoyants (1998). The literature on estimation for Lévy processes is still very sparse, but there are several examples of estimation techniques for particular models in, e.g., Schoutens (2003), Cont and Tankov (2004) and Raible (2000). The above list is not exhaustive but a good starting point for further reading.

References

Abramowitz, M. and Stegun, I. (1964). *Handbook of Mathematical Functions*. Dover Publications, New York.

Aït-Sahalia, Y. (1996a). Nonparametric pricing of interest rate derivative securities. *Econometrica* **64**, 527–560.

Aït-Sahalia, Y. (1996b). Testing continuous-time models of the spot interest rate. *Rev. Financial Stud* **9**, 2, 385–426.

Barndorff-Nielsen, O. and Shepard, N. (2001). Normal modified stable processes. *Theory of Probability and Mathematical Statistics* **65**, 1–19.

Barndorff-Nielsen, O. E. (1997). Normal inverse gaussian distributions and stochastic volatility modelling. *Scand. J. Statist.* **24**, 1–13.

Black, F. and Scholes, M. (1973). The pricing of options and corporate liabilities. *Journal of Political Economy* **81**, 637–654.

Brennan, M. and Schwartz, E. (1980). Analyzing convertible securities. *J. Financial Quant. Anal.* **15**, 4, 907–929.

Carr, P., Geman, H., Madan, D., and Yor, M. (2002). The fine structure of asset returns: an empirical investigation. *J. Business* **75**, 305–332.

Chan, K., Karolyi, G., Longstaff, F., and Sanders, A. (1992). An empirical investigation of alternative models of the short-term interest rate. *J. Finance* **47**, 1209–1227.

Cont, R. and Tankov, P. (2004). *Financial Modelling With Jump Processes*. Chapman & Hall/CRC, Boca Raton.

Cox, J., Ingersoll, J., and Ross, S. (1980). An analysis of variable rate loan contracts. *J. Finance* **35**, 2, 389–403.

Cox, J., Ingersoll, J., and Ross, S. (1985). A theory of the term structure of interest rates. *Econometrica* **53**, 385–408.

De Gregorio, A. and Iacus, S. M. (2008). Parametric estimation for the standard and the geometric telegraph process observed at discrete times. *Statistical Inference for Stochastic Processes* **11**, 249–263.

Di Crescenzo, A. and Pellerey, F. (2002). On prices' evolutions based on geometric telegrapher's process. *Applied Stochastic Models in Bussiness and Industry* **18**, 171–184.

Dothan, U. (1978). On the term structure of interest rates. *J. Financial Econ.* **6**, 59–69.

Genon-Catalot, V. and Jacod, J. (1993). On the estimation of the diffusion coefficient for multidimensional diffusion processes. *Ann. Inst. Henri Poincaré* **29**, 119–151.

Iacus, S. (2008). *Simulation and Inference for Stochastic Differential Equations. With R Examples*. Springer Series in Statistics, Springer, New York.

Iacus, S. and Yoshida, N. (2008). Estimation for the discretely observed telegraph process. *Theory of Probability and Mathematical Statistics* **78**, 33–43.

Jacod, J. (1997). On continuous conditional gaussian martingales and stable convergence in law. *Séminaire de probabilités (Strasbourg)* **31**, 232–246.

Jacod, J. (2002). On processes with conditional independent increments and stable convergence in law. *Séminaire de probabilités (Strasbourg)* **36**, 383–401.

Kou, S. G. (2002). A jump diffusion model for option pricing. *Manag. Sci.* **48**, 1086–1101.

Kutoyants, Y. (1998). *Statistical inference for spatial Poisson processes*. Lecture Notes in Statistics, Springer-Verlag, New York.

Kutoyants, Y. (2004). *Statistical Inference for Ergodic Diffusion Processes*. Springer-Verlag, London.

Madan, D., Carr, P., and Change, E. (1998). The variance gamma process and option pricing. *European Finance Review* **2**, 79–105.

Madan, D. and Seneta, E. (1990). The variance gamma (v.g.) model for share market returns. *Journal of Business* **64**, 4, 511–524.

Merton, R. C. (1973). Theory of rational option pricing. *Bell Journal of Economics and Management Science* **4**, 1, 141–183.

Merton, R. C. (1976). Option pricing with discontinuous returns. *Bell J. Financ. Econ.* **3**, 145–166.

Phillips, P. and Yu, J. (2009). Maximum likelihood and gaussian estimation of continuous time models in finance. *Handbook of Financial Time Series, Springer* 497–530.

Prakasa Rao, B. (1999). *Statistical Inference for Diffusion Type Processes*. Oxford University Press, New York.

Raible, S. (2000). *Lévy Processes in Finance: Theory, Numerics, and Empirical Facts*. PhD Thesis, University of Freiburg, http://www.freidok.uni-freiburg.de/volltexte/51/, Freiburg.

Schoutens, W. (2003). *Lévy Processes in Finance*. John Wiley & Sons, Ltd, Chichester.

Schoutens, W. and Teugels, J. (1998). Lévy processes, polynomials and martingales. *Commun. Statist.- Stochastic Models* **14**, 1, 335–349.

Sørensen, M. (2009). Parametric inference for discretely sampled stochastic differential equations. *Handbook of Financial Time Series, Springer* 531–553.

Tweedie, M. (1984). An index which distinguishes between some important exponential families. *In Statistics: Applications and New Directions: Proc. Indian Statistical Institute Golden Jubilee International Conference (ed. J. Gosh and J. Roy)* 579–604.

Vasicek, O. (1977). An equilibrium characterization of the term structure. *J. Financial Econ.* **5**, 177–188.

Yoshida, N. (1992). Estimation for diffusion processes from discrete observation. *J. Multivar. Anal.* **41**, 2, 220–242.

6

European option pricing

6.1 Contingent claims

In this chapter we will focus on the determination of the fair price of a derivative, like options. We will assume geometric Brownian motion for asset price dynamics and pricing formulas of the Black and Scholes (1973) and Merton (1973) model.

Although we mainly discuss European call options we will discuss pricing of generic *contingent claims* which include call and put but also Asian options or options with barriers. Options of American type require a special attention and hence will be treated separately in Chapter 7.

In Chapter 8 we will consider deviations from the theory presented here. The main ingredient to be replaced in the Black and Scholes theory will be the model which describes assets prices to admit jumps or non-Gaussian returns.

We assume that there is a probability space $(\Omega, \mathcal{A}, P)$, and a filtration $\mathcal{F} = \{\mathcal{F}_t, t \geq 0\}$. When not explicitly mentioned, all processes are adapted to the filtration $\mathcal{F}$.

Definition 6.1.1 *A T-contingent claim or T-claim is a contract which pays to the holder a stochastic amount X at time T. The random variable X is $\mathcal{F}_T$-measurable and T is called exercise time of the contingent claim.*

The definition of T-claim explicitly requires $\mathcal{F}_T$ measurability of X because of the fundamental principle of finance in which only the information up to some given T (and not future information) determines the price. In this setup, the information is the one generated by Brownian motion paths up to time T.

All contingent claims have an associated *payoff function* $f(\cdot)$ usually calculated at point S_T (the final value of the asset price at time T). If $X = f(S_T)$,

Option Pricing and Estimation of Financial Models with R, First Edition. Stefano M. Iacus.
© 2011 John Wiley & Sons, Ltd. Published 2011 by John Wiley & Sons, Ltd.

then X depends on S_T which in turn depends on B_T. Therefore, X is clearly $\mathcal{F}_T$ adapted.

European call options have a payoff function of the form $f(x) = \max(x - K, 0)$, while for put options we have $f(x) = \max(K - x, 0)$. Here K is the strike price. Indeed, for a call option the payoff is not null only if S_T is higher than the strike price K, i.e. $S_T - K > 0$. Hence, its payoff can be written as $f(S_T) = \max(S_T - K, 0)$ and similarly for the put option. The next code shows how to plot the payoff functions for the call and put options represented in Figure 6.1.

```
R> f.call <- function(x) sapply(x, function(x) max(c(x - K, 0)))
R> f.put <- function(x) sapply(x, function(x) max(c(K - x, 0)))
R> K <- 1
R> curve(f.call, 0, 2, main = "Payoff functions", col = "blue",
+       lty = 1, lwd = 1, ylab = expression(f(x)))
R> curve(f.put, 0, 2, col ="black", add = TRUE, lty = 2, lwd = 2)
R> legend(0.9, 0.8, c("call", "put"), lty = c(1, 2),
        col = c("blue",
+       "black"), lwd = c(2, 2))
```

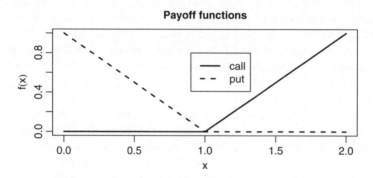

Figure 6.1   The payoff function of call and put options.

A barrier option is characterized by the fact that some threshold β is given and and the contingent claim pays (like a put or a call) only if, in $[0, T]$, the underlying asset never passes the threshold β. Imagine a call option with barrier β and strike price K. The payoff will be of the form:

$$X = \mathbf{1}_{\{S_t < \beta, t \leq T\}} \max(S_T - K, 0)$$

and the payoff X depends on the whole trajectory S_t, $t \leq T$, and hence by B_t, $t \leq T$. So it is still $\mathcal{F}_T$-measurable.

An Asian option is another contingent claim which pays a payoff that is a function of the average of the values of the underlying asset calculated on the whole time interval $[0, T]$. An example of Asian option is a claim which pays only if the average price is higher than the strike price

$$X = \max\left(\frac{1}{T}\int_0^T S_t dt - K, 0\right)$$

which satisfies all the hypotheses.

One powerful technique to find fair price of options is the equivalent martingale measure approach. We will discuss in detail this technique later. Other opportunity rely on the rare case in which Itô formula can be used to solve $f(S_T)$ explicitly. We will see that this not always trivial because payoff functions are usually not differentiable.

Before discussing the continuous time Black and Scholes model, we start with the analysis of market with one single period. This simplified exposition allows us to explain other basic concepts such as: risk neutral measure, hedging and arbitrage.

6.1.1   The main ingredients of option pricing

Before going into detail, it is useful to get the general idea of what is reasonable to ask of an option price theory. We do not discuss technical assumptions in detail, but just the general idea for the moment.

One of the natural assumptions to ask is that the market under which the prices are fixed do not admit *arbitrage*. Arbitrage free markets are markets such that it is not possible to start without capital and gain a positive amount of money with certainty. On real markets and in the short run, this assumption may not always be verified due to asymmetry of information among the actors, but in the (not so) long run the market actors tend to behave in a way that eliminates arbitrage opportunity. If there are no arbitrage opportunities, then the prices on the market are fair. We are particularly interested here in the price of derivatives. Denote by p_0 the (fair) price of a derivative like a contingent T-claim at time 0.

As mentioned in Chapter 1, the writer who sells the contract should guarantee himself from potential loss, so he has to set up a hedging portfolio at time 0 such that at time T the risk associated with the right of exercise of the option right is covered by the value of the portfolio. We denote by H_t the value of the hedging portfolio at time $t \in [0, T]$. When this hedging portfolio exists for all kind of derivatives exchanged on the market, we say that the market is also *complete*. Completeness is rarely observed in real markets for several reasons which we will discuss later (for example, transaction costs have an impact on the management of the hedging portfolio) but it is not unreasonable to ask that the writer protect himself and, at the same time, the holder of the derivative is guaranteed to keep the right to exercise the option at maturity. So hedging means that $H_T = X$, where X is the payoff of the contingent claim. This is clearly an equality between random variables.

Non-arbitrage means, we cannot obtain a positive gain with probability one if we start without capital and that the price of derivative is fair. This is written as

$P_0 = H_0$. Clearly, the hedging portfolio should contain a part of risk and a part of risk free elements in it but it cannot also contain the derivative itself. So we can write the value of the portfolio at time t as $H_t = aS_t + bR_t$, where a is the quantity of assets and b a quantity of risk free activities, e.g. bonds, R_t. Notice that, in general, a and b may also be function of t but this implies a continuous adjusting of the composition of the portfolio. This continuous adjustment is in practice prevented by transaction costs, and this motivates the above comment about absence of completeness of real markets.

Finally, because the derivatives can be exchanged at any time up to maturity on the market (even if the holder can exercise it at maturity) and because we assume arbitrage-free markets, we also have to ask $P_t = H_t$, with P_t the price of the derivative at any time $t \in [0, T]$. So we would like to solve this problem by considering the sequence of equalities:

$$P_0 = H_0 = aS_0 + bR_0 \qquad P_t = H_t = aS_t + bR_t \qquad \text{no arbitrage}$$

$$H_T = X \qquad\qquad\qquad\qquad\qquad\qquad\qquad\qquad\qquad \text{completeness}$$

Now notice that P_0 and hence H_0 are nonstochastic, while both X and H_T are, but we need to fix P_0 (and H_0). The idea is then to consider the expected payoff $\mathbb{E}X = \mathbb{E}H_T$ and then discount the value of $\mathbb{E}H_T$ as if X were constant in order to obtain H_0 and then fix both P_0 and the composition of the hedging portfolio. Loosely speaking, we would like to write $H_0 = (1 + r)^{-1}\mathbb{E}X$. This is not possible in this simple way, but to explain how it may become possible we start with the simple one period market.

6.1.2   One period market

Imagine a market where only a single risky asset and a risk free bond are exchanged. We assume that the asset at time 0 has an initial value of s_0, i.e. $S_0 = s_0$, and that the bond has a value of 1, i.e. $b_0 = 1$. We further assume that, at time T, only two possible events may occur $\Omega = \{\omega_1, \omega_2\}$ with probability $P(\omega_1) = p$ and $P(\omega_2) = 1 - p$, $p \in (0, 1)$. Therefore, either $S(T, \omega_1) = s_1$ or $S(T, \omega_2) = s_2$. For simplicity we assume $s_1 > s_0 > s_2$. Given the assumptions, we have $P(S_T = s_1) = p$ and $P(S_T = s_2) = 1 - p$. We are in a one-period market and we assume that the interest rate is r, then at time T, the value of the bond is $b_T = 1 + r$. Let us now construct a portfolio H buying a assets and b bonds. At time 0, our investment has a cost of $H(0)$ given by

$$H_0 = a \cdot s_0 + b \cdot 1,$$

while at time T, either

$$H(T, \omega_1) = a \cdot s_1 + b \cdot (1 + r)$$

with probability p or

$$H(T, \omega_2) = a \cdot s_2 + b \cdot (1 + r)$$

with probability $1 - p$. So the expected value of our investment at time T is given by

$$\mathbb{E}_P H_T = p H(T, \omega_1) + (1 - p) H(T, \omega_2)$$

where we explicitly use the notation $\mathbb{E}_P$ to denote that the expected value is calculated under the measure P.

Now the question is the following: can we change the probability measure from $P := \{P(\omega_1) = p, P(\omega_2) = 1 - p\}$ to another measure $Q := \{Q(\omega_1) = q, Q(\omega_2) = 1 - q\}$, $q \in (0, 1)$, such that we can write

$$H_0 = (1 + r)^{-1} \mathbb{E}_Q H_T \ ?$$

Which means, can we find a constant q, such that

$$H_0 = (1 + r)^{-1} (q H(T, \omega_1) + (1 - q) H(T, \omega_2)) \ ?$$

If this new probability measure Q exists, then we can discount the final value of the portfolio H_T and obtain the initial amount H_0 of investment needed at time zero in order to obtain that exact expected value H_T at the end, i.e. we are completely eliminating the risk of our investment and treat the risk portfolio as if it were a bond. If this Q exists, then we call the new measure Q the *risk neutral measure*. Let us try to verify if such a constant q exists. We need to find q such that

$$a \cdot s_0 + b = (1 + r)^{-1} (q(a \cdot s_1 + b \cdot (1 + r)) + (1 - q)(a \cdot s_2 + b \cdot (1 + r)))$$

which we can manipulate as follows:

$$(a \cdot s_0 + b)(1 + r) = q(a \cdot s_1 + b \cdot (1 + r)) + a \cdot s_2 + b \cdot (1 + r)$$
$$- q(a \cdot s_2 + b \cdot (1 + r))$$
$$a \cdot s_0(1 + r) + b(1 + r) - a \cdot s_2 - b \cdot (1 + r) = q(a \cdot s_1 + b \cdot (1 + r)$$
$$- a \cdot s_2 - b \cdot (1 + r))$$
$$a \cdot s_0(1 + r) - a \cdot s_2 = q(a \cdot s_1 - a \cdot s_2)$$

and finally obtain

$$q = \frac{s_0(1 + r) - s_2}{s_1 - s_2}.$$

We now have to check that $q \in (0, 1)$ in order to obtain the new probability measure $Q := \{Q(\omega_1) = q, Q(\omega_2) = 1 - q\}$, but this is indeed true because $s_0(1 + r) > s_2$ by construction. Notice that we expect $p \neq q$, so that, while p is the probability that the asset increases its value from s_0 to s_1, for q there is no such direct interpretation. In this sense, P is called the true/historical/physical measure while Q is an instrumental measure that we only use to perform calculations. So, using non-arbitrage and completeness, we can now write

$$P_0 = H_0 = (1+r)^{-1}\mathbb{E}_Q H_T = (1+r)^{-1}\mathbb{E}_Q X.$$

From the above, one can now obtain the value of $H(0)$ which is necessary to put in the hedging portfolio. There is still the problem on how to allocate a and b inside the portfolio. In particular, there may be many or one solution. Let us determine the optimal allocation of the hedging portfolio. Denote by $x_1 = X(\omega_1)$ and $x_2 = X(\omega_2)$. If we want to have $H_T = X$, we have to solve the following system

$$\begin{cases} as_1 + b(1+r) = x_1 \\ as_2 + b(1+r) = x_2 \end{cases}$$

from which

$$a = \frac{x_1 - x_2}{s_1 - s_2} \qquad b = \frac{x_2 s_1 - x_1 s_2}{s_1 - s_2}(1+r)^{-1}.$$

We now look at other properties of the new measure Q. Let us set $a = 1$ and $b = 0$, i.e. we construct a portfolio of only assets. Then, from the above equalities, the relation

$$H(0) = (1+r)^{-1}\mathbb{E}_Q H_T$$

reduces to

$$s_0 = (1+r)^{-1}\mathbb{E}_Q S_T.$$

This means that investing in risky assets has the same return of a bond if we perform our calculations under the measure Q instead of using the physical measure P. Again, Q removes, in a sense, the risk from the evaluation of market activities.

So, we have seen that if a measure Q exists, it allows along with completeness and non-arbitrage, to have fair prices, hedging portfolios (and their composition) and, moreover, the randomness of the underlying process is neutralized (the asset price process behaves as a bond). We will see that, in general, the measure Q maintains these properties in the general setup with some differences like, for example, that all risky activities behave like martingales (not simply as bonds).

To conclude this discussion, we need a more in depth discussion of why the requirement of a fair price is equivalent to ask that $P_0 = H_0$. We have to prove that this requirement is unavoidable. Let us assume that on the market the contingent claim is offered at a price $\tilde{P}_0$ which is lower than the fair price $P_0 = H_0$ and let us prove that we can make arbitrage, i.e. make profit without risk. Of course, we have to assume that we know P_0.

We can short-sell a hedging portfolio H_0 at a price of $N H_0$. Given that $\tilde{P}_0 < P_0$ we can buy N of the above claims at the cost of $N\tilde{P}_0$ and we still have a surplus value of $N(H_0 - \tilde{P}_0)$. Let us buy bonds for that amount. Notice that we don't need to really own the claims, we just need to sell a portfolio in order to obtain a value of H_0 to start our speculation, without putting one of our pennies. At maturity T our portfolio will have a value of $N H_T$ and because it is a hedging portfolio $N H_T = N X$ where $N X$ is the true value of the contingent claims. So we have enough money for the contingent claims. But also our bonds are at maturity and we get a strictly positive profit of $N(H_0 - \tilde{P}_0)(1 + r)$. Clearly, this risk-less profit is null if $\tilde{P}_0 = H_0$ i.e. if $\tilde{P}_0 = P_0$. A similar argument holds for the symmetric case $\tilde{P}_0 > P_0$.

So, the one-period market is a very simple set up but gives an overview of the general approach to the option pricing theorem.

Notice that in reality markets are incomplete because not all claims are completely replicable. If it were the case, instead of buying a derivative one could just buy the equivalent hedging portfolio. In some cases, especially if the underlying asset follows a Lévy model, there will be more than one way to replicate a claim (more than one measure Q) and we have to choose one among the others. Conversely, notice that the price p_0 and the hedging portfolio are obtained via the risk-neutral measure, so both are independent of the physical measure P. But to obtain a risk neutral measure we need to assume non-arbitrage and hedging. So we cannot construct the portfolio without passing through Q.

6.1.3   The Black and Scholes market

We now consider the continuous time case. We assume to have a financial market with only one asset (risky investment), a bond and a contingent claim. The dynamics of the underlying asset is represented by the geometric Brownian motion

$$dS_t = \mu S_t dt + S_t dB_t$$

while interest rates follow the following nonstochastic differential equation

$$dR_t = r R_t dt.$$

For simplicity we assume $R_0 = 1$. The constant rate r is, for example, the one year interest rate on the bonds. The solution to the differential equation for R_t is $R_t = \exp(rt)$, which can be easy obtained as follows:

$$dR_t = r R_t dt \quad \Leftrightarrow \quad \frac{dR_t}{dt} = r R_t \quad \Leftrightarrow \quad \frac{\frac{dR_t}{dt}}{R_t} = r$$

taking the limit as $dt \to 0$

$$\frac{R'_t}{R_t} = r \quad \Leftrightarrow \quad \frac{d}{dt} \log R_t = r \quad \Leftrightarrow \log R_t - \log$$

$$R_0 = \int_0^t r du = rt \quad \Leftrightarrow \quad \frac{R_t}{R_0} = e^{rt}$$

the result follows from the fact that $R_0 = 1$. The discount rate r_a can be obtained as the solution of the equation $(1 + r_a)^t = \exp(rt)$ in the following manner:

$$(1 + r_a)^t = \exp(rt) \Leftrightarrow \log(1 + r_a)^t = rt \Leftrightarrow t\log(1 + r_a) = rt \Leftrightarrow 1$$
$$+ r_a = e^r \Leftrightarrow r_a = e^r - 1.$$

Let $P(t) = P_t$ the price of the contingent claim at time t. We need to derive the dynamics of P_t to complete the description of how the market evolves in continuous time. We assume that all stochastic processes are now adapted to the filtration generated by the Brownian motion without explicitly mentioning it.

6.1.4   Portfolio strategies

In the Black and Scholes market a portfolio can be composed of assets, bonds and claims. We denote by $a(t)$ the amount of assets in the portfolio at time t; by $b(t)$ the amount of bonds and by $c(t)$ the amount of claims.

Definition 6.1.2 *A portfolio strategy is a triplet (a, b, c) of stochastic processes and the value of the corresponding portfolio at time t is given by*

$$V_t = a_t S_t + b_t R_t + c_t P_t.$$

The processes (a, b, c) are supposed to be adapted, though in the simplest case, as we will see, they are supposed to be constant. In both cases, the process V_t is also an adapted stochastic process. This means that an investor owning such a portfolio does not have more information than what the market reveals up to time t. In the Black and Scholes market, portfolios have an additional constraint.

Definition 6.1.3 *A portfolio is said to be a self-financing portfolio if and only if its value at time t depends only on the variation of the processes and the triplet (a, b, c), i.e. it is not possible to add value or remove value from the portfolio but it is only possible to adjust the strategy (a, b, c). Therefore, a self-financing portfolio verifies*

$$dV_t = a_t dS_t + b_t dR_t + c_t dP_t.$$

Remember that the writing

$$dV_t = a_t dS_t + b_t dR_t + c_t dP_t$$

is interpreted as

$$V_t = V_0 + \int_0^t a_u dS_u + \int_0^t b_u dR_u + \int_0^t c_u dP_u.$$

In order to complete the description we have to discuss the dynamics of P_t to make sense of the last stochastic integral. To this end we will make use of the Itô formula.

6.1.5 Arbitrage and completeness

We now need to clarify the meaning of non-arbitrage and completeness of the market in the continuous time framework. Remember that arbitrage means the possibility of starting with a negative investment and ending with a positive one, almost surely. Technically, this notion is defined as follows:

Definition 6.1.4 *A self-financing strategy is called an arbitrage opportunity if it satisfies*

$$V_0 \leq 0, \qquad V_T \geq 0, \qquad \mathbb{E}V_T > 0.$$

Completeness requires the existence of a replicating or hedging portfolio composed only by assets and bonds which are able to cover the risk of the payoff associated with the contingent claim. We denote the hedging portfolio by H_t and the corresponding strategy by (a^H, b^H). At time t, the value of the hedging portfolio is

$$H_t = a_t^H S_t + b_t^H R_t.$$

Definition 6.1.5 *A portfolio strategy* (a^H, b^H)

$$H_t = a_t^H S_t + b_t^H R_t$$

is called hedging portfolio for a contingent claim c if

$$H_T = X$$

where X is the payoff at exercise time of the contingent claim c.

We say that the market is complete if for all contingent claims there exists a hedging portfolio.

Under some circumstances, it is possible to prove that, given the above conditions, there can still exists portfolio strategies such that starting with an initial value of zero, produce positive investments. These are called *doubling strategies* and in order to exclude them from the Black and Scholes market, it is necessary to assume the following assumption: there exists $K > 0$ such that $V_t \geq K$ for all $t \in [0, T]$.

6.1.6 Derivation of the Black and Scholes equation

We remember that we have denoted the payoff of the contingent claim as

$$X = f(S_T).$$

We are interested in the price of the contingent claim at time $t = 0$ and we know that the payoff is the value of the contingent claim at time $t = T$. Further, the market is free to sell and buy derivatives even before maturity, so there is the need to consider the price of the claim for all time $t \in [0, T]$. We denote by $C(t, S_t) = P_t$ the price of the contingent claim at time t, which of course depends on the payoff function f and the underlying asset S. Notice that, at time T, $P_T = C(T, S_T) = f(S_T) = X$. If we want to derive a stochastic differential equation for P_t we need to apply the Itô formula to $f(S(T))$ so we need to require that $\mathbb{E} f(S_T)^2 < \infty$.

Now our task will be to derive an equation for $P_t = C(t, S_t)$ for $t \leq T$ which will provide the price of the option $P_0 = C(0, S_0)$. We denote by $C(t, x)$ the generic function which represents the price of the contingent claim.

Theorem 6.1.6 *Let $C(t, x)$ be the function that describes the price $P_t = C(t, S_t)$ of the contingent claim. Assume that $C(t, x)$ is differentiable in t and two times differentiable in x. Then, $C(t, x)$ solves the following (nonstochastic) partial differential equation:*

$$rC(t, x) = C_t(t, x) + rxC_x(t, x) + \frac{1}{2}\sigma^2 x^2 C_{xx}(t, x). \tag{6.1}$$

Equation (6.1) is called the Black and Scholes equation.

Proof. We start by applying the Itô formula to $P_t = C(t, S_t)$

$$
\begin{aligned}
dP_t &= C_t(t, S_t)dt + C_x(t, S_t)dS_t + \frac{1}{2}C_{xx}(t, S_t)(dS_t)^2 \\
&= C_t(t, S_t)dt + C_x(t, S_t)(\mu S_t dt + \sigma S_t dB_t) \\
&\quad + \frac{1}{2}C_{xx}(t, S_t)(\mu S_t dt + \sigma S_t dB_t)^2 \\
&= C_t(t, S_t)dt + C_x(t, S_t)\mu S_t dt + \sigma C_x(t, S_t)S_t dB_t \\
&\quad + \frac{1}{2}C_{xx}(t, S_t)\mu^2 S_t^2(dt)^2 + \frac{1}{2}C_{xx}(t, S_t)S_t^2\sigma^2(dB_t)^2 + C_{xx}(t, S_t) \\
&\quad \mu\sigma S_t^2 dt dB_t
\end{aligned}
$$

and by dropping the terms of order $(dt)^2$ we obtain

$$dP_t = \left\{ C_t(t, S_t) + \mu C_x(t, S_t)S_t + \frac{1}{2}\sigma^2 C_{xx}(t, S_t)S_t^2 \right\} dt + \sigma C_x(t, S_t)S_t dB_t \tag{6.2}$$

We now make use of the assumption of completeness and non-arbitrage. Completeness of the market means that there exists a hedging portfolio strategy (a^H, b^H) such that $H_T = X$ and non-arbitrage implies that $H_t = P_t$ for all

$t \leq T$. Therefore, we have the following equality

$$
\begin{aligned}
dP_t &= dH_t \\
&= a_t^H dS_t + b_t^H dR_t \\
&= a_t^H \{\mu S_t dt + \sigma S_t dB_t\} + b_t^H r R_t dt \\
&= \{\mu a_t^H S_t + r b_t^H R_t\} dt + \sigma a_t^H S_t dB_t.
\end{aligned}
\tag{6.3}
$$

Comparing the last line of (6.3) and (6.2) we realize that, necessarily, we have

$$
a_t^H = C_x(t, S_t).
\tag{6.4}
$$

Therefore, from

$$
C(t, S_t) = P_t = H_t = a_t^H S_t + b_t^H R_t
$$

we also obtain b_t^H

$$
b_t^H = \frac{C(t, S_t) - a_t^H S_t}{R_t}.
\tag{6.5}
$$

Now replacing (6.4) and (6.5) into (6.3) we obtain

$$
\begin{aligned}
dP_t &= \mu C_x(t, S_t) S_t dt + r R_t^{-1} \{C(t, S_t) - a_t^H S_t\} R_t dt + \sigma C_x(t, S_t) S_t dB_t \\
&= \{\mu C_x(t, S_t) S_t - r C_x(t, S_t) S_t + r C(t, S_t)\} dt + \sigma C_x(t, S_t) S_t dB_t
\end{aligned}
\tag{6.6}
$$

Finally, equating (6.6) and (6.2) we obtain

$$
\begin{aligned}
\mu C_x(t, S_t) S_t - r C_x(t, S_t) S_t + r C(t, S_t) &= C_t(t, S_t) + \mu C_x(t, S_t) S_t \\
&\quad + \frac{1}{2}\sigma^2 C_{xx}(t, S_t) S_t^2
\end{aligned}
$$

and then

$$
r C(t, S_t) = C_t(t, S_t) + r S_t C_x(t, S_t) + \frac{1}{2}\sigma^2 S_t^2 C_{xx}(t, S_t).
$$

Therefore, $C(t, x)$ satisfies (6.1).

Theorem 6.1.6 saying nothing about the explicit price $C(t, x) = P_t$. To derive the expression of the price we need to solve the Black and Scholes equation

$$
r C(t, x) = C_t(t, x) + r x C_x(t, x) + \frac{1}{2}\sigma^2 x^2 C_{xx}(t, x)
$$

for all $x \geq 0$. Notice that, at time T, $C(T, S_T) = f(S_T)$, which implies that we have to find solutions for $C(T, x) = f(x)$, i.e. only in terms of the variable x. We will discuss this problem in the next section. Before that, we look again at the necessity to impose the non-arbitrage conditions. Suppose again that the market sells the claim at a price $\tilde{P}_0 < H_0$. We assume that both the real price P_t and the wrong price $\tilde{P}_t$ are adapted processes. Given that the claim costs less than it should, we buy one unit and put it in our portfolio, i.e. we set $c_t = 1$, for all $t \in [0, T]$. Let us short-sell a portfolio H_0 which is a hedging portfolio for the claim and with the surplus $H_0 - \tilde{P}_0 > 0$ we buy bonds. Our portfolio strategy is then

$$(-a_t^H, -b_t^H + H_0 - \tilde{P}_0, 1).$$

This portfolio has an initial value of 0, i.e. $V_0 = 0$, while at time T we get

$$\begin{aligned} V_T &= -H_T + (H_0 - \tilde{P}_0)R_T + P_T \\ &= -X + (H_0 - \tilde{P}_0)R_T + X \\ &= (H_0 - \tilde{P}_0)R_T > 0 \end{aligned}$$

thus a strictly positive value. It is an arbitrage opportunity if we also show that the portfolio is self-financing, i.e. if

$$dV_t = a_t dS_t + b_t dR_t + c_t dP_t.$$

In our case we have

$$V_t = -H_t + (H_0 - \tilde{P}_0)R_t + \tilde{P}_t$$

or

$$\begin{aligned} dV_t &= -dH_t + (H_0 - \tilde{P}_0)dR_t + d\tilde{P}_t \\ &= -a_t^H dS_t + (-b_t^H + H_0 - \tilde{P}_0)dR_t + d\tilde{P}_t \end{aligned}$$

so, it is a real self-financing portfolio, and we have realized arbitrage.

6.2   Solution of the Black and Scholes equation

Let B_t be a standard Brownian motion and define the following new process

$$B_t^x = x + B_t$$

with x some constant. Then, B_t^x is still a Brownian motion which starts from x at time 0. If we want a Brownian motion which is at x at time t we need to introduce the process

$$B_s^{t,x} = x + B_s - B_t, \quad s \geq t,$$

which is called *translated Brownian motion*. In a similar manner we can define the *translated geometric Brownian motion* as follows:

$$Z_s^{t,x} = x + \int_t^s r Z_u^{t,x} du + \int_t^s \sigma Z_u^{t,x} dB_u.$$

which is a geometric Brownian motion which is at x at time t. The process $\{Z_s^{t,x}, s \geq t\}$ satisfies the stochastic differential equation

$$dZ_s^{t,x} = r Z_s^{t,x} ds + \sigma Z_s^{t,x} dB_s$$

with the following explicit solution:

$$Z_s^{t,x} = x \exp\left\{\left(r - \frac{1}{2}\sigma^2\right)(s-t) + \sigma(B_s - B_t)\right\}. \tag{6.7}$$

Before proving that $C(t, x)$ in Theorem 6.1.6 is a solution of the Black and Scholes equation (6.1) we show a similar fact for the translated Brownian motion. Notice that (6.1) is a standard partial differential equation without any random term, so no stochastic calculus is needed to get the result. But let us first consider the process $B_t^x = x + B_t$. The random variable B_t^x is normally distributed with mean x and variance t. Thus, for any function $g(x)$, we can write $\mathbb{E}g(B_t^x)$ as follows:

$$\mathbb{E}g(B_t^x) = \int_{-\infty}^{\infty} g(y) p(t, x - y) dy$$

where $p(t, z) = \frac{1}{\sqrt{2\pi t}} e^{-\frac{z^2}{2t}}$ is the density of the $N(0, t)$ random variable. We want to prove that $v(t, x) = \mathbb{E}g(B_T^{t,x})$ is a solution of a particular partial differential equation subject to the final condition $v(T, x) = g(x)$. It is an easy exercise to prove that $p(t, x)$ solves the following partial differential equation:

$$\frac{\partial}{\partial t} p(t, z) = \frac{1}{2} \frac{\partial^2}{\partial z^2} p(t, z)$$

called the *heat equation*. Indeed,

$$\frac{\partial}{\partial t} p(t, z) = p(t, z) \frac{1}{2}\left(\frac{z^2}{t^2} - \frac{1}{t}\right),$$

$$\frac{\partial}{\partial x} p(t, z) = -p(t, z) \frac{z}{t},$$

$$\frac{\partial^2}{\partial z^2} p(t, z) = -\frac{1}{t} p(t, z) + \frac{z^2}{t^2} p(t, z).$$

Now, let us define $u(t, x) = \mathbb{E}g(B_t^x)$. By direct differentiation (we assume it is possible to exchange integration and derivation) we also get that

$$\frac{\partial}{\partial t} u(t, x) = \frac{1}{2} \frac{\partial^2}{\partial x^2} u(t, x)$$

with initial condition

$$u(0, x) = \mathbb{E}g(B_0^x) = \mathbb{E}g(x) = g(x).$$

The next step is to reverse time in order to get the translated Brownian motion. We define a new function $v(t, x) = u(T - t, x)$ so that $\frac{\partial}{\partial t}v(t, x) = -\frac{\partial}{\partial t}u(t, x)$. Therefore

$$\frac{\partial}{\partial t}v(t, x) + \frac{1}{2}\frac{\partial^2}{\partial x^2}v(t, x) = 0$$

with terminal condition $v(T, x) = u(0, x) = g(x)$. Now we recall that

$$v(t, x) = u(T - t, x) = \mathbb{E}g(B_{T-t}^x) = \mathbb{E}g(B_T^{t,x})$$

because $B_{T-t}^x = x + B_{T-t} \sim x + B_T - B_t = B_T^{t,x}$. So we have proved that $v(t, x) = \mathbb{E}g(B_T^{t,x})$ with final condition $v(T, x) = g(x)$ is a solution of the partial differential equation

$$\frac{\partial}{\partial t}v(t, x) + \frac{1}{2}\frac{\partial^2}{\partial x^2}v(t, x) = 0. \tag{6.8}$$

Theorem 6.2.1 *The solution $C(t, x)$ of (6.1) is*

$$C(t, x) = e^{-r(T-t)}\mathbb{E}f(Z_T^{t,x}). \tag{6.9}$$

Proof. We start considering that $Y = \log Z_T^{t,x}$

$$Y \sim N\left(\log x + \left(r - \frac{1}{2}\sigma^2\right)(T - t) + \sigma(B_T - B_t)\right)$$

with density

$$p(t, y; x) = \frac{1}{\sqrt{2\pi\sigma^2(T - t)}}\exp\left\{-\frac{\left(y - \log x - \left(r - \frac{1}{2}\sigma^2\right)(T - t)\right)^2}{2\sigma^2(T - t)}\right\}$$

and we put in evidence the variable x in the notation $p(t, z; x)$. Now we can write $C(t, x)$ in this form:

$$C(t, x) = e^{-r(T-t)}\int_{-\infty}^{+\infty} f(e^y)p(t, y; x)\mathrm{d}y.$$

In order to prove that $C(T, x)$ solves (6.1) we first need to evaluate the partial derivatives of $p(t, y; x)$.

$$\frac{\partial}{\partial t} p(t, y; x) = p(t, y; x) \left\{ \frac{1}{2(T - t)} - \frac{\left(r - \frac{\sigma^2}{2}\right)\left(y - \log x - \left(r - \frac{\sigma^2}{2}\right)(T - t)\right)}{\sigma^2(T - t)} \right.$$

$$\left. - \frac{\left(y - \log x - \left(r - \frac{\sigma^2}{2}\right)(T - t)\right)^2}{2\sigma^2(T - t)} \right\}$$

$$\frac{\partial}{\partial x} p(t, y; x) = p(t, y; x) \frac{y - \log x - \left(r - \frac{\sigma^2}{2}\right)(T - t)}{x\sigma^2(T - t)},$$

$$\frac{\partial^2}{\partial x^2} p(t, y; x) = p(t, y; x)$$

$$\left\{ \frac{\left(y - \log x - \left(r - \frac{\sigma^2}{2}\right)(T - t)\right)^2}{x^2\sigma^4(T - t)^2} - \frac{y - \log x - \left(r - \frac{\sigma^2}{2}\right)(T - t)}{x^2\sigma^2(T - t)} \right\}.$$

It is easy to see that

$$\frac{\partial}{\partial t} p(t, y; x) + rx\frac{\partial}{\partial x} p(t, y; x) + \frac{1}{2}\sigma^2 x^2 \frac{\partial^2}{\partial x^2} p(t, y; x) = 0.$$

Now, direct calculations show the following

$$\frac{\partial}{\partial t} C(t, x) = rC(t, x) + e^{-r(T-t)}\frac{\partial}{\partial t}\int_{-\infty}^{+\infty} f(e^y)p(t, y; x)dy$$

$$= rC(t, x) + e^{-r(T-t)}\int_{-\infty}^{+\infty} f(e^y)\frac{\partial}{\partial t} p(t, y; x)dy$$

$$= rC(t, x) - e^{-r(T-t)}\int_{-\infty}^{+\infty} f(e^y)$$

$$\left\{ rx\frac{\partial}{\partial x} p(t, y; x) + \frac{1}{2}\sigma^2 x^2 \frac{\partial^2}{\partial x^2} p(t, y; x) \right\} dy$$

$$= rC(t, x) - rx\frac{\partial}{\partial x} C(t, x) - \frac{1}{2}\sigma^2 x^2 \frac{\partial^2}{\partial x^2} C(t, x).$$

So we have proved that $C(t, x)$ solves the Black and Scholes equation (6.1). To finish the proof, we need to show that the boundary condition is also fulfilled. Indeed,

$$C(T, x) = e^{-r(T-T)}\mathbb{E}f\left(Z_T^{T,x}\right) = \mathbb{E}f\left(e^{\log x}\right) = \mathbb{E}f(x) = f(x)$$

which is what we needed.

6.2.1   European call and put prices

The above calculation of the Black and Scholes price can be made exact in the case of the European call.

Theorem 6.2.2 *Let $f(x) = \max(0, x - K)$, the payoff function of the European call with strike price K. Then, the solution of*

$$C(t, x) = e^{-r(T-t)} \mathbb{E} f(Z_T^{t,x}).$$

is

$$P_t = C(t, S_t) = S_t \Phi(d_1) - e^{-r(T-t)} K \Phi(d_2)$$

with

$$d_1 = d_2 + \sigma \sqrt{T - t}$$

$$d_2 = \frac{\ln \frac{S_t}{K} + \left(r - \frac{1}{2}\sigma^2\right)(T - t)}{\sigma \sqrt{T - t}}.$$

Proof. Let us rewrite $\sigma(B_T - B_t)$ as

$$\sigma(B_T - B_t) = \sigma \sqrt{T - t} \cdot Y$$

with $Y = (B_T - B_t)/\sqrt{T - t} \sim N(0, 1)$. In order to calculate $\mathbb{E} f(Z_T^{t,x})$ we rewrite the quantity of interest as follows:

$$\mathbb{E}\left\{\max(0, Z_T^{t,x} - K)\right\} = \mathbb{E}\left\{\max\left(0, e^{\ln Z_T^{t,x}}\right)\right\}$$

$$= \mathbb{E}\left\{\max\left(0, e^{\ln x + \left(r - \frac{1}{2}\sigma^2\right)(T-t) + \sigma\sqrt{T-t}\cdot Y} - K\right)\right\}$$

The payoff is zero if $Z_T^{t,x}$ is lower than the strike price K and hence the expected value above will be zero as well. Let us restrict the calculation of the expected value to those trajectories such that we have a positive payoff. We have that

$$\max\left(0, e^{\ln x + \left(r - \frac{1}{2}\sigma^2\right)(T-t) + \sigma\sqrt{T-t}\cdot Y} - K\right) = 0$$

if

$$e^{\ln x + \left(r - \frac{1}{2}\sigma^2\right)(T-t) + \sigma\sqrt{T-t}\cdot Y} < K = e^{\log K}$$

or, better, if

$$\log x + \left(r - \frac{1}{2}\sigma^2\right)(T - t) + \sigma\sqrt{T - t} \cdot Y < \log K.$$

Therefore, we have

$$Y < \frac{\log K - \log x - \left(r - \frac{1}{2}\sigma^2\right)(T-t)}{\sigma\sqrt{T-t}}$$

which we can rewrite as $Y < -d_2$

$$d_2 = \frac{\ln \frac{x}{K} + \left(r - \frac{1}{2}\sigma^2\right)(T-t)}{\sigma\sqrt{T-t}}.$$

Thus

$$\begin{aligned}
\mathbb{E}\left\{\max(0, Z_T^{t,x} - K)\right\} &= \mathbb{E}\left\{Y\mathbf{1}_{\{Y > -d_2\}}\right\} \\
&= \int_{-d_2}^{\infty} \left(e^{\ln x + \left(r - \frac{1}{2}\sigma^2\right)(T-t) + \sigma\sqrt{T-t}\cdot y} - K\right)\phi(y)dy \\
&= xe^{r(T-t)} \int_{-d_2}^{\infty} e^{-\frac{\sigma^2(T-t)}{2} + \sigma\sqrt{T-t}y}\phi(y)dy \\
&\quad - K \int_{-d_2}^{\infty} \phi(y)dy
\end{aligned}$$

where $\phi(y)$ is the density function of the standard Gaussian random variable, i.e. $\phi(y) = e^{-\frac{y^2}{2}}/\sqrt{2\pi}$. By symmetry of the Gaussian density, we have that

$$\int_{-d_2}^{\infty} \phi(y)dy = P(Y > -d_2) = P(Y < d_2) = \Phi(d_2)$$

and then

$$\mathbb{E}\left\{\max(0, Z_T^{t,x} - K)\right\} = xe^{r(T-t)} \int_{-d_2}^{\infty} e^{-\frac{\sigma^2(T-t)}{2} + \sigma\sqrt{T-t}y}\phi(y)dy - K\Phi(d_2).$$

We now change the variable of integration in the first integral as $z = y - \sigma\sqrt{T-t}$

$$\int_{-d_2}^{\infty} e^{-\frac{\sigma^2(T-t)}{2} + \sigma(T-t)y}\phi(y)dy$$

and obtain

$$\begin{aligned}
&\int_{-d_2-\sigma\sqrt{T-t}}^{\infty} e^{-\frac{\sigma^2(T-t)}{2} + \sigma\sqrt{T-t}(z+\sigma\sqrt{T-t})} \frac{1}{\sqrt{2\pi}} e^{-\frac{(z+\sigma\sqrt{T-t})^2}{2}} dz \\
&= \int_{-d_2-\sigma\sqrt{T-t}}^{\infty} \frac{1}{\sqrt{2\pi}} e^{-\frac{1}{2}\sigma^2(T-t) + \sigma\sqrt{T-t}z + \sigma^2(T-t) - \frac{1}{2}z^2 - \frac{1}{2}\sigma^2(T-t) - z\sigma\sqrt{T-t}} dz \\
&= \int_{-d_2-\sigma\sqrt{T-t}}^{\infty} \frac{e^{-\frac{1}{2}z^2}}{\sqrt{2\pi}} dz \\
&= P(Y > -d_2 - \sigma\sqrt{T-t}) = P(Y < d_2 + \sigma\sqrt{T-t}) = \Phi(d_1)
\end{aligned}$$

with $d_1 = d_2 + +\sigma\sqrt{T-t}$. Putting all steps together we obtain

$$\mathbb{E}\left\{\max(0, Z_T^{t,x} - K)\right\} = xe^{r(T-t)}\int_{-d_2}^{\infty} e^{-\frac{1}{2}\sigma^2(T-t)+\sigma(T-t)y}\phi(y)\mathrm{d}y - K\Phi(d_2)$$
$$= xe^{r(T-t)}\Phi(d_1) - K\Phi(d_2).$$

Therefore, noting that

$$C(t,x) = e^{-r(T-t)}\mathbb{E}f(0, Z^{t,x}(T))$$

we obtain the statement of the theorem

$$C(t,x) = x\Phi(d_1) - e^{-r(T-t)}K\Phi(d_2).$$

Similar calculations can be obtained for the price of the European put option

Theorem 6.2.3 *Let $f(x) = \max(0, K - x)$, the payoff function of the European put with strike price K. Then, the solution of*

$$C(t,x) = e^{-r(T-t)}\mathbb{E}f(Z_T^{t,x}).$$

is

$$P_t = C(t, S_t) = e^{-r(T-t)}K\Phi(-d_2) - S_t\Phi(-d_1) \qquad (6.10)$$

with

$$d_1 = d_2 + \sigma\sqrt{T-t}$$
$$d_2 = \frac{\ln\frac{S_t}{K} + \left(r - \frac{1}{2}\sigma^2\right)(T-t)}{\sigma\sqrt{T-t}}.$$

The proof follows exactly the same steps as the one for the price of the call and it is omitted. The price of the put option can be obtained also via the put-call parity.

6.2.2 Put-call parity

Let us consider a portfolio Π with only puts and calls on the same underlying asset and the asset itself. Assume that the put and the call have the same maturity date T and strike price K. Let us denote by P and C the value of the put and call respectively. We short sell a call, so we put a minus in our portfolio

$$\Pi_t = S_t + P_t - C_t.$$

The value of Π at time T is a function of the payoff of the two options. Remember that the payoff of a call is $f(x) = \max(S(T) - K, 0)$ while the payoff of the put is $f(x) = \max(K - S(T), 0)$. Then, the final value of the portfolio will be

$$\Pi_T = S_T + (K - S_T) - 0 = K, \quad \text{if} \quad K > S_T,$$

$$\Pi_T = S_T + 0 - (S_T - K) = K, \quad \text{if} \quad K \leq S_T.$$

In this case, the final value of the portfolio is always K which is deterministic and we can make arbitrage out of it if we don't set some constraints on the value of the put P_t. So, we can discount this final value K and construct a proper hedging portfolio whose value at time $t = 0$ is H_0 and impose the non-arbitrage condition

$$K e^{-rT} = S_0 + P_0 - C_0.$$

Therefore, given the price C_0 from Theorem 6.2.2 we obtain the value of P_0 by the *put-call parity* formula

$$P_0 = C_0 - S_0 + K e^{-rT}.$$

In general we have that

$$P_t = C_t - S_t + K e^{-r(T-t)}, \quad 0 \leq t \leq T. \tag{6.11}$$

Exercise 6.1 *Prove that for the put option $P_t = e^{-r(T-t)} K \Phi(-d_2) - S_t \Phi(-d_1)$ as in Theorem 6.2.3 using the put-call parity formula (6.11).*

6.2.3 Option pricing with R

It is quite easy to implement put and call prices with R. The next code illustrates the idea for the call option

```
R> call.price <- function(x = 1, t = 0, T = 1, r = 1, sigma = 1,
+        K = 1) {
+        d2 <- (log(x/K) + (r - 0.5 * sigma^2) * (T - t))/(sigma *
+            sqrt(T - t))
+        d1 <- d2 + sigma * sqrt(T - t)
+        x * pnorm(d1) - K * exp(-r * (T - t)) * pnorm(d2)
+ }
```

and for the put option

```
R> put.price <- function(x = 1, t = 0, T = 1, r = 1, sigma = 1,
+        K = 1) {
+        d2 <- (log(x/K) + (r - 0.5 * sigma^2) * (T - t))/(sigma *
+            sqrt(T - t))
+        d1 <- d2 + sigma * sqrt(T - t)
+        K * exp(-r * (T - t)) * pnorm(-d2) - x * pnorm(-d1)
+ }
```

We can now calculate the price of a constract with $S_0 = 100$, strike price $K = 110$, interest rate $r = 0.05$ with maturity 3 months. In this case $T = 1/4$, i.e. one fourth of the year, if we consider daily data. We assume a volatility of $\sigma = 0.25$.

```
R> S0 <- 100
R> K <- 110
R> r <- 0.05
R> T <- 1/4
R> sigma <- 0.25
R> C <- call.price(x = S0, t = 0, T = T, r = r, K = K,
    sigma = sigma)
R> C
```

```
[1] 1.980506
```

and for the price of the put

```
R> P <- put.price(x = S0, t = 0, T = T, r = r, K = K,
    sigma = sigma)
R> P
```

```
[1] 10.61406
```

and check the put-call parity formula

```
R> C - S0 + K * exp(-r * T)
```

```
[1] 10.61406
```

Another solution is to use the **fOptions** package from the **Rmetrics** suite (see Appendix B.1.1). We have to use the function GBSOption from the **fOptions** which calculates several exact formulas for options of the Generalized Black and Scholes model

```
R> require(fOptions)
```

For the call option we use the call price we need to write

```
R> GBSOption(TypeFlag = "c", S = S0, X = K, Time = T, r = r,
    b = r,
+       sigma = sigma)
```

```
Title:
 Black Scholes Option Valuation

Call:
 GBSOption(TypeFlag = "c", S = S0, X = K, Time = T, r = r, b = r,
     sigma = sigma)

Parameters:
         Value:
 TypeFlag c
```

```
S          100
X          110
Time       0.25
r          0.05
b          0.05
sigma      0.25
```

```
Option Price:
1.980509
```

```
Description:
Wed Nov 17 00:08:45 2010
```

Notice that the function produces extensive output. If we just want a numeric value we need to access the slot price as follows:

```
R> GBSOption(TypeFlag = "c", S = S0, X = K, Time = T, r = r,
    b = r,
+      sigma = sigma)@price
```

[1] 1.980509

Note further that the generalized Black and Scholes formula includes an additional parameter b which is the *cost of carry*. In order to obtain the standard formulas one has to put b = r, as in our examples. For the put options, we need to change the argument TypeFlag from c to p

```
R> GBSOption(TypeFlag = "p", S = S0, X = K, Time = T, r = r,
    b = r,
+      sigma = sigma)@price
```

[1] 10.61407

We can make a final remark about the formula of the price of the European call option. For the formula

$$P_t = C(t, S_t) = S_t \Phi(d_1) - e^{-r(T-t)} K \Phi(d_2)$$

we see immediately that, if we add the strike price K to the notation of P_t, i.e. $P_t^K = C(t, S_t, K)$, we have that

$$a P_t^K = C(t, a S_t, a K).$$

Indeed, we can see it also numerically

```
R> a <- 5
R> a * GBSOption(TypeFlag = "c", S = S0, X = K, Time = T, r = r,
+      b = r, sigma = sigma)@price
```

[1] 9.902546

```
R> GBSOption(TypeFlag = "c", S = a * S0, X = a * K, Time = T,
    r = r,
+       b = r, sigma = sigma)@price
```

[1] 9.902546

6.2.4   The Monte Carlo approach

When explicit formulas for the calculations of the option price do not exist, it
is still possible to rely on the Monte Carlo method of Section 4.1 given that the
general price formula is given in the following form:

$$C(t, x) = e^{-r(T-t)} \mathbb{E} f(Z_T^{t,x}).$$

In order to evaluate the option price we need to be able to simulate independent
copies of the random variable $Z_T^{t,x}$ which is extremely simple due to the fact that

$$Z_T^{t,x} = x \exp\left\{\left(r - \frac{1}{2}\sigma^2\right)(T - t) + \sigma\sqrt{T - t}u\right\}$$

with $u \sim N(0, 1)$. So we need to simulate M copies of the the random variable
u and apply the transform above to obtain the value of $Z_T^{t,x}$. To each value of
the simulated $Z_T^{t,x}$ we apply the payoff function and finally we calculate the
average of these values and discount by the factor $e^{-r(T-t)}$. The next code
implements the above algorithm in the function MCPrice assuming that M is the
number of Monte Carlo replications and f a generic payoff function. To speed
up the simulation, we use antithetic sampling, because the simulation involves
the simulation of symmetric random variables.

```
R> MCPrice <- function(x = 1, t = 0, T = 1, r = 1, sigma = 1,
    M = 1000,
+       f) {
+       h <- function(m) {
+           u <- rnorm(m/2)
+           tmp <- c(x * exp((r - 0.5 * sigma^2) * (T - t) + sigma *
+               sqrt(T - t) * u), x * exp((r - 0.5 * sigma^2) * (T -
+               t) + sigma * sqrt(T - t) * (-u)))
+           mean(sapply(tmp, function(xx) f(xx)))
+       }
+       p <- h(M)
+       p * exp(-r * (T - t))
+ }
```

We now compare the Monte Carlo estimate of the price and the exact price of
a European call option.

```
R> S0 <- 100
R> K <- 110
R> r <- 0.05
R> T <- 1/4
```

```
R> sigma <- 0.25
R> GBSOption(TypeFlag = "c", S = S0, X = K, Time = T, r = r,
    b = r,
+      sigma = sigma)@price

[1] 1.980509

R> f <- function(x) max(0, x - K)
R> set.seed(123)
R> M <- 1000
R> MCPrice(x = S0, t = 0, T = T, r = r, sigma, M = M, f = f)

[1] 1.872703

R> set.seed(123)
R> M <- 50000
R> MCPrice(x = S0, t = 0, T = T, r = r, sigma, M = M, f = f)

[1] 1.991576

R> set.seed(123)
R> M <- 1e+06
R> MCPrice(x = S0, t = 0, T = T, r = r, sigma, M = M, f = f)

[1] 1.984967
```

As seen from previous example, the Monte Carlo approach is not very precise
in the calculation unless we greatly increase with the number of simulations as
any Monte Carlo estimate. The advantage of the Monte Carlo method is that it
can be easily parallelized and its precision is not affected by the dimensionality
of the problem but only by the number of replications. The next code prepares
a parallelized version of the previous MCPrice function which simply splits the
task of simulation of the M random variables between the nodes of the cluster.
We make use of the package **foreach** to distribute the task between the nodes
of a cluster. The **foreach** package assumes that a cluster is already in place,
if not, the code is executed sequentially. We modify our MCPrice function to
work with the dopar operator from the **foreach** package

```
R> MCPrice <- function(x = 1, t = 0, T = 1, r = 1, sigma = 1,
    M = 1000,
+      f) {
+      require(foreach)
+      h <- function(m) {
+          u <- rnorm(m/2)
+          tmp <- c(x * exp((r - 0.5 * sigma^2) * (T - t) + sigma *
+              sqrt(T - t) * u), x * exp((r - 0.5 * sigma^2) * (T -
+              t) + sigma * sqrt(T - t) * (-u)))
+          mean(sapply(tmp, function(xx) f(xx)))
+      }
+      nodes <- getDoParWorkers()
+      p <- foreach(m = rep(M/nodes, nodes), .combine = "c") %dopar%
```

```
+          h(m)
+       p <- mean(p)
+       p * exp(-r * (T - t))
+  }
```

Notice that we added the `foreach` command to distribute the simulations and the command `mean(p)` because, in general, the result will be a vector (one entry per node of the cluster). Before executing the new version of the parallelized code, we need to adjust the definition of the payoff function `f` which requires the parameter `K`. This parameter, is not passed to the nodes of the cluster, because it is not visible in the code inside the function `MCPrice`. So we redefine the function and then call `MCPrice` again

```
R> f <- function(x) max(0, x - 110)
R> set.seed(123)
R> M <- 50000
R> MCPrice(x = S0, t = 0, T = T, r = r, sigma, M = M, f = f)
```

```
[1] 1.991576
```

We do not give here too much detail on how to set up a cluster with R because an extensive explanation is given in Section A.6 and we not mean to be efficient in the code above, but just show that it is a feasible approach. In our example below, we will set up a cluster using the `snowfall` package, but a similar cluster can be set up using the `multicore` package. Now, to test this parallelized version we set up a cluster of two cpus (assuming, e.g., a multicore processor) and assign it to an object `cl` and start up a random number generator for parallelized tasks

```
R> require(snowfall)
R> sfInit(parallel = TRUE, cpus = 2)
```

```
R Version:   R version 2.10.1 (2009-12-14)
```

```
R> cl <- sfGetCluster()
R> clusterSetupRNG(cl, seed = rep(123, 2))
```

```
[1] "RNGstream"
```

next we inform the **foreach** package that it can make use of the **snow** cluster

```
R> require(foreach)
R> require(doSNOW)
R> registerDoSNOW(cl)
```

Just to be sure that everything is working correctly, we check how many nodes are there in our cluster using the command `getDoParWorkers`

```
R> getDoParWorkers()
```

```
[1] 2
```

Now we are ready to call MCPrice, but this time the execution will use the nodes of the cluster

```
R> M <- 50000
R> MCPrice(x = S0, t = 0, T = T, r = r, sigma, M = M, f = f)

[1] 1.992274
```

When finished with our simulations, we should not forget to stop the cluster with sfStop, but before doing that we give an empirical proof about the speed of convergence of the naive Monte Carlo method. The experiment is as follows: we make the number M of Monte Carlo replications vary in the set $m = (10, 50, 100, 150, 200, 250, 500, 1000)$. For each value of M=m, we replicate the experiment repl = 100 times. In this way, we can evaluate the empirical variability of the Monte Carlo estimate and compare with the theoretical confidence bands. The result is shown in Figure 6.2. The next code executes the double Monte Carlo runs

```
R> set.seed(123)
R> m <- c(10, 50, 100, 150, 200, 250, 500, 1000)
R> p1 <- NULL
R>·err <- NULL
R> nM <- length(m)
R> repl <- 100
R> mat <- matrix(, repl, nM)
R> for (k in 1:nM) {
+       tmp <- numeric(repl)
+       for (i in 1:repl) tmp[i] <- MCPrice(x = S0, t = 0, T = T,
+           r = r, sigma, M = m[k], f = f)
+       mat[, k] <- tmp
+       p1 <- c(p1, mean(tmp))
+       err <- c(err, sd(tmp))
+ }
R> colnames(mat) <- m
```

The next portion of code is less interesting because it is only about the graphical representation of the results of the experiment, so we present it without additional comments.

```
R> p0 <- GBSOption(TypeFlag = "c", S = S0, X = K, Time = T, r = r,
+       b = r, sigma = sigma)@price
R> minP <- min(p1 - err)
R> maxP <- max(p1 + err)
R> plot(m, p1, type = "n", ylim = c(minP, maxP), axes = F,
    ylab = "MC price",
+       xlab = "MC replications")
R> lines(m, p1 + err, col = "blue")
R> lines(m, p1 - err, col = "blue")
R> axis(2, p0, "B&S price")
R> axis(1, m)
R> boxplot(mat, add = TRUE, at = m, boxwex = 15, col = "orange",
```

```
+        axes = F)
R> points(m, p1, col = "blue", lwd = 3, lty = 3)
R> abline(h = p0, lty = 2, col = "red", lwd = 3)
```

Finally, we close the cluster with `sfStop` and inform the **foreach** package that the cluster is no longer available using the command `registerDoSEQ`.

```
R> sfStop()
R> registerDoSEQ()
```

6.2.5   Sensitivity of price to parameters

We now analyze how the theoretical price reacts as a function of the parameters of the contract. For simplicity we consider the price of the European options only. The results which we obtain from the formula should be consistent from what we expect from the price of the option. We have seen that the price of a European call is available in the following explicit form:

$$P_t = S_t \Phi(d_1) - e^{-r(T-t)} K \Phi(d_2)$$

with

$$d_1 = d_2 + \sigma \sqrt{T - t}$$
$$d_2 = \frac{\ln \frac{S_t}{K} + \left(r - \frac{1}{2}\sigma^2\right)(T - t)}{\sigma \sqrt{T - t}}$$

while for the put it reads as

$$P_t = e^{-r(T-t)} K \Phi(-d_2) - S_t \Phi(-d_1).$$

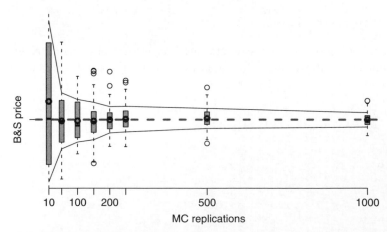

Figure 6.2   The speed of convergence of the naive Monte Carlo price toward the true Black and Scholes price.

Note that the price P_t is a function of the interest rate but not of the drift μ of the underlying geometric Brownian motion. And it is also a function of the underlying asset S_t, the strike price K, the volatility σ and the time T. We will now discuss individually the effect of each of these parameters on the price P_t.

6.2.5.1   Sensitivity to the value of underlying asset S_t

Now, consider all the rest of the parameters as fixed and assume that, at a given instant t, the value S_t is much higher than the exercise price K. The ratio S_t/K will be very large and thus both d_2 and d_1 will be large as well. In this case $\Phi(d_1) \simeq \Phi(d_2) \simeq 1$, so the price of call option becomes

$$S_t - e^{-r(T-t)}K$$

which means, the value of the underlying asset S_t is so large that we will probably exercise the option and thus we have a nonrandom payoff. So the fair price of the option is just the different between the present value and the discounted strike price K. Conversely, for the put we have $\Phi(-d_1) \simeq \Phi(-d_2) \simeq 0$, therefore the price of the option will be zero. Indeed, we will almost surely not exercise the option.

6.2.5.2   Sensitivity to the volatility σ

Suppose now that the volatility σ is zero, all the rest being fixed. The option becomes a nonrisky contract, thus the value of the option at time T is nothing but $S_0 e^{rT}$. Exactly like a bond. Then, the payoff of the call will be

$$\max(S_0 e^{rT} - K, 0)$$

which, discounted at time $t = 0$, becomes

$$e^{-rT}\max(S_0 e^{rT} - K, 0) = \max(S_0 - K e^{-rT}, 0).$$

The same conclusion can be derived directly from the Black and Scholes formula. Indeed, suppose that $S_T > K$ i.e. $S_0 > e^{-rT}K$, then $\log(S_0/K) + rT > 0$. As σ converges to 0, we have that both d_1 and d_2 diverge and then $\Phi(d_1) \sim P(d_2) \to 1$. Therefore, the final payoff will be $S_0 - K e^{-rT}$. Conversely, assume that $S_T < K$. In this case we have that $\log(S_0/K) + rT < 0$ and then $d_1, d_2 \to -\infty$ as $\sigma \to 0$. Consequently, $\Phi(d_1) \sim \Phi(d_2) \to 0$ and thus the payoff is zero.

6.2.5.3   Sensitivity to the strike price K

Figure 6.3 shows that, as expected, the price of the call option is a decreasing function of the strike price K. The plot represents the price of a call option with $S_0 = 100$, $T = 100$, $r = 0.01$ given, and presents P_t as a function of both K and the volatility σ. The following code has been used to produce Figure 6.3.

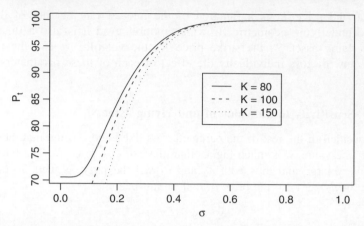

Figure 6.3   The sensitivity of the price of a call option with respect to the strike price K.

```
R> S0 <- 100
R> r <- 0.01
R> T <- 100
R> p <- function(sigma) call.price(x = S0, t = 0, T = T, r = r,
+      K = K, sigma = sigma)
R> K <- 80
R> curve(p, 0, 1, xlab = expression(sigma),
     ylab = expression(P[t]))
R> K <- 100
R> curve(p, 0, 1, add = TRUE, lty = 2)
R> K <- 150
R> curve(p, 0, 1, add = TRUE, lty = 3)
R> legend(0.5, 90, c("K=80", "K=100", "K=150"), lty = 1:3)
```

6.2.5.4   Sensitivity to the expiry time *T*

Figure 6.3 shows that the price of the call option is a increasing function of the expiry date T. The plot represents the price of a call option with $S_0 = 100$, $K = 100$, $r = 0.01$ given, and presents P_t as a function of both T and the volatility σ. The following code has been used to produce Figure 6.4.

```
R> S0 <- 100
R> r <- 0.01
R> K <- 100
R> T <- 10
R> curve(p, 0, 1, xlab = expression(sigma),
     ylab = expression(P[t]),
+      ylim = c(0, 100))
R> T <- 50
R> curve(p, 0, 1, add = TRUE, lty = 2)
R> T <- 100
R> curve(p, 0, 1, add = TRUE, lty = 3)
R> legend(0.5, 40, c("T=10", "T=50", "T=100"), lty = 1:3)
```

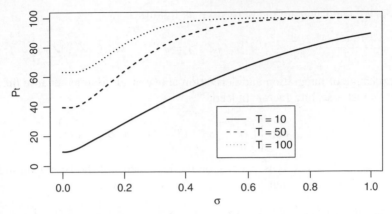

Figure 6.4   The sensitivity of the price of a call option with respect to the strike price K.

6.3   The δ-hedging and the Greeks

We have seen that the hedging portfolio (a^H, b^H) is a key ingredient of the Black and Scholes model. Remember also that the model derives an explicit formula of both a^H and b^H. In particular,

$$a^H(t) = \frac{\partial}{\partial x} C(t, S_t)$$

which can be rewritten as

$$a^H(t) = \frac{\partial}{\partial S_t} P_t$$

as a derivative, it can be interpreted as the sensitivity of the option price P_t with respect to the value of the underlying asset S_t. This derivative is also called δ and the corresponding portfolio strategy is called δ-*hedging*. The quantity δ is also one of the Greeks of the option. The attribute 'Greek' is only due to the fact that originally Greek letters were used to define these quantities. We will consider other Greeks at the end of this section. Here we just analyze the δ and, in particular, we consider the problem of evaluation of this Greek which is straightforwardly applicable to all the other Greeks.

The first thing we note is that the Greek δ is a function of time, i.e. $\delta = \delta(t) = a^H(t)$. This means that the hedging portfolio should be adapted continuously with time. This is practically impossible for several good reasons, including transaction costs (i.e. continuous adjusting will have a total transaction cost equal to ∞!) This means that completeness cannot be attained at any instant.

We now derive the explicit formula of the δ in the case of the European call. The hedging portfolio (a^H, b^H) is such that $a^H(t) = \partial C(t, x)/\partial x$ evaluated at

$x = S_t$. We can calculate the derivative as follows:

$$\frac{\partial C(t, x)}{\partial x} = \frac{\partial}{\partial x} e^{-r(T-t)} \mathbb{E} f(Z_T^{t,x}) = e^{-r(T-t)} \mathbb{E} \left\{ f'(Z_T^{t,x}) Z_T^{t,1} \right\}.$$

The exchange of integration and derivation above is admissible because the variable x is just a scaling factor. Indeed,

$$Z_T^{t,x} = x \exp \left\{ \left(r - \frac{1}{2}\sigma^2 \right) (T - t) + \sigma (B_T - B_t) \right\}$$

so the law of Z_T depends essentially on the increments of the Brownian motion. Further, $Z_T^{t,1}$ is nothing but $\partial/\partial_x f(Z_T^{t,x})$. In fact, we have

$$\frac{\partial}{\partial_x} f(Z_T^{t,x}) = \frac{\partial}{\partial_x} x \exp \left\{ \left(r - \frac{1}{2}\sigma^2 \right) (T - t) + \sigma (B_T - B_t) \right\}$$

$$= 1 \cdot \exp \left\{ \left(r - \frac{1}{2}\sigma^2 \right) (T - t) + \sigma (B_T - B_t) \right\}$$

$$= Z_T^{t,1}.$$

The critical point is now the derivative of the payoff function.

Theorem 6.3.1 *Let $f(x) = \max(0, x - K)$ be the payoff function of the European call option. Then*

$$a^H(t) = \delta = \Phi(d_1)$$

with d_1 from Theorem 6.2.2.

   Proof. We have that $f(x) = \max(0, x - K)$, therefore

$$f'(x) = \begin{cases} 1, & x > K \\ 0, & x < K \end{cases}$$

and hence

$$\frac{\partial C(t, x)}{\partial x} = e^{-r(T-t)} \mathbb{E} \left\{ f'(Z_T^{t,x}) Z^{t,1}(T) \right\}$$

$$= e^{-r(T-t)} \mathbb{E} \left\{ Z_T^{t,1} 1_{\{Z_T^{t,x} > K\}} \right\}$$

$$= e^{-r(T-t)} \int_{-d_2}^{\infty} \exp \left\{ \left(r - \frac{1}{2}\sigma^2 \right) (T - t) + \sigma \sqrt{T - t} y \right\} \phi(y) dy$$

$$= \int \exp \left\{ -\frac{1}{2}\sigma^2 (T - t) + \sigma \sqrt{T - t} y \right\} \phi(y) dy$$

with $y \sim N(0, 1)$ and $\phi(\cdot)$ the density of the same Gaussian distribution. As in the proof of Theorem 6.2.2, after the change of variable $z = y - \sigma\sqrt{T - t}$

we get

$$\frac{\partial C(t,x)}{\partial x} = \int_{-d_2-\sigma\sqrt{T-t}}^{\infty} \phi(z)dz = 1 - \Phi(-d_2 - \sigma\sqrt{T-t})$$

$$= 1 - \Phi(-d_1) = \Phi(d_1).$$

Previous results say that the hedging portfolio is such that the δ is a value in $[0, 1]$ given that it is just the cumulative distribution function of a Gaussian random variable evaluated at a particular point

$$0 \le a^H(t) = \delta = \Phi(d_1) \le 1.$$

The δ is then also called the *hedge ratio* for this reason, even though it is not strictly a ratio. Notice also that the δ is the quantity of underlying assets we have to include in the hedging portfolio, which should amount at least to one unit. In this view, the hedge ratio is interpreted as a proportion but in practice this implies that in the Black and Scholes model all the quantities involved are infinitely divisible even though we didn't mention this assumption explicitly earlier.

We finally note that the δ of a put option is given by the formula

$$\delta_{\text{put}} = -\Phi(-d_1) = \Phi(d_1) - 1 \le 0$$

which implies that while the δ of the call is always non-negative, the δ of the European put option is always negative. This is also intuitive in that, if the price of underlying asset increases, the δ of the call increases, which means that the hedging portfolio must include more assets than bonds. Vice versa for the put option.

6.3.1 The hedge ratio as a function of time

It is also interesting to see how the δ varies along the duration of the contract. Let us consider a European option with $T = 100$, $K = 100$, and let $r = 0.01$, $\sigma = 0.05$. Figure 6.5 represents the value of the hedge ratio $a^H(t)$ as a function of time for different values of the underlying asset S_t. The plot shows that, as the time approaches maturity $t = 99.5$, and the underlying assets S_t approaches the strike price K, the hedging portfolio should contain essentially only assets rather than bonds because δ is almost equal to 1. Conversely if the underlying asset S_t is smaller than K. In this case, the hedging portfolio should contain only bonds, i.e. $\delta \simeq 0$. For all times before maturity t, it is also true that that if the value of the underlying asset is much higher than the strike price K, the δ approaches 1 and vice versa if S_t is much lower than K.

```
R> delta <- function(x) {
+       d2 <- (log(x/K) + (r - 0.5 * sigma^2) * (T - t))/(sigma *
+           sqrt(T - t))
+       d1 <- d2 + sigma * sqrt(T - t)
```

```
+      pnorm(d1)
+ }
R> r <- 0.01
R> K <- 100
R> T <- 100
R> sigma <- 0.05
R> t <- 1
R> curve(delta, 0, 200, xlab = expression(S[t]),
   ylab = expression(delta = a^H(t)))
R> t <- 50
R> curve(delta, 0, 200, lty = 2, add = TRUE)
R> t <- 99.5
R> curve(delta, 0, 200, lty = 3, add = TRUE)
R> legend(150, 0.6, c("t=1", "t=50", "t=99.5"), lty = 1:3)
```

6.3.2   Hedging of generic options

There are other cases in which the payoff function of the claim is not differentiable or when its derivative is meaningless. For example, the *digital* or *binary option* option is a contract which pays a constant amount, say c, if $S_T > K$ and zero otherwise. In this case, the payoff function can be written as $f(x) = c\mathbf{1}\{x > K\}$ but its derivative is zero for all values of x different from K, which is a discontinuity point. This means, that we cannot apply the previous formula, because we obtain

$$\delta = a^H(t) = \frac{\partial C(t, x)}{\partial_x} = e^{-r(T-t)}\mathbb{E}\left\{f'(Z_T^{t,x})Z_T^{t,1}\right\} = 0.$$

This clearly implies an unrealistic allocation of the portfolio with only bonds and no assets. We now present another widely used method to overcome the problem with differentiation of the payoff function, which is the density method.

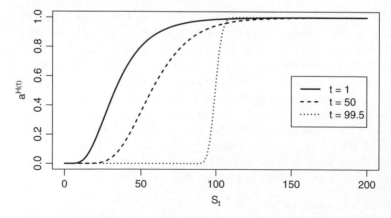

Figure 6.5   The sensitivity of the δ of a call option with respect to time t.

6.3.3   The density method

Theorem 6.3.2 *Let f be the payoff function of a contingent claim. Then*

$$C_x(t, x) = e^{-r(T-t)}\mathbb{E}\{g(t, x, Z_T^{t,x})\}$$

with

$$g(t, x, z) = f(z)\frac{\log\frac{z}{x} - \left(r - \frac{1}{2}\sigma^2\right)(T - t)}{x\sigma^2(T - t)}.$$

Notice that in the above formula, the payoff function $f(\cdot)$ now appears as is, without the need of even the existence of the first derivative. The proof is very elementary.

Proof. As before, let us define

$$Y = \log x + \left(r - \frac{1}{2}\sigma^2\right)(T - t) + \sigma(B_T - B_t), \qquad Z^{t,x} = \exp(Y).$$

Thus

$$Y \sim N\left(\log x + \left(r - \frac{1}{2}\sigma^2\right)(T - t), \sigma^2(T - t)\right).$$

Denote by $p_Y(y; x)$ the density of the random variable Y, keeping the explicit dependence on x in our notation. Then

$$C(t, x) = e^{-r(T-t)}\mathbb{E}f\left(Z_T^{t,x}\right) = e^{-r(T-t)}\int f(e^y)p_Y(y; x)dy.$$

Now we proceed with differentiation

$$\begin{aligned}
C_x(t, x) &= \frac{\partial}{\partial x}C(t, x) = e^{-r(T-t)}\int f(e^y)\frac{\partial}{\partial x}p_Y(y; x)dy \\
&= e^{-r(T-t)}\int f(e^y)\frac{\partial}{\partial x}p_Y(y; x)\frac{y - \log x - \left(r - \frac{1}{2}\sigma^2\right)(T - t)}{x\sigma^2(T - t)}dy \\
&= e^{-r(T-t)}\mathbb{E}\left\{f(Z_T^{t,x})\frac{\log Z_T^{t,x} - \log x - \left(r - \frac{1}{2}\sigma^2\right)(T - t)}{x\sigma^2(T - t)}\right\} \\
&= e^{-r(T-t)}\mathbb{E}\{g(t, x, Z_T^{t,x})\}
\end{aligned}$$

because $Y = \log Z_T^{t,x}$ and

$$p_Y(y; x) = \frac{1}{\sqrt{2\pi\sigma^2(T - t)}}\exp\left\{-\frac{\left(y - \left(\log x + \left(r - \frac{1}{2}\sigma^2\right)(T - t)\right)\right)^2}{2\sigma^2(T - t)}\right\}.$$

6.3.4 The numerical approximation

Because $C_x(t, x)$ a derivative, it is possible to evaluate the δ using the incremental ratio as seen in Section 4.2. Indeed, we can evaluate the derivate as

$$\frac{\partial}{\partial x} C(t, x) = \lim_{h \to 0} \frac{C(t, x + h) - C(t, x)}{h}$$

which we can approximate numerically as follows:

$$\frac{\partial}{\partial x} C(t, x) \sim \frac{C(t, x + h) - C(t, x)}{h}$$

with very small h. For example,

```
R> r <- 0.01
R> K <- 100
R> T <- 100
R> sigma <- 0.05
R> t <- 10
R> St <- 70
R> h <- 0.01
R> delta.num <- function(x) (call.price(x = x + h, t = t, T = T,
+      sigma = sigma, r = r, K = K) - call.price(x = x, t = t,
    T = T,
+      sigma = sigma, r = r, K = K))/h
R> delta(St)

[1] 0.9166063

R> delta.num(St)

[1] 0.9166294
```

Another option is to use the centered derivative which is better than the simple incremental ratio

$$\frac{\partial}{\partial x} C(t, x) \sim \frac{C(t, x + h) - C(t, x - h)}{2h}$$

```
R> delta.num2 <- function(x) (call.price(x = x + h, t = t, T = T,
+      sigma = sigma, r = r, K = K) - call.price(x = x - h, t = t,
+      T = T, sigma = sigma, r = r, K = K))/(2 * h)
R> delta(St)

[1] 0.9166063

R> delta.num2(St)

[1] 0.9166063
```

If we decrease h, the approximation converges to the true value quickly also for the simple approximation as shown in the next numerical example.

```
R> delta(St)

[1] 0.9166063

R> h <- 0.001
R> delta.num(St)

[1] 0.9166086

R> h <- 1e-04
R> delta.num(St)

[1] 0.9166066

R> h <- 1e-05
R> delta.num(St)

[1] 0.9166063
```

6.3.5   The Monte Carlo approach

When the formula of the price is not available in explicit form because, for example, the underlying process is not a geometric Brownian motion, the Monte Carlo method is useful in the evaluation of the Greeks. As usual, being a Monte Carlo method, it is always necessary to take into consideration the speed of convergence of the method. The Monte Carlo method can be used to calculate the δ directly using the density method or inside the numerical differentiation. We first consider the density method approach because it involves a simple modification of the formula of the option price. In this case, the δ is given by direct evaluation of the expected value

$$\delta = e^{-r(T-t)} \mathbb{E}\{g(t, x, Z_T^{t,x})\}$$

with

$$g(t, x, z) = f(z) \frac{\log \frac{z}{x} - \left(r - \frac{1}{2}\sigma^2\right)(T - t)}{x\sigma^2(T - t)}$$

and we already know that, in order to generate the random variable $Z_T^{t,x}$, we just need to generate Gaussian random variables and then apply the proper transform.

```
R> MCdelta <- function(x = 1, t = 0, T = 1, r = 1, sigma = 1,
    M = 1000,
+       f) {
+       h <- function(m) {
+           u <- rnorm(M/2)
+           tmp <- c(x * exp((r-0.5 * sigma^2) * (T - t) + sigma *
+               sqrt(T - t) * u), x * exp((r-0.5 * sigma^2) * (T -
```

```
+                    t) + sigma * sqrt(T - t) * (-u)))
+            g <- function(z) f(z) * (log(z/x) - (r - 0.5 * sigma^2) *
+                    (T - t))/(x * sigma^2 * (T - t))
+            mean(sapply(tmp, function(z) g(z)))
+        }
+        nodes <- getDoParWorkers()
+        p <- foreach(m = rep(M/nodes, nodes), .combine = "c")
+            %dopar%
+                h(m)
+        p <- mean(p)
+        p * exp(-r * (T - t))
+ }
```

We now test the new function `MCdelta`

```
R> r <- 0.01
R> K <- 100
R> T <- 100
R> t <- 10
R> sigma <- 0.05
R> S0 <- 70
R> f <- function(x) max(0, x - 100)
R> delta(S0)

[1] 0.9166063

R> set.seed(123)
R> M <- 10000
R> MCdelta(x = S0, t = 0, T = T, r = r, sigma, M = M, f = f)

[1] 0.9260659
```

6.3.6   Mixing Monte Carlo and numerical approximation

There are cases in which it is not possible to use the density method due to the fact that the underlying process is not the geometric Brownian motion. In this case, Monte Carlo is required to simulate paths of the underlying process but Monte Carlo alone is not enough. The idea is then to evaluate the derivative of Monte Carlo price path wisely. So for example, the following approach:

```
R> r <- 0.01
R> K <- 100
R> T <- 100
R> t <- 0
R> sigma <- 0.05
R> S0 <- 70
R> delta(S0)

[1] 0.9378105

R> h <- 0.001
R> p1 <- MCPrice(x = S0 + h, t = 0, T = T, r = r, sigma, M = M,
```

```
+       f = f)
R> p2 <- MCPrice(x = S0 - h, t = 0, T = T, r = r, sigma, M = M,
+       f = f)
R> p1
```

```
[1] 34.4279
```

```
R> p2
```

```
[1] 34.3844
```

```
R> (p1 - p2)/(2 * h)
```

```
[1] 21.74834
```

is wrong because the two Monte Carlo prices $C(0, S_0 + h)$ (p1) and $C(0, S_0)$ (p2) are calculated on different trajectories. But, if set the same seed for the random number generator, we actually use the same trajectories. Indeed, we have

```
R> delta(S0)
```

```
[1] 0.9378105
```

```
R> set.seed(123)
R> p1 <- MCPrice(x = S0 + h, t = 0, T = T, r = r, sigma, M = M,
+       f = f)
R> set.seed(123)
R> p2 <- MCPrice(x = S0 - h, t = 0, T = T, r = r, sigma, M = M,
+       f = f)
R> p1
```

```
[1] 34.26478
```

```
R> p2
```

```
[1] 34.26291
```

```
R> (p1 - p2)/(2 * h)
```

```
[1] 0.9368775
```

But, as mentioned, we are using the same trajectories, so there is no need to generate them twice and we can modify the code of the MCPrice as follows:

```
R> MCdelta2 <- function(x = 1, t = 0, T = 1, r = 1, sigma = 1,
    M = 1000,
+       f, dx = 0.001) {
+       h <- function(m) {
+           u <- rnorm(M/2)
+           tmp1 <- c((x + dx) * exp((r - 0.5 * sigma^2) * (T - t) +
+               sigma * sqrt(T - t) * u), (x + dx) * exp((r - 0.5 *
+               sigma^2) * (T - t) + sigma * sqrt(T - t) * (-u)))
+           tmp2 <- c((x - dx) * exp((r - 0.5 * sigma^2) * (T - t) +
```

```
+              sigma * sqrt(T - t) * u), (x - dx) * exp((r - 0.5 *
+              sigma^2) * (T - t) + sigma * sqrt(T - t) * (-u)))
+          mean(sapply(tmp1, function(x) f(x)) - sapply
+              (tmp2, function(x) f(x)))/(2 *
+              dx)
+      }
+      nodes <- getDoParWorkers()
+      p <- foreach(m = rep(M/nodes, nodes), .combine = "c")
+          %dopar%
+          h(m)
+      p <- mean(p)
+      p * exp(-r * (T - t))
+ }
```

which is faster but gives, clearly, the same result as in the above

```
R> set.seed(123)
R> MCdelta2(x = S0, t = 0, T = T, r = r, sigma = sigma, f = f,
    M = M,
+      dx = h)

[1] 0.9368775

R> delta(S0)

[1] 0.9378105
```

6.3.7 Other Greeks of options

We have seen that the first greek δ represents the sensitivity of the price of the derivative with respect to the variation of the underlying value of the asset. It is possible to study the sensitivity with respect to the other input of the formula.

6.3.7.1 The Greek theta

The Greek θ (*theta*) represents the sensitivity of the price with respect to time or, better, how the price varies as a function of time. It is defined as the derivative of $C(t, x)$ with respect to variable t. For the European call option it corresponds to

$$\text{theta} = \theta = C_t(t, x) = -rKe^{-r(T-t)}\Phi(d_2) - \frac{\sigma x}{2\sqrt{T-t}}\phi(d_1) < 0$$

where $\phi(\cdot)$ is the density of the standard Gaussian random variable. Notice that it is always a negative value, which implies that as time passes by, the value of the option tends to decrease. For the put option the corresponding formula is

$$\text{theta} = \theta_{\text{put}} = C_t(t, x) = rKe^{-r(T-t)}\Phi(-d_2) - \frac{\sigma x}{2\sqrt{T-t}}\phi(d_1)$$

and this value is not necessarily negative, which means that the price of the put may also increase with time.

6.3.7.2   The Greek gamma

Another interesting Greek is the so-called γ (*gamma*) which is the second derivative of the option price with respect to argument x or, in other words, the derivative of the δ with respect to x. So it represents the sensitivity of the hedge ratio with respect to the variation of the underlying asset. For the European call and put options it has the following form:

$$\text{gamma} = \gamma = C_{xx}(t, x) = \frac{\partial}{\partial x}C_x(t, x) = \frac{1}{\sigma x\sqrt{T - t}}\phi(d_1) > 0.$$

In this case there is no difference between put and call options.

6.3.7.3   The Greek rho

The Greek ρ (*rho*) represents the sensitivity of the options price with respect to the interest rate. For the European call option has the following form:

$$\text{rho} = \rho = \frac{\partial C(t, x)}{\partial r} = K(T - t)\Phi(d_2)e^{-r(T-t)} > 0$$

while for the European put it has the form:

$$\text{rho} = \rho_{put} = \frac{\partial C(t, x)}{\partial r} = -K(T - t)\Phi(-d_2)e^{-r(T-t)} < 0.$$

6.3.7.4   The Greek kappa

The Greek *kappa* represents the sensitivity of the option price with respect to the strike price K. For the European call option it has the form:

$$\text{kappa} = \kappa = \frac{\partial C(t, x)}{\partial K} = e^{-r(T-t)}(\Phi(-d_2) - 1) < 0$$

and for the put option:

$$\text{kappa} = \kappa_{put} = \frac{\partial C(t, x)}{\partial K} = e^{-r(T-t)}\Phi(-d_2) > 0.$$

6.3.7.5   The Greek vega

The Greek *vega* represents the sensitivity of the option price with respect to the volatility. For the European call and put options it has the form:

$$\text{vega} = \frac{\partial C(t, x)}{\partial \sigma} = x\sqrt{T - t}\phi(d_1) > 0.$$

Clearly *vega* is not a letter of the Greek alphabet but all kind of derivatives of $C(t, x)$ are called *Greeks* in mathematical finance. There are an enormous

amount of Greeks with funny names (volga, vanna, charm, speed, color) that can
be defined. For a vast review one can see Haug (1997). A last remark about the
Greeks is that they are fundamentally linked to the Black and Scholes equation
(6.1). Indeed, we can rewrite

$$rC(t, x) = C_t(t, x) + rxC_x(t, x) + \frac{1}{2}\sigma^2 x^2 C_{xx}(t, x)$$

in terms of the Greeks as follows:

$$rC(t, S_t) = \theta_t + rS_t\delta_t + \frac{1}{2}\sigma^2 S_t^2 \gamma_t.$$

6.3.8   Put and call Greeks with **Rmetrics**

The package **fOptions** which we have already encountered, allows direct calculations of the Greeks for put and call options via the function GBSCharacteristics. Here we give a self-explanatory example for the European call option.

```
R> r <- 0.01
R> K <- 100
R> T <- 100
R> t <- 10
R> sigma <- 0.05
R> S0 <- 70
R> GBSCharacteristics(TypeFlag = "c", S = S0, X = K, Time = T -
+      t, r = r, b = r, sigma = sigma)

$premium
[1] 30.89978

$delta
[1] 0.9166063

$theta
[1] -0.360923

$vega
[1] 101.8671

$rho
[1] 2993.639

$lambda
[1] 2.076469

$gamma
[1] 0.004619824
```

The function GBSCharacteristics calculates other parameters of interest (Haug
1997) about which we do not comment here. Similar output can be obtained for
the put option.

```
R> GBSCharacteristics(TypeFlag = "p", S = S0, X = K, Time = T -
+     t, r = r, b = r, sigma = sigma)

$premium
[1] 1.556748

$delta
[1] -0.08339374

$theta
[1] 0.04564667

$vega
[1] 101.8671

$rho
[1] -665.4879

$lambda
[1] -3.749842

$gamma
[1] 0.004619824
```

6.4 Pricing under the equivalent martingale measure

In Section 6.1.2 we have seen that it is possible to price options as if they were nonrisky contracts by changing the measure in the expected value of the payoff X from the physical measure P to a risk-free measure Q. We now derive the same approach of option pricing in the continuous case under the Black and Scholes market.

Definition 6.4.1 *A probability measure Q on $(\Omega, \mathcal{A})$ is called equivalent martingale measure if there exist a positive random variable $Y > 0$ such that*

$$Q(A) = \mathbb{E}\{\mathbf{1}_A Y\} = P(Y \in A), \qquad \forall A \subset \mathcal{A}$$

and the discounted price process $S_t^d = e^{-rt} S_t$ is a martingale with respect to Q.

Contrary to the one-period market in which the measure Q transform quantities in risk-free financial activities, in the above definition Q transforms the discounted stock price into a martingale. Such a measure, when it exists, it is still called *risk neutral measure*.

Theorem 6.4.2 *Let $\lambda \in \mathbb{R}$ and define*

$$M_T = \exp\left\{-\lambda B_T - \frac{1}{2}\lambda^2 T\right\}.$$

Define further a new probability measure Q as

$$Q(A) = \mathbb{E}\left(1_A M_T\right), \quad A \in \mathcal{A}.$$

Then, $W_t = B_t + \lambda t$, $0 \leq t \leq T$, is a Brownian motion with respect to Q.

Proof. To prove that W_t is a Brownian motion under the new measure Q, we have to show that its characteristic function is the one of the random variable $N(0, t)$ for each t. First, notice that $Q(A) = \mathbb{E}\left(1_A M_T\right) = \int_A M_T(\omega) P(d\omega)$. Therefore, we can rewrite integrals in this way $\mathbb{E}_Q(X) = \mathbb{E}(M_T X)$. We now calculate the characteristic function of W_t.

$$
\begin{aligned}
\varphi_{W_t}(u) &= \mathbb{E}_Q\left(e^{iuW_t}\right) = \mathbb{E}\left(M_T e^{iuW_t}\right) \\
&= e^{iu\lambda t - \frac{1}{2}\lambda^2 T} \mathbb{E}\left(\exp\left\{-\lambda B_T + iu B_t\right\}\right) \\
&= e^{iu\lambda t - \frac{1}{2}\lambda^2 T} \mathbb{E}\left(e^{-iu(B_T - B_t)}\right) \mathbb{E}\left(e^{(iu-\lambda)B_T}\right) \\
&= e^{iu\lambda t - \frac{1}{2}\lambda^2 T} \mathbb{E}\left(e^{-iuY}\right) \mathbb{E}\left(e^X\right)
\end{aligned}
$$

with $-Y \sim Y \sim N(0, T - t)$ and $X \sim N(0, (iu - \lambda)^2 T)$. So $\mathbb{E}(\exp\{iu(-Y)\})$ is the characteristic function of the Gaussian random variable Y and $\mathbb{E}(\exp\{X\})$ is the mean of the log-Normal random variable X. If we replace the quantities we have

$$\varphi_{W_t}(u) = e^{iu\lambda t - \frac{1}{2}\lambda^2 T} e^{\frac{1}{2}(T-t)u^2} e^{\frac{1}{2}(iu-\lambda)^2 T} = e^{-\frac{1}{2}u^2 t},$$

which is the characteristic function of the Gaussian random variable $N(0, t)$.

Clearly, M_T is nothing but the Radon-Nikodým derivative in the Girsanov's theorem 3.16.1 for W_t against B_t. Notice that W_t is not a Brownian motion under the physical measure P. Indeed,

$$\mathbb{E}(W_t) = \mathbb{E}(B_t + \lambda t) = \lambda t \neq 0.$$

On the other hand, $M_t = \exp\{\lambda B_t - \frac{1}{2}\lambda^2 t\}$ is a martingale under P. Indeed, $M_t = f(t, B_t)$ with $f(t, x) = e^{\lambda x - 0.5\lambda^2 t}$, thus by direct application of Itô formula, we get

$$dM_t = \lambda M_t dB_t + \frac{1}{2}\lambda^2 M_t dt - \frac{1}{2}\lambda^2 M_t dt = \lambda M_t dB_t, \qquad M_0 = 1$$

or $M_t = M_0 \lambda \int_0^t M_s dB_s$ which proves that M_t is a martingale (see Section 3.13.1). Notice that M_t is log-Normal distributed, so $\mathbb{E}|M_t| = \mathbb{E} M_t < \infty$.

Exercise 6.2 *Prove that Q defined in Theorem 6.4.2 is a probability measure.*

We can now state that the measure Q defined in Theorem 6.4.2 is the equivalent martingale measure we were aiming at.

Theorem 6.4.3 *Let* $\lambda = \frac{\mu-r}{\sigma}$. *Then,* Q *from Theorem 6.4.2 is the equivalent martingale measure in the Black and Scholes model, i.e.* $S_t^d = e^{-rt}S_t$ *is a martingale. Further,* S_t^d *solves the following stochastic differential equation:*

$$dS_t^d = \sigma S_t^d dW_t.$$

Proof. We have to prove that $S_t^d = e^{-rt}S_t$ is a martingale under the new measure Q, because clearly P and Q are equivalent. So we start by writing the stochastic differential equation for S_t^d using Itô formula for $f(t,x) = e^{-rt}x$:

$$dS_t^d = e^{-rt}dS_t - re^{-rt}S_t dt = e^{-rt}\mu S_t dt + e^{-rt}\sigma S_t dB_t - re^{-rt}S_t dt$$

$$= (\mu-r)S_t^d + S_t^d\sigma\left(dW_t - \frac{\mu-r}{\sigma}dt\right) = \sigma S_t^d dW_t.$$

Clearly $S_0^d = S_0$, thus we also have that

$$S_t^d = S_0 + \sigma\int_0^t S_u^d dW_u.$$

Finally, by the martingale representation theorem (see Section 3.13.1) and the fact that W_t is a Brownian motion under the measure Q, we can conclude that S_t^d is a martingale.
A further remark is that the process

$$Z_s^{t,x} = x\exp\left\{\left(r - \frac{1}{2}\sigma^2\right)(s-t) + \sigma(B_s - B_t)\right\}, \quad s \geq t,$$

of Equation 6.7 is replaced by the new translated Brownian motion under the Q measure

$$S_s^{t,x} = x\exp\left\{\left(r - \frac{1}{2}\sigma^2\right)(s-t) + \sigma(W_s - W_t)\right\}, \quad s \geq t.$$

Theorem 6.4.4 *Let* Q *be the risk-neutral measure of Theorem 6.4.2 and* $\lambda = (\mu-r)/\sigma$. *Then, the price of the contingent claim with payoff* $X = f(S_T)$ *is given by*

$$C(t,x) = e^{-r(T-t)}\mathbb{E}_Q\left\{f(S_T^{t,x})\right\}$$

and the delta-hedge ratio is

$$\frac{\partial}{\partial x}C(t,x) = e^{-r(T-t)}\mathbb{E}_Q\left\{g(t,x,S_T^{t,x}\right\}$$

with

$$g(t,x,s) = f(s)\frac{\log\frac{s}{x} - \left(r - \frac{1}{2}\sigma^2\right)(T-t)}{x\sigma^2(T-t)}.$$

The proof is the same as the one under the physical measure P, thus we do not replicate it. Apparently we did not gain too much with the introduction of the new measure Q, but it is not the case, in that we are now able to replicate the ideas presented in Section 6.1.2 for the one-period market.

Exercise 6.3 *Let $X_t = \exp\{\mu t + \sigma B_t\}$. Prove that X_t is a martingale with respect to the natural filtration of the Brownian motion if and only if $\mu = -\frac{1}{2}\sigma^2$.*

6.4.1   Pricing of generic claims under the risk neutral measure

Given a contingent claim with payoff X, there must exist a hedging portfolio H such that $H(T) = X$ and, due to non-arbitrage requirements, $H(t) = P(t)$. Due to the fact that S_t^d is a martingale and R_t is as well, one can easily see that $H_t^d = e^{-rT} H_t$ is a martingale under the measure Q. Thus, it is also true that

$$H_T^d = e^{-rT} H_T = e^{-rT} X.$$

For the martingale property of H_t^d we have that

$$H_t^d = \mathbb{E}_Q\{H_T^d|\mathcal{F}_t\} = \mathbb{E}_Q\{e^{-rT} X|\mathcal{F}_t\}$$

and, by the non-arbitrage property $P_t = H_t$, we also have that

$$H_t^d = e^{-rt} P_t.$$

Therefore,

$$e^{-rt} P_t = H_t^d = e^{-rT} \mathbb{E}_Q\{X|\mathcal{F}_t\}$$

and thus

$$P_t = e^{-r(T-t)} \mathbb{E}_Q\{X|\mathcal{F}_t\}.$$

Finally, at time $t = 0$

$$P_0 = e^{-r(T-0)} \mathbb{E}_Q\{X|\mathcal{F}_0\} = e^{-rT} \mathbb{E}_Q(X).$$

So, the fair price of the contingent claim at time $t = 0$ is nothing but the discounted value of the expected payoff, where expectation is calculated under the new measure Q.

6.4.2   Arbitrage and equivalent martingale measure

Notice that, not only S_t becomes a martingale under Q (after discounting by the factor e^{-rt}) but all related quantities behave as such and in particular the price of the option $P_t^d = e^{-rt} P_t$ is also a martingale (bonds are always martingale in this approach). The main effect of this new 'martingale world' is that arbitrage is prevented by construction. Indeed, assume for a while that, there exists an arbitrage opportunity, i.e. there exists a portfolio strategy (a, b, c) such that

$$V_0 \leq 0, \quad V_T \geq 0 \quad \text{and} \quad \mathbb{E}(V_T) > 0.$$

From the fact that $V_t^d = e^{-rt} V_t$ is also a martingale under Q, we have that

$$0 \geq V_0 = e^{-rT} \mathbb{E}_Q(V_T) = e^{-rT} \mathbb{E}_Q(V_T | \mathcal{F}_0)$$

or, better, $\mathbb{E}_Q(V_T) \leq 0$ and hence $V_T \leq 0$. So, we have that

$$Q(V_T > 0) = 0.$$

Given that P and Q are equivalent, we will also have that $P(V_T > 0) = 0$ and hence $P(V_T \leq 0) = 1$ which makes our arbitrage assumption $\mathbb{E}(V_T > 0)$ impossible!

We have just shown that, if there exists an equivalent martingale measure, there are no arbitrage opportunities. It turns out that the contrary is also true, i.e. a market is arbitrage free if and only if there exists a martingale measure. In this simple framework it is possible to state that there is only one martingale measure, but in general (e.g. when the underlying process is not the geometric Brownian motion) there may be multiple martingale measures, say Q_1, Q_2, etc. In this case, $\mathbb{E}_{Q_1}(X)$, $\mathbb{E}_{Q_2}(X)$, etc., might all be different numbers, so a unique fair price may not exist. In this situation hedging is a bit questionable but still possible. We will consider a general statement of this fact in Chapter 6.7.

6.5   More on numerical option pricing

Remember that for a contingent claim with payoff $X = f(S_T)$, we denote its price at time t, $P_t = C(t, S_t)$, with

$$C(t, x) = e^{-r(T-t)} \mathbb{E}_Q \{ f(S_T^{t,x}) \}$$

and Q the equivalent martingale measure. The process $S_u^{t,x}$ is a translated geometric Brownian motion under Q with explicit expression

$$S_T^{t,x} = x \exp \left\{ \left(r - \frac{1}{2}\sigma^2 \right) (T - t) + \sigma (W_T - W_t) \right\}.$$

As for the $Z_T^{t,x}$ under P, in order to simulate $S_T^{t,x}$ we just need to simulate the increments of a standard Brownian motion. So, from the point of view of the Monte Carlo approach, there is essentially no difference because $B_t - B_t \sim W_T - W_t$. Similarly for the delta hedge ratio $\delta = a_t^H$

$$a_t^H = \frac{\partial}{\partial x} C(t, S_t)$$

and b_t^H

$$b_t^H = \frac{(C(t, S_t) - a_t^H S_t)}{R_t}.$$

By Theorem 6.4.4 we have

$$C_x(t, x) = e^{-r(T-t)}\mathbb{E}_Q\{g(t, x, S_T^{t,x})\}$$

with

$$g(t, x, s) = f(s)\frac{\ln\frac{s}{x} - \left(r - \frac{1}{2}\sigma^2\right)(T - t)}{x\sigma^2(T - t)}.$$

Notice that

$$g(t, x, S_T^{t,x}) = f(S_T^{t,x})\frac{W_T - W_t}{x\sigma(T - t)}$$

$$= f(S_T^{t,x})\frac{Z\sqrt{T - t}}{x\sigma(T - t)}$$

$$= f(S_T^{t,x})\frac{Z}{x\sigma\sqrt{T - t}}$$

with $Z \sim N(0, 1)$. Thus, the Monte Carlo algorithm should just calculate

$$e^{-r(T-t)}\frac{1}{M}\sum_{i=1}^{M} f\left(x\exp\left\{\left(r - \frac{1}{2}\sigma^2\right)(T - t) + \sigma\sqrt{T - t}z_i\right\}\right)\frac{z_i}{x\sigma\sqrt{T - t}}$$

after the simulation of M pseudo-random numbers $z_1, z_2, \ldots, z_M$ from the standard Gaussian distribution. But, in general, in the standard Black and Scholes model, there is no need to simulate under the measure Q. We only need to know that this measure exists in order to know that the price calculated as in Equation (6.9) is fair.

6.5.1 Pricing of path-dependent options

For path dependent options, there rarely exist closed form formulas like the simple European and call options. In this case, one possibility is to make use of the general Equation (6.9) inside a Monte Carlo algorithm but we need to modify it because we need to simulate and keep the whole path of the process.

6.5.1.1 Barrier options

These contracts are such that their payoff is zero if, during the time $0 \leq t \leq T$, the event $S_t > \beta$ occurs, where β is the barrier. For a European call with barrier, the payoff function is given by the formula

$$X = \max\left(0, (S_T - K)\mathbf{1}_{\{S_t \leq \beta, t \in [0, T]\}}\right).$$

The price formula is

$$P(0) = e^{-rT} \mathbb{E}_Q \left[\max \left(0, (S(T) - K) \mathbf{1}_{\{S(t) \leq b, t \in [0,T]\}} \right) \right]$$

and, for $t = 0$, $S_t^{0,S_0} = S_t$. Thus,

$$P_0 = e^{-rT} \mathbb{E}_Q \left\{ \max \left(0, (S_T - K) \mathbf{1}_{\{S_t \leq \beta, t \in [0,T]\}} \right) \right\}.$$

We need to simulate a full path up to the time instant when S_t eventually crosses the barrier β. To simulate S_t we need to create a grid of time $t_i = i\Delta t$, with $t_{i+1} - t_i = \Delta_t = T/N$ with N large enough. Then, we can simulate $S_{t_{i+1}}$ conditionally on S_{t_i} as follows:

$$s_{i+1} = s_i \exp \left\{ \left(r - \frac{1}{2}\sigma^2 \right) \Delta t + \sigma \sqrt{\Delta t} Z \right\}$$

with $Z \sim N(0, 1)$ and $s_i = S(t_i)$. Figure 6.6 gives a representation of the algorithm.

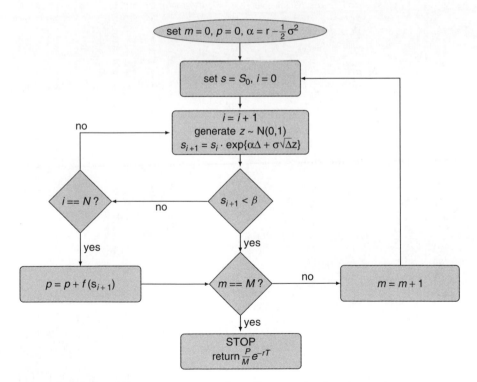

Figure 6.6   Monte Carlo algorithm to price barrier options.

6.5.1.2   Asian options

The Asian option has a payoff of the form:

$$X = \max\left(0, \frac{1}{T}\int_0^T S_t dt - K\right)$$

In practice, the integral in most cases is discretized with formulas like the following

$$X = \max\left(0, \frac{1}{N+1}\sum_{i=0}^{N} S(t_i) - K\right)$$

where $t_i = i\Delta t$, $i = 0, \ldots, N$, $\Delta t = \frac{T}{N}$. One way is to proceed as follows: generate several trajectories, evaluate the average of S_t over $t \in [0, T]$, apply the payoff function and finally discount the average payoff. The function MCAsian implements this using the usual **foreach** approach.

```
R> MCAsian <- function(S0 = 100, K = 100, t = 0, T = 1, mu = 0.1,
+       sigma = 0.1, r = 0.1, N = 100, M = 1000) {
+       require(foreach)
+       h <- function(x) {
+           require(sde)
+           z <- colMeans(sde.sim(X0 = S0, model = "BS", theta = c(mu,
+               sigma), M = x, N = N))
+           f <- function(x) max(x - K, 0)
+           p0 <- mean(sapply(z, f))
+       }
+       nodes <- getDoParWorkers()
+       p <- foreach(m = rep(M/nodes, nodes),.combine = "c") %dopar%
+           h(m)
+       p <- mean(p)
+       p * exp(-r * (T - t))
+ }
```

We can now test the function

```
R> M <- 5000
R> mu <- 0.1
R> s <- 0.5
R> K <- 110
R> r <- 0.01
R> T <- 1
R> S0 <- 100
R> set.seed(123)
R> p0 <- MCAsian(S0 = S0, K = K, t = 0, T = T, mu = mu, sigma = s,
+       r = r, N = 250, M = M)

R> p0

[1] 9.835132
```

An alternative approach is to use another recursive approach (Benth 2004). Let us define $\tilde{Z}_i = (r - 0.5\sigma^2)\Delta t + \sigma\sqrt{\Delta t}Z_i$, with Z_i, $i = 1, \ldots, N$, independent draws from N(0, 1). Then

$$s_{i+1} = s_i \exp\left\{\left(r - \frac{1}{2}\sigma^2\right)\Delta t + \sigma\sqrt{\Delta t}Z_i\right\} = s_{i+1}\exp(\tilde{Z}_i).$$

To explain the idea, let us set $N = 3$.

$$\frac{1}{4}\sum_{i=0}^{3} s_i = \frac{1}{4}(s_0 + s_1 + s_2 + s_3)$$

$$= \frac{1}{4}(s_0 + s_0 \exp \tilde{Z}_1 + s_{2+3})$$

$$= \frac{1}{4}(s_0 + s_0 \exp \tilde{Z}_1 + s_1 \exp \tilde{Z}_2 + s_3)$$

$$= \frac{1}{4}(s_0 + s_0 \exp \tilde{Z}_1 + s_0 \exp \tilde{Z}_1 \exp \tilde{Z}_2 + s_3)$$

$$= \frac{1}{4}(s_0 + (s_0 \exp \tilde{Z}_1(1 + \exp \tilde{Z}_2) + s_3)$$

$$= \frac{1}{4}(s_0 + (s_0 \exp \tilde{Z}_1(1 + \exp \tilde{Z}_2) + s_0 \exp \tilde{Z}_1 \exp \tilde{Z}_2 \exp \tilde{Z}_3)$$

$$= \frac{s_0}{4}(1 + \exp \tilde{Z}_1(1 + \exp \tilde{Z}_2(1 + \exp \tilde{Z}_3)))$$

Now let us define

$$Y_N = 1 + \exp \tilde{Z}_N$$
$$Y_{i-1} = 1 + \exp(\tilde{Z}_{i-1})Y_i, \quad i = N, N-1, \dots, 2$$

Then

$$\frac{1}{N+1}\sum_{i=0}^{N} S(t_i) = \frac{s_0}{N+1}Y_1.$$

Figure 6.7 contains the above algorithm for Monte Carlo option pricing.

6.5.2 Asian option pricing via asymptotic expansion

The geometric Brownian motion process can also be interpreted in the framework of small diffusion processes solution to the stochastic differential equations

$$dX_t^\varepsilon = a(X_t^\varepsilon, \varepsilon)dt + b(X_t^\varepsilon, \varepsilon)dW_t, \quad \varepsilon \in (0, 1].$$

The geometric Brownian motion is the particular case when $a(x, \varepsilon) = \mu x$ and $b(x, \varepsilon) = \varepsilon x$ with $\sigma = \varepsilon \in (0, 1]$. Hence, another way to estimate the price of

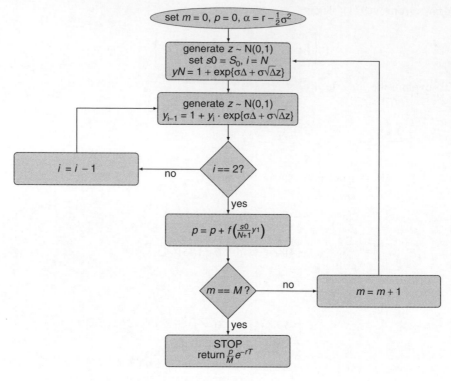

Figure 6.7   Monte Carlo algorithm to price Asian options.

the Asian option is to recognize that the payoff function contains the evaluation of the functional

$$\frac{1}{T}\int_0^T X_t \mathrm{d}t$$

which is a particular case of the general functional

$$F^\varepsilon(X_t^\varepsilon) = \sum_{\alpha=0}^r \int_0^T f_\alpha(X_t^\varepsilon, \mathrm{d})\mathrm{d}W_t^\alpha + F(X_t^\varepsilon, \varepsilon)$$

setting $W_t^0 = t$ by definition and taking

$$f_0(x, \varepsilon) = \frac{x}{T}, \quad f_1(x, \varepsilon) = 0, \quad F(x, \varepsilon) = 0.$$

Asymptotic expansion via Malliavin calculus is implemented in the **yuima** package. The next code offers an example of use via the function AEAsian

```
R> AEAsian <- function(S0 = 100, K = 100, t = 0, T = 1, mu = 0.1,
+       e = 0.1, r = 0.1, N = 1000) {
+       require(yuima)
+       diff.matrix <- matrix(c("x*e"), 1, 1)
+       model <- setModel(drift = c("mu*x"), diffusion = diff.matrix)
+       xinit <- S0
+       f <- list(expression(x/T), expression(0))
+       F <- 0
+       yuima <- setYuima(model = model, sampling =
+           setSampling(Terminal = T,
+           n = 1000))
+       yuima <- setFunctional(yuima, f = f, F = F, xinit = xinit,
+           e = e)
+       F0 <- F0(yuima)
+       rho <- expression(0)
+       get_ge <- function(x, epsilon, K, F0) {
+           tmp <- (F0 - K)+(epsilon * x)
+           tmp[(epsilon * x)<(K - F0)] <- 0
+           return(tmp)
+       }
+       epsilon <- e
+       g <- function(x) {
+           tmp <- (F0 - K) + (epsilon * x)
+           tmp[(epsilon * x) < (K - F0)] <- 0
+           tmp
+       }
+       asymp <- asymptotic_term(yuima, block = 10, rho, g)
+       exp(-r * T) * (asymp$d0+e * asymp$d1)
+ }
```

The AEAsian constructs a geometric Brownian model first, then specifies the functional and finally calculate the asymptotic expansion up to the first term. For more details we suggest to look into the **yuima** manual. The use is similar to MCAsian, but there is no simulation step and evaluation is almost instantaneous

```
R> p1 <- AEAsian(S0 = S0, K = K, t = 0, T = T, mu = mu, e = s,
   r = r)

YUIMA: Solution variable (lhs) not specified. Trying to use
   state variables.

YUIMA: 'delta' (re)defined.

YUIMA: Get variables...

YUIMA: Done.

YUIMA: Initializing...

YUIMA: Done.

YUIMA: Calculating d0...

YUIMA: Done
```

```
YUIMA: Calculating d1 term...

YUIMA: Done

R> p1

[1] 10.26708
```

6.5.3   Exotic option pricing with `Rmetrics`

We conclude mentioning that the package **fExoticOptions** contains the implementation of several exact of expansion formulas for exotic options including
Asian and barrier options. Most formulas are taken from Haug (1997). For
example, if we want to replicate previous example we can use the function
TurnbullWakemanAsianApproxOption

```
R> require(fExoticOptions)
R> p2 <- TurnbullWakemanAsianApproxOption("c", S = S0, SA = S0,
+      X = K, Time = T, time = T, tau = 0, r = r, b = r,
    sigma = s)©price
R> p2

[1] 7.96567
```

Alternatively, one can use the **fAsianOptions** which includes several specific
approximations for Asian options as well as functions to evaluate upper bounds
and lower bounds for the prices of several types of Asian options. For example,

```
R> require(fAsianOptions)
R> p3 <- GemanYorAsianOption("c", S = S0, X = K, Time = T, r = r,
+      sigma = s, doprint = FALSE)$price
R> p3

[1] 7.920475

R> p4 <- VecerAsianOption("c", S = S0, X = K, Time = T, r = r,
    sigma = s,
+      table = NA, nint = 800, eps = 1e-08, dt = 1e-10)
R> p4

[1] 7.920546

R> p5 <- ZhangAsianOption("c", S = S0, X = K, Time = T, r = r,
    sigma = s,
+      table = NA, correction = TRUE, nint = 800, eps = 1e-08,
    dt = 1e-10)
R> p5

[1] 8.294416
```

Table 6.1 summarizes the different results.

Table 6.1   Different approximations of Asian option price.

Monte Carlo	Asymp. Exp.	Turnbull-Wakeman	Geman-Yor	Vecer	Zhang
9.84	10.27	7.97	7.92	7.92	8.29

6.6   Implied volatility and volatility smiles

If we look at the market price of e.g., a call option at a given time instant, we can compare it with the price predicted by the Black and Scholes formula. Let us denote by p the price observed on the market. Now, consider the Black and Scholes price of a call option at time $t = 0$

$$p_0 = S_0\Phi(d_1) - e^{-rT}K\Phi(d_2)$$

with

$$d_1 = d_2 + \sigma\sqrt{T}, \qquad d_2 = \frac{\ln\frac{S_0}{K} + \left(r - \frac{1}{2}\sigma^2\right)T}{\sigma\sqrt{T}}.$$

Given the strike price K, the time to maturity T, the interest rate r, the current price of the asset S_0 and its volatility σ, we are able to calculate the predicted price p_0 by the above formula. We can compare this price p_0 with the market price p. The only delicate matter is which value of σ we should plug in the formula. One should think at taking the historical volatility estimated on the log-returns (see Section 5.1.2). In the next example we consider the data for the Atlantia (ATL.MI) asset for the period from 23 July 2004 to 13 May 2005. We download the data from the Yahoo server using the function `yahooSeries` from the package **fImport**

```
R> require(fImport)
R> S <- yahooSeries("ATL.MI", from = "2004-07-23",
   to = "2005-05-13")
R> head(S)

GMT
           ATL.MI.Open ATL.MI.High ATL.MI.Low ATL.MI.Close ATL.MI.Volume
2005-05-13       20.82       20.87      20.50        20.55       5944700
2005-05-12       20.88       21.10      20.63        20.77       3324700
2005-05-11       20.81       21.01      20.66        20.86       7415700
2005-05-10       20.65       20.98      20.61        20.80       2357700
2005-05-09       20.40       20.66      20.23        20.60       4171500
2005-05-06       20.30       20.68      20.08        20.50       3038800
           ATL.MI.Adj.Close
2005-05-13            16.96
2005-05-12            17.14
2005-05-11            17.22
2005-05-10            17.17
2005-05-09            17.00
2005-05-06            16.92

R> Close <- S[, "ATL.MI.Close"]
```

and then look at the data with `chartSeries` from the **quantmod** package. The result is in Figure 6.8.

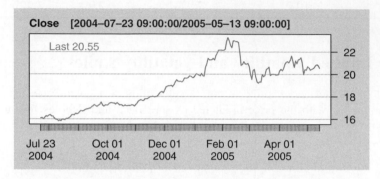

Figure 6.8   Close values of the ATL.MI asset.

```
R> require(quantmod)
R> chartSeries(Close, theme = "white")
```

We now calculate the variance of the log returns in order to obtain the historical volatility setting $\Delta = 1/252$ because we use daily data

```
R> X <- returns(Close)
R> Delta <- 1/252
R> sigma.hat <- sqrt(var(X)/Delta)[1, 1]
R> sigma.hat
```

```
[1] 0.1933289
```

We have used the R function `returns` from the **timeSeries** package. In order to use the Black and Scholes formula we need to identify all quantities. We consider a call option priced on 13 May 2005. The market price was $p = 0.0004$, the strike price $K = 23$, $S_0 = 20.55$. The expiry date was 3 June 2005 which corresponds to 15 days, thus we set $T = 15 \cdot \Delta$. The annual interest rate was $r = 0.02074$. On 13 May 2005, ATL.MI call option was priced

```
R> S0 <- Close[1]
R> K <- 23
R> T <- 15 * Delta
R> r <- 0.02074
R> sigma.hat <- as.numeric(sigma.hat)
R> require(fOptions)
R> p0 <- GBSOption("c", S = S0, X = K, Time = T, r = r, b = r,
    sigma = sigma.hat)@price
R> p0
```

```
[1] 0.003125474
```

We notice here that there is a difference in the theoretical price p_0 and the market price p. Apart from the fact that the market price is influenced by many factors (including the fact that most of the Black and Scholes hypotheses are not satisfied), one can interpret this saying that the market expectation on the exercise of this call is very small. From another point of view, one can instead consider the Black and Scholes formula replacing p_0 with p

$$p = S_0\Phi(d_1) - e^{-rT}K\Phi(d_2)$$

and solve it with respect to σ. The value of σ which satisfies the equality is called the *implied volatility*. We can use the function GBSVolatility to solve this problem

```
R> p <- 4e-04
R> sigma.imp <- GBSVolatility(p, "c", S = S0, X = K, Time = T,
       r = r,
+      b = r)
R> sigma.imp
```

```
[1] 0.1557277
```

As we see, the implied volatility is lower than the historical volatility. This is interpreted again to mean that the market expects low probability of exercising the contract. The historical probability and the implied volatility rarely match. One reason is that the Black and Scholes model assumes a fixed volatility σ over time, while market actors know that volatility is far from being stable and try to predict its trend and levels. So, implied volatility incorporates the expectation of market actors on the options and the underlying assets. If one looks at the plot of the returns for this asset (Figure 6.9) one can see the volatility changing a lot around the beginning of 2005. To be sure about this change, we make use of the cpoint function in the **sde** package. This function will allow for the discovery of a structural change point in the structure of the volatility of a generic stochastic differential equation following De Gregorio and Iacus (2008). In Section 9.1 we will discuss this in detail in a more general approach to the problem of change point in the volatility.

```
R> require(sde)
R> cp <- cpoint(as.ts(X))
R> cp
R> time(X)[cp$k0]
R> plot(X)
R> abline(v = time(X)[cp$k0], lty = 3)
```

Using the second part of the series will make the theoretical Black and Scholes price p_0 even more far than the current price market p. Similar evidence occurs for most of the standard options priced in the market.

```
$k0
[1] 123

$tau0
[1] 123

$theta1
[1] 0.00580984

$theta2
[1] 0.01649384

GMT
[1] [2005-01-12]
```

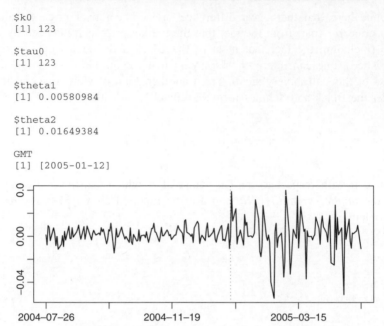

Figure 6.9   Returns of the ATL.MI asset with change point estimation.

6.6.1   Volatility smiles

The same analysis on the volatility can be done on the same asset for options with different strike prices or expiry dates. What happens in general is that the implied volatility changes for given maturity T but different values of the strike price K, but in a nonlinear way. Plotted as a function of the strike price K, the implied volatility designs a curve, sometimes u-shaped and this curve is called *volatility smile*. Consider for example, the price of the call options for the asset Apple, Inc. (AAPL). We have collected the prices of the options for different strike prices in Table 6.2. They refer to the same expiry date 17 July 2009. Data were collected on 23 April 2009, so T is about 60 working days. The current value of the assets was $S_0 = 123.90$. For each price we calculate the implied volatility:

Table 6.2   Call options prices p for Apple, Inc. for different strikes K. In all cases expiry date is 17 July 2009.

p	22.20	18.40	15.02	11.90	9.20	7.00	5.20	3.60	2.62
K	105	110	115	120	125	130	135	140	145
p	1.76	1.28	0.80	0.53	0.34	0.23	0.15	0.09	0.10
K	150	155	160	165	170	175	180	185	190

```
R> S <- yahooSeries("AAPL", from = "2009-01-02", to = "2009-04-23")
R> Close <- S[, "AAPL.Close"]
R> X <- returns(Close)
R> Delta <- 1/252
R> sigma.hat <- sqrt(var(X)/Delta)
R> sigma.hat


              AAPL.Close
AAPL.Close   0.4495024


R> Pt <- c(22.2, 18.4, 15.02, 11.9, 9.2, 7, 5.2, 3.6, 2.62, 1.76,
+       1.28, 0.8, 0.53, 0.34, 0.23, 0.15, 0.09, 0.1)
R> K <- c(105, 110, 115, 120, 125, 130, 135, 140, 145, 150, 155,
+       160, 165, 170, 175, 180, 185, 190)
R> S0 <- 123.9
R> nP <- length(Pt)
R> T <- 60 * Delta
R> r <- 0.056
R> smile <- sapply(1:nP, function(i) GBSVolatility(Pt[i], "c",
    S = S0,
+       X = K[i], Time = T, r = r, b = r))
```

and then plot the values as a function of the strike price K. Figure 6.10 shows
this almost u-shaped volatility smile picture.

```
R> vals <- c(smile, sigma.hat)
R> plot(K, smile, type = "l", ylim = c(min(vals, na.rm = TRUE),
+       max(vals, na.rm = TRUE)), main = "")
R> abline(v = S0, lty = 3, col = "blue")
R> abline(h = sigma.hat, lty = 3, col = "red")
R> axis(2, sigma.hat, expression(hat(sigma)), col = "red")
```

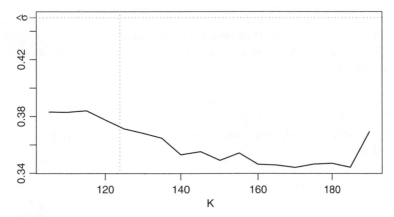

*Figure 6.10   Example of volatility smile: implied volatility as a function of the
strike price K for given expiry date T. The vertical dotted line is the current price
of the underlying asset and $\hat{\sigma}$ is the value of the historical volatility.*

6.7 Pricing of basket options

We end this chapter with a simplified approach to pricing of derivatives written on more than one asset under the multidimensional Black and Scholes model. Assume we have n assets and denote their prices by $S_1(t), S_2(t), \ldots, S_n(t)$. Assume further that m independent Brownian motions $B_1(t), \ldots, B_m(t)$ act on the market as source of noise. These noises are independent but the dynamics of asset prices may depend on one or more of them. Let Σ be the *volatility matrix* defined as

$$
\Sigma = \begin{bmatrix}
\sigma_{11} & \sigma_{12} & \cdots & \sigma_{1m} \\
\sigma_{21} & \sigma_{22} & \cdots & \sigma_{2m} \\
\vdots & \vdots & & \vdots \\
\sigma_{n1} & \sigma_{n2} & \cdots & \sigma_{nm}
\end{bmatrix}
$$

The constants σ_{ij} represent covariances between log-returns of asset prices and Brownian motions. Let further μ_i, $i = 1, \cdots, n$ be n constant. The i-th asset price satisfies the following stochastic differential equation

$$
dS_i(t) = \mu_i S_i(t)dt + S_i(t) \sum_{j=1}^{m} \sigma_{ij} dB_j(t)
$$

with solution

$$
dS_i(t) = S_i(0) \exp \left\{ \left(\mu_i - \frac{1}{2} \sum_{j=1}^{m} \sigma_{ij}^2 \right) + \sum_{j=1}^{m} \sigma_{ij} B_j(t) \right\}.
$$

The vector process $(S_1(t), S_2(t), \ldots, S_n(t))$ is called *multidimensional geometric Brownian motion* and will be the underlying process of options based on multiple assets. These kind of options are called *basket options*. Example of basket options are contracts which pay only when the difference between two assets is positive, i.e.

$$
X = f(S_1, S_2) = \max(S_1(T) - S_2(T), 0)
$$

or options which pay only when the maximal value of all underlying pass some threshold K

$$
X = f(S_1, S_2) = \max \left(\max(S_1(T), S_2(T)) - K, 0 \right)
$$

Generally speaking, the payoff $f(\cdot)$ of basket options or multidimensional T-contingent claims are

$$
X = f(S_1(T), S_2(T), \ldots, S_n(T))
$$

When an equivalent martingale measure Q exists (we will see that this is true when $n = m$) we obtain

$$P(t) = e^{-r(T-t)} \mathbb{E}_Q(X|\mathcal{F}_t)$$

where $\mathcal{F}_t$ is the σ-algebra (all the information) generated by the m Brownian motion up to time t.

In order to calculate $P(T)$ we need, as in the one-dimensional case, the equivalent martingale measure. One way to obtain it is to require the existence of the inverse of the volatility matrix Σ, but the inverse[1] exists only for square matrixes therefore one necessary, but not sufficient, condition for the existence of an equivalent martingale measure is that $m = n$. So assume $m = n$. Introduce the vector $\lambda = \Sigma^{-1}(\mu - r\,\mathbf{1})$ and define the following vector process $M(t)$

$$M(t) = \exp\left\{-\lambda' \mathbf{B}_t - \frac{1}{2}\lambda\lambda't\right\}$$

with $\mathbf{B}_t = (B_1(t), B_2(t), \dots, B_n(t))'$, $\mathbf{1} = (1, 1, \dots, 1)'$, $\mu = (\mu_1, \mu_2, \dots, \mu_n)$ and r is the interest rate. The martingale measure is obtained as in the one-dimensional case. For each $A \subset \Omega$ we define

$$Q(A) = \mathbb{E}\mathbf{1}_A M(T)$$

then Q is equivalent to P and

$$\mathbf{W}_t = \mathbf{B}_t + \lambda t$$

is a n-dimensional Brownian motion with respect to measure Q. Similarly to the one-dimensional case it can be proved that $\mathbf{W}_t$ is a Brownian motion under Q and not under P and further that $M(t)$ is a martingale under Q.

Under the measure Q the dynamics of the underlying prices are as follows:

$$dS_i(t) = rS_i(t)dt + S_i(t) \sum_{j=1}^{n} \sigma_{ij} dW_j(t)$$

and the discounted prices $S_i^d(t) = e^{-rt} S_i(t)$ satisfy the stochastic differential equations

$$dS_i^d(t) = S_i^d(t) \sum_{j=1}^{n} \sigma_{ij} dW_j(t).$$

It can be proved that the discounted process is a martingale, i.e.

$$\mathbb{E}_Q[S_i^d(t)|\mathcal{F}_s] = S_i^d(s)$$

---

[1] The inverse of a matrix Σ is denoted by Σ^{-1} and it is such that $\Sigma^{-1}\Sigma = \mathbf{I}$ where $\mathbf{I}$ is the identity matrix.

Further, the payoff of a T-contingent claim we need to calculate the conditional expected value under Q

$$P(t) = e^{-(T-t)} = \mathbb{E}_Q[X|\mathcal{F}_t] = C(t, S_1(t), S_2(t), \ldots, S_n(t))$$

with $X = f(S_1(T), S_2(T), \ldots, S_n(T))$.

6.7.1 Numerical implementation

In order to use Monte Carlo method we need to consider translated geometric Brownian motions $S_i^{t,x}$. Indeed, we can write

$$C(t, x_1, x_2, \ldots, x_n) = e^{-r(T-t)} \mathbb{E}_q[f(S_1^{t,x_1}(T), S_2^{t,x_2}(T), \ldots, S_n^{t,x_n}(T))]$$

with

$$S_i^{t,x_i}(T) = x_i \exp\left(\left(r - \frac{1}{2}\sum_{j=1}^{n}\sigma_{ij}^2\right)(T-t) + \sum_{j=1}^{n}\sigma_{ij}(W_j(T) - W_j(t))\right)$$

The hedging portfolio is then

$$H(t) = \sum_{i=1}^{n} a_i^H(t)S_i(t) + b^H(t)R(t)$$

where

$$a_i^H(t) = \frac{\partial}{\partial x_i}C(t, S_1(t), S_2(t), \ldots, S_n(t))$$

6.7.2 Completeness and arbitrage

Under the above model

- the market does not allow for arbitrage opportunities if and only if $n \leq m$
- the market is complete if and only if $n \geq m$

therefore $m = n$ is a necessary condition for complete and arbitrage free markets. This result is due to Harrison and Pliska (1981). See also references therein.

6.7.3 An example with two assets

Assume we have two assets only

$$dS_1(t) = \mu_1 S_1(t)dt + S_1(t)(\sigma_{11}dB_1(t) + \sigma_{12}dB_2(t))$$
$$dS_2(t) = \mu_2 S_2(t)dt + S_2(t)(\sigma_{21}dB_1(t) + \sigma_{22}dB_2(t))$$

with $B_1(t)$ and $B_2(t)$ two independent Brownian motions. The log-return are defined as

$$X(t) = \ln(S_1(t)/S_1(t-1)) \qquad Y(t) = \ln(S_2(t)/S_2(t-1)).$$

Let us calculate the correlation between $X(t)$ and $Y(t)$. Denote by $\Delta B_i(t) = B_i(t) - B_i(t-1) \sim N(0,1)$. Replace S_1 and S_2 in the expressions of the log-return

$$X(t) = \left(\mu_1 - \frac{1}{2}(\sigma_{11}^2 + \sigma_{12}^2)\right) + \sigma_{11}\Delta B_1(t) + \sigma_{12}\Delta B_2(t)$$

$$Y(t) = \left(\mu_2 - \frac{1}{2}(\sigma_{21}^2 + \sigma_{22}^2)\right) + \sigma_{21}\Delta B_1(t) + \sigma_{22}\Delta B_2(t).$$

Given that B_1 e B_2 are independent, so are ΔB_1 and ΔB_2, then $X(t)$ and $Y(t)$ are two Gaussian random variables.

$$X(t) \sim N\left(\mu_1 - \frac{1}{2}(\sigma_{11}^2 + \sigma_{12}^2), \sigma_{11}^2 + \sigma_{12}^2\right)$$

$$Y(t) \sim N\left(\mu_2 - \frac{1}{2}(\sigma_{21}^2 + \sigma_{22}^2), \sigma_{21}^2 + \sigma_{22}^2\right).$$

The covariance is given by

$$\begin{aligned}
\mathrm{Cov}(X(t), Y(t)) &= \mathbb{E}X(t)Y(t) - \mathbb{E}X(t)\mathbb{E}Y(t) \\
&= \sigma_{11}\sigma_{21}\mathbb{E}(\Delta B_1(t))^2 + \sigma_{12}\sigma_{22}\mathbb{E}(\Delta B_2(t))^2 \\
&\quad + (\sigma_{11}\sigma_{22} + \sigma_{12}\sigma_{21})\mathbb{E}(\Delta B_1(t)\Delta B_2(t)) \\
&= \sigma_{11}\sigma_{21} + \sigma_{12}\sigma_{22}
\end{aligned}$$

$$\begin{aligned}
\mathrm{Cor}(X(t), Y(t)) &= \frac{\mathrm{Cov}(X(t), Y(t))}{\sqrt{\mathrm{Var}X(t)}\sqrt{\mathrm{Var}Y(t)}} \\
&= \frac{\sigma_{11}\sigma_{21} + \sigma_{21}\sigma_{22}}{\sqrt{\sigma_{11}^2 + \sigma_{12}^2}\sqrt{\sigma_{21}^2 + \sigma_{22}^2}}.
\end{aligned}$$

We now set $\sigma_{11} = \sigma_1$, $\sigma_{21} = \sigma_2$, $\sigma_{12} = \sigma_1\rho$ and $\sigma_{22} = 0$, i.e.

$$\Sigma = \begin{bmatrix} \sigma_{11} & \sigma_{12} \\ \sigma_{21} & \sigma_{22} \end{bmatrix} = \begin{bmatrix} \sigma_1 & \rho\sigma_1 \\ \sigma_2 & 0 \end{bmatrix}$$

then

$$\mathrm{Cor}(X(t), Y(t)) = \frac{\sigma_1\sigma_2}{\sqrt{\sigma_1^2 + \sigma_1^2\rho^2}\sqrt{\sigma_2^2}} = \frac{\pm 1}{\sqrt{1+\rho^2}}.$$

Hence, from the above, given the sign of $\sigma_1\sigma_2$ and the value of ρ^2 it is possible to model a two assets option very simply.

6.7.4   Numerical pricing

Consider n assets $(S_1, S_2, \ldots, S_n)$ and n independent Brownian motions $(W_1, W_2, \ldots, W_n)$. The stochastic differential equations are

$$dS_i(t) = rS_i(t)dt + S_i(t) \sum_{j=1}^{n} \sigma_{ij} dW(t).$$

The payoff is $X = f(t, S_1(T), S_2(T), \ldots, S_n(T))$ and

$$P(t) = C(t, S_1(t), S_2(t), \ldots, S_n(t))$$

where

$$C(t, x_1, x_2, \ldots, x_n) = e^{-r(T-t)} \mathbb{E}_Q \left[S_1^{t,x_1}(T), \ldots, S_n^{t,x_n}(T) \right].$$

Further

$$S_i^{t,x_i}(T) = x_1 \exp \left\{ \left(r - \frac{1}{2} \sum_{j=1}^{n} \sigma_{ij} \right)(T-t) + \sum_{j=1}^{n} \sigma_{ij}(W(T) - W(t) \right\}.$$

The algorithm is as follows:

- simulate N samples of size n from the $N(0,1)$ distribution

$$(z_1^1, \ldots, z_n^1) \cdots (z_1^N, \ldots, z_n^N)$$

- calculate

$$e^{-r(T-t)} \frac{1}{N} \sum_{k=1}^{N} f(s_1^k, s_2^k, \ldots, s_n^k)$$

where

$$s_i^k = x_i \exp \left\{ \left(r - \frac{1}{2} \sum_{j=1}^{n} \sigma_{ij} \right)(T-t) + \sum_{j=1}^{n} \sigma_{ij} z_j^k \right\}.$$

6.8   Solution to exercises

Solution 6.1 (to Exercise 6.1) *We start from* $P_t = e^{-r(T-t)} K \Phi(-d_2) - S_t \Phi(-d_1)$ *in (6.10) and remind that* $\Phi(-z) = 1 - \Phi(z)$. *Therefore*

$$P_t = e^{-r(T-t)} K(1 - \Phi(d_2)) - S_t(1 - \Phi(d_1))$$
$$= S_t \Phi(d1) - e^{-r(T-t)} K \Phi(d_2) + e^{-r(T-t)} K - S_t$$
$$= C_t - S_t + e^{-r(T-t)} K.$$

which is (6.11).

Solution 6.2 (to Exercise 6.2) *First we notice that M_T is positive, and this implies that $Q(A) = \mathbb{E}(1_A M_T) \geq 0$ for all $A \in \mathcal{A}$. Further*

$$Q(\Omega) = \mathbb{E}(1_\Omega M_T) = \mathbb{E}M_T = e^{-\frac{1}{2}\lambda^2 T}\mathbb{E}\left(e^{-\lambda B_T}\right) = e^{-\frac{1}{2}\lambda^2 T}e^{\frac{1}{2}\lambda^2 T} = 1$$

because $e^{-\lambda B_T}$ is the log-Normal distribution. Now, let us consider two disjoint sets A and B in $\mathcal{A}$. Clearly, $1_{A \cup B}(\omega) = 1_A(\omega) + 1_B(\omega)$. Then

$$Q(A \cup B) = \mathbb{E}(1_{A \cup B} M_T) = \mathbb{E}(1_A M_T) + \mathbb{E}(1_B M_T) = Q(A) + Q(B).$$

Finally, let A_i, $i = 1, 2, \dots$ such that $A_i \cap A_j = \emptyset$ for every $i \neq j$.

$$Q\left(\bigcup_{i=1}^{\infty} A_i\right) = \mathbb{E}\left(1_{\bigcup_{i=1}^{\infty} A_i} M_T\right) = \mathbb{E}\left(\sum_{i=1}^{\infty} 1_{A_i} M_T\right) = \sum_{i=1}^{\infty} \mathbb{E}\left(1_{A_i} M_T\right) = \sum_{i=1}^{\infty} Q(A_i)$$

if we admit that we can exchange integration and summation. This has to be verified, but we don't do it here.

Solution 6.3 (to Exercise 6.3) *Clearly. $X_0 = 1$ and it is $\mathcal{F}_t$ measurable. Assume that X_t is a martingale. Then, X_t being a martingale is such that its expected value is constant. Hence*

$$\mathbb{E}X_t = \mathbb{E}\exp\{\mu t + \sigma B_t\} = e^{\mu t}\mathbb{E}e^{\sigma B_t} = e^{\mu t}e^{\frac{1}{2}\sigma^2 t} = e^{t(\mu + \frac{1}{2}\sigma^2)}.$$

Then, $\mathbb{E}X_t$ is constant and independent of t only if $\mu = -\frac{1}{2}\sigma^2$. Now, let $\mu = -\frac{1}{2}\sigma^2$. Let us verify that X_t is a martingale.

$$\mathbb{E}|X_t| = \mathbb{E}X_t = e^{t(\mu + \frac{1}{2}\sigma^2)} = e^0 = 1 < \infty.$$

Further,

$$\mathbb{E}\{X_t | \mathcal{F}_s\} = \mathbb{E}\left\{e^{\mu t + \sigma B_t} \,\middle|\, \mathcal{F}_s\right\} = e^{\mu t}\mathbb{E}\left\{\exp(\sigma B_s) + \exp(\sigma(B_t - B_s))| \mathcal{F}_s\right\}$$

$$= e^{\mu t}e^{\sigma B_s}\mathbb{E}\left\{\exp(\sigma(B_t - B_s))\right\} = e^{\mu t}e^{\sigma B_s}e^{\frac{1}{2}\sigma^2(t-s)}$$

$$= e^{-\frac{1}{2}\sigma^2 t}e^{\sigma B_s}e^{\frac{1}{2}\sigma^2(t-s)} = e^{\mu s + \sigma B_s} = X_s.$$

6.9   Bibliographical notes

The standard reference for option pricing is the famous book of Hull (2000). This book contains all the basic theory and a lot of insights about option pricing in practice. Advanced books for mathematically educated readers are Shreve (2004a,b) and Musiela and Rutkowski (2005). An intermediate approach is contained in Benth (2004), Mikosch (1998), Wilmott et al. (1995) and Ross (2003). The book of Haug (1997) contains an extensive list of exact formulas for pricing including non-European options. The above list of references is largely incomplete, but those books are easily found in any library of practitioners and researcher.

References

Benth, F. (2004). *Option Theory with Stochastic Analysis. An introduction to Mathematical Finance.* Springer-Verlag Berlin, Heidelberg.

Black, F. and Scholes, M. (1973). The pricing of options and corporate liabilities. *Journal of Political Economy* **81**, 637–654.

De Gregorio, A. and Iacus, S. M. (2008). Least squares volatility change point estimation for partially observed diffusion processes. *Communications in Statistics, Theory and Methods* **37**, 15, 2342–2357.

Harrison, M. and Pliska, S. (1981). Martingales and stochastic integrals in the theory of continuous trading. *Stoch. Proc. and Their App.* **11**, 215–260.

Haug, E. (1997). *The Complete Guide to Option Pricing Formulas.* McGraw-Hill, New York.

Hull, J. (2000). *Options, Futures and Other Derivatives.* Prentice-Hall, Englewood Cliffs, N.J.

Merton, R. C. (1973). Theory of rational option pricing. *Bell Journal of Economics and Management Science* **4**, 1, 141–183.

Mikosch, T. (1998). *Elementary stochastic calculus with finance in view.* World Scientific, Singapore.

Musiela, M. and Rutkowski, M. (2005). *Martingale Methods in Financial Modeling. Second Edition.* Springer-Verlag Berlin, Heidelberg.

Ross, S. (2003). *An Elementary Introduction to Mathematical Finance. Options and Other Topics.* Cambridge University Press, New York.

Shreve, S. (2004a). *Stochastic Calculus for Finance I. The Binomial Asset Pricing Model.* Springer, New York.

Shreve, S. (2004b). *Stochastic Calculus for Finance II. Continuous-Time Models.* Springer, New York.

Wilmott, P., Howison, S., and Dewynne, J. (1995). *The Mathematics of Financial Derivatives. A Student Introduction.* Cambridge University Press, New York.

7

American options

7.1 Finite difference methods

American options are similar to European options with the peculiarity that they can be exercised during the whole time interval $[0, T]$.

Assume that we own an American call contract with a strike price of 40 €, expiry date one month and the current price of the underlying asset is 50 €.

If this is the case, it will appear very profitable to exercise the option immediately and gain 10 €. But, if we also own a portfolio which includes the underlying asset of this option and we want to keep it for at least one month, then exercising now is not the best strategy. So we want to wait a little more.

Another good reason to wait is that there is still some probability (even if very small) that the price of the underlying asset decreases below 40 €; in such a case the American call option plays the role of a warranty against the decrease in value of the assets in our portfolio.

On the contrary, if we think that the asset in our portfolio is overvalued by the market it becomes interesting to understand when it is more convenient to exercise our option or sell the underlying asset.

Notice that the value of an American call is always higher than the value of the corresponding European option, while this is untrue for put options or in general in the case of dividends.

Consider now an example of an American put option: assume that the exercise price is fixed to 10 € and today's value of the underlying asset is almost zero. If we exercise immediately we have a payoff of about 10 €. If we keep waiting, the payoff can decrease below 10 €. In this case, the American put option should be exercised immediately. In general, the value of an American put is a function of the initial price of the asset S_0: the lower, the higher the value of the option.

Option Pricing and Estimation of Financial Models with R, First Edition. Stefano M. Iacus.
© 2011 John Wiley & Sons, Ltd. Published 2011 by John Wiley & Sons, Ltd.

American options which can be exercised only at prescribed dates before the expiry date are called *Bermudan*.

The price of an American put must satisfy the following Black and Scholes inequality:

$$rC(t, x) \leq C_t(t, x) + rxC_x(t, x) + \frac{1}{2}\sigma^2 x^2 C_{xx}(t, x) \qquad (7.1)$$

under additional and technical regularity conditions which ensures the continuity of the quantities involved. This function $c(t, x)$ should be evaluated at each time instant and each value of the underlying assets $x = S_t$ to decide whether is better to keep the option live or exercise. It is in general an optimal stopping time problem. An American put if the current value of the underlying asset is below the exercise frontier, say f_t, i.e. if $S_t < f_t$.

Exact formulas for this problem do not exist even in the simple case but several approximation techniques have been proposed. The first class of techniques is called finite difference method which consists in the search of the solution of the above Black and Scholes inequality (7.1) using numerical arguments. We will present two standard methods in the following (see also Hull 2000).

7.2 Explicit finite-difference method

Suppose we have a put option with final maturity date T. The idea is to partition the time interval $[0, T]$ into N intervals of the same length, say $\Delta t = T/N$. So we have $N + 1$ time instants $t_i = i\Delta t$, $i = 0, 1, \ldots, N$. We also need to discretize the potential support of the process S_t for $t \in [0, T]$. We denote by $[S_{\min}, S_{\max}]$ this support. Those two values are chosen in a substantive way and case by case for each option. Then, we divide this interval into M subintervals of the same length and denote by $\Delta S = (S_{\max}] - S_{\min}])/M$ the size of the increments of the price process S_t. We denote by $x_j = S_{\min} + j\Delta S$, $j = 0, 1, \ldots, M$, the $M + 1$ points of the grid. So we now have a grid of $(N + 1) \times (M + 1)$ points in which we evaluate numerically the solution $C(t, x)$ of (7.1). We assume that the strike price K is in between $S_{\max}$ and $S_{\min}$, otherwise the problem of pricing of the option is not particularly interesting. To shorten the notation we define the following quantities:

$$C_{i,j} = C(t_i, x_j)$$
$$= C(i\Delta t, S_{\min} + j\Delta S), \ldots i = 0, 1, \ldots, N, \quad j = 0, 1, \ldots, M.$$

Figure 7.1 gives a representation of the grid in which each point of the grid $C_{i,j}$ corresponds to the above quantity.

We now use different approximations of the partial derivatives of the function $C(t, x)$. We use the centered derivative

$$\frac{\partial}{\partial x}C(t, x) = \frac{C_{i,j+1} - C_{i,j-1}}{2\Delta S} \qquad (7.2)$$

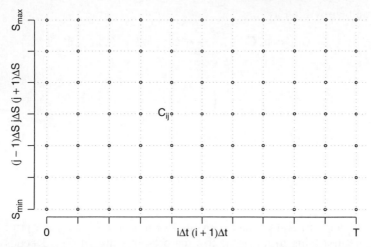

Figure 7.1   Grid of $(N+1) \times (M+1)$ point in which the function is being evaluated numerically.

and, in order to construct the second partial derivative with respect to x^2, we introduce these two approximations

$$\frac{\partial}{\partial x}C(t,x) = \frac{C_{i,j+1} - C_{i,j}}{\Delta S} \quad \text{forward approximation} \qquad (7.3)$$

and

$$\frac{\partial}{\partial x}C(t,x) = \frac{C_{i,j} - C_{i,j-1}}{\Delta S} \quad \text{backward approximation} \qquad (7.4)$$

Finally, mixing both (7.3) and (7.4) we obtain

$$\frac{\partial^2 C(t,x)}{\partial x^2} = \frac{\frac{C_{i,j+1}-C_{i,j}}{\Delta S} - \frac{C_{i,j}-C_{i,j-1}}{\Delta S}}{\Delta S} = \frac{C_{i,j+1} + C_{i,j-1} - 2C_{i,j}}{\Delta S^2} \qquad (7.5)$$

while for the time derivative we use the forward approximation

$$\frac{\partial}{\partial t}C(t,x) = \frac{C_{i+1,j} - C_{i,j}}{\Delta t}. \qquad (7.6)$$

Now, by replacing (7.2), (7.5) and (7.6) in the Black and Scholes equation we get

$$rjC_{i,j} = \frac{C_{i+1,j} - C_{i,j}}{\Delta t} + rj\Delta S\frac{C_{i,j+1} - C_{1,j-1}}{2\Delta S}$$
$$+ \frac{1}{2}\sigma^2 j^2 (\Delta S)^2 \frac{C_{i,j+1} + C_{i,j-1} - 2C_{i,j}}{2\Delta S}$$

which can be rewritten as

$$C_{i,j} = a_j^* C_{i+1,j-1} + b^* C_{i+1,j} + c_j^* C_{i+1,j+1} \qquad (7.7)$$

with

$$a_j^* = \frac{1}{1+r\Delta t} \left(-\frac{1}{2} r j \Delta t + \frac{1}{2}\sigma^2 j^2 \Delta t \right)$$

$$b_j^* = \frac{1}{1+r\Delta t} \left(1 - \sigma^2 j^2 \Delta t \right)$$

$$c_j^* = \frac{1}{1+r\Delta t} \left(\frac{1}{2} r j \Delta t + \frac{1}{2}\sigma^2 j^2 \Delta t \right)$$

which provides an iterative approximation formula for $C(t, x)$, the price of the option. The first thing to remark is that the present value of $C_{i,j}$ depends on the future values of $C_{i+1,j-1}$, $C_{i+1,j}$ and $C_{i+1,j+1}$ and our interest is in $C(0, S_0)$ in order to first price the American put option. Therefore, in order to get the solution we need to start from the rightmost part of the grid. Figure 7.2 shows this fact and also put in evidence that, luckily enough, we know the values of the edges of this grid. Indeed, at time T, we know that the payoff is given by $\max(K - S_T, 0)$ like the usual put option. Hence

$$C_{N,j} = \max(K - j\Delta S, 0), \quad j = 0, 1, \dots, M. \qquad (7.8)$$

This gives us the rightmost dots (●) in Figure 7.2. We now look at the upper edge of the figure. This corresponds to the case when $S_t = S_{max}$ at each given

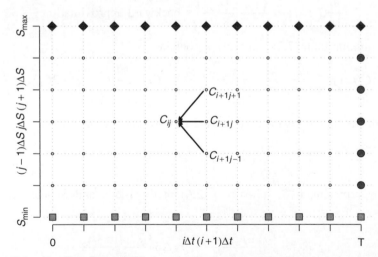

Figure 7.2 Explicit finite difference method. The value $C(t, x)$ depends on future values of the same function. Values on the edges are known by construction, so the algorithm proceeds from right to left iteratively to determine $C(0, S_0)$.

time t. Therefore, given that $S_{\max} > K$, the payoff is always zero. Then

$$C_{i,M} = 0, \quad i = 0, 1, \ldots, N, \tag{7.9}$$

which correspond to the upper diamonds ($\blacklozenge$) in Figure 7.2. Finally, if $S_t = S_{\min}$, the corresponding payoff at each given time t is the maximal payoff, i.e.

$$C_{i,0} = K - S_{\min}, \quad i = 0, 1, \ldots,, N, \tag{7.10}$$

i.e. the squares ($\blacksquare$) at the bottom of Figure 7.2. We are now able to write the equation for the $i = N - 1$

$$C_{N-1,j} = a_j^* C_{N,j-1} + b_j^* C_{N,j} + c_j^* C_{N,j+1}, \quad j = 1, 2, \ldots, M - 1$$

with

$$C_{N-1,0} = K - S_{\min} \qquad C_{N-1,M} = 0.$$

The above is a system of $M - 1$ equations with M_1 unknowns $C_{N-1,1}$, $C_{N-1,2}, \ldots, C_{N-1,M-1}$ but, each equation has an explicit solution, so there is no need to invert the matrix of the system. Once these values are obtained, we need to compare the value of $C_{N-1,j}$ with the corresponding value of European put in order to satisfy the Black and Scholes inequality (7.1). If this value $C_{N-1,j}$ is smaller than the corresponding value $C_{N,j}$ it is convenient to exercise. Therefore $N - 1$ will be the optimal time. In practice this means that, if at time $t = (N - 1)$ Δt the underlying value of the asset is $S_t = x_j = S_{\min} + j\Delta S$ and $C_{N-1,j} < C_{N,j}$, the value of the American put is lower than the value of the value of corresponding European put (i.e. the present payoff), so it is no longer convenient to keep the contract alive. We need to repeat the analysis for each value of j in order to determine which is the value of j which identifies the frontier of exercise. Once this is done, the last two columns are completely determined and we can proceed with the $i = N - 2$ back to $i = 0$. Once all the points of the grid have been determined, the first column, $C_{0,1}, C_{0,2}, \ldots, C_{0,M-1}$, corresponds to the price of the option at time $t = 0\Delta$ (or $i = 0$) we were looking for.

Example 7.2.1 *Suppose we have an American put with the following parameters: strike price $K = 30$, time to maturity $T = 1$, volatility $\sigma = 0.4$, interest rate $r = 0.05$ and current price of the underlying asset $S_0 = 36$. Let us set $S_{\min} = 0$ and $S_{\max} = 60$. We further choose $N = 10$ so that $\Delta t = 1/10 = 0.1$ and $M = 10$, with $\Delta S = 60/10 = 6$. We now fix all needed quantities. The rightmost column of Figure 7.2 is, top to bottom, as follows:*

$$\max(30 - j\Delta S, 0) = 0, 0, 0, 0, 0, 0, 6, 12, 18, 24, 30.$$

The topmost row is always zero and the bottom line is constantly equal to $K - S_{\min} = 30 - 0 = 30$. Now the explicit difference method is ready to start. We do not do it manually for all entries of the grid but instead only calculate one single *point. Let use calculate $C(N-1, j=3)$ using the recursive formula (7.7). We need the three constants a_j^*, b_j^* and c_j^**

$$a_3^* = \frac{-\frac{1}{2}r3\Delta t + \frac{1}{2}\sigma^2 3^2 \Delta t}{1 + r\Delta t} \simeq 0.0642, \quad b_3^* \simeq 0.8517, \quad c_3^* \simeq 0.0791$$

and then

$$C_{N-1,3} = a_3^* C_{N,2} + b_3^* C_{N,3} + c_3^* C_{N,4}$$
$$= 0.0642 \cdot 18 + 0.8517 \cdot 12 + 0.0791 \cdot 6 \simeq 11.85$$

The next algorithm is a simple implementation of the finite difference method which we will use later to calculate the grid of Example 7.2.1

```
R> AmericanPutExp <- function(Smin = 0, Smax, T = 1,
     N = 10, M = 10,
+      K, r = 0.05, sigma = 0.01) {
+      Dt = T/N
+      DS = (Smax - Smin)/M
+      t <- seq(0, T, by = Dt)
+      S <- seq(Smin, Smax, by = DS)
+      A <- function(j) (-0.5 * r * j * Dt + 0.5 * sigma^2 * j^2 *
+          Dt)/(1 + r * Dt)
+      B <- function(j) (1 - sigma^2 * j^2 * Dt)/(1 + r * Dt)
+      C <- function(j) (0.5 * r * j * Dt + 0.5 * sigma^2 * j^2 *
+          Dt)/(1 + r * Dt)
+      P <- matrix(, M + 1, N + 1)
+      colnames(P) <- round(t, 2)
+      rownames(P) <- round(rev(S), 2)
+      P[M + 1, ] <- K
+      P[1, ] <- 0
+      P[, N + 1] <- sapply(rev(S), function(x) max(K - x, 0))
+      optTime <- matrix(FALSE, M + 1, N + 1)
+      optTime[M + 1, ] <- TRUE
+      optTime[which(P[, N + 1] > 0), N + 1] <- TRUE
+      for (i in (N - 1):0) {
+          for (j in 1:(M - 1)) {
+              J <- M + 1 - j
+              I <- i + 1
+              P[J, I] <- A(j) * P[J + 1, I + 1] + B(j) * P[J, I +
+                  1] + C(j) * P[J - 1, I + 1]
+              if (P[J, I] < P[J, N + 1])
+                  optTime[J, I] <- TRUE
+          }
+      }
+      colnames(optTime) <- colnames(P)
+      rownames(optTime) <- rownames(P)
+      ans <- list(P = P, t = t, S = S, optTime = optTime, N = N,
+          M = M)
+      class(ans) <- "AmericanPut"
+      return(invisible(ans))
+ }
```

The function `AmericanPutExp` is a simple implementation of the finite different method, so it is by no means optimal and, given its iterative nature, it should be coded in C, but for reasons we will explain shortly, it is not worth doing that. The only thing to note is that the function also calculates the frontier of optimal exercise and stores this information in the object before exiting. Before using the function `AmericanPutExp` we also define a `plot` method for the object of class `AmericanPut` that this function creates.

```
R> plot.AmericanPut <- function(obj) {
+       plot(range(obj$t), range(obj$S), type = "n", axes = F,
        xlab = "t", ylab = "S")
+       axis(1, obj$t, obj$t)
+       axis(2, obj$S, obj$S)
+       abline(v = obj$t, h = obj$S, col = "darkgray",
        lty = "dotted")
+       for (i in 0:obj$N) {
+           for (j in 0:obj$M) {
+               J <- obj$M + 1 - j
+               I <- i + 1
+               cl <- "grey"
+               if (obj$optTime[J, I])
+                   cl <- "black"
+               text(obj$t[i + 1], obj$S[j + 1], round(obj$P[J, I],
+                   2), cex = 0.75, col = cl)
+           }
+       }
+       DS <- mean(obj$S[1:2])
+       y <- as.numeric(apply(obj$optTime, 2, function(x)
        which(x)[1]))
+       lines(obj$t, obj$S[obj$M + 2 - y] + DS, lty = 2)
+ }
```

We are now ready to replicate Example 7.2.1 with R.

```
R> put <- AmericanPutExp(Smax = 60, sigma = 0.4, K = 30)
R> round(put$P, 2)
```

	0	0.1	0.2	0.3	0.4	0.5	0.6	0.7	0.8	0.9	1
60	0.00	0.00	0.00	0.00	0.00	0.00	0.00	0.00	0.00	0.00	0
54	0.25	0.21	0.16	0.11	0.07	0.03	0.00	0.00	0.00	0.00	0
48	0.60	0.50	0.41	0.31	0.22	0.12	0.06	0.00	0.00	0.00	0
42	1.16	1.02	0.87	0.72	0.56	0.41	0.25	0.11	0.00	0.00	0
36	2.15	1.98	1.79	1.60	1.38	1.15	0.89	0.61	0.30	0.00	0
30	3.86	3.70	3.52	3.32	3.10	2.85	2.56	2.21	1.76	1.12	0
24	6.71	6.62	6.52	6.42	6.31	6.20	6.08	5.97	5.88	5.85	6
18	11.03	11.08	11.13	11.19	11.27	11.35	11.45	11.57	11.70	11.85	12
12	16.58	16.71	16.85	16.98	17.12	17.26	17.41	17.55	17.70	17.85	18
6	22.58	22.71	22.85	22.99	23.13	23.27	23.41	23.56	23.70	23.85	24
0	30.00	30.00	30.00	30.00	30.00	30.00	30.00	30.00	30.00	30.00	30

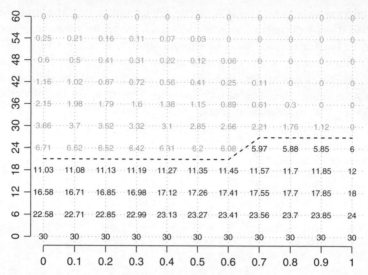

Figure 7.3   Explicit finite difference method grid for the American put option of Example 7.2.1.

and eventually plot the whole grid as in Figure 7.3 with a simple `plot` command.

```
R> plot(put)
```

Now, looking at Figure 7.3 and given that $S_0 = 36$, the price of the option is 2.15. We should keep the option until the price S_t crosses the frontier marked as a dashed line in the picture.

7.2.1   Numerical stability

Now, let us return to the expression of the solution the finite difference method of Equation 7.7. We have seen that the present price $C_{i,j}$ is the average of future prices when the underlying stock decreases its values $C_{i+1,j-1}$, remains unchanged $C_{i+1,j}$, or increases its values $C_{i+1,j+1}$. The coefficients, up to the scaling factor $1/(1 + r\Delta t) > 0$, are such that

$$(1 + r\Delta t) \times (a_j^* + b_j^* + c_j^*) = 1$$

so, if they are also positive, they can be interpreted as the probability of decrease, no change, increase of the value of the underlying asset. The problem is that in some cases this is not true. In particular, if $j < r/\sigma^2$, then $a_j^* < 0$ and if $j^2\sigma^2\Delta t > 1$ then $b_j^* < 0$. This fact also causes numerical instability because, for some iteration, the value of the average $C_{i,j}$ based on those negative coefficients may be negative as well. This negative values then propagates backward in the algorithm leading to unrealistic (actually wrong) negative prices of the option. For example, if we change M without increasing N in our example we get this result

```
R> put.bad <- AmericanPutExp(Smax = 60, sigma = 0.4, K = 30, M = 15)
R> round(put.bad$P, 2)
```

	0	0.1	0.2	0.3	0.4	0.5	0.6	0.7	0.8	0.9	1
60	0.00	0.00	0.00	0.00	0.00	0.00	0.00	0.00	0.00	0.00	0
56	-119.08	27.64	-5.54	0.98	0.00	0.00	0.00	0.00	0.00	0.00	0
52	153.67	-39.56	10.40	-2.26	0.64	0.00	0.00	0.00	0.00	0.00	0
48	-120.15	36.96	-10.89	3.97	-0.89	0.49	0.00	0.00	0.00	0.00	0
44	70.11	-22.99	9.50	-2.75	1.84	-0.23	0.44	0.00	0.00	0.00	0
40	-28.14	12.91	-3.59	3.21	-0.17	1.23	0.20	0.47	0.00	0.00	0
36	11.85	-2.02	3.84	0.76	2.07	0.97	1.27	0.62	0.61	0.00	0
32	1.01	4.24	2.42	3.10	2.37	2.46	1.96	1.84	1.28	0.98	0
28	5.21	4.45	4.63	4.27	4.21	3.92	3.75	3.42	3.15	2.67	2
24	6.78	6.78	6.64	6.57	6.44	6.35	6.21	6.10	5.95	5.85	6
20	9.50	9.49	9.49	9.49	9.50	9.52	9.55	9.61	9.70	9.85	10
16	12.80	12.88	12.97	13.06	13.17	13.29	13.41	13.55	13.70	13.85	14
12	16.58	16.71	16.84	16.98	17.12	17.26	17.41	17.55	17.70	17.85	18
8	20.54	20.69	20.83	20.97	21.12	21.26	21.41	21.55	21.70	21.85	22
4	24.57	24.71	24.85	24.99	25.13	25.27	25.41	25.56	25.70	25.85	26
0	30.00	30.00	30.00	30.00	30.00	30.00	30.00	30.00	30.00	30.00	30

7.3   Implicit finite-difference method

The implicit finite difference method has been designed to overcome the problem of numerical instability of the previous one. Finite difference method is also used in European option pricing (Brennan and Schwartz 1978). The idea is to approximate the future value $C_{i+1,j}$ with present values $C_{i,j-1}$, $C_{i,j}$ and $C_{i,j+1}$. Although it appears to be a more natural approach then the previous one, the solution of this method is more involved. The first step consists in the approximation of the partial derivates as follows:

$$\frac{\partial}{\partial t}C(t,x) = \frac{C_{i+1,j} - C_{i,j}}{\Delta t}$$

$$\frac{\partial}{\partial x}C(t,x) = \frac{C_{i,j+1} - C_{i,j-1}}{2\Delta S}$$

$$\frac{\partial}{\partial x^2}C(t,x) = \frac{C_{i,j+1} + C_{i,j-1} - 2C_{i,j}}{\Delta S^2}$$

and inserting these derivates in the Black and Scholes equation we get this new approximated solution for $C_{i,j}$:

$$C_{i+1,j} = a_j C_{i,j-1} + b_j C_{i,j} + c_j C_{i,j+1}$$

with

$$a_j = \frac{1}{2}rj\Delta t - \frac{1}{2}\sigma^2 j^2 \Delta t$$

$$b_j = 1 + \sigma^2 j^2 \Delta t + r\Delta t$$

$$c_j = -\frac{1}{2}rj\Delta t - \frac{1}{2}\sigma^2 j^2 \Delta t$$

and the same conditions on the edges of the $(N+1) \times (M+1)$ grid as in (7.8), (7.9), and (7.10). In this case the solution is not explicit. Indeed, let $i = N - 1$, then

$$C_{N,j} = a_j C_{N-1,j-1} + b_j C_{N-1,j} + c_j C_{N-1,j+1} \quad j = 1, \ldots, M-1,$$

which is the following system of $M - 1$ unknowns $C_{N-1,j}$, $j = 1, \ldots, M-1$,

$$\begin{cases} a_1 C_{N-1,0} + b_1 C_{N-1,1} + c_1 C_{N-1,2} = C_{N,1} \\ a_2 C_{N-1,1} + b_2 C_{N-1,2} + c_2 C_{N-1,3} = C_{N,2} \\ \vdots \\ a_{M-1} C_{N-1,M-2} + b_{M-1} C_{N-1,M-1} + c_{M-1} C_{N-1,M} = C_{N,M-1} \end{cases}$$

where the known quantities are the terms $C_{N,j}$, a_j, b_j e c_j. To solve this system we write it in matrix form:

$$Ax = b$$

with

$$A = \begin{bmatrix} b_1 & c_1 & 0 & 0 & 0 & 0 & \cdots & 0 \\ a_2 & b_2 & c_2 & 0 & 0 & 0 & \cdots & 0 \\ 0 & a_3 & b_3 & c_3 & 0 & 0 & \cdots & 0 \\ \vdots & & & \vdots & 0 & 0 & 0 & \vdots \\ 0 & 0 & 0 & 0 & 0 & a_{M-2} & b_{M-2} & c_{M-2} \\ 0 & 0 & 0 & 0 & 0 & 0 & a_{M-1} & b_{M-1} \end{bmatrix}$$

the vector of known terms b is

$$b = \left(C_{N,1} - a_1 C_{N-1,0}, C_{N-2}, \ldots, C_{N,M-1} - c_{M-1} C_{N-1,M} \right)'$$

and the unknowns x are

$$x' = \left(C_{N-1,1}, C_{N-1,2}, \ldots, C_{N-1,M-1} \right).$$

At this point the system can be easily solved within R with `solve(A,b)` in order to obtain the values of $C_{N-1,j}$, $j = 1, \ldots, M-1$. At this point, we need to compare the value of $C_{N-1,j}$ with the payoff $K - j\Delta S$. If $C_{N-1,j} < K - j\Delta S$

we set $C_{N-1,j} = K - j\Delta S$ and the time $i = N - 1$ corresponds to the optimal exercise time. Once the columns of the values $C_{N-1,j}$, $j = 1, \ldots, M - 1$, are available we can proceed with the calculation of $C_{N-2,j}$ by updating the above system. And so forth, back to the columns of $C_{0,j}$, $j = 1, \ldots, M - 1$. Although this method is computationally more intensive than the explicit method, it converges if both Δt and ΔS are sufficiently small. The next code implements the implicit difference method in R.

```
R> AmericanPutImp <- function(Smin = 0, Smax, T = 1, N = 10,
    M = 10,
+      K, r = 0.05, sigma = 0.01) {
+      Dt = T/N
+      DS = (Smax - Smin)/M
+      t <- seq(0, T, by = Dt)
+      S <- seq(Smin, Smax, by = DS)
+      A <- function(j) 0.5 * r * j * Dt - 0.5 * sigma^2 * j^2 *
+          Dt
+      B <- function(j) 1 + sigma^2 * j^2 * Dt + r * Dt
+      C <- function(j) -0.5 * r * j * Dt - 0.5 * sigma^2 * j^2 *
+          Dt
+      a <- sapply(0:M, A)
+      b <- sapply(0:M, B)
+      c <- sapply(0:M, C)
+      P <- matrix(, M + 1, N + 1)
+      colnames(P) <- round(t, 2)
+      rownames(P) <- round(rev(S), 2)
+      P[M + 1, ] <- K
+      P[1, ] <- 0
+      P[, N + 1] <- sapply(rev(S), function(x) max(K - x, 0))
+      AA <- matrix(0, M - 1, M - 1)
+      for (j in 1:(M - 1))  {
+          if (j > 1)
+              AA[j, j - 1] <- A(j)
+          if (j < M)
+              AA[j, j] <- B(j)
+          if (j < M - 1)
+              AA[j, j + 1] <- C(j)
+      }
+      optTime <- matrix(FALSE, M + 1, N + 1)
+      for (i in (N - 1):0) {
+          I <- i + 1
+          bb <- P[M:2, I + 1]
+          bb[1] <- bb[1] - A(1) * P[M + 1 - 0, I + 1]
+          bb[M - 1] <- bb[M - 1] - C(M - 1) * P[M + 1 - M, I +
+              1]
+          P[M:2, I] <- solve(AA, bb)
+          idx <- which(P[, I] < P[, N + 1])
+          P[idx, I] <- P[idx, N + 1]
+          optTime[idx, I] <- TRUE
+      }
+      optTime[M + 1, ] <- TRUE
+      optTime[which(P[, N + 1] > 0), N + 1] <- TRUE
```

```
+          colnames(optTime) <- colnames(P)
+          rownames(optTime) <- rownames(P)
+          ans <- list(P = P, t = t, S = S, optTime = optTime, N = N,
+             M = M)
+          class(ans) <- "AmericanPut"
+          return(invisible(ans))
+ }
```

We don't need to change the `plot` method because the resulting structure of the output of `AmericanPutImp` is the same as `AmericanPutExp`. The only difference between the two is that in the implicit method the value of the option under the exercise frontier is replaced by the payoff according to the rule explained in the above. We now again replicate Example 7.2.1.

```
R> put <- AmericanPutImp(Smax = 60, sigma = 0.4, K = 30)
R> round(put$P, 2)
```

	0	0.1	0.2	0.3	0.4	0.5	0.6	0.7	0.8	0.9	1
60	0.00	0.00	0.00	0.00	0.00	0.00	0.00	0.00	0.00	0.00	0
54	0.25	0.21	0.17	0.14	0.10	0.07	0.04	0.02	0.01	0.00	0
48	0.59	0.50	0.42	0.34	0.26	0.19	0.12	0.07	0.03	0.01	0
42	1.13	0.99	0.86	0.72	0.58	0.44	0.31	0.20	0.10	0.03	0
36	2.09	1.91	1.72	1.52	1.30	1.08	0.84	0.60	0.36	0.15	0
30	3.84	3.65	3.44	3.21	2.95	2.66	2.33	1.93	1.44	0.82	0
24	6.96	6.84	6.71	6.58	6.45	6.32	6.20	6.10	6.03	6.00	6
18	12.00	12.00	12.00	12.00	12.00	12.00	12.00	12.00	12.00	12.00	12
12	18.00	18.00	18.00	18.00	18.00	18.00	18.00	18.00	18.00	18.00	18
6	24.00	24.00	24.00	24.00	24.00	24.00	24.00	24.00	24.00	24.00	24
0	30.00	30.00	30.00	30.00	30.00	30.00	30.00	30.00	30.00	30.00	30

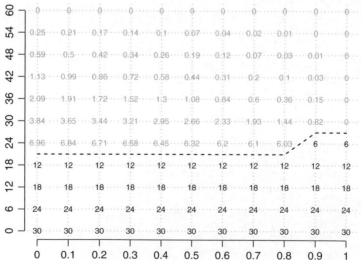

Figure 7.4   Implicit finite difference method grid for the American put option of Example 7.2.1.

and again plot the grid with

```
R> plot(put)
```

in Figure 7.4 to see the difference with the explicit difference method.

7.4 The quadratic approximation

The idea due to McMillan (1986) and Barone-Adesi and Whaley (1987) was to consider an approximation of the value of an option with a constant dividend q and a constant rate r. We did not consider dividends previously, but it does not make a substantial difference to the derivation of the partial differential equation of the price of the option. For simplicity we explain the method for the American call, and then finally give the formula also for the American put. When we have constant dividends, the Black and Scholes equation (6.1) is simply transformed into the following one

$$\frac{\partial}{\partial t}C(t,x) = \frac{1}{2}\sigma^2 x^2 \frac{\partial^2}{\partial x^2}C(t,x) + (r-q)x\frac{\partial}{\partial x}C(t,x) - rC(t,x) = 0. \quad (7.11)$$

We can introduce the so-called *early exercise premium* defined as

$$e(t,x) = C^a(t,x) - C(t,x)$$

where $C^a(t,x)$ is the price of the American call and $C(t,x)$ is the price of the European call. Both $C^a(t,x)$ and $C(t,x)$ solve (7.11) because of non-arbitrage (the difference between the two is in the boundary conditions) so also $e(t,x)$ solves the same partial differential equation. Let $k_1 = \frac{2r}{\sigma^2}$, $k_2 = \frac{2(r-q)}{\sigma^2}$, $\tau = T_t$ and $h(t) = 1 - e^{-r\tau}$. Then rewrite $e(t,x)$ as follows:

$$e(x,t) = h(\tau)\eta(h,x)$$

then Equation (7.11) can be transformed into

$$x^2\frac{\partial^2}{\partial x^2}\eta + k_2 x\frac{\partial}{\partial x}\eta - k_1(1-h)\frac{\partial}{\partial h}\eta - \frac{k_1}{h}\eta = 0.$$

Now, notice that, as $\tau \to 0$, then $\frac{\partial}{\partial h}\eta \simeq 0$ but also when $\tau = 1$ the term $(1-h) = 0$. Thus the term $1-h$ can be dropped from the partial differential equation. Hence, we are left with the simple ordinary differential equation

$$x^2\frac{\partial^2}{\partial x^2}\eta + k_2 x\frac{\partial}{\partial x}\eta - \frac{k_1}{h}\eta = 0. \quad (7.12)$$

We now set $\eta = bx^\gamma$, thus (7.12) becomes a simple quadratic equation

$$bx^\gamma\left(\gamma^2 + (k_2-1)\gamma - \frac{k_1}{h}\right) = 0$$

with solutions

$$\gamma_1 = \frac{-(k_2 - 1) - \sqrt{(k_2 - 1)^2 - 4\frac{k_1}{h}}}{2} < 0,$$

$$\gamma_2 = \frac{-(k_2 - 1) + \sqrt{(k_2 - 1)^2 - 4\frac{k_1}{h}}}{2} > 0.$$

We need to exclude $\gamma_1 < 0$ because for the call option we necessarily have $\frac{\partial}{\partial x}\eta > 0$, i.e. the early exercise premium increases with S. So our solution is $\eta = bx^{\gamma_2}$, hence

$$e(t, x) = h(\tau)\eta(h, x) = (1 - e^{-r\tau})bx^{\gamma_2}$$

or, better,

$$C^a(t, x) = C(t, x) + (1 - e^{-r\tau})bx^{\gamma_2}.$$

Now, if S^* is the value of S_t above the strike price K such that it is possible to exercise the American call option, we can write

$$S^* - K = C(t, S^*) + (1 - e^{-r\tau})b(S^*)^{\gamma_2}. \tag{7.13}$$

Further, if we assume the continuity of the hedge ratio δ for the option at point S^*, i.e. we derive the above Equation (7.13) with respect to x and evaluate it at $x = S^*$, we also obtain

$$1 = e^{-q\tau}\Phi(d_1^*) + (1 - e^{-r\tau})b\gamma_2(S^*)^{\gamma_2-1} \tag{7.14}$$

where d_1^* corresponds to d_1 for the European call option when $x = S^*$. We can now explicit b from Equation (7.14) as a function of S^*

$$b = \frac{1 - e^{-q\tau}\Phi(d_1^*)}{(1 - e^{-r\tau})\gamma_2(S^*)^{\gamma_2-1}}$$

and replace it in (7.13) which gives

$$S^* - K = C(t, S^*) + \frac{1 - e^{-q\tau}\Phi(d_1^*)}{\gamma_2}S^*$$

which can be solved numerically. Hence, finally we have that

$$C^a(t, S_t) = \begin{cases} C(t, S_t) + \frac{1-e^{-q\tau}\Phi(d_1^*)}{\gamma_2}S^*\left(\frac{S_t}{S^*}\right)^{\gamma_2}, & S_t < S^* \\ S_t - K, & S_t \geq S^*. \end{cases}$$

With the same steps, one arrives at the equation:

$$P^a(t, x) = P^a(t, x) + (1 - e^{-r\tau})cx^{\gamma_1}$$

with $P^a(t, x)$ being the price of the American put and $P(t, x)$ the price of the European put. Therefore, as before, we can define S^{**} the price of the underlying stock such that the American put is exercised and write

$$K - S^{**} = P(t, S^{**}) + (1 - e^{-r\tau})c(S^{**})^{\gamma_1}.$$

Again, imposing the continuity of the delta, we obtain a second equation:

$$-1 = -e^{-q\tau}\Phi(-d_1^{**}) + (1 - e^{-r\tau})c\gamma_1(S^{**})^{\gamma_1 - 1}$$

with d_1^{**} is d_1 for the European put with $x = S^{**}$. Now, we can get an explicit formula for c in terms of S^{**} and write

$$K - S^{**} = P(t, S^{**}) - \frac{1 - e^{-q\tau}\Phi(-d_1^{**})}{\gamma_1}S^{**}$$

and solve it numerically for S^{**}. Finally, the price of the American put option is given as follows:

$$P^a(t, S_t) = \begin{cases} K - S_t, & S_t \le S^{**}, \\ P(t, S_t) - \frac{1 - e^{-q\tau}\Phi(-d_1^{**})}{\gamma_1}S^{**}\left(\frac{S_t}{S^{**}}\right)^{\gamma_1}, & S_t > S^{**}. \end{cases}$$

Barone-Adesi and Whaley (1987) propose an efficient nonlinear algorithm to solve the above problem. We do not treat this algorithm here but rather propose the use of **fOption** package to solve this problem. In particular, the function BAWAmericanApproxOption implements this functionality. We consider the same setup of Example 7.2.1 and solve it using the quadratic approximation

```
R> require(fOptions)
R> T <- 1
R> sigma = 0.4
R> r = 0.05
R> S0 <- 36
R> K <- 30
R> BAWAmericanApproxOption("p", S = S0, X = K, Time = T, r = r,
+       b = r, sigma = sigma)@price

[1] 2.293530
```

and we can compare with the solution given by the implicit method

```
R> put <- AmericanPutImp(0, 100, T = T, K = K, r = r,
         sigma = sigma,
+        M = 100, N = 100)
R> put$P["36", 1]

[1] 2.264198
```

A quick remark about this method is that the above method is a bit unstable for very long maturities. Ju and Zhong (1999) provide an efficient modification of the quadratic approximation which is also stable.

7.5   Geske and Johnson and other approximations

Geske and Johnson (1984) proposed a method which consists in the approximation of the true American option with a sequence of Bermudan options (i.e. options which can be exercised only at given fixed dates). Let P_1 be the price of a European put option which can be exercised at time T; P_2 the price a Bermudan option which can be exercised only at times $T/2$ and T and let P_3 be the price of a Bermudan option which can be exercised at times $T/3$, $2T/3$ and T. The so-called Richardson's extrapolation method allows us to calculate the approximate value of an American option in this way:

$$P^a = P_3 + \frac{7}{2}(P_3 - P_1) - \frac{1}{2}(P_2 - P_1).$$

The problem with this approximation is that the calculation of P_2 and P_3 requires evaluation of bivariate and trivariate Gaussian integrals. It was also noticed by some authors that the approximation may not be uniform, so a modified version of this approach was later proposed by Bunch and H.E. (1992). This modification involves only two time instants but those have be chosen in an optimal way. Formally, the approximation formula is the following

$$P^a = P_{2\max} + (P_{2\max} - P_1)$$

where $P_{2\max}$ is a Bermudan option where the two time instants are chosen in a way to maximize the value of the option.

There are a number of other approximation methods which are not discussed in this book. Some of these approximations admit explicit formulas but they are quite long and the derivation is not that interesting. The package fOptions implements the method of Bjerksund and Stensland (1993) via the function BSAmericanApproxOption and the method known as Roll-Geske-Whaley (R. 1979, 1977; Whaley 1981) via the function RollGeskeWhaleyOption. The reader is invited to check the corresponding code.

7.6   Monte Carlo methods

All the above mentioned methods work quite well for one-dimensional options but closed or approximated formulas rarely exist in the case of American options written on more than one asset. As we have seen in Chapter 4. one of the advantages of the Monte Carlo method is that the computational complexity does not increase (in general) with the dimensionality of the problem. Simulation methods for pricing American options tend to exploit this particular feature of the Monte Carlo method.

7.6.1   Broadie and Glasserman simulation method

In their famous paper Broadie and Glasserman (1997) claim that it is not possible to obtain an unbiased estimator of the value of an American option via simulation.

The main reason is that all simulation schemes are by their nature discrete and hence, depending on the configuration of each simulated path, there is still a non-negative probability to reverse the decision (exercise or not) at each step (stopping time) of the simulation. As usual, r is the risk-free interest rate, T the maturity, K the strike price S_T and S_0 respectively the terminal and initial values of the stock price. We remember that the objective here is to estimate the value

$$C = \max_{\tau} \mathbb{E} \left\{ e^{-r\tau} \max(S_\tau - K, 0) \right\}$$

over all possible sets of stopping times $\tau \leq T$. In practice this is only a finite set of times on a grid as for the other numerical methods seen previously. Therefore, given a grid of times $0 = t_0 < t_1 < \cdots < t_d = T$, the idea is to simulate $S_1 = S_{t_1}$, $S_2 = S_{t_2}$, ..., $S_T = S_{t_d}$ starting from a given S_0 and then estimate the price of the option using the discretized version of the above formula

$$C = \max_{i=0,\ldots,d} \mathbb{E} \left\{ e^{-rt_i} \max(S_i - K, 0) \right\}.$$

This simulation method generates a tree where at each step b different new branches are created. Thus, for example, starting from S_0, b new values of S_1 are simulated, say $S_1^1, S_1^2, \ldots, S_1^b$, for the possible future prices of the stock at time t_1. Then, for each of the S_1^j values at time t_1, b new branches are generated at time t_2. Therefore, starting from S_1^1 at time t_1 we generate b new future values S_2^{11}, $S_2^{12}, \ldots, S_2^{1b}$. And so forth for the other $b - 1$ nodes at time t_1. For simplicity we consider $b = 3$ as in the original paper of Broadie and Glasserman (1997). Figure 7.5 represents a simulated three. It is important to notice that the tree is ordered only horizontally with time and not vertically with the asset price as in the numerical methods. This means that (see Figure 7.5) there is no ordering between S_T^{11}, S_T^{12} and S_T^{13} but they all depend on S_1^1. Similarly, S_1^1, S_1^2 and S_1^3 are not ordered. Once such a simulated tree is available, it is possible to introduce

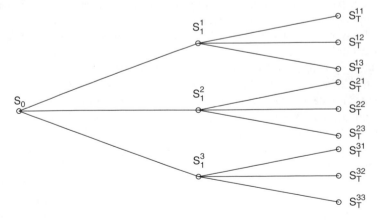

Figure 7.5   Example of simulated path with $b = 3$ branches per node.

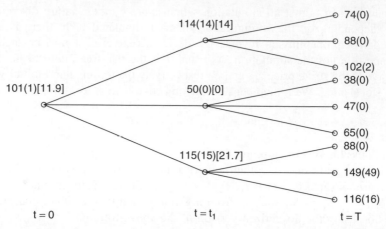

114(14)[14]

101(1)[11.9]

50(0)[0]

115(15)[21.7]

74(0)

88(0)

102(2)
38(0)

47(0)

65(0)
88(0)

149(49)

116(16)

t = 0   t = t₁   t = T

Figure 7.6   Example of simulated path from Broadie and Glasserman (1997). In parentheses the payoff at the node and in squared brackets the value of the upper estimator Θ of the American call option.

two estimators of the option prices. both biased (one from above Θ and one from below θ) but asymptotically (with the number of Monte Carlo simulations and the number of branches b in the tree) unbiased. This leads to the construction of confidence intervals for the real price of the American option. We start the description of the upper estimator Θ. Consider the bottom node in Figure 7.6 at time t_1. For this node the asset price is $S = 115$ and the payoff of the call with a strike price of $K = 100$ is 15 (in parentheses in the plot). At time T, the last three nodes at time $t = T$ have payoff 0, 49 and 16. The expected payoff is then $21.7 = (0 + 49 + 16)/3$. We compare this value with current payoff which is 15. Then the upper estimator is the maximum between 15 and 21.7. We put this number in squared brackets. For the middle node at time t_1 both the expected and current payoff are null, thus the present value of the option is zero as well. For the upper node we need to compare the present payoff, which is 14, with the expected payoff $(0 + 0 + 2)/3 = 0.7$, therefore we put 14 in squared brackets. Then, the final estimate is obtained by comparing the payoff at time 0, which is 1, and the expected payoff at time $t = 1$, i.e. $(14 + 0 + 21.7)/3 = 11.9 = \Theta$.

For the lower estimator the procedure is more involved. Consider again the same bottom node at time $t = 1$. The expected payoff at time $t = T$ is calculated using branch 2 and 3, i.e. $(49 + 16)/2 = 32.5$. This value is compared with the current payoff which is 15. In this case it is worth to continue, and the continuation value is 0, the payoff of branch 1. The same procedure is obtained considering only branch 1 and 3 to calculate the expected payoff, i.e. $(0 + 16)/2 = 8$. In this case $8 < 15$ then present value of option is 15. Finally, the expected payoff using branches 1 and 2 is $(0 + 49)/2 = 24.5$ which is bigger than 15. So the continuation value is 16, the payoff of branch 3. Now we have three different values: 0, 15 and 16. The estimated value for this node is the mean of these values, i.e. $(0 + 15 + 16)/3 = 10.3$. The procedure is iterated on the remaining nodes back to the initial node. The final lower estimate $\theta = 8.1$. Thus, for this option we have an upper bound $\Theta = 11.9$ and a lower bound $\theta = 8.1$. These are upper and lower estimates of the option in one single Monte Carlo replication. In order to get the convergence, we need to iterate for this procedure a sufficient number of Monte Carlo replications. We now present a script which implements this procedure on a single trajectory. The reader can easily embed this code into a Monte Carlo procedure. This code is designed to be sufficiently flexible, it accepts any number of periods d and any number of branches b. In order to save memory, the tree is stored on a vector where the initial node occupies position 0, the first b nodes occupy position 1 to b; the first b nodes exiting from node 1 occupy position $b + 1$ to $b + 2b$; the b subnodes of node 2 occupy positions $2b + 1$ to $2b + 2b$ and so forth. Though flexible, this code has been written only for didactic reasons and does not include dividends and discounting factor. The next code defines a simulator for the tree.

```
R> simTree <- function(b, d, S0, sigma, T, r) {
+       tot <- sum(b^(1:(d - 1)))
+       S <- numeric(tot + 1)
+       S[1] <- S0
+       dt <- T/d
+       for (i in 0:(tot - b^(d - 1))) {
+           for (j in 1:b) {
+               S[i * b + j + 1] <- S[i + 1] *
                  exp((r - 0.5 * sigma^2) *
+                   dt + sigma * sqrt(dt) * rnorm(1))
+           }
+       }
+       S
+ }
```

The next functions calculate the upper and lower estimators of the American call option:

```
R> upperBG <- function(S, b, d, f) {
+       tot <- sum(b^(1:(d - 1)))
+       start <- tot - b^(d - 1) + 1
+       end <- tot + 1
+       P <- S
+       P[start:end] <- f(S[start:end])
+       tot1 <- sum(b^(1:(d - 2)))
+       for (i in tot1:0) {
+           m <- mean(P[i * b + 1:b + 1])
+           v <- f(S[i + 1])
+           P[i + 1] <- max(v, m)
+       }
+       P
+ }
R> lowerBG <- function(S, b, d, f) {
+       tot <- sum(b^(1:(d - 1)))
```

```
+      start <- tot - b^(d - 1) + 1
+      end <- tot + 1
+      p <- S
+      p[start:end] <- f(S[start:end])
+      tot1 <- sum(b^(1:(d - 2)))
+      m <- numeric(b)
+      for (i in tot1:0) {
+          v <- f(S[i + 1])
+          for (j in 1:b) {
+              m[j] <- mean(p[i * b + (1:b)[-j] + 1])
+              m[j] <- ifelse(v > m[j], v, p[i * b + (1:b)[j] +
+                  1])
+          }
+          p[i + 1] <- mean(m)
+      }
+      p
+ }
```

Now we can test these functions on the example of Figure 7.6, so we prepare
the vector *S* according to the data in the picture.

```
R> b <- 3
R> d <- 3
R> S0 <- 101
R> S <- c(101, 114, 50, 115, 74, 88, 102, 38, 47, 65, 88, 149,
      116)
R> K <- 100
R> f <- function(x) sapply(x, function(x) max(x - K, 0))
```

where f is the payoff of the call option. We now call the two functions to obtain
the lower and upper estimates:

```
R> lowerBG(S, b, d, f)

 [1]  8.111111 14.000000  0.000000 10.333333  0.000000  0.000000
      2.000000
 [8]  0.000000  0.000000  0.000000  0.000000 49.000000 16.000000

R> upperBG(S, b, d, f)

 [1] 11.88889 14.00000  0.00000 21.66667  0.00000  0.00000
      2.00000  0.00000
 [9]  0.00000  0.00000  0.00000 49.00000 16.00000
```

Notice that the estimates of the values of the option correspond with the first
element of the returned vector but we plot the entire sequence just to show the
correspondence between the values returned by these routines and the values
in Figures 7.6 and 7.7 taken from Broadie and Glasserman (1997). In general,
one has to simulate a trajectory and extract the values of the lower and upper
estimates as follows:

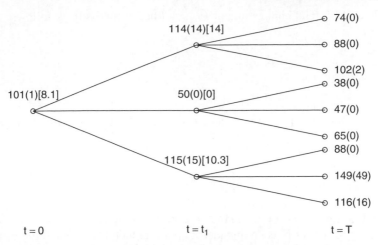

Figure 7.7   Example of simulated path from Broadie and Glasserman (1997). In parentheses the payoff at the node and in squared brackets the value of the lower estimator θ of the the American call option.

```
R> set.seed(123)
R> b <- 3
R> d <- 3
R> K <- 100
R> f <- function(x) sapply(x, function(x) max(x - K, 0))
R> T <- 1
R> r <- 0.05
R> sigma <- 0.4
R> S0 <- 101
R> S <- simTree(b, d, S0, sigma, T, r)
R> lowerBG(S, b, d, f)[1]

[1] 17.68876

R> upperBG(S, b, d, f)[1]

[1] 22.681
```

and embed the last three lines of code in a Monte Carlo script.

To understand the asymptotic results, we give a precise definition of the estimator. The random tree with b branches per node is represented by the array

$$\{S_t^{i_1 i_2 \cdots i_t} : t = 0, 1, \ldots, T; i_j = 1, \ldots, b; j = 1, \ldots, t\}$$

A simulated path is then the sequence

$$S_0 S_1^{i_1} S_2^{i_1 i_2} \cdots S_T^{i_1 i_2 \cdots i_T}$$

as in Figure 7.6. The high estimator Θ is defined recursively as follows:

$$\Theta_T^{i_1 \cdots i_T} = f_T \left(S_T^{i_1 \cdots i_T} \right)$$

and

$$\Theta_t^{i_1 \cdots i_t} = \max \left\{ h_t \left(S_t^{i_1 \cdots i_t} \right), \frac{1}{b} \sum_{j=1}^{b} e^{-R_{t+1}^{i_1 \cdots i_t j}} \Theta_{t+1}^{i_1 \cdots i_t j} \right\}, \quad t = 0, \ldots, T-1,$$

where $h_t(x) = \max\{f_t(x), g_t(x)\}$ is the value of the option at time t when $S_t = x$, $g_t(x) = \mathbb{E}\{e^{-R_{t+1}} h_{t+1}(S_{t+1}) | S_t = x\}$ is the continuation value at time t, $f_t(x)$ is the payoff at time t, $h_T(x) = f_T(x)$ and e^{-R_t} is the discount factor from $t-1$ to t, with $R_t \geq 0$. In our R code this factor is not included. The final high estimator is Θ_0 and $\bar{\Theta}_{0,n}$ is the average value of Θ_0 in a Monte Carlo experiment with n replications. Under regularity conditions the following result holds true.

Theorem 7.6.1 (Broadie and Glasserman (1997)) *For a given number of Monte Carlo replications n, not necessarily diverging to ∞, then*

$$\mathbb{E}(\bar{\Theta}_{0,n}(b)) \overset{b \to \infty}{\to} f_0(S_0)$$

and for finite b the bias is always positive:

$$\mathbb{E}(\Theta_0(b)) \geq f_0(S_0).$$

The low estimator θ is defined recursively as follows. Let

$$\theta_T^{i_1 \cdots i_T} = f_T \left(S_T^{i_1 \cdots i_T} \right)$$

and define

$$\eta_t^{i_1 \cdots i_t j} = \begin{cases} f_t \left(S_t^{i_1 \cdots i_t} \right) & \text{if } f_t \left(S_t^{i_1 \cdots i_t} \right) \geq \frac{1}{b} \sum_{i=1 i \neq j}^{b} e^{-R_{t+1}^{i_1 \cdots i_t i}} \theta_{t+1}^{i_1 \cdots i_t i} \\ e^{-R_{t+1}^{j}} \theta_{t+1}^{i_1 \cdots i_t j} & \text{otherwise} \end{cases}$$

for $j = 1, \ldots, b$. Then let

$$\theta_t^{i_1 \cdots i_t} = \frac{1}{b} \sum_{j=1}^{b} \eta_t^{i_1 \cdots i_t j}, \quad t = 0, \ldots, T-1.$$

As before θ_0 is the low estimate in a single simulation and $\bar{\theta}_{0,n}$ is the Monte Carlo estimate after n replications. Then, the following result holds true.

Theorem 7.6.2 (Broadie and Glasserman 1997) *For a given number of Monte Carlo replications n, unnecessarily diverging to ∞, then*

$$\mathbb{E}(\bar{\theta}_{0,n}(b)) \overset{b\to\infty}{\to} f_0(S_0)$$

and for finite b the bias is always negative

$$\mathbb{E}(\theta_0(b)) \leq f_0(S_0).$$

The approximate $(1-\alpha)\%$ Monte Carlo confidence interval for the true price of the American option is defined as

$$\left[\max\left\{ \max(S_0 - K, 0), \bar{\theta}_{0,n} - z_{\frac{\alpha}{2}} \frac{s(\bar{\theta}_{0,n})}{n} \right\}, \bar{\Theta}_{0,n} + z_{\frac{\alpha}{2}} \frac{s(\bar{\Theta}_{0,n})}{n} \right]$$

where $s(\bar{\theta}_{0,n})$ is the standard deviation of $\bar{\theta}_{0,n}$, $s(\bar{\Theta}_{0,n})$ the standard deviation $\bar{\Theta}_{0,n}$ and $z_{\frac{\alpha}{2}}$ the $1 - \frac{\alpha}{2}$ quantile of the Gaussian distribution. Notice that the interval is cut from below by the payoff of immediate exercise. The point estimate in a given replication is given by the following formula:

$$\hat{C} = \frac{1}{2} \max \{\max(S_0 - K, 0), \theta_0\} + \frac{1}{2}\Theta_0.$$

7.6.2   Longstaff and Schwartz Least Squares Method

After the paper of Broadie and Glasserman (1997) many other variations of the algorithm and new solutions have been proposed in the literature. The most notable one is the Least Squares Method (LSM) developed by Longstaff and Schwartz (2001). This method is extremely easy to understand and implement. The key argument is the estimation of the continuation value by a simple regression via Least Squares. This method requires a simulation of a single path, rather than the construction of a tree, on a grid of times t_i, $i = 0, 1, \ldots, d$. Let $V_i(x)$ and $f_i(x)$ denote respectively the value of the option and the payoff function at time t_i given $S_{t_i} = x$. The continuation value at time t_i given $S_{t_i} = x$ is

$$C_i(x) = \mathbb{E}\left\{ V_{i+1}(S_{t_{i+1}}) \,\middle|\, S_{t_i} = x \right\} = \sum_{r=1}^{M} \beta_{ir} \psi_r(x)$$

for some basis functions ψ_r and coefficients β_{ir}, $r = 1, \ldots, M$. This representation comes from the fact that the conditional expectation lives in a L^2 space where some basis exists, so it can be represented in term of the element of the basis. This coefficient β can be estimated via simple regression using the values $(S_{t_i}, V_{i+1}(S_{t_{i+1}}))$. Thus, the continuation value is estimated as

$$\hat{C}_i(x) = \hat{\beta}_i' \psi(x) \tag{7.15}$$

with

$$\hat{\beta}_i' = \left(\hat{\beta}_{i1}, \ldots, \hat{\beta}_{iM}\right), \quad \psi(x) = (\psi_1(x), \ldots, \psi_M(x))'.$$

Before going into the details of the algorithm, we present the working example found in the original paper of Broadie and Glasserman (1997) because it is very instructive. In this example it is assumed that the strike price is $K = 1.10$, there are only times 0, 1, 2 and $3 = T$, the interest rate is $r = 0.06$ and the initial value of the asset is $S_0 = 1$. Only 8 paths have been generated and Table 7.1 reports each path by row. At time $t = 2$ only options which are in the money are considered. These are indicated with an asterisk in Table 7.1. The holder of the option decides to exercise immediately or wait till expiration comparing the present payoff and the discounted expected payoff. In our case, the discount factor is $\exp(-r) = 0.94176$. The expected payoff is estimated via regression taking into account all the trajectories using this matrix:

$$\begin{bmatrix} \text{Path} & Y & X \\ 1 & .00 \cdot 0.94176 & 1.08 \\ 2 & -- & -- \\ 3 & .07 \cdot 0.94176 & 1.07 \\ 4 & .18 \cdot 0.94176 & 0.97 \\ 5 & -- & -- \\ 6 & .20 \cdot 0.94176 & 0.77 \\ 7 & .09 \cdot 0.94176 & 0.84 \\ 8 & -- & -- \end{bmatrix}$$

From these data the following simple regression model can be estimated $\mathbb{E}(Y|X) = -1.070 + 2.983X - 1.812X^2$. Putting S_2 in place of X, we obtain the continuation values $\hat{C}$ which are given in Table 7.1. For example, for the first trajectory we have $-1.070 + 2.983 \cdot 1.08 - 1.813 \cdot (1.08)^2 = 0.0369$.

Table 7.1   Numerical example for the LSM algorithm from Longstaff and Schwartz (2001). In parentheses the payoff at expiry date. The asterisk indicates paths such that the option is in-the-money at time $t = 2$.

Path	$t = 0$	$t = 1$	$t = 2$	$t = 3$	Continuation $\hat{C}$	payoff at $t = 2$
1*	1.00	1.09	1.08	1.34 (.00)	0.0369	0.02
2	1.00	1.16	1.26	1.54 (.00)	–	–
3*	1.00	1.22	1.07	1.03 (.07)	0.0461	0.03
4*	1.00	0.93	0.97	0.92 (.18)	0.1176	0.13
5	1.00	1.11	1.56	1.52 (.00)	–	–
6*	1.00	0.76	0.77	0.90 (.20)	0.1520	0.33
7*	1.00	0.92	0.84	1.01 (.09)	0.1565	0.26
8	1.00	0.88	1.22	1.34 (.00)	–	–

This value has to be compared with the value of immediate exercise which, for the first path, is $K - S_2 = 1.10 - 1.08 = 0.02$. For paths 1 and 3 it is worth waiting, while for paths 4, 6 and 7 it is more convenient to exercise immediately. The algorithm proceeds backward for $t = 1$, using the same rule and only considering trajectories which are in the money at time $t = 1$. Formally the algorithm proceeds as follows:

(i) simulate n independent paths on the grid of times

(ii) at $t = T$ set $\hat{V}_{t_j} = f_d(S_{t_j})$, $j = 1, \ldots, n$,

(iii) for $i = d - 1, \ldots, 1$

    – specify the I of paths in the money

    – discount the value $\hat{V}_{i+1,j}$, $j \in I$ to input in the regression

    – given the estimates $\hat{V}_{i+1,j}$, run regression to get $\hat{\beta}_j$

    – estimate the continuation value as in (7.15)

    – if $f_i(S_i^j) \geq \hat{C}_i(S_i^j)$, set $\hat{V}_{ij}(S_i^j) = f_i(S_i^j)$ else $\hat{V}_{ij} = \hat{V}_{i+1,j}$

(iv) calculate $(\hat{V}_{11} + \cdots + \hat{V}_{1n})/n$ and discount it to get $\hat{V}_0$

As anticipated, the discounted value can be written as an expansion in a proper L^2 space. Longstaff and Schwartz (2001) propose the following approximation formula:

$$F(t_{k-1}) = \sum_{j=0}^{M} a_j L_j(X)$$

where a_i are constants coefficients and the basis is formed by the Laguerre polynomials

$$L_0(x) = e^{-\frac{x}{2}}$$
$$L_1(x) = e^{-\frac{x}{2}}(1 - x)$$
$$L_2(x) = e^{-\frac{x}{2}}\left(1 - 2x + \frac{x^2}{2}\right)$$
$$\vdots$$
$$L_n(x) = e^{-\frac{x}{2}}\frac{e^x}{n!}\frac{d^n}{dx^n}\left(x^n e^{-x}\right)$$

The next code is taken almost identical from the corresponding R code by Coşkan (2008) and implements the algorithm using the first three Laguerre polynomials.

```
R> LSM <- function(n, d, S0, K, sigma, r, T) {
+      s0 <- S0/K
+      dt <- T/d
```

```
+       z <- rnorm(n)
+       s.t <- s0 * exp((r - 1/2 * sigma^2) * T + sigma * z * (T^0.5))
+       s.t[(n + 1):(2 * n)] <- s0 * exp((r - 1/2 * sigma^2) * T -
+           sigma * z * (T^0.5))
+       CC <- pmax(1 - s.t, 0)
+       payoffeu <- exp(-r * T) * (CC[1:n] + CC[(n + 1):(2 * n)])/2 *
+           K
+       euprice <- mean(payoffeu)
+       for (k in (d - 1):1) {
+           z <- rnorm(n)
+           mean <- (log(s0) + k * log(s.t[1:n]))/(k + 1)
+           vol <- (k * dt/(k + 1))^0.5 * z
+           s.t.1 <- exp(mean + sigma * vol)
+           mean <- (log(s0) + k * log(s.t[(n + 1):(2 * n)]))/(k +
+               1)
+           s.t.1[(n + 1):(2 * n)] <- exp(mean - sigma * vol)
+           CE <- pmax(1 - s.t.1, 0)
+           idx <- (1:(2 * n))[CE > 0]
+           discountedCC <- CC[idx] * exp(-r * dt)
+           basis1 <- exp(-s.t.1[idx]/2)
+           basis2 <- basis1 * (1 - s.t.1[idx])
+           basis3 <- basis1 * (1 - 2 * s.t.1[idx] + (s.t.1[idx]^2)/2)
+           p <- lm(discountedCC ~ basis1 + basis2 + basis3)$coefficients
+           estimatedCC <- p[1] + p[2] * basis1 + p[3] * basis2 +
+               p[4] * basis3
+           EF <- rep(0, 2 * n)
+           EF[idx] <- (CE[idx] > estimatedCC)
+           CC <- (EF == 0) * CC * exp(-r * dt) + (EF == 1) * CE
+           s.t <- s.t.1
+       }
+       payoff <- exp(-r * dt) * (CC[1:n] + CC[(n + 1):(2 * n)])/2
+       usprice <- mean(payoff * K)
+       error <- 1.96 * sd(payoff * K)/sqrt(n)
+       earlyex <- usprice - euprice
+       data.frame(usprice, error, euprice)
+ }
```

The function LSM returns the estimated value of the American option, the radius
of the Monte Carlo confidence interval at 99% level, and the corresponding value
of the European option. We compare this value with the approximation formulas
of Sections 7.5 and 7.4 as implemented in the **Rmetrics** package.

```
R> S0 <- 36
R> K <- 30
R> T <- 1
R> r <- 0.05
R> sigma <- 0.4
R> LSM(10000, 3, S0, K, sigma, r, T)

    usprice      error  euprice
1 2.230353 0.04077382 2.201232

R> require(fOptions)
R> BSAmericanApproxOption("p", S0, K, T, r, r, sigma)@price
```

```
[1] 2.250694

R> BAWAmericanApproxOption("p", S0, K, T, r, r, sigma)@price

[1] 2.293530
```

7.7   Bibliographical notes

The present review of methods is not exhaustive, but it is a summary of what can be reasonably understood without introducing more advanced topics. An introduction to finite difference methods as explained here can be found in Hull (2000). Monte Carlo methods and their extensions can be found in Glasserman (2004). In Coşkan (2008) the author also introduces the RSA approach, an efficient modification of the LSM algorithm for the one-dimensional case. He also provides all relevant R code for special two-dimensional American basket options.

References

Barone-Adesi, G. and Whaley, R. (1987). Efficient analytic approximation of American option values. *Journal of Finance* **42**, 2, 301–320.

Bjerksund, P. and Stensland, G. (1993). Closed-form approximation of American options. *Scandinavian Journal of Management* **9**, 87–99.

Brennan, M. and Schwartz, E. (1978). Finite difference methods and jump processes arising in the pricing of contingent claims: A synthesis. *J. Financial Quant. Anal.* **13**, 3, 461–474.

Broadie, M. and Glasserman, P. (1997). Pricing American-style securities using simulation. *Journal of Economic Dynamics and Control* **21**, 8-9, 1323–1352.

Bunch, D. and H.E., J. (1992). A simple and numerically efficient valuation method for Aamerican put using a modified geske-johnson approach. *Journal of Finance* **47**, 809–816.

Coşkan, C. (2008). *Pricing American Options by Simulation*. Master Thesis, Istanbul Technical University, http://www.ie.boun.edu.tr/~hormannw/BounQuantitiveFinance/Thesis/coskan%.pdf, Istambul.

Geske, R. (1979). A note on an analytical formula for unprotected American call options on stocks with known dividends. *Journal of Financial Economics* **7**, 63–81.

Geske, R. and Johnson, E. (1984). The American put option valued analytically. *Journal of Finance* **39**, 1511–1524.

Glasserman, P. (2004). *Monte Carlo Methods in Financial Engineering*. Springer, New York.

Hull, J. (2000). *Options, Futures and Other Derivatives*. Prentice-Hall, Englewood Cliffs, N.J.

Ju, N. and Zhong, R. (1999). An approximate formula for pricing American options. *Journal of Derivatives* **7**, 2, 31–40.

Longstaff, F. and Schwartz, E. (2001). Valuing American options by simulation: A simple least-squares approach. *The Review of Financial Studies* **14**, 1, 113–147.

McMillan, L. (1986). Analytic approximation for the American put option. *Advances in Futures and Options Research* **1**, A, 119–139.

Roll, R. (1977). An analytic valuation formula for unprotected American call options on stocks with known dividends. *Journal of Financial Economics* **5**, 251–258.

Whaley, R. E. (1981). On the valuation of American call options on stocks with known dividends. *Journal of Financial Economics* **9**, 207–211.

8

Pricing outside the standard Black and Scholes model

The standard Black and Scholes model relies on several assumptions which allow for explicit formulas and easy calculations in most cases. Unfortunately some of these hypotheses like the constant volatility, Gaussianity of the returns and the continuity of the paths of the geometric Brownian motion process are unlikely to hold for many observed prices. Indeed, prices often show jumps, changes in volatility (see e.g. Section 6.6.1) and the distribution of the returns is usually skewed and with high tails (see e.g. Section 5.4). Most of these stylized facts are indeed captured by Lévy processes as we have seen. In this chapter we mainly consider the problem of pricing under the assumption that the dynamic of the financial prices includes some kind of jump process and/or non-Gaussian behaviour. We start with the simpler and most studied Lévy process.

8.1 The Lévy market model

Consider again the simple exponential Lévy model analyzed in Sections 4.5.5 and 5.4

$$S_t = S_0 e^{Z_t} \tag{8.1}$$

where Z_t is a Lévy process with triplet (b, c, ν) and canonical decomposition

$$Z_t = bt + \sqrt{c}B_t + \int_0^t \int_{\mathbb{R}} x(\mu^Z - \nu^Z)(\mathrm{d}s, \mathrm{d}x)$$

such that $\mathbb{E}|L_1| < \infty$. We have seen that it is possible to estimate the infinitely distribution of the Lévy processes quite easily from the log-returns of real financial data due to the fact that the increments of the Lévy process are independent.

Option Pricing and Estimation of Financial Models with R, First Edition. Stefano M. Iacus.
© 2011 John Wiley & Sons, Ltd. Published 2011 by John Wiley & Sons, Ltd.

The property of the Lévy process Z_t naturally propagates to the asset price process S_t thus, if Z_t is a pure jump process, then also S_t is a pure jump process. We have seen in Example 3.18.10 that applying Itô formula (3.56) to (8.1) we can see that S_t solves the stochastic differential equation

$$dS_t = S_{t-}\left\{dZ_t + \frac{c}{2}dt + \int_{\mathbb{R}}\left(e^x - 1 - x\right)\mu^Z(dt, dx)\right\}.$$

As usual, to develop a theory of option pricing, one way is to identify a change of measure which makes the discounted process $S_t^d = e^{-rt S_t}$ a martingale under this new measure. In this section we can consider the general case of dividends and hence we will focus on the discounted process $S_t^d = e^{-(r-\delta)t} S_t$ where δ is the continuous dividend yield.

The first task is then to construct a proper change of measure such that the discounted process is a martingale. We have seen already under which conditions a Lévy process can be transformed into a martingale in Section 3.18.6 and we now exploit this approach.

8.1.1   Why the Lévy market is incomplete?

The problem with Lévy markets is that a whole set of equivalent martingale measures exist and this, in turn, implies that the market is incomplete as seen. To understand why multiple martingale measures exist in the Lévy market we follow Papapantoleon (2008). Let us denote by P and Q the true physical measure of Z_t and the equivalent martingale measure which makes $S_t^d = e^{-(r-\delta)t} S_t$ a martingale. We denote the corresponding characteristic triplets of P and Q as (b, c, v) and $(\tilde{b}, \tilde{c}, \tilde{v})$ respectively. The two measures are related, via the Girsanov theorem, in this way

$$\tilde{c} = c, \quad \tilde{v} = Yv, \quad \tilde{b} = b + c\beta + x(Y - 1) * v. \tag{8.2}$$

with (β, Y) the two processes from Theorem 3.18.13. Under Q. Z_t has the canonical decomposition

$$Z_t = \tilde{b}t + \sqrt{\tilde{c}}\tilde{W}_t + \int_0^t \int_{\mathbb{R}} x \left(\mu^Z - \tilde{v}^Z\right)(ds, dx)$$

where $\tilde{W}_t$ is a Brownian motion under Q and $\tilde{v}^Z$ is a compensator of the jump measure μ^Z under Q (see again Theorem 3.18.13.) But, in order to have a martingale, the following condition must hold

$$\tilde{b} = r - \delta - \frac{\tilde{c}}{2} - \int_{\mathbb{R}}\left(e^x - 1 - x\right)\tilde{v}(dx). \tag{8.3}$$

Therefore, equating (8.2) and (8.3) with $\tilde{c} = c$ and $v = Y\tilde{v}$, we obtain

$$0 = b + c\beta + x(Y - 1) * v - r - \frac{\tilde{c}}{2} + \left(e^x - 1 - x\right) * \tilde{v}$$

$$= b - r + c\left(\beta + \frac{1}{2}\right) + \left((e^x - 1)Y - x\right) * v. \tag{8.4}$$

This means that, in order to obtain a martingale, we need to solve an equation with two unknown β and Y and each solution of (8.4) in terms of the couple (β, Y) provides an equivalent martingale measure. As noticed in Papapantoleon (2008), this result is completely in line with what we know about market completeness in the Black and Scholes model. Indeed, for the geometric Brownian motion the Lévy process takes the form $Z_t = bt + \sqrt{c}W_t$. Then Equation (8.4) has a unique solution, which is:

$$\beta = \frac{r - b}{c} - \frac{1}{2}.$$

So the martingale measure is unique and the market complete.

In their seminal paper Eberlein and Jacod (1997) have characterized the range of possible fair price of options under the exponential Lévy model. Let us denote by Q a generic equivalent martingale measure and let X be the payoff of a contingent claim. Then, at time zero

$$P_0 = P_0^Q = e^{-rT}\mathbb{E}_Q(X).$$

Then $P_0^Q \in [m, M]$ where

$$m = \inf\left\{e^{-rT}\mathbb{E}_Q(X) | Q \text{ an equivalent martingale measure}\right\}$$

$$M = \inf\left\{e^{-rT}\mathbb{E}_Q(X) | Q \text{ an equivalent martingale measure}\right\}$$

Without any additional information, this interval is usually very large, but for options with payoff $X = f(S_T)$, then $m = e^{-rT}f\left(e^{rT}S_0\right)$ and $M = S_0$. In addition, it is possible to prove that if the price falls outside this interval, there are arbitrage opportunities, so arbitrage free option prices should lie in that interval and the result is independent of the particular Lévy model chosen to construct the process S_t. So now the problem is how to choose one among all equivalent martingale measures. We will consider a few examples from the different alternative proposals available in the literature.

8.1.2   The Esscher transform

Gerber and Shiu (1994) introduced for the first time the so-called *Esscher transform* to find one possible equivalent martingale measure. Let $f(x)$ be the density of Z_1 under the real measure P. Let θ be a real number such that $\theta \in \{\theta \in \mathbb{R} : \int_{\mathbb{R}} \exp(\theta x)f(x)dx < \infty\}$. Then, we can construct a new density

$$f_\theta(x) = \frac{\exp(\theta x)f(x)}{\int_{\mathbb{R}} \exp(\theta x)f(x)dx}$$

and choose a value of θ such that the discounted price process, $S_t^d = e^{-(r-\delta)t}S_t$, is a martingale, i.e.

$$S_0 = e^{-(r-\delta)t}\mathbb{E}_\theta(S_t)$$

where $\mathbb{E}_\theta$ is the expectation under $f_\theta(\cdot)$. Now let $\varphi(u) = \mathbb{E} \exp(iuZ_1)$, the characteristic function of Z_1, then in order to have a martingale θ must solve the following equation:

$$\exp(r - \delta) = \frac{\varphi(-i(\theta + 1))}{\varphi(-i\theta)}. \tag{8.5}$$

We denote this solution as θ^*, thus under $f_{\theta*}(\cdot)$ the discounted process is a martingale and the corresponding equivalent martingale measure is the one with density $f_{\theta*}$. Gerber and Shiu (1996) justify the choice of this particular equivalent martingale measure in terms of utility-maximization theory. The nice properties of this approach is that the new density is still infinitely divisible. Moreover, if Z_1 has characteristic triplet (b, c, ν) then, the characteristic function $\varphi_\theta(u) = \mathbb{E}_\theta \exp(iuZ_1)$ is given by

$$\varphi_\theta(u) = \frac{\varphi(u - i\theta)}{\varphi(-i\theta)}$$

and has triplet $(b_\theta, c, \nu_\theta)$, where

$$b_\theta = b + c^2\theta + \int_{-1}^{1} \left(e^{\theta x} - 1\right) \nu(\mathrm{d}x), \quad \nu_\theta(\mathrm{d}x) = e^{\theta x} \nu(\mathrm{d}x).$$

Last but not least, this transform is very easy to obtain for many models.

Example 8.1.1 *Let us consider the Black and Scholes model. In this case* $Z_1 \sim \mathrm{N}\left(\mu - \frac{1}{2}\sigma^2, \sigma^2\right)$, *hence the characteristic function is* $\varphi(u) = \exp\left(iu(\mu - \frac{1}{2}\sigma^2) - \frac{\sigma^2 u^2}{2}\right)$. *The martingality condition* (8.5) *becomes*

$$r - \delta = \mu - \frac{1}{2}\sigma^2 + \frac{1}{2}\sigma^2(2\theta + 1)$$

thus

$$\theta^* = \frac{r - \delta - \mu}{\sigma^2}.$$

The density of the equivalent martingale measure is

$$f_{\theta*}(x) = \frac{\exp\left(\theta^* x - \frac{\left(x - \mu + \frac{1}{2}\sigma^2\right)^2}{2\sigma^2}\right)}{\int_{\mathbb{R}} \exp\left(\theta^* x - \frac{\left(x - \mu + \frac{1}{2}\sigma^2\right)^2}{2\sigma^2}\right) \mathrm{d}x}$$

$$= \frac{\exp\left(-\frac{x(r - \delta - \mu)}{\sigma^2} - \frac{\left(x - \mu + \frac{\sigma^2}{2}\right)}{2\sigma^2}\right)}{\int_{\mathbb{R}} \exp\left(-\frac{x(r - \delta - \mu)}{\sigma^2} - \frac{\left(x - \mu + \frac{\sigma^2}{2}\right)}{2\sigma^2}\right) \mathrm{d}x}$$

$$= \frac{\exp\left(-\frac{x(r-\delta-\mu)}{\sigma^2} - \frac{\left(x-\mu+\frac{\sigma^2}{2}\right)}{2\sigma^2}\right)}{\exp\left(\frac{(r-\delta-\mu)(r-\delta+\mu-\sigma^2)}{2\sigma^2}\right)\sqrt{2\pi\sigma^2}}$$

$$= \frac{1}{\sqrt{2\pi\sigma^2}} e^{-\frac{\left(x-\left(r-\delta-\frac{1}{2}\sigma^2\right)\right)^2}{2\sigma^2}}$$

which is the density of the Gaussian law $\mathrm{N}\left(r - \delta - \frac{1}{2}\sigma^2, \sigma^2\right)$ *as expected for this model from Section 6.4.*

For the Meixner process with $Z_1 \sim \text{Meixner}(\alpha, \beta, \gamma)$, the Esscher transform gives a new martingale measure with parameters $\text{Meixner}(\alpha, \alpha\theta^* + \beta, \gamma)$ with

$$\theta^* = \frac{-1}{\alpha}\left(\beta + 2\arctan\left(\frac{-\cos\left(\frac{\alpha}{2}\right) + \exp\left(\frac{\delta-r}{2\gamma}\right)}{\sin\left(\frac{\alpha}{2}\right)}\right)\right)$$

while for the $\text{NIG}(\alpha, \beta, \gamma)$ the resulting equivalent martingale measure is $\text{NIG}(\alpha + \theta^* + \beta, \gamma)$ with θ^* the solution of

$$r - \delta = \gamma\left(\sqrt{\alpha^2 - (\beta + \theta)^2} - \sqrt{\alpha^2 - (\beta + \theta + 1)^2}\right).$$

In many other cases, the solution is not explicit but can be obtained numerically. For more details see Schoutens (2003).

8.1.3   The mean-correcting martingale measure

We have seen that adjusting for the mean is a way to transform the discounted process into a martingale. Assume that $\varphi(u)$ is the characteristic function of a Lévy process Z_t. It is possible to construct a new process $\tilde{Z}_t$ by adding a constant drift mt into a new process such that the distribution of Z_t is translated by the same amount. This is done directly setting $\tilde{Z}_t = Z_t + mt$ and, in terms of characteristic functions,

$$\tilde{\varphi}(u) = \varphi(u)\exp(ium).$$

In this way the characteristic triplet changes simply from (b, c, ν) to $(\tilde{b} = b + m, \tilde{c} = c, \tilde{\nu} = \nu)$ and the two densities are related by the formula $f(x) = \tilde{f}(x - m)$. In this setup, all the processes of Section 5.4.1 are transformed as follows: $\text{Meixner}(\alpha, \beta, \delta)$ into $\text{Meixner}(\alpha, \beta, \delta, m)$; $\text{NIG}(\alpha, \beta, \delta)$ into $\text{NIG}(\alpha, \beta, \delta, m)$; $\text{VG}(C, G, M)$ into $\text{VG}(C; G, M, m)$; etc. Each model requires a different value of m. According to Schoutens (2003), the solution requires the following identity:

$$m' = m + r - \delta - \log\varphi(-i).$$

The initial m can be zero. For example, the Black and Scholes model is such that $\varphi(-i) = \mu$. The physical measure has $m = \mu - \frac{1}{2}\sigma^2$ and the new measure has $m' = r - \delta - \frac{1}{2}\sigma^2$, and we obtain again the standard result. Table 6.2 in Schoutens (2003) provides a summary of the mean transformation for most of the processes of Section 5.4.1.

8.1.4   Pricing of European options

Once a martingale measure Q is available, it is possible to price options under the Lévy market. If we have a contingent T-claim such that the payoff function is $f(\cdot)$ depends only on S_T, the usual formula applies

$$\mathbb{E}_Q\left\{e^{-rT}f(S_T)\right\}.$$

When Q has been obtained with the Esscher transform, then the above expected value is calculated as

$$\mathbb{E}_Q\left\{e^{-rT}f(S_T)\right\} = \mathbb{E}_{\theta*}\left\{e^{-rT}f(S_T)\right\} = e^{-\delta T}S_0\int_c^\infty g_{\theta*}(x)\mathrm{d}x$$
$$- e^{-rT}K\int_c^\infty g_{\theta*}(x)\mathrm{d}x$$

where $g_{\theta*}(\cdot)$ is the density of Q and $c = \log(K/S0)$.

8.1.5   Option pricing using Fast Fourier Transform method

Another way is to use the FFT algorithm to invert the density of the transformed process. The idea dates back to Scott (1997), Carr and Madan (1998) and Bakshi and Madan (2000). For simplicity we assume no dividends, i.e. $\delta = 0$ and we describe the approach under the physical measure. Let $\varphi(u) = \mathbb{E}(\exp(iu\log(S_T)))$, the characteristic function of the random variable $\log(S_T)$. For a European call option the general formula for the price is

$$C(K, T) = S_0\Pi_1 - Ke^{-rT}\Pi_2$$

where $\Pi_2 = P(S_t > K)$ is the probability of finishing in the money and Π_1 is the delta. These two values can be obtained from the inversion of the characteristic function $\varphi(u)$ as follows:

$$\Pi_1 = \frac{1}{2} + \frac{1}{\pi}\int_0^\infty \mathrm{Re}\left(\frac{\exp(-iu\log K)\mathbb{E}(\exp(i(u-i)\log S_T))}{iu\mathbb{E}(S_T)}\right)\mathrm{d}u$$

$$= \frac{1}{2} + \frac{1}{\pi}\int_0^\infty \mathrm{Re}\left(\frac{\exp(-iu\log K)\varphi(u-i)}{iu\varphi(-i)}\right)\mathrm{d}u,$$

$$\Pi_2 = \frac{1}{2} + \frac{1}{\pi}\int_0^\infty \mathrm{Re}\left(\frac{\exp(-iu\log K)\mathbb{E}(\exp(iu\log S_T))}{iu}\right)\mathrm{d}u$$

$$= \frac{1}{2} + \frac{1}{\pi}\int_0^\infty \mathrm{Re}\left(\frac{\exp(-iu\log K)\varphi(u)}{iu}\right)\mathrm{d}u.$$

The above two integrals can be evaluated by numerical integration when the characteristic function is known as in the exponential Lévy models. Another way to see this problem is to use the FFT algorithm to obtain the price of the option. Let $k = \log K$, the logarithm of the strike price and let $C_T(k)$ be the price at maturity T of the call option with strike $\exp(k)$. Let $s_T = \log(S_T)$ and $q_T(\cdot)$ the density of the equivalent martingale measure. The characteristic function of the density $q_T(\cdot)$ is defined as

$$\varphi_T(u) = \int_{\mathbb{R}} e^{ius} q_T(s) ds.$$

Then

$$C_T(k) = \int_k^\infty e^{-rT} \left(e^s - e^k \right) q_T(s) ds.$$

This function of k is not square integrable because it does not converge to zero but to S_0 as k goes to $-\infty$, therefore Carr and Madan (1998) propose to introduce an exponential dampening factor α into the function $C_T(k)$. The new function is defined as

$$c_T(k) = e^{\alpha k} C_T(k)$$

which is possibly square integrable for all k on same range of values of α. The characteristic function of $c_T(k)$, denoted by $\psi_T(\cdot)$, is given by

$$\psi_T(u) = \int_{\mathbb{R}} e^{iuk} c_T(k) dk.$$

The idea is to express $C_T(k)$ as a function of $\psi_T(\cdot)$

$$C_T(k) = \frac{e^{-\alpha k}}{2\pi} \int_{\mathbb{R}} e^{-iuk} \psi_T(u) du = \frac{e^{-\alpha k}}{\pi} \int_0^\infty e^{-iuk} \psi_T(u) du \qquad (8.6)$$

where the last equality follows from the fact that $C_T(k)$ is a real number. The last step is to establish a relationship between $\psi_T(\cdot)$ and $\varphi_T(\cdot)$ and then apply the FFT algorithm. Luckily, this is quite straightforward. In fact,

$$\psi_T(u) = \int_{\mathbb{R}} e^{iuk} \int_k^\infty e^{\alpha k} e^{-rT} \left(e^s - e^k \right) ds dk$$

$$= \int_{\mathbb{R}} e^{-rT} q_T(s) \int_{-\infty}^s \left(e^{s+\alpha k} - e^{(1+\alpha)k} \right) e^{iuk} dk ds$$

$$= \int_{\mathbb{R}} e^{-rT} q_T(s) \left\{ \frac{e^{(\alpha+1+iu)s}}{\alpha + iu} - \frac{e^{(\alpha+1+iu)s}}{\alpha + 1 + iu} \right\} ds$$

$$= e^{-rT} \left(\frac{1}{\alpha + iu} - \frac{1}{\alpha + 1 + iu} \right) \int_{\mathbb{R}} e^{(\alpha+1+iu)s} q_T(s) ds$$

$$= \frac{e^{-rT} \varphi_T(u - (\alpha + 1)i)}{\alpha^2 + \alpha - u^2 + i(2\alpha + 1)u}.$$

Now this version of $\psi_T(\cdot)$ can be directly plugged into (8.6).

The unsolved issue is how to choose α. The first thing to notice is that when $\alpha = 0$ also $\psi_T(u) = 0$, thus the factor $\exp(\alpha k)$ is required in a sense. In order to have integrability, we need to check whether $\psi(0) < \infty$ and $\phi(0)$ is finite only if $\varphi_T(-(\alpha + 1)i)$ if finite. This requirement is a condition on the following expected value

$$\mathbb{E}\left(S_T^{\alpha+1}\right) < \infty.$$

In practice, one has to find the maximal value of α such that the above moment condition is satisfied. Carr and Madan (1998) propose to choose α to be one fourth of the maximal α.

8.1.6   The numerical implementation of the FFT pricing

Using the trapezoidal rule to approximate the integral, Equation (8.6) can be rewritten in approximate form as follows:

$$C_T(k) \approx \frac{e^{-\alpha k}}{\pi} \sum_{j=1}^{N} e^{-iv_j k} \psi_T(v_j)\eta, \tag{8.7}$$

where $v_j = \eta(j-1)$. According to Carr and Madan (1998), one should keep in mind that the effective upper bound of integration is $a = N\eta$. We set up a regular grid for the argument k (the log strike level) in the following form:

$$k_u = -b + \lambda(u - 1), \quad u = 1, \ldots, N,$$

where λ is the step size of the regular grid. Thus, this grid spans the interval from $-b$ to b. where $b = \frac{1}{2}N\lambda$. Now replacing the quantities into (8.7) we have:

$$C_T(k_u) \approx \frac{e^{-\alpha k_u}}{\pi} \sum_{j=1}^{N} e^{-i\lambda\eta(j-1)(u-1)} e^{ibv_j} \psi_T(v_j)\eta.$$

To apply the FFT transform we need to ask for this condition:

$$\lambda\eta = \frac{2\pi}{N}.$$

Finally, replacing this last argument and applying Simpson's rule to increase the approximation of the integral (8.6) we obtain the following formula:

$$C_T(k_u) \approx \frac{e^{-\alpha k}}{\pi} \sum_{j=1}^{N} e^{-i\frac{2\pi}{N}} e^{ibv_j} \psi_T(v_j)\frac{\eta}{3}\left(3 + (-1)^j - \delta_{j-1}\right). \tag{8.8}$$

where δ_n is 1 for $n = 0$ and zero otherwise. The above calculations assume that the initial value S_0 is one. In practice we should always set $k = \log(K/S_0)$ rather than $k = \log K$ and thus, the option price for the general case $S_0 \neq 1$ is given by the formula $C_T^*(k_u) = S_0 C_t(k_u)$. This is how the script is implemented. A further note is that all calculations should be made under the characteristic function of an equivalent martingale measure. As seen, one of the most effective ways is to use the mean-correcting martingale measure approach in Section 8.1.3. This means that, given the characteristic function φ of Z_1, one transforms it into

$$\tilde{\varphi}(u) = \varphi(u) \cdot \exp(ium)$$

and the new location parameter is $m' = r - \log \varphi(-i)$. The next code takes as argument the strike price K, the initial value of the underline asset S_0, the interest rate r, the expiry date T and the characteristic function of Z_1. Then it applies the mean-correcting transform, calculates ψ_T from φ_T, inverts the call price via the FFT algorithm and finally rescales the price in Equation (8.8) by S_0 to obtain C_T^*. Internally the algorithm applies the spline smoothing before returning the final value. This algorithm is a version of what is available on http://quantcode.com/ and in Sengul (2008).

```
R> FFTcall.price <- function(phi, S0, K, r, T, alpha = 1,
    N = 2^12,
+      eta = 0.25) {
+      m <- r - log(phi(-(0+1i)))
+      phi.tilde <- function(u) (phi(u) * exp((0+1i) * u * m))^T
+      psi <- function(v) exp(-r * T) * phi.tilde((v - (alpha +
+          1) * (0+1i)))/(alpha^2 + alpha - v^2 + (0+1i) * (2 *
+          alpha + 1) * v)
+      lambda <- (2 * pi)/(N * eta)
+      b <- 1/2 * N * lambda
+      ku <- -b + lambda * (0:(N - 1))
+      v <- eta * (0:(N - 1))
+      tmp <- exp((0+1i) * b * v) * psi(v) * eta *
    (3 + (-1)^(1:N) -
+          ((1:N) - 1 == 0))/3
+      ft <- fft(tmp)
+      res <- exp(-alpha * ku) * ft/pi
+      inter <- spline(ku, Re(res), xout = log(K/S0))
+      return(inter$y * S0)
+ }
```

Now, recall that we know the exact formula of the Black and Scholes price of the European call option and it is implemented in the GBSOption function in package **fOptions**. We test the FFT method against the exact formula for the geometric Brownian motion model for which the characteristic function of Z_1 is

$$\varphi(u) = \exp\left(iu\left(\mu - \frac{1}{2}\sigma^2\right) - \frac{\sigma^2 u^2}{2}\right)$$

which we code as the phiBS function below:

```
R> phiBS <- function(u) exp((0+1i) * u * (mu - 0.5 * sigma^2) -
+      0.5 * sigma^2 * u^2)
```

We now price the European call option with

```
R> S0 <- 100
R> K <- 110
R> r <- 0.05
R> T <- 1/4
R> sigma <- 0.25
R> mu <- 1
R> require(fOptions)
R> GBSOption(TypeFlag = "c", S = S0, X = K, Time = T, r = r,
    b = r,
+      sigma = sigma)@price

[1] 1.980509

R> FFTcall.price(phiBS, S0 = S0, K = K, r = r, T = T)

[1] 1.984243
```

the two prices look quite close. Figure 8.1 show a plot of the difference between the price given by the exact formula and with the FFT approximation. The variation is very small and mostly due to numerical factors. The plot has been generated using the following code:

```
R> K.seq <- seq(100, 120, length = 100)
R> exactP <- NULL
R> fftP <- NULL
R> for (K in K.seq) {
+      exactP <- c(exactP, GBSOption(TypeFlag = "c", S = S0, X = K,
+          Time = T, r = r, b = r, sigma = sigma)@price)
+      fftP <- c(fftP, FFTcall.price(phiBS, S0 = S0, K = K, r = r,
+          T = T))
+ }
R> plot(K.seq, exactP - fftP, type = "l", xlab = "strike price K")
```

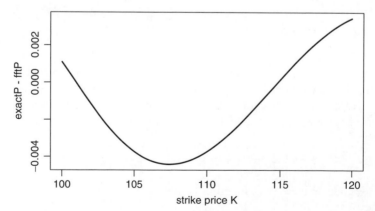

Figure 8.1   Difference in the price of the European call option between the exact Black and Scholes formula and the FFT approximation.

As a different example, we consider now a Variance Gamma process. This process is obtained by evaluating the Brownian motion with drift θ and volatility σ at a random time given by the gamma process with mean rate 1 and variance v, i.e.

$$B_t(\theta, \sigma) = \theta t + \sigma W_t$$

where W_t is the standard Brownian motion and t is a random variable distributed as $\Gamma(t/v, 1/v)$. The exponential Lévy process S_t is then written in this form under the risk neutral measure:

$$S_t = S_0 e^{rt + Z_t + \omega t}$$

where

$$\omega = \frac{1}{v} \log \left(1 - \theta v - \frac{\sigma^2 v}{2} \right)$$

where ω is the compensator which ensures a martingale property. The characteristic function of the logarithm of the process S_t is the following:

$$\varphi_T(u) = \exp \left(\log \left(S_0 + (r + \omega)T \right) (1 - i\theta v u + \sigma^2 u^2 \frac{v}{2})^{-T/nu} \right)$$

which we implement in the code `phiVG` below.

```
R> theta <- -0.1436
R> nu <- 0.3
R> r <- 0.1
R> sigma <- 0.12136
R> T <- 1
R> K <- 101
R> S <- 100
R> alpha <- 1.65
R> phiVG <- function(u) {
+       omega <- (1/nu) * (log(1 - theta * nu - sigma^2 * nu/2))
+       tmp <- 1 - (0+1i) * theta * nu * u + 0.5 * sigma^2 * u^2 *
+          nu
+       tmp <- tmp^(-1/nu)
+       exp((0+1i) * u * log(S0) + u * (r + omega) * (0+1i)) * tmp
+ }
```

We now need to pass this characteristic function to our `FFTcall.price` function

```
R> FFTcall.price(phiVG, S0 = S0, K = K, r = r, T = T)
```

```
[1] 10.98145
```

and obtain the result. We now want to compare the price given by the FFT approach with a Monte Carlo price, so we need to simulate the terminal value of S_T. For simulation purposes, one can see that conditional on the gamma time change t, the Variance Gamma process Z_t over an interval of length t is normally

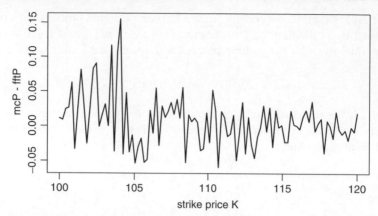

Figure 8.2   Difference in the price of the European call option between the Monte Carlo price and the FFT approximation for Variance Gamma process.

distributed with mean θt and variance $\sigma\sqrt{t}$ and thus one can simulate the process using this representation

$$Z_t = \theta t + \sigma\sqrt{t}z$$

where $z \sim N(0, 1)$ independent of $t \sim \text{Gamma}(t/\nu, 1/\nu)$. The next code simulates n terminal values of the VG process and calculates the Monte Carlo value of the option.

```
R> n <- 50000
R> t <- rgamma(n, shape = T/nu, scale = nu)
R> N <- rnorm(n, 0, 1)
R> X <- theta * t + N * sigma * sqrt(t)
R> omega <- (1/nu) * (log(1 - theta * nu - sigma^2 * nu/2))
R> S <- S0 * exp(r * T + omega * T + X)
R> payoff <- sapply(S, function(x) max(x - K, 0))
R> mean(payoff) * exp(-r * T)

[1] 11.00923
```

As seen these prices are quite close; we can run an experiment similar to what we did for the Black and Scholes model and the results are given in Figure 8.2 without any comment on the code.

```
R> K.seq <- seq(100, 120, length = 100)
R> mcP <- NULL
R> fftP <- NULL
R> for (K in K.seq) {
+       t <- rgamma(n, shape = T/nu, scale = nu)
+       N <- rnorm(n, 0, 1)
+       X <- theta * t + N * sigma * sqrt(t)
+       S <- S0 * exp(r * T + omega * T + X)
+       payoffvec <- sapply(S, function(x) max(x - K, 0))
+       tmp <- mean(payoffvec) * exp(-r * T)
```

```
+        mcP <- c(mcP, tmp)
+        fftP <- c(fftP, FFTcall.price(phiVG, S0 = S0, K = K, r = r,
+           T = T))
+ }
R> plot(K.seq, mcP - fftP, type = "l", xlab = "strike price K")
```

These results show that, at least for this configuration of the parameters, the FFT approach provides a reasonable approximation of the price and FFT is almost instantaneous compared to the Monte Carlo approach. Unfortunately this is not always the case as noticed by several authors and several modifications to the FFT algorithm have been proposed. One notable variation is the so-called Lewis regularization method which we do not present here. For further reading see Lewis (2001).

8.2 Pricing under the jump telegraph process

Following Ratanov (2007a), let $\{\sigma_t, t \geq 0\}$ be a Markov process with values ± 1 and transition probability intensities $\lambda_\pm$ such that, as $\Delta t \to 0$,

$$P(\sigma(t + \Delta t) = +1 | \sigma(t) = -1) = \lambda_- \Delta + o(\Delta t),$$

$$P(\sigma(t + \Delta t) = -1 | \sigma(t) = +1) = \lambda_+ \Delta + o(\Delta t).$$

From this process, we build the new processes $c_{\sigma_t} = c_\pm$, $h_{\sigma_t} = h_\pm > -1$ and $r_{\sigma_t} = r_\pm > 0$. Further we denote by $X_t^\sigma = \int_0^t c_{\sigma_s} ds$ the telegraph process and we introduce a new jump process J_t^σ, with alternating jumps of sizes $h_\pm$. The risk-free asset $\{B_t, t \geq 0\}$, is given by the exponent of the process $Y_t^\sigma = \int_0^t r_{\sigma_s} ds$, i.e. $B_t = \exp(Y_t^\sigma)$, which means that the current interest rate depends on the market state. Everywhere the exponent σ indicates the starting value of the process σ_t, i.e. $\sigma = \sigma_0$. The process r_{σ_t} captures the movement of interest rates and the jump process J_t^σ the market jumps or crashes. The driving process σ_t is the only source of randomness in this model and represents contractions of expansions of the market. In this setup the risky asset follows this stochastic differential equation

$$dS_t^\sigma = S_{t-}^\sigma d(X_t^\sigma + J_t^\sigma)$$

According to Ratanov (2007a), the solution of this equation has the following exponential form:

$$S_t^\sigma = S_0 \mathcal{E}_t(X^\sigma + J^\sigma) = S_0 e^{X_t^\sigma} \kappa_t^\sigma, \quad S_0 = S_0^\sigma,$$

where

$$\kappa_t^\sigma = \prod_{\tau \leq t}(1 + \Delta J_\tau^\sigma) = \prod_{j=1}^{N_t^\sigma} \left(1 + h_{\sigma_{\tau_j^-}}\right)$$

and τ_j, $j \geq 1$ are the jump time instants of the process and N_t^σ the Poisson process with alternating probabilities λ_+ and λ_-. The symbol $\mathcal{E}_t(\cdot)$ is called the *stochastic exponential* and widely used in the analysis of jump processes (see Jacod and Shiryaev 2003). Ratanov (2007a) proved that $X^\sigma + J^\sigma$ is a martingale if and only if the following relationships are realized:

$$h_- = -\frac{c_-}{\lambda_-}, \quad h_+ = -\frac{c_+}{\lambda_+}$$

and since the model is driven by a single source of noise, the market can only have a unique martingale measure.

Theorem 8.2.1 (Ratanov (2007a)) *Let* $Z_t = \mathcal{E}_t(X^* + J^*)$, $t \geq 0$, *with* $h_\sigma^* = -c_\sigma^*/\lambda_\sigma$ *be the Radon-Nikodym density of the probability P^* with respect of P, i.e.*

$$Z_t = \frac{dP^*}{dP} = \mathcal{E}_t(X^* + J^*) = e^{X_t^*}\kappa_t^*, \quad t \geq 0.$$

Then, the process

$$B_t^{-1} S_t^\sigma$$

is a martingale with respect to the measure P^* if and only if

$$c_\sigma^* = \lambda_\sigma + \frac{c_\sigma - r_\sigma}{h_\sigma}, \quad \sigma = \pm 1.$$

Under P^*, the Markov process has the new intensity

$$\lambda_\sigma^* = \frac{r_\sigma - c_\sigma}{h_\sigma} > 0, \quad \sigma = \pm 1$$

A pricing formula to complete the theory is available as well. Let $f(\cdot)$ denote the payoff function of T-claim and introduce the pricing function

$$F(t, x, \sigma) = \mathbb{E}^* \left\{ e^{-Y(T-t)} f\left(xe^X(T-t)\kappa(T-t)\right) \big| \sigma_0 = \sigma \right\},$$
$$\sigma = \pm 1, \quad 0 \leq t \leq T.$$

Then $F(\cdot, \cdot, \cdot)$ solves the following difference-differential equation:

$$F_t(t, x, \sigma) + c_\sigma x F_x(t, x, \sigma) = \left(r_\sigma + \frac{r_\sigma - c_\sigma}{h_\sigma}\right) F(t, x, \sigma)$$
$$- \frac{r - \sigma - c_\sigma}{h_\sigma} F(t, x(1 + h_\sigma), -\sigma), \quad \sigma = \pm 1$$

with terminal condition $F_{t \to T} = f(x)$. The above equation is the analogue of the fundamental equation (6.1) in the Black and Scholes model. This equation has the following explicit form

$$F(t, x, \sigma) = e^{b_r(T-t)} \sum_{n=0}^{\infty} \int_{\mathbb{R}} e^{-a_r y} f\left(xe^y \kappa^\sigma\right) p(y, T - t : n) dy,$$

where $p(\cdot, \cdot : n)$ is the transition density of telegraph process after n jumps and

$$a_r = \frac{r_+ - r_-}{c_+ - c_-}, \qquad b_r = \frac{c_+ r_- - c_- r_+}{c_+ - c_-}.$$

Notice that the equation do not depend on $\lambda_\pm$ as the original Black and Scholes equation does not depend on the drift μ. When $\lambda_- = \lambda_+ = \lambda$ the formula simplifies and the price of the call can be written in a way similar to the standard Black and Scholes formula. We do not investigate this model further, but more in depth analysis, as the convergence of the model to the Black and Scholes model, can be found in the following series of papers Ratanov (2005a,b, 2007a,b). Earlier works which use the telegraph process as the basis of financial models are Di Masi *et al.* (1994) and Di Crescenzo and Pellerey (2002).

8.3 Markov switching diffusions

In Section 3.20 we introduced stochastic differential equations with Markov switching regimes. We remind briefly the components of this model. A Markov switching diffusion is a process X_t solution to

$$\mathrm{d}X_t = f(X_t, \alpha_t)\mathrm{d}t + g(X_t, \alpha_t)\mathrm{d}\mathbf{B}_t, \qquad X_0 = x_0, \alpha(0) = \alpha, \qquad (8.9)$$

where $\{\mathbf{B}_t, t \geq 0\}$ is an n-dimensional Brownian motion, α_t is a finite-state Markov chain in continuous time with state space $\mathcal{S}$ and generator $Q = [q_{ij}]$, $f(\cdot, \cdot) : \mathbb{R}^r \times \mathcal{S} \to \mathbb{R}^n$ and $g(\cdot, \cdot) : \mathbb{R}^n \times \mathcal{S} \to \mathbb{R}^{n \times n}$. The initial value x, the Brownian motion $\mathbf{B}$ and the Markov chain α_t are all mutually independent. An example of this model in $\mathbb{R}$ is the geometric Brownian motion with switching

$$\mathrm{d}S_t = \mu(\alpha_t)S_t \mathrm{d}t + \sigma(\alpha_t)S_t \mathrm{d}\mathbf{B}_t, \qquad S_0 = s_0$$

where $\mu(i)$, $i \in \mathcal{S}$ is the expected return and $\sigma(i) > 0$, $i \in \mathcal{S}$ represents the stock volatility with $f(x, y) = \mu(y)x$ and $g(x, y) = \sigma(y)x$ in the general stochastic differential equation (8.9). This model is intended to capture macro fluctuations of the market which change the sets of parameters μ and σ temporarily during the period. So it is a slight generalization of the standard Black and Scholes model in which μ and σ are constant. For example, in a Markov chain α_t with only two states, i.e. $\mathcal{S} = \{0, 1\}$, $\alpha_t = 1$ may indicate that the market is in expansion and $\mu(1) = \mu_1$ and $\sigma(1) = \sigma_1$ are the trend and volatility parameters of the standard Black and Scholes model during market expansion; $\alpha_t = 0$ implies $\mu(0) = \mu_0$ and $\sigma(0) = \sigma_0$ another set of parameters during market contraction. The Markov process α_t can also be of the type $\alpha_t = (\alpha_t^\mu, \alpha_t^\sigma)$ where α_t^μ models the trend at time t and α_t^σ the volatility at time t. For simplicity we now write $\mu_{\alpha_t} = \mu(\alpha_t)$ and $\sigma_{\alpha_t} = \sigma(\alpha_t)$ so that we have the following system of differential equations:

$$\begin{cases} \mathrm{d}S_t = S_t \mu_{\alpha_t} \mathrm{d}t + S_t \sigma_{\alpha_t} \mathrm{d}\mathbf{B}_t, \\ \mathrm{d}R_t = r R_t \mathrm{d}t \end{cases}$$

for $t \geq 0$, where R_t is bond-type risk-free asset. We assume that $S = \{0, 1\}$ so there are only the four parameters μ_i, σ_i, $i = 0, 1$. Clearly $\mu_1 = \mu_0$ and $\sigma_1 = \sigma_0$ correspond to the standard Black and Scholes model. We assume that the continuous time Markov chain changes its state from i to j with rate λ_i and the waiting time is denoted by τ_i. Then

$$P(\tau_i > t) = e - \lambda_i t, \quad i = 0, 1.$$

To live in a market without arbitrage, we need to find an equivalent martingale measure such that the discounted process $S_t^d = e^{-rt} S_t$ is a martingale under the new measure. We now describe how to get the martingale measure which is similar to what we see in Section 6.4. Let $f(t, x) = e^{-rt} x$ and apply the Itô formula to $f(t, S_t)$. Clearly we have that

$$f_f(t, x) = -re^{-rt} x, \quad f_x(t, x) = e^{-rt}, \quad f_{xx}(t, x) = 0$$

therefore, by Itô formula we have

$$f(t, S_t) = f(0, S_0) + \int_0^t \left\{ f_t(u, S_u) + f_x(u, S_u) \mu_{\alpha_u} S_u \right\} du + \int_0^t f_x(u, S_u) \sigma_{\alpha_u} S_u dB_u$$

$$= S_0 + \int_0^t \left\{ -re^{-ru} S_u + e^{-ru} \mu_{\alpha_u} S_u \right\} du + \int_0^t e^{-ru} \sigma_{\alpha_u} S_u dB_u$$

$$= S_0 + \int_0^t \left(\mu_{\alpha_u} - r \right) e^{-ru} S_u du + \int_0^t e^{-ru} \sigma_{\alpha_u} S_u dB_u$$

thus

$$S_t^d = S_t^0 + \int_0^t \left(\mu_{\alpha_u} - r \right) S_u^d du + \int_0^t \sigma_{\alpha_u} S_u^d dB_u$$

and then

$$dS_t^d = (\mu_{\alpha_t} - r) S_t^d dt + \sigma_{\alpha_t} S_t^d dB_t.$$

Now we can define a new Brownian motion under the new measure Q:

$$W_t = \int_0^t \frac{\mu_{\alpha_u} - r}{\sigma_{\alpha_u}} du + B_t.$$

Now replacing B_t with W_t in the stochastic differential equation of S_t^d we get

$$dS_t^d = S_t^d \left\{ (\mu_{\alpha_u} - r) dt + \sigma_{\alpha_u} dB_t \right\}$$

$$= S_t^d \left\{ (\mu_{\alpha_u} - r) dt + \sigma_{\alpha_u} \left(dW_t - \frac{\mu_{\alpha_t} - r}{\sigma_{\alpha_t}} dt \right) \right\}$$

$$= \sigma_{\alpha_t} S_t^d dW_t$$

which shows, by the properties of the Itô integral, that S_t^d is a martingale. In this setup

$$M_T = \frac{dQ}{dP} = \exp\left\{-\int_0^T \frac{\mu_{\alpha_t} - r}{\sigma_{\alpha_t}} dW_t - \frac{1}{2}\int_0^T \left(\frac{\mu_{\alpha_t} - r}{\sigma_{\alpha_t}}\right)^2 dt\right\}$$

is the Radon-Nikodým derivative of Q over P given by the Girsanov's theorem 3.16.1, so clearly P and Q are equivalent measures. Under the new measure we also have that

$$dS_t = rS_t dt + \sigma_{\alpha_t} dW_t,$$

by replacing S_t^d by $e^{-rt}S_t$. Now that we have a martingale measure we need to ensure completeness of the market. This market is not complete because we have an additional noise process α_t which is not adapted to the filtration of the Brownian motion, say $\{\mathcal{F}_t, t \geq 0\}$, but only to the filtration generated by S_t, say $\{\mathcal{G}_t, t \geq 0\}$, thus according to Harrison and Pliska (1981) the market is not complete as it is. In order to have a complete market Guo (2001) suggests a new security be introduced into the market. That is, at each time t, there exists a security that pays one unit of account (say \$) the next time $\tau(t) = \inf\{u > t | \alpha_u \neq \alpha_t\}$ that the continuous-time Markov chain switches its state. One can think of this as an insurance contract that compensates its holder for any losses that occur when the next state change occurs. This security is called COS 'change-of-state' contract. According to Guo (2001), the COS contract should be traded for a price of

$$V_t = \mathbb{E}\left\{\left\{e^{-(r+k(\alpha_t))(\tau(t)-t)}\middle| \mathcal{G}_t\right\}\right.$$

where $k : \{0, 1\} \to \mathbb{R}$ is a given value that can be considered as a risk-premium coefficient. More precisely, the COS price is given by the formula:

$$V_t = J(\alpha_t), \quad \text{with } J_i = \frac{\lambda(i)}{r + k(i) + \lambda(i)}$$

and $\lambda(\alpha_t)$ is the intensity of the point process N that counts the changes of states in the Markov chain. Under the measure Q, the price V_t takes this form:

$$V_t = \mathbb{E}_Q\left\{e^{-r(\tau_t - t)}\middle| \mathcal{G}_t\right\}$$

and the price of the COS security is zero after the next change of state. Under the martingale measure Q, the counting process N has intensity $\lambda_Q(\alpha_t)$, thus

$$V_t = J_Q(\alpha_t), \quad \text{with } J_Q(i) = \frac{\lambda_Q(i)}{r + \lambda_Q(i)}.$$

Now, due to the fact that $J = J_Q$, we also have that

$$\lambda_Q(i) = \frac{r\lambda(i)}{r + k(i)}.$$

Similarly, the underlying risky-asset under the new measure Q solves the following stochastic differential equation

$$dS_t = (r - d_{\alpha_t})S_t dt + S_t \sigma_{\alpha_t} dW_t \qquad (8.10)$$

where $d_{\alpha_t} = r - \mu_{\alpha_t}$. The solution of the above stochastic differential equation is

$$S_t = S_0 \exp\left\{ \int_0^t \left(r - d_{\alpha_s} - \frac{1}{2}\sigma_{\alpha_s}^2 \right) ds + \int_0^t \sigma_{\alpha_s} dW_s \right\}$$

with initial condition S_0. As usual the proof can be done using Itô formula as in the standard case of the geometric Brownian motion process. We are now ready to express the pricing formula for the European call option. Remember that the payoff function for the call is $f(x) = \max(x - K, 0)$.

Theorem 8.3.1 (Guo 2001) *Let α_T be a continuous time Markov process with state space $S = \{0, 1\}$ and T_i be the random variable which measures the total time between 0 and T during which $\alpha_t = 0$ starting from state $i = 0, 1$. Assume we have a COS security, a risk-free interest rate r a martingale measure Q such that the price S_t satisfies (8.10). Then the arbitrage-free price of a European call option with maturity T and strike price K is*

$$C_i(T, K, r) = \mathbb{E}_Q \left\{ e^{-rT} f(S_T) \big| \alpha_0 = i \right\}$$

$$= e^{-rT} \int_0^\infty \int_0^T \frac{y}{y + K} \varphi(\log(y + K); m_t, v_t) \psi_i(dt, T) dy \quad (8.11)$$

where $\varphi(x; m, v)$ is the density of the Gaussian random variable $N(m, v)$, with

$$m_t = \log(S_0) + \left(d_1 - d_0 - \frac{\sigma_0^2 - \sigma_1^2}{2} \right) t + \left(r - d_1 \frac{1}{2}\sigma_1^2 \right) T$$

$$v_t = \left(\sigma_0^2 - \sigma_1^2 \right) t + \sigma_1^2 T$$

and $\psi_i(\cdot, T)$ is the probability distribution of T_i, $i = 0, 1$

$$\psi_0(t, T) = e^{-\lambda_0 T} \delta_0(T - t)$$

$$+ e^{-\lambda_1(T-t) - \lambda_0 t} \left\{ \lambda_0 I_0 \left(2\sqrt{\lambda_0 \lambda_1 t(T - t)} \right) \right.$$

$$\left. + \sqrt{\frac{\lambda_0 \lambda_1 t}{T - t}} I_1 \left(2\sqrt{\lambda_0 \lambda_1 t(T - t)} \right) \right\}$$

$$\psi_1(t, T) = e^{-\lambda_1 T} \delta_0(T - t)$$

$$+ e^{-\lambda_1(T-t) - \lambda_0 t} \left\{ \lambda_1 I_0 \left(2\sqrt{\lambda_0 \lambda_1 t(T - t)} \right) \right.$$

$$\left. + \sqrt{\frac{\lambda_0 \lambda_1 t}{T - t}} I_1 \left(2\sqrt{\lambda_0 \lambda_1 t(T - t)} \right) \right\}$$

with λ_0, λ_1 are the total rate of leaving out of state 0 and 1 respectively and $I_v(x)$ is the modified Bessel function of order v (see (Abramowitz and Stegun 1964)) and δ is the Dirac delta function.

Notice that if $\mu_0 = \mu_1$ and $\sigma_0 = \sigma_1$, and $\lambda_0 = \lambda_1 = 0$, then m_t and v_t do not depend on t and the pricing equation falls back to the standard Black and Scholes solution. Although Equation (8.11) is explicit, even numerically it is quite hard to compute. Fuh *et al.* (2002) proved an approximation formula based on the following asymptotic argument. The random variable $T_i/T \to \pi_i$ in probability as $T \to \infty$ where π_i is the element of the stationary distribution of α_t corresponding to state $i \in S$. Then T_i can be replaced by the quantity $\pi_i T$ in Equation (8.11) and, as $T \to \infty$, we have that

$$\psi_0(t, T) = \psi_1(t, T) = \begin{cases} 0, & \text{when } t = T, \\ 1, & \text{when } t \neq T. \end{cases}$$

Then, Equation (8.11) becomes

$$\begin{aligned}
\tilde{V}_i = e^{-rT} \int_0^\infty \frac{y}{y + K} &\Big\{ \varphi(\log(y + K); m(\pi_0 T), v(\pi_0 T)) \\
&\times \left(1 - e^{-\lambda_0 T} \delta_0(i - 0) - e^{-\lambda_1 T} \delta_0(1 - i) \right) + \varphi(\log(y + K); m(T), v(T)) \\
&\times e^{-\lambda_0 T} \delta_0(i - 0) \varphi(\log(y + K); m(0), v(0)) e^{-\lambda_1 T} \delta_0(1 - i) \Big\} dy
\end{aligned}$$

which gives the two solutions $\tilde{V}_0$ and $\tilde{V}_1$ below:

$$\begin{aligned}
\tilde{V}_0 = e^{-rT} \int_0^\infty \frac{y}{y + K} &\Big\{ \varphi(\log(y + K); m(\pi_0 T), v(\pi_0 T)) \left(1 - e^{-\lambda_0 T} \right) \\
&+ \varphi(\log(y + K); m(T), v(T)) e^{-\lambda_0 T} \Big\} dy
\end{aligned} \tag{8.12}$$

$$\begin{aligned}
\tilde{V}_1 = e^{-rT} \int_0^\infty \frac{y}{y + K} &\Big\{ \varphi(\log(y + K); m(\pi_0 T), v(\pi_0 T)) \left(1 - e^{-\lambda_1 T} \right) \\
&+ \varphi(\log(y + K); m(T), v(T)) e^{-\lambda_1 T} \Big\} dy
\end{aligned}$$

These two approximations require only one-dimensional integration which is more accurate and faster. We present the code for $\tilde{V}_0$ and $\tilde{V}_1$ below.

```
R> V0 <- function(S0, K, T, r, s0, s1, L0, L1, p0) {
+       m <- function(t) log(S0) + (d1 - d0 - 0.5 *
      (s0^2 - s1^2)) *
+               t + (r - d1 - 0.5 * s1^2) * T
+       v <- function(t) (s0^2 - s1^2) * t+s1^2 * T
+       f <- function(y) {
+           y/(y + K) * (dnorm(log(y + K), m(p0 * T), sqrt(v(p0 *
+               T))) * (1 - exp(-L0 * T))+dnorm(log(y + K), m(T),
```

```
+                sqrt(v(T))) * exp(-L0 * T))
+        }
+        integrate(f, 0, Inf, subdivisions = 1000)$value * exp(-r *
+            T)
+ }
R> V1 <- function(S0, K, T, r, s0, s1, L0, L1, p0) {
+        m <- function(t) log(S0) + (d1 - d0 - 0.5 *
+   (s0^2 - s1^2)) *
+            t + (r - d1 - 0.5 * s1^2) * T
+        v <- function(t) (s0^2 - s1^2) * t + s1^2 * T
+        f <- function(y) {
+            y/(y + K) * (dnorm(log(y + K),  m(p0 * T), sqrt(v(p0 *
+                T))) * (1 - exp(-L1 * T)) + dnorm(log(y + K), m(0),
+                sqrt(v(0))) * exp(-L1 * T))
+        }
+        integrate(f, 0, Inf, subdivisions = 1000)$value * exp(-r *
+            T)
+ }
```

Clearly, if $\lambda_i = 0$, the value $\tilde{V}_i$ converges to the corresponding Black and Scholes price under state $i = 0$. Figure 8.3 shows this behaviour for $\sigma_0 = 0.2$, $\sigma_1 = 0.4$, $r = 0.1$, $d_0 = d_1 = 0$, $S_0 = 100$, $K = 110$ and $\pi_0 = .5$ for different maturity dates $T \in (0, 1]$ when $\lambda_0 = \lambda_1 = 1$.

```
R> require(fOptions)
R> r <- 0.1
R> s0 <- 0.2
R> s1 <- 0.4
R> L0 <- 1
R> L1 <- 1
R> K <- 110
R> S0 <- 100
R> d1 <- 0
R> d0 <- 0
R> p0 <- 0.5
R> tt <- seq(0, 1, length = 50)
R> pV0 <- NULL
R> pV1 <- NULL
R> pBS0 <- NULL
R> pBS1 <- NULL
R> for (T in tt) {
+        pV0 <- c(pV0, V0(S0, K, T, r, s0, s1, L0, L1, p0))
+        pV1 <- c(pV1, V1(S0, K, T, r, s0, s1, L0, L1, p0))
+        pBS0 <- c(pBS0, GBSOption(TypeFlag = "c", S = S0, X = K,
+            Time = T, r = r, b = r, sigma = s0)@price)
+        pBS1 <- c(pBS1, GBSOption(TypeFlag = "c", S = S0, X = K,
+            Time = T, r = r, b = r, sigma = s1)@price)
+ }
R> matplot(tt, cbind(pV0, pV1, pBS0, pBS1), type = "o",
   lty = rep(1,
+        4), pch = c(0, 1, 15, 16), cex = 0.5, col = rep(1, 4),
   main = "",
```

```
+        xlab = expression(T))
R> legend(0.1, 12, c(expression(tilde(V)[0]),
     expression(tilde(V)[1]),
+        expression(BS[0]), expression(BS[1])), lty = rep(1, 4),
     pch = c(0,
+        1, 15, 16), cex = 0.75, col = rep(1, 4))
```

From Figure 8.3 we notice that the prices under the Markov switching model are always between the corresponding Black and Scholes prices. When λ_0 increases, the price $\tilde{V}_0$ moves away from the Black and Scholes price BS_0, while when λ_1 decreases to zero, $\tilde{V}_1$ converges to BS_1 (see Figure 8.4). Finally, in Figure 8.5, we show that when both λ_0 and λ_1 increase, the two prices $\tilde{V}_0$ and $\tilde{V}_1$ move away from the Black and Scholes prices and both converge to a common limit.

This model and the approximation formula can be generalized to the case of a Markov chain α_t with state space $\mathcal{S} = \{0, 1, \ldots, N\}$. In this case the pricing formula of the call option is given by the following expression:

$$
\begin{aligned}
\tilde{V}_i = e^{-rT} \int_0^\infty \Bigg\{ &\varphi(\log(y + K); m(t_0, t_1, \ldots, T_n), v(t_0, t_1, \ldots, t_n)) \\
&\left(1 - e^{-\lambda_0 T} \delta_0(0 - i) - \cdots - e^{-\lambda_N T} \delta_0(N - i)\right) \\
&+ \varphi(\log(y + K); m(T, 0, \ldots, 0), v(T, 0, \ldots, 0)) e^{-\lambda_0 T} \delta_0(0 - i) \\
&+ \varphi(\log(y + K); m(0, 0, \ldots, T), v(0, 0, \ldots, T)) e^{-\lambda_N T} \delta_0(N - i) \Bigg\} dy
\end{aligned}
$$

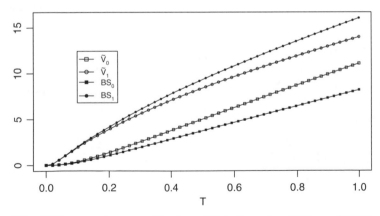

Figure 8.3 *Difference between Black and Scholes price (BS_0 and BS_1) and the price under the Markov switching diffusion ($\tilde{V}_0$ and $\tilde{V}_1$). Parameters: $\sigma_0 = 0.2$, $\sigma_1 = 0.4$, $r = 0.1$, $d_0 = d_1 = 0$, $S_0 = 100$, $K = 110$ and $\pi_0 = .5$ for different maturity dates $T \in (0, 1]$ when $\lambda_0 = \lambda_1 = 1$.*

where $t_i = \pi_i T$, $i = 0, \ldots, N$ and

$$m(t_0, t_1, \ldots, t_N) = \log(S_0) + \sum_{i=0}^{N} \left(r - d_i - \frac{1}{2}\sigma_i^2 \right) t_i,$$

$$v(t_0, t_1, \ldots, t_N) = \sum_{i=0}^{N} \sigma_i^2 t_i.$$

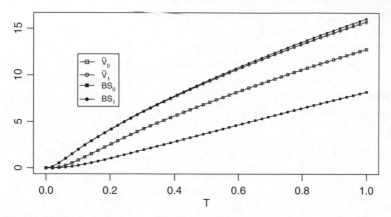

Figure 8.4 *Difference between Black and Scholes price (BS_0 and BS_1) and the price under the diffusion Markov switching ($\tilde{V}_0$ and $\tilde{V}_1$). Parameters: $\sigma_0 = 0.2$, $\sigma_1 = 0.4$, $r = 0.1$, $d_0 = d_1 = 0$, $S_0 = 100$, $K = 110$ and $\pi_0 = .5$ for different maturity dates $T \in (0, 1]$ when $\lambda_0 = 10$, $\lambda_1 = 0.1$. The price $\tilde{V}_0$ moves up while the price $\tilde{V}_1$ converges to BS_1.*

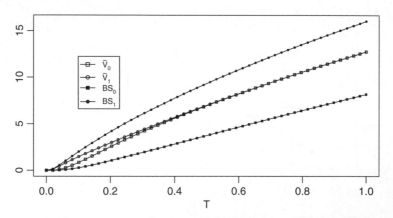

Figure 8.5 *Difference between Black and Scholes price (BS_0 and BS_1) and the price under the diffusion Markov switching ($\tilde{V}_0$ and $\tilde{V}_1$). Parameters: $\sigma_0 = 0.2$, $\sigma_1 = 0.4$, $r = 0.1$, $d_0 = d_1 = 0$, $S_0 = 100$, $K = 110$ and $\pi_0 = .5$ for different maturity dates $T \in (0, 1]$ when $\lambda_0 = 10$, $\lambda_1 = 10$. The price $\tilde{V}_0$ moves up and $\tilde{V}_1$ moves down and both converge to the same curve.*

8.3.1 Monte Carlo pricing

As for the other models, it is possible to price options using the Monte Carlo approach. In our case, we make use of the code `simMSdiff` presented in Section 4.5.7.

```
R> simMarkov <- function(x0, n, x, P) {
+       mk <- numeric(n + 1)
+       mk[1] <- x0
+       state <- which(x == x0)
+       for (i in 1:n) {
+           mk[i + 1] <- sample(x, 1, prob = P[state, ])
+           state <- which(x == mk[i + 1])
+       }
+       return(ts(mk))
+ }
R> simMSdiff <- function(x0, a0, S, delta, n, f, g, Q) {
+       require(msm)
+       P <- MatrixExp(delta * Q)
+       alpha <- simMarkov(a0, n, S, P)
+       x <- numeric(n + 1)
+       x[1] <- x0
+       for (i in 1:n) {
+           A <- f(x[i], alpha[i]) * delta
+           B <- g(x[i], alpha[i]) * sqrt(delta) * rnorm(1)
+           x[i + 1] <- x[i] + A + B
+       }
+       ts(x, deltat = delta, start = 0)
+ }
```

After preparing all the code, we now need to specify the infinitesimal generator Q, which in the two-state cases is

$$Q = \begin{bmatrix} -\lambda_0 & \lambda_0 \\ \lambda_1 & -\lambda_1 \end{bmatrix}$$

and simulate the diffusion with Markov switching under the risk-neutral measure, i.e. the process

$$dS_t = (r - d_{\alpha_t})S_t dt + \sigma_{\alpha_t} S_t dW_t$$

where W_t is the Brownian motion under the martingale measure. The parameters of the model are as follows: $T = 0.75$, $r = 0.05$, $\sigma_0 = 0.2$, $\sigma_1 = 0.4$, $\lambda_0 = \lambda_1 = 1$, $d_0 = d_1 = 0$, $\pi_0 = 0.5$ and $K = 80$, $S_0 = 85$.

```
R> T <- 0.75
R> r <- 0.05
R> s0 <- 0.2
R> s1 <- 0.4
R> L0 <- 1
R> L1 <- 1
R> K <- 80
```

```
R> S0 <- 85
R> d1 <- 0
R> d0 <- 0
R> p0 <- 0.5
R> Q <- matrix(c(-L0, L1, L0, -L1), 2, 2)
```

Notice that for the infinitesimal generator we always have $\pi Q = 0$, indeed

```
R> c(p0, p0) %*% Q
```

```
     [,1] [,2]
[1,]    0    0
```

Now we prepare the drift and diffusion coefficients to pass to the function simMS-diff.

```
R> f <- function(x, a) ifelse(a == 0, (r - d0) * x, (r - d1) * x)
R> g <- function(x, a) ifelse(a == 0, s0 * x, s1 * x)
```

and we simulate with a time mesh of $\Delta = 1/n$ with $n = 1000$. Remember that the price of the call option is given by

$$V_0 = e^{-rT}\mathbb{E}_Q\{\max(S_T - K, 0)\}.$$

The algorithm is very simple then

```
R> n <- 1000
R> set.seed(123)
R> nsim <- 1000
R> ST0 <- numeric(nsim)
R> for (i in 1:nsim) {
+       X <- simMSdiff(x0 = S0, a0 = 0, S = 0:1, delta = 1/n,
     n = T *
+            n, f, g, Q)
+       ST0[i] <- X[length(X)]
+ }
R> v0 <- exp(-r * T) * pmax(ST0 - K, 0)
R> mean(v0)
```

```
[1] 11.93979
```

which we compare with price given by the approximation formula for $\tilde{V}_0$

```
R> mean(v0)
```

```
[1] 11.93979
```

```
R> V0(S0, K, T, r, s0, s1, L0, L1, p0)
```

```
[1] 11.97513
```

Similarly we proceed with V_1. The only difference is the starting state $\alpha_0 = i$ of the Markov chain α_t. Therefore

```
R> set.seed(123)
R> ST1 <- numeric(nsim)
R> for (i in 1:nsim) {
+      X <- simMSdiff(x0 = S0, a0 = 1, S = 0:1, delta = 1/n,
     n = T *
+            n, f, g, Q)
+        ST1[i] <- X[length(X)]
+ }
R> v1 <- exp(-r * T) * pmax(ST1 - K, 0)
R> mean(v1)
```

```
[1] 13.88842
```

```
R> V1(S0, K, T, r, s0, s1, L0, L1, p0)
```

```
[1] 14.39285
```

Although the number of replications in the Monte Carlo pricing are small, the prices are reasonably close. These algorithms are sufficiently general so one can price Markov switching diffusion processes with any number of states but also with nonlinear drift and diffusion coefficients $f(\cdot, \cdot)$ and $g(\cdot, \cdot)$.

8.3.2   Semi-Monte Carlo method

According to Buffington and Elliott (2002), a full Monte Carlo method is not always needed if it is possible to generate a continuous Markov Chain α_t. Indeed, it is possible to use a Black and Scholes formula based on the fact that

$$S_t = S_0 \exp\{X_t\} = S_0 \exp\left\{ \int_0^t \left(r - d_{\alpha_s} - \frac{1}{2}\sigma_{\alpha_s}^2 \right) ds + \int_0^t \sigma_{\alpha_s} dW_s \right\}$$

with W_t the Brownian motion under the risk-free measure. The call option price can be written as

$$C = \mathbb{E}\left\{ e^{-rT} \max(S_T - K, 0) \right\} = \mathbb{E}\left(\mathbb{E}\left\{ e^{-rT} \max(S_T - K, 0) \big| \mathcal{F}_T \right\} \right)$$

where $\{\mathcal{F}_t, 0 \leq t \leq T\}$ is the σ-algebra generated by the Markov chain α_t. Now, as noticed by Liu $et\ al.$ (2006), the conditional expectation can be calculated as follows:

$$\mathbb{E}\left\{ e^{-rT} \max(S_T - K, 0) \big| \mathcal{F}_T \right\} = S_0 e^{-(rT - L_T)} \Phi(d1) - K e^{-rT} \Phi(d2)$$

where

$$d_1 = \frac{\log\left(\frac{S_0}{K} \right) + L_T + \frac{1}{2}V_T}{\sqrt{V_T}}, \quad d_2 = d_1 - \sqrt{V_T}$$

and

$$L_T = \int_0^T (r - d_{\alpha_t})dt, \quad V_T = \int_0^T \sigma_{\alpha_t}^2 dt.$$

Once the trajectory α_T is available, it is possible to calculate L_T and V_T directly by discretization of the integrals. We construct now the function Vsmc which gives the semi-Monte Carlo price of the option for a Markov chain α with state space S, generator Q and initial state a0. This time we use package **msm** to simulate a full path of the continuous time Markov chain instead of using the ϵ-skeleton approach.

```
R> Vsmc <- function(S0, K, T, r, a0, S, Q, mu, sigma, M = 1000) {
+       require(msm)
+       C <- numeric(M)
+       for (i in 1:M) {
+           MC <- sim.msm(Q, T, start = which(S == a0))
+           alpha <- S[MC$states]
+           t <- diff(MC$times)
+           s <- sapply(alpha[-length(alpha)], sigma)^2
+           m <- sapply(alpha[-length(alpha)], mu)
+           VT <- sum(t * s)
+           LT <- sum(t * m)
+           d1 <- (log(S0/K)+ LT + 0.5 * VT)/sqrt(VT)
+           d2 <- d1 - sqrt(VT)
+           RT <- r * T
+           C[i] <- S0 * exp(-(RT - LT)) * pnorm(d1) - K *
     exp(-RT) *
+               pnorm(d2)
+       }
+       mean(C)
+ }
```

Now we compare the semi-Monte Carlo price with the approximation formulas (8.12) of $\tilde{V}_i$, $i = 0, 1$. We use the following parameters: $T = 1, r = 0.1, \sigma_0 = 0.2,$ $\sigma_1 = 0.3, \lambda_0 = \lambda_1 = 1, d_0 = d_1 = 0, K = 90, S_0 = 100$ as in the original work of Liu *et al.* (2006).

```
R> T <- 1
R> r <- 0.1
R> s0 <- 0.2
R> s1 <- 0.3
R> L0 <- 1
R> L1 <- 1
R> K <- 90
R> S0 <- 100
R> d1 <- 0
R> d0 <- 0
R> p0 <- 0.5
R> Q <- matrix(c(-L0, L1, L0, -L1), 2, 2)
R> sigma <- function(a) ifelse(a == 0, s0, s1)
R> mu <- function(a) ifelse(a == 0, r - d0, r - d1)
```

Now we test the result for an initial sate $\alpha_0 = 0$

```
R> Vsmc(S0, K, T, r, a0 = 0, 0:1, Q, mu, sigma)
```

```
[1] 20.71375
```

```
R> V0(S0, K, T, r, s0, s1, L0, L1, p0)
```

```
[1] 20.81154
```

and for an initial state $\alpha_0 = 1$

```
R> Vsmc(S0, K, T, r, a0 = 1, 0:1, Q, mu, sigma)
```

```
[1] 21.81279
```

```
R> V1(S0, K, T, r, s0, s1, L0, L1, p0)
```

```
[1] 21.73914
```

both results are close. Notice that this method is much faster than the pure Monte Carlo method of previous section, although it is limited to the standard linear model, while the full Monte Carlo method works with any Markov switching diffusion.

8.3.3 Pricing with the Fast Fourier Transform

In the same work of Liu *et al.* (2006), it is possible to find the explicit formulas for the characteristic functions of V_0 and V_1 which are need to perform option pricing using the FFT algorithm. We remind briefly only the formulas and the basic idea because pricing by FFT is the same as what has already been treated in Section 8.1.5. Let $\rho > 0$ denote the dampening factor and $k = \log(K)$. By Carr and Madan (1998) we know that the call price formula can be given as

$$c(k) = e^{\rho k} \frac{C(K)}{S_0}$$

and the objective is to derive a formula for the characteristic function of $\psi(u)$ of $c(k)$. The formula is the following:

$$\psi(u) = \frac{\mathbb{E}\left(e^{-rT}\phi_T(u - i(1 + \rho))\right)}{\rho^2 + \rho - u^2 + i(1 + 2\rho)u}$$

where $\phi_T(u) = \mathbb{E}\left\{e^{iuX_T} \mid \mathcal{F}_T\right\}$ is the conditional expectation of X_T given $\{\mathcal{F}_t, 0 \leq t \leq T\}$, the σ-algebra generated by α_t and X_t as in previous section. X_T is a Gaussian random variable with mean $L_T - \frac{1}{2}V_T$ and variance V_T, hence:

$$\phi_T(u) = \exp\left\{iu\left(L_T - \frac{1}{2}V_T\right) - \frac{1}{2}u^2 V_T\right\}.$$

Putting all terms together we obtain

$$\psi(u) = \frac{\mathbb{E}\left(\exp\left\{(1+\rho)\left(L_T - \frac{1}{2}\rho V_T\right) - rT - \frac{1}{2}u^2 V_T + iu\left(L_T + \left(\frac{1}{2}+\rho\right)V_T\right)\right\}\right)}{\rho^2 + \rho - u^2 + i(1+2\rho)u}$$

Taking into account a Markov chain with only two states, it is possible to obtain the following explicit formula for $\psi(u)$

$$\psi(u) = \frac{\exp\{B(u)T\phi_{\alpha_0}(A(u), T)\}+}{\rho^2 + \rho - u^2 + i(1+2\rho)u}$$

where

$$A(u) = \left\{(d_1 - d_0) + \left(\frac{1}{2}+\rho\right)(\sigma_0^2 - \sigma_1^2)\right\}u + \frac{1}{2}u^2\left(\sigma_0^2 - \sigma_1^2\right)i$$

$$+ \left\{(1+\rho)(d_0 - d_1) - \frac{1}{2}\rho(1+\rho)\left(\sigma_0^2 - \sigma_1\right)^2\right\}i$$

$$B = iu\left\{r - d_1 + \left(\frac{1}{2}+\rho\right)\sigma_1^2\right\} - \frac{1}{2}u^2\sigma_1^2 + (1+\rho)(r - d_1) - r + \frac{1}{2}\rho(1+\rho)\sigma_1^2$$

and

$$\phi_0(\theta, T) = \frac{1}{s_1 - s_2}\left\{(s_1 + \lambda_0 + \lambda_1)e^{s_1 T} - (s_2 + \lambda_0 + \lambda_1)e^{s_2 T}\right\}$$

$$\phi_1(\theta, T) = \frac{1}{s_1 - s_2}\left\{(s_1 + \lambda_0 + \lambda_1 - i\theta)e^{s_1 T} - (s_2 + \lambda_0 + \lambda_1 - i\theta)e^{s_2 T}\right\}$$

with s_1 and s_2 the two roots of the equation

$$s^2 + (\lambda_0 + \lambda_1 - i\theta)s - i\theta\lambda_1 = 0.$$

Although it appears complicated, these formulas are explicit and FFT algorithm can be applied taking into account (Liu *et al.* 2006) that

$$c(k_l) \approx \frac{\Delta_u}{\pi}\sum_{j=0}^{N-1}e^{-ijl\frac{2\pi}{N}}e^{ij\pi}\psi(j\Delta_u)w(j), \quad l = 0, 1, \ldots, N-1,$$

with $k_l = \left(l - \frac{N}{2}\right)\Delta_k, l = 0, 1, \ldots, N-1$, Δ_u the grid size for variable u, with $u_j = j\Delta_u, j = 0, 1, \ldots, N-1$, $\Delta_u\Delta_k = 2\pi/N$ and a set of weights

$$w(j) = \begin{cases} \frac{1}{3}, & j = 0, \\ \frac{4}{3}, & j \text{ is odd}, \\ \frac{2}{3}, & j \text{ is even}. \end{cases}$$

Finally, apply `fft` to the sequence $e^{ij\pi}\psi(j\Delta_u)w(j), j = 0, \ldots, N-1$.

8.3.4   Other applications of Markov switching diffusion models

Di Masi *et al.* (1994) considered a model similar to (8.9) where only the volatility is affected by Markov switching. The model presented in this section is more general in that the drift is also controlled by the underlying Markov process. The model in Di Masi *et al.* (1994) falls in the category of stochastic volatility models which are not treated in this book. A good starting point for reading is Shephard (2005). Under the diffusion model with Markov switching it is also possible to do pricing of other types of options, such as perpetual lookback options, Russian options, perpetual American options as shown in Guo (1999) and Buffington and Elliott (2002). Optimal stock trading rules under this model have been developed in Zhang (2001). Estimation of this model has been considered under different situations and approaches. In Elliott *et al.* (2008) moment type estimation is considered while Fearnhead *et al.* (2008) considered particle filter and Hahn and Sass (2009) the Bayesian approach.

8.4   The benchmark approach

We have seen so far that the main ingredients of option pricing are the existence of a martingale measure, which implies non-arbitrage, and the ability to hedge risk, which implies market completeness. In some cases, like the Lévy market, there exist an infinite set of equivalent martingale measures and thus there exist more than one fair price depending on which measure the market chooses. In other cases, like market governed by the jump telegraph process or Markov switching diffusions, there are additional sources of noise so that completeness is ensured only after the addition of special additional conditions like the introduction of COS securities. But in all models, the pricing requires the existence of a martingale measure, which is more a technical tool than an economic need. Other approaches exist like the so-called *benchmark* approach by Platen and David (2006) and Platen and Bruti-Liberati (2010). We cannot go into too much detail, but it is at least worth mentioning the basic facts of this approach. The main object in this theory is not the martingale measure but the benchmark. The benchmark has the function of *numeraire* with respect to which all derivates and other financial products should be priced. All pricing occur under the real measure P and not under a martingale measure Q and this links more closely the pricing task with the part of the theory related to estimation of the model's parameters from historical data. The key ingredient is the *growth optimal portfolio* GOP which plays the role of the reference unit or numeraire in this market. The GOP is a portfolio that maximizes the expected logarithmic utility from the terminal wealth, see Kelly (1956) and Long (1990). The idea is to target the market on a long run period. With this in mind, the search is for a strictly positive process, like a market index, which when used as numeraire or benchmark, generates benchmarked price processes that become martingales under the physical measure. This implies that the benchmarked derivative process represents the best forecast of their future benchmarked values. Let us now see what happens in the standard Black and Scholes market under this approach. For this market we already know how to derive the fair (non-arbitrage) price under the equivalent martingale measure. As usual, we take the geometric Brownian motion for the asset price dynamics

$$dS_t = \mu S_t dt + \sigma S_t dB_t$$

with some initial condition $S_0 > 0$, and $\sigma > 0$. For the savings account nonrisky process we consider

$$dR_t = r R_t dt, \ R_0 = 1.$$

We denote the self-financing strategy as $\delta(a_t, b_t)$, $t \geq 0$ where a_t is the unit of S_t invested at time t and b_t the units invested in savings account. The value of the portfolio under this strategy is

$$V_t = a_t S_t + b_t R_t$$

with the the self-financing structure

$$
\begin{aligned}
dV_t^\delta &= a_t dS_t + b_t dR_t \\
&= a_t \mu S_t dt + a_t \sigma S_t dB_t + b_t r R_t dt \\
&= (b_t r R_t + a_t \mu S_t)dt + a_t \sigma S_t dB_t \\
&= V_t^\delta (\pi_t^0 r + \pi_t^1 \mu)dt + \pi_t^1 \sigma dB_t
\end{aligned}
$$

where $\pi_\delta^0(t)$ and $\pi_\delta^1(t)$ are the *fractions* that are held in the respective securites

$$\pi_\delta^0(t) = b_t \frac{R_t}{V_t^\delta}, \quad \pi_\delta^1(t) = a_t \frac{S_t}{V_t^\delta}$$

and $\pi_\delta^0(t) + \pi_\delta^1(t) = 1$, for $t \in [0, T]$. Clearly the fractions makes sense as long as the value of the portfolio V_t^δ is not zero. The *growth optimal portfolio* (GOP) is the portfolio that maximizes the drift of the logarithm of the value of the portfolio for any time horizon. By Itô formula we can write the stochastic differential equation for $\log V_t^\delta$ which reads as

$$d \log V_t^\delta = g_t^\delta dt + \pi_\delta^1(t)\sigma dB_t$$

with *growth rate*

$$g_t^\delta = r + \pi_\delta^1(t)(\mu - r) - \frac{1}{2}\left(\pi_\delta^1(t)\right)^2 \sigma^2, \quad t \in [0T].$$

Definition 8.4.1 *Under the Black and Scholes model, the GOP is the portfolio $V_t^{\delta*} = \{V_t^{\delta*}, 0 \leq t \leq T\}$ with optimal growth rate $g_t^{\delta*}$ at time t such that*

$$g_t^\delta \leq g_t^{\delta*}$$

almost surely for all $t \in [0, T]$ and all strictly positive portfolio processes V_t^δ.

For this model is easy to obtain the optimal growth rate, which is indeed the portfolio with optimal fraction $\pi_\delta^1(t)$. We apply simple derivation to find the maximum

$$\frac{\partial}{\partial \pi_\delta^1(t)} g_t^\delta = \mu - r\pi_{\delta*}^1(t)\sigma^2 = 0, t \in [0, T].$$

Therefore, the *optimal fraction* is given by

$$\pi_{\delta*}^1(t) = \frac{\mu - r}{\sigma^2}$$

and

$$\pi_{\delta*}^0(t) = 1 - \pi_{\delta*}^1(t), \quad t \in [0, T].$$

Finally, the optimal growth rate is given by the formula

$$g_t^{\delta*} = r + \frac{1}{2}\left(\frac{\mu - r}{\sigma}\right)^2, \quad t \in [0, T].$$

Now we can replace this expression in the stochastic differential equation of dV_t^δ and obtain

$$dV_t^{\delta*} = V_t^{\delta*}\left((r + \theta_t^2)dt + \theta_t dB_t\right), \quad S_0^{\delta*} > 0$$

and GOP volatility

$$\theta_t = \pi_{\delta*}^1(t)\sigma = \frac{\mu - r}{\sigma}, \quad t \in [0, T].$$

The quantity θ_t is called the *market price of risk* at time t. Now, given all the above, the optimal growth rate for the Black and Scholes model is

$$g_t^{\delta*} = r + \frac{1}{2}\theta_t^2, \quad t \in [0, T].$$

We now introduce the discounted GOP

$$\bar{V}_t^{\delta*} = \frac{V_t^{\delta*}}{R_t}$$

which, again using Itô formula, is the solution of the following stochastic differential equation

$$d\bar{V}_t^{\delta*}\theta_t(\theta_t dt + dB_t), \quad t \in [0, T].$$

Notice that for this model the drift is determined by the square of its diffusion coefficient.

8.4.1 Benchmarking of the savings account

We now benchmark the saving account and the risky asset. The benchmarked savings account $\hat{R}^0 = \{\hat{R}_t^0, 0 \leq t \leq T\}$ where

$$\hat{R}_t^0 = \frac{R_t}{V_t^{\delta*}}, \quad t \in [0, T].$$

Again, by Itô formula we get

$$d\hat{R}_t^0 = -\theta_t \hat{R}_t^0 dB_t.$$

This process is clearly a martingale because its solution is just an Itô integral. So the first effect we notice is that benchmarking the savings account returns a martingale under the physical measure.

8.4.2 Benchmarking of the risky asset

We now perform benchmarking of the underlying asset price process:

$$\hat{S}_t^1 = \frac{S_t}{S_t^{\delta*}}, \quad t \in [0, T].$$

Again, by Itô formula and recalling that $\theta_t = (\mu - r)/\sigma$, we get:

$$d\hat{S}_t^1 = \hat{S}_t^1 (\mu - r - \sigma\theta_t)dt + \hat{S}_t^1 (\sigma - \theta_t)dB_t$$
$$= \hat{S}_t^1 (\sigma - \theta_t)dB_t$$

which is again a martingale under the physical measure.

8.4.3 Benchmarking the option price

Now, let $P_t = P(t, S_t)$ be the price of, e.g., a European option. We can introduce the benchmarked price

$$\hat{P}_t = \frac{P_t}{V_t^{\delta*}} = \frac{\bar{P}(t, \bar{S}_t)}{V_t^{\delta*}} \hat{R}_t^0$$

where $\bar{P}(t, \bar{S}_t) = P(t, S_t)/R_t$ and $\bar{S}_t = S_t/R_t$. Therefore, the benchmarked payoff has the form

$$\hat{P}_T = \frac{H(S_T)}{V_t^{\delta*}}.$$

We now apply Itô formula to derive an expression for the stochastic differential equation of $\bar{P}_t$.

$$d\bar{P}(t, \bar{S}_t) = \left(\frac{\partial}{\partial t} \bar{P}(t, \bar{S}_t) + (\mu - r)\bar{S}_t \frac{\partial}{\partial x} \bar{P}(t, \bar{S}_t) + \frac{1}{2}\sigma^2 \bar{P}(t, \bar{S}_t)^2 \frac{\partial^2}{\partial x^2} \bar{P}(t, \bar{S}_t) \right) dt$$

$$+ \sigma \bar{P}(t, \bar{S}_t) \frac{\partial}{\partial x} \bar{P}(t, \bar{S}_t) dB_t$$

$$= \frac{\partial}{\partial x} \bar{P}(t, \bar{S}_t) \bar{S}_t \left((\mu - r)dt + \sigma dB_t \right).$$

Similar calculations lead to the following stochastic differential equation for $\hat{P}_t$

$$d\hat{P}_t = \hat{R}_t^0 \left(\sigma \bar{S}_t \frac{\partial}{\partial x} \bar{P}(t, \bar{S}_t) - \theta_t \bar{P}(t, \bar{S}_t) \right) dB_t, \quad t \in [0, T].$$

Notice that the benchmarked option price has no drift term.

In summary, the GOP has the property that, if used as numeraire or benchmark, transforms all related processes into martingale. For this reason the GOP is called in the literature the numeraire portfolio, see Long (1990). Now we can introduce the notion of fair price under this setup.

Definition 8.4.2 *A price process $\{P_t, 0 \le t \le T\}$ is called fair if its benchmarked value $\hat{P}_t = P_t / V_t^{\delta *}$ is a martingale under the physical measure.*

Therefore, the following price formula is available

$$P_t = V_t^{\delta *} \mathbb{E} \left\{ \frac{P_T}{S_T^{\delta *}} \middle| \mathcal{F}_t \right\}$$

where $\mathcal{F} = \{\mathcal{F}_t, 0 \le t \le T\}$ is the filtration under the real work measure P.

8.4.4   Martingale representation of the option price process

The stochastic differential equation of the benchmarked option price process $\hat{P}_t$ can be rewritten in this integral form:

$$\hat{P}_T = \hat{P}_t \int_t^T \left(\sigma \frac{S_u}{V_u^{\delta *}} \frac{\partial}{\partial x} \bar{P}(u, \bar{S}_u) - \theta_u \hat{P}_u \right) dB_u$$

and then the benchmarked payoff can be written also in this way

$$\hat{P}_T = \frac{H(S_T)}{V_T^{\delta *}}.$$

Since $\hat{P}$ is a martingale, we have that

$$\mathbb{E} \left\{ \hat{P}_T \middle| \mathcal{F}_t \right\} = \hat{P}_t$$

now multiplying both sides of the above equation by $V_t^{\delta*}$ we get

$$P_t = V_t^{\delta*} \hat{P}_t = V_t^{\delta*} \mathbb{E}\left\{\hat{P}_T \,\middle|\, \mathcal{F}_t\right\}$$

therefore

$$P_t = V_t^{\delta*} \mathbb{E}\left\{\frac{H(S_T)}{V_T^{\delta*}} \,\middle|\, \mathcal{F}_t\right\}, \quad t \in [0, T].$$

And this is a pricing formula ready to be used once the growth optimal portfolio (GOP) strategy $V_t^{\delta*}$ is available. Platen and David (2006) show how to determine the GOP in practice and extend the above result to a variety of financial applications (not just option pricing) and also under jump models.

8.5 Bibliographical notes

There is a vast literature on pricing models out of the Black and Scholes method. In this chapter we presented few examples where some of them are more used than others in practice like the exponential Lévy market model. In particular, for this model one should mention at least the following references: Schoutens (2003), Cont and Tankov (2004), Joundeau *et al.* (2007), Di Nunno *et al.* (2009) and Platen and Bruti-Liberati (2010). Models based on the telegraph process have been studied in Di Masi *et al.* (1994), Di Crescenzo and Pellerey (2002), Ratanov (2005a,b, 2007a,b). For the Markov switching diffusion process one can start from the following papers Guo (1999), Buffington and Elliott (2002), Zhang (2001) and references there in. The benchmark approach is mostly treated in Platen and David (2006) and Platen and Bruti-Liberati (2010).

References

Abramowitz, M. and Stegun, I. (1964). *Handbook of Mathematical Functions*. Dover Publications, New York.

Bakshi, G. and Madan, D. (2000). Spanning and derivative security valuation. *Journal of Financial Economics* **55**, 2, 205–238.

Buffington, J. and Elliott, R. J. (2002). American options with regime switching. *International Journal of Theoretical and Applied Finance* **5**, 497–514.

Carr, P. and Madan, D. (1998). Option valuation using the fast fourier transform. *Journal of Computational Finance* **2**, 61–73.

Cont, R. and Tankov, P. (2004). *Financial Modelling With Jump Processes*. Chapman & Hall/CRC, Boca Raton.

Di Crescenzo, A. and Pellerey, F. (2002). On prices' evolutions based on geometric telegrapher's process. *Applied Stochastic Models in Bussiness and Industry* **18**, 171–184.

Di Masi, G., Kabanov, Y., and Runggaldier, W. (1994). Mean-variance hedging of options on stocks with Markov volatilities. *Theory of Probability and its Applications* **39**, 172–182.

Di Nunno, G., Øksendal, B., and Proske, F. (2009). *Malliavin Calculus for Lévy Processes with Applications to Finance*. Springer, New York.

Eberlein, E. and Jacod, J. (1997). On the range of options prices. *Finance Stoch*. **1**, 131–140.

Elliott, R., Krishnamurthy, V., and Saass, J. (2008). Moment based regression algorithms for drift and volatility estimation in continuous-time Markov switching models. *Econometrics Journal* **11**, 244–270.

Fearnhead, P., Papaspiliopoulos, O., and Roberts, G. O. (2008). Particle filters for partially observed diffusions. *Journal of the Royal Statistical Society: Series B* **70**, 4, 755–777.

Fuh, C.-D., Wang, R.-H., and Cheng, J.-C. (2002). Option pricing in a Black-Scholes model with Markov switching. *Working Paper C-2002-10, Institute of Statistical Science, Academia Sinica, http://www3.stat.sinica.edu.tw/library/c_tec_rep/c-2002-10.pdf*.

Gerber, H. and Shiu, E. (1994). Option pricing by Esscher-transforms. *Transactions of the Society of Actuaries* **46**, 99–191.

Gerber, H. and Shiu, E. (1996). Actuarial bridges to dynamic hedging and option pricing. *Insurance: Mathematics and Economics* **18**, 3, 183–218.

Guo, X. (1999). *Inside Information and Stock Fluctuations*. Ph.D. dissertation, Department of Mathematics, Rutgers University, New Jersey.

Guo, X. (2001). Information and option pricings. *Quant. Finance* **1**, 1, 38–44.

Hahn, M. and Sass, J. (2009). Parameter estimation in continuous time Markov switching models: A semi-continuous Markov chain Monte Carlo approach. *Bayesian Analysis* **4**, 1, 63–84.

Harrison, M. and Pliska, S. (1981). Martingales and stochastic integrals in the theory of continuous trading. *Stoch. Proc. and Their App*. **11**, 215–260.

Jacod, J. and Shiryaev, A. (2003). *Limit Theorems for Stochastic Processes (2nd ed.)*. Springer, New York.

Joundeau, E., Poon, S., and Rockinger, M. (2007). *Financial Modeling under Non-Gaussian Distributions*. Springer, New York.

Kelly, J. R. (1956). A new interpretation of information rate. *Bell Syst. Techn. J*. **35**, 917–926.

Lewis, A. L. (2001). A simple option formula for general jump-diffusion and other exponential lévy processes. In *Other Exponential Lévy Processes, Environ Financial Systems and OptionCity.net*. http://optioncity.net/pubs/ExpLevy.pdf.

Liu, R., Zhang, Q., and Yin, G. (2006). Option pricing in a regime-switching model using the Fast Fourier Transform. *Journal of Applied Mathematics and Stochastic Analysis* 1–22.

Long, J. B. (1990). The numeraire portfolio. *J. Financial Economics* **26**, 29–69.

Papapantoleon, A. (2008). *An Introduction to Lévy Processes with Applications in Finance*. Lecture notes, TU Vienna, http://arxiv.org/abs/0804.0482, Vienna.

Platen, E. and Bruti-Liberati, N. (2010). *Numerical Solution of Stochastic Differential Equations with Jumps in Finance*. Springer, New York.

Platen, E. and David, H. (2006). *A Benchmark Approach to Quantitative Finance*. Springer, New York.

Ratanov, N. (2005a). Pricing options under telegraph processes. *Revista de Economia del Rosario* **8**, 2, 131–150.

Ratanov, N. (2005b). Quantile hedging for telegraph markets and its applications to a pricing of equity-linked life insurance contracts. *Borradores de Investigatión* **62**, 1–30. http://www.urosario.edu.co/FASE1/economia/documentos/pdf/bi62.pdf.

Ratanov, N. (2007a). A jump telegraph model for option pricing. *Quantitative Finance* **7**, 5, 575–583.

Ratanov, N. (2007b). Telegraph models of financial markets. *Revista Colombiana de Mathemáticas* **41**, 247–252.

Schoutens, W. (2003). *Lévy Processes in Finance*. John Wiley & Sons, Ltd, Chichester.

Scott, L. (1997). Pricing stock options in a jump diffusion model with stochastic volatility and interest rates: Application of fourier inversion methods. *Mathematical Finance* **7**, 413–426.

Sengul, B. (2008). Lamperti stable processes for modeling of financial data. *Working Paper, Department of Mathematics, University of Bath* 1–15.

Shephard, N. (2005). *Stochastic Volatility: Selected Readings*. Oxford University Press, New York.

Zhang, Q. (2001). Stock trading: An optimal selling rule. *SIAM J. Control Optim.* **40**, 1, 64–87.

9

Miscellanea

9.1 Monitoring of the volatility

We have seen that volatility of the market or of the asset prices plays a crucial role in many aspects. One of the underlying assumptions in the standard Black and Scholes market of Chapter 6 is that volatility is supposed to be constant. We have seen many examples (see, e.g. Section 6.6) about the fact that this assumption is simply unrealistic when we go to analyze real financial data. We have also seen deviations from the standard geometric Brownian motion model which allow for nonconstant volatility. Change point analysis was initially introduced in the framework of independent and identically distributed data by these authors: Hinkley (1971), Csörgő and Horváth (1997), Inclan and Tiao (1994), Bai (1994, 1997) and quickly applied to the analysis of time series: Kim *et al.* (2000), Lee *et al.* (2000), Chen *et al.* (2005). For continuous time diffusion models Kutoyants (1994, 2004) and Lee *et al.* (2006) studied change point in the drift term from continuous time observations. Due to the fact that volatility can be estimated without error in continuous time, the change point analysis in this setup is not very interesting. Recently, De Gregorio and Iacus (2008) considered least squares estimation for the volatility of a one-dimensional stochastic differential equation and later Iacus and Yoshida (2009) consider the problem under the general setup of multidimensional Itô processes also observed in discrete time. Although we will only consider the problem of change point estimation, we mention that Song and Lee (2009) proposed the CUSUM test statistics to discover structural change points for one-dimensional diffusion processes. We briefly recall both least squares and maximum likelihood estimation approach.

Option Pricing and Estimation of Financial Models with R, First Edition. Stefano M. Iacus.
© 2011 John Wiley & Sons, Ltd. Published 2011 by John Wiley & Sons, Ltd.

9.1.1   The least squares approach

In Section 6.6 we have used, without details, the cpoint function to discover change points in the volatility of geometric Brownian motion. We now describe the precise setup. Let $X = \{X_t, 0 \leq t \leq T\}$ be a diffusion process, with state space $\mathcal{X} = (l, r)$, $-\infty \leqslant l \leqslant r \leqslant +\infty$, such that

$$X_t = \begin{cases} X_0 + \int_0^t b(X_s)ds + \int_0^t \sqrt{\theta_1}\sigma(X_s)dB_s, & 0 \leq t \leq \tau^*, \\ X_{\tau^*} + \int_{\tau^*}^t b(X_s)ds + \int_{\tau^*}^t \sqrt{\theta_2}\sigma(X_s)dB_s, & \tau^* < t \leq T, \end{cases} \tag{9.1}$$

with $X_0 = x_0$, $0 < \theta_1, \theta_2 < \infty$ and $\{B_t, t \geq 0\}$ a standard Brownian motion. The value $\tau^* \in (0, T)$ is the change point instant. The parameters θ_1 and θ_2 belong to Θ, a compact set of $\mathbb{R}^+$. The coefficients $b : \mathcal{X} \to \mathbb{R}$ and $\sigma : \mathcal{X} \to (0, \infty)$ are supposed to be known, continuous with continuous derivatives and regular so that (9.1) is well defined and the process X is unique and such that the process possesses the ergodic property. Let $s(x) = \exp\{-\int_{x_0}^x 2b(u)/\sigma^2(u)du\}$ be the scale function (where x_0 is an arbitrary point inside $\mathcal{X}$). The following condition will be required throughout this section:

$$\lim_{x_1 \to l} \int_{x_1}^x s(u)du = +\infty, \quad \lim_{x_2 \to r} \int_x^{x_2} s(u)du = +\infty, \tag{9.2}$$

where $l < x_1 < x < x_2 < r$. Condition (9.2) guarantees that the exit time from $\mathcal{X}$ is infinite (Karatzas and Shreve 1991). We later assume that $b(\cdot)$ is unknown and we will make use of nonparametric estimators. The process X is supposed to be observed at $n + 1$ equidistant discrete times $0 = t_0 < t_1 < ... < t_n = T$, with $t_i = i\Delta_n$. For the sake of simplicity we assume $T = 1$ and with little abuse of notation, we will write X_i instead of X_{t_i} and B_i instead of B_{t_i}. The asymptotic framework is a high frequency scheme: $n \to \infty$, $\Delta_n \to 0$ with $n\Delta_n = T$ with T fixed. Given the observations $X_i, i = 0, 1, ..., n$, the aim of this work is to estimate the change time τ^* as well as the two parameters θ_1, θ_2. The solution of this problem is an adaptation of the least squares approach of Bai (1994) for autoregressive models. By Euler scheme, we first construct the standardized residuals

$$Z_i = \frac{X_{i+1} - X_i - b(X_i)\Delta_n}{\sqrt{\Delta_n}\sigma(X_i)}, \quad i = 1, ..., n.$$

Let us denote by $k_0 = [n\tau^*]$ and $k = [n\tau]$, $\tau, \tau^* \in (0, 1)$, where $[x]$ is the integer part of the real value x. The least squares estimator of the change point is given by

$$\hat{k}_0 = \arg\min_k \left(\min_{\theta_1,\theta_2} \left\{ \sum_{i=1}^k (Z_i^2 - \theta_1)^2 + \sum_{i=k+1}^n (Z_i^2 - \theta_2)^2 \right\} \right)$$

$$= \arg\min_k \left\{ \sum_{i=1}^k (Z_i^2 - \bar{\theta}_1)^2 + \sum_{i=k+1}^n (Z_i^2 - \bar{\theta}_2)^2 \right\}, \tag{9.3}$$

where

$$\bar{\theta}_1 = \frac{1}{k}\sum_{i=1}^{k} Z_i^2 =: \frac{S_k}{k} \quad \text{and} \quad \bar{\theta}_2 = \frac{1}{n-k}\sum_{i=k+1}^{n} Z_i^2 =: \frac{S_{n-k}^*}{n-k}.$$

It is easy to show that the problem (9.3) is equivalent to the following

$$\hat{k}_0 = \arg\max_k |D_k| \tag{9.4}$$

where

$$D_k = \frac{k}{n} - \frac{S_k}{S_n}.$$

Once $\hat{k}_0$ has been obtained, the following estimator of the parameters θ_1 and θ_2 can be used:

$$\hat{\theta}_1 = \frac{S_{\hat{k}_0}}{\hat{k}_0}, \quad \hat{\theta}_2 = \frac{S_{n-\hat{k}_0}^*}{n-\hat{k}_0}.$$

Denote by $\mathcal{W}(v)$ the *two-sided Brownian motion*, i.e.

$$\mathcal{W}(u) = \begin{cases} B_1(-u), & u < 0 \\ B_2(u), & u \geq 0 \end{cases} \tag{9.5}$$

where $B_1(t)$, $B_2(t)$, $t \geq 0$, are two independent Brownian motions. De Gregorio and Iacus (2008) show the following asymptotic results. Let $\vartheta_n = |\theta_2 - \theta_1| \neq 0$ for finite n. Under the additional condition that $\vartheta_n \to 0$ and $\frac{\sqrt{n}\vartheta_n}{\sqrt{\log n}} \to \infty$, the change point estimator $\hat{\tau}_n^* = \hat{k}_0/n$ is also consistent and such that

$$\frac{n\vartheta_n^2(\hat{\tau}_n^* - \tau^*)}{2\tilde{\theta}^2} \xrightarrow{d} \arg\max_v \left\{ \mathcal{W}(v) - \frac{|v|}{2} \right\} \tag{9.6}$$

for any consistent estimator $\tilde{\theta}$ for the common limiting value θ_0 of θ_1 and θ_2. The condition $\vartheta_n \to 0$ corresponds to the setup of contiguous alternatives, i.e. the two parameters $\theta_1 = \theta_1(n)$ and $\theta_2 = \theta_2(n)$ are allowed to be closer and closer as the sample size increases. Thus, in order to discriminate the two regimes a sufficiently large number of observations n (or rate of convergence ϑ_n) is required. Under the conditions above, the estimators $\hat{\theta}_1$, $\hat{\theta}_2$ are $\sqrt{n}$-consistent and such that

$$\sqrt{n}\begin{pmatrix} \hat{\theta}_1 - \theta_1 \\ \hat{\theta}_2 - \theta_2 \end{pmatrix} \xrightarrow{d} N(0, \Sigma), \quad \text{where} \quad \Sigma = \begin{pmatrix} 2\frac{\theta_0^2}{\tau^*} & 0 \\ 0 & 2\frac{\theta_0^2}{1-\tau^*} \end{pmatrix}.$$

The above results hold in the high frequency case $\Delta_n \to 0$ with $n \to \infty$ and $n\Delta = T$ fixed, but also for the rapidly increasing design case, i.e. $n\Delta_n^2 \to 0$, $n\Delta_n = T \to \infty$.

In most cases, the drift coefficient is not known or treated as a *nuisance* term in the statistical model. If we assume $T \to \infty$, then it is possible to estimate consistently the drift coefficient, see, e.g. Iacus (2008). The `cpoint` function in the **sde** package implements the nonparametric drift estimator proposed in Bandi and Phillips (2003) although similar alternative nonparametric estimators can be used in this setup. The reader might want to consider the results in earlier papers by Pham (1981), Florens-Zmirou (1993) or Stanton (1997). Let $K \geqslant 0$ be a kernel function, i.e. K is symmetric and continuously differentiable, with $\int_{\mathbb{R}} u K(u) du = 0$, $\int_{\mathbb{R}} K^2(u) du < \infty$ and such that $\int_{\mathbb{R}} K(u) du = 1$. We plug into the object function (9.3) these new standardized residuals

$$\hat{Z}_i = \frac{X_{i+1} - X_i - \hat{b}(X_i)\Delta_n}{\sqrt{\Delta_n}}$$

where

$$\hat{b}(x) = \frac{\sum_{i=1}^{n} K\left(\frac{X_i - x}{h_n}\right) \frac{X_{i+1} - X_i}{\Delta_n}}{\sum_{i=1}^{n} K\left(\frac{X_i - x}{h_n}\right)}$$

is our nonparametric estimator of the drift. Then, the change point estimator is obtained by maximizing D_k as in the case of known drift. This mixed result of parametric and nonparametric estimation is quite useful in applications because in practice there is no need to fully specify the data generating model for the observed data. The change point analysis for this reduced model identifies a change in the scale (or intensity) in the volatility levels. De Gregorio and Iacus (2008) also discuss the choice of the bandwidth selection problem for h_n which is implemented in the `cpoint` function.

9.1.2   Analysis of multiple change points

Usually, change point statistics identify the largest change point, but it may be that other change points exist in the series. One possibility to solve this problem is that once the large change point has been detected, the time series is split into two subseries and the analysis of structural breaks is pursued on the subseries separately. The next code analyzes the volatility of the AAPL stock in this way. We first discover the largest change point

```
R> require(sde)

R> require(fImport)
R> Delta <- 1/252
R> S <- yahooSeries("AAPL", from = "2009-01-01",
    to = "2009-12-31")
R> Close <- S[, "AAPL.Close"]
R> sqrt(var(returns(Close))/Delta)

        AAPL.Close
```

```
AAPL.Close   0.3327552

R> Close <- rev(Close)
R> cp <- cpoint(as.ts(Close))
R> cp

$k0
[1] 201

$tau0
[1] 201

$theta1
[1] 2.510733

$theta2
[1] 3.486225
```

then we split the time series in two parts and evaluate the volatility individually assuming that the underlying model is the geometric Brownian motion:

$$dX_t = \mu X_t dt + \sigma X_t dB_t$$

thus we do not consider the estimates of θ_1 and θ_2 provided by the software for the model
$$dX_t = b(X_t)dt + \theta dB_t.$$

```
R> Close1 <- Close[time(Close)[1:cp$k0], ]
R> Close2 <- Close[time(Close)[-(1:cp$k0)], ]
R> sqrt(var(returns(Close1))/Delta)

           AAPL.Close
AAPL.Close   0.3442099

R> sqrt(var(returns(Close2))/Delta)

           AAPL.Close
AAPL.Close   0.2704417
```

Now we repeat the analysis on left-hand time series:

```
R> cp2 <- cpoint(as.ts(Close1))
R> cp2

$k0
[1] 99

$tau0
[1] 99

$theta1
```

```
[1] 2.756492

$theta2
[1] 2.240476
```

and we plot the change points values against the plot of the returns of the asset value in Figure 9.1.

```
R> plot(returns(Close), theme = "white", ylab = "AAPL Returns",
+    main = "", xlab = "")
R> abline(v = time(Close)[cp$k0], lty = 3)
R> abline(v = time(Close1)[cp2$k0], lty = 3)
```

9.1.3   An example of real-time analysis

The change point statistic presented is a retrospective tool, i.e. once the data are collected we look retrospectively to the change point instant. This is very useful to calibrate financial models in the sense that, instead of taking the whole trajectory of an asset price, this tool permits us to extract only the last relevant observations with more homogeneous volatility. In other applications, it is instead interesting to use change point statistics to do real time monitoring. Figure 9.2 contains the results of a real time change point analysis of the last recent financial crisis. By real time it is meant that the analysis proceeds as follows: only data up to the last week of June 2008 are considered and change points are estimated; then data are increased by one week and again a new change point analysis is performed; then the third week is added, and so forth. As soon as change points were found, they were kept (and added to the plot in Figure 9.2). The 'official' week of the total collapse is represented in a different color. The experiment in Smaldone (2009)

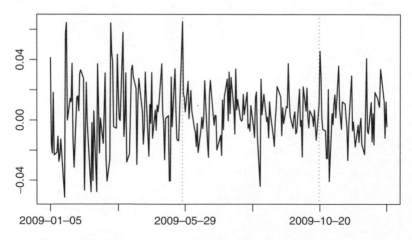

Figure 9.1   First two change points for the AAPL time series.

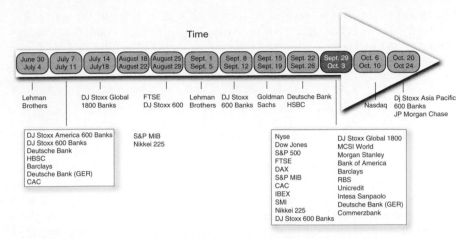

Figure 9.2   Summary of a crude real-time structural change point analysis of several financial markets and assets. Change-point analysis seems to be able to capture structural breakpoints not only retrospectively but also in real time. Source: Smaldone (2009).

involved the change point analysis of several market indexes including Nyse, Dow Jones, S&P 500 and Nasdaq; Dow Jones Stoxx 600, Nikkei, Dow Jones Global 1800, MSCI World, FTSE (UK), DAX (Germany), S&P/Mib (Italy), CAC (France), Ibex (Spain) and SMI (Switzerland). Then several indexes for the bank compartment like Dow Jones Stoxx Global 1800 Banks (worldwide), Dow Jones Stoxx Americas 600 Banks (USA), Dow Jones Stoxx Asia-Pacific 600 Banks (Asia), Dow Jones Stoxx 600 Banks (Europe). Further, a bunch of individual stock prices from the USA, UK, Italy, France, Germany, Spain and Japan stock exchanges. The evidence from this analysis is interesting in showing how the crisis (measure here by uncertainty) affects the different markets at different dates (e.g. more protected markets, like the Italian case, reacts slowly to the crisis). Although these real-time approach can't be considered to be reliable in general, our experience says that real time monitoring by change point tools is worth considering, possibly along with the monitoring of other indexes.

9.1.4   More general quasi maximum likelihood approach

Iacus and Yoshida (2009) extended the result on change point analysis for the volatility to general d-dimensional Itô processes described by stochastic differential equations of this form:

$$dY_t = b_t dt + \sigma(X_t, \theta)dB_t, \quad t \in [0, T], \tag{9.7}$$

where B_t is an r-dimensional standard Brownian motion, b_t and X_t are vector valued processes, and $\sigma(x, \theta)$ is a matrix valued function. The parameter θ

belongs to Θ which is a bounded domain in $\mathbb{R}^{d_0}$, $d_0 \geq 1$. As in the above, it is assumed that there is a time τ^* across which the diffusion coefficient changes from $\sigma(x, \theta_1)$ to $\sigma(x, \theta_2)$. More precisely, (Y, X) satisfy the following stochastic integral equation:

$$Y_t = \begin{cases} Y_0 + \int_0^t b_s \mathrm{d}s + \int_0^t \sigma(X_s, \theta_1) \mathrm{d}B_s & \text{for } t \in [0, \tau^*) \\ Y_{\tau^*} + \int_{\tau^*}^t b_s \mathrm{d}s + \int_{\tau^*}^t \sigma(X_s, \theta_2) \mathrm{d}B_s & \text{for } t \in [\tau^*, T]. \end{cases}$$

The change point $\tau^* \in (0, T)$ is unknown and is to be estimated from the observations sampled from the path of (X, Y). As before, $\mathcal{X}$ denotes the state space of X. The coefficient $\sigma(x, \theta)$ is assumed to be known up to the parameter θ, while b_t is completely unknown and unobservable, therefore possibly depending on θ and τ^*. In this framework the interest is purely on τ^* and the estimation of θ is secondary. We assume that consistent estimators exist for θ_k and later describe how to obtain it. The results are not affected by this assumption. Note that diffusion models are included in this framework by simply taking $Y = X$ in Equation (9.7). The sample consists of (X_{t_i}, Y_{t_i}), $i = 0, 1, \ldots, n$, where $t_i = i\Delta_n$ for $\Delta = \Delta_n = T/n$. The time T is fixed, so this scheme is purely high frequency and asymptotic results are of mixed normal type because the asymptotic Fisher information for the model will be a random matrix. Denote by θ_i^* the true value of θ_k for $k = 1, 2$ and $\vartheta_n = |\theta_1^* - \theta_2^*|$. In order to obtain the asymptotic results some conditions should be further assumed on the regularity of the coefficient σ and on the continuity of the trajectories of the coordinate process X and the behaviour of the process b_t, i.e. the process itself cannot be too irregular in order to discover changes in the structure. The process b_t can have jumps but must be controlled to avoid that those jumps are interpreted as change points. The conditions to express the above remarks on the regularity of the processes involved are rather technical and we do not present them here but the reader can refer to the original paper. The only condition which is necessary to explicit concerns, as in the one dimensional case of Section 9.1.1, the contiguity of the two parameters θ_1 and θ_2. In particular, we assume that θ_1^* and θ_2^* depend on n, and as $n \to \infty$,

$$\theta_1^* \to \theta_0^* \in \Theta, \quad \vartheta_n \to 0, \quad \text{and} \quad n\vartheta_n^2 \to \infty. \tag{9.8}$$

We need some additional notation. For a matrix A, we denote by $A^\top$ the transpose of A, $\mathrm{Tr}(A)$ the trace of A, $A^{\otimes 2} = A \cdot A^\top$ and by A^{-1} the inverse of A. Let $S(x, \theta) = \sigma(x, \theta)^{\otimes 2}$.

9.1.5   Construction of the quasi-MLE

We now explain how to construct the quasi-MLE (i.e., quasi maximum likelihood estimator) of the change point. Let $\Delta_i Y = Y_{t_i} - Y_{t_{i-1}}$ and define

$$\Phi_n(t; \theta_1, \theta_2) = \sum_{i=1}^{[nt/T]} G_i(\theta_1) + \sum_{i=[nt/T]+1}^{n} G_i(\theta_2), \tag{9.9}$$

with

$$G_i(\theta) = \log \det S(X_{t_{i-1}}, \theta) + \Delta_n^{-1}(\Delta_i Y)' S(X_{t_{i-1}}, \theta)^{-1}(\Delta_i Y). \qquad (9.10)$$

The contrast function in (9.9) is a version of the one in Genon-Catalot and Jacod (1993), Suppose that there exists an estimator $\hat{\theta}_k$ for each θ_k, $k = 1, 2$. In case θ_k^* are known, we define $\hat{\theta}_k$ just as $\hat{\theta}_k = \theta_k^*$. The change point estimator of τ^* is

$$\hat{\tau}_n = \arg\min_{t \in [0,T]} \Phi_n(t; \hat{\theta}_1, \hat{\theta}_2).$$

The estimator $\hat{\tau}_n$ has the same structure of the estimator $\hat{k}_0$ in Equation (9.3). It is possible to prove that $\hat{\tau}_n$ is a consistent for τ^* and the rate of convergence is of order $n\vartheta_n^2$. Now, let us assume that the following limit random variable

$$\eta = \lim_{n \to \infty} \vartheta_n^{-1}(\theta_2^* - \theta_1^*)$$

exists. Let Ξ be the positive-definite matrix

$$\Xi(x, \theta) = \left(\mathrm{Tr}((\partial_{\theta^{(i_1)}} S) S^{-1} (\partial_{\theta^{(i_2)}} S) S^{-1})(x, \theta) \right)_{i_1, i_2 = 1}^{d_0}, \qquad \theta = (\theta^{(i)}).$$

Define further,

$$\mathbb{H}(v) = -2 \left(\Gamma_\eta^{\frac{1}{2}} \mathcal{W}(v) - \frac{1}{2} \Gamma_\eta |v| \right)$$

for $\Gamma_\eta = (2T)^{-1} \eta' \Xi(X_{\tau^*}, \theta_0^*) \eta$, where $\mathcal{W}$ is the multidimensional version of the two-sided Brownian motion in (9.5) and independent of X_{τ^*}. Then,

$$n\vartheta_n^2(\hat{\tau}_n - \tau^*) \to_s^d \arg\min_{v \in \mathbb{R}} \mathbb{H}(v)$$

as $n \to \infty$. This result is equivalent to (9.6) although in the present case, the double-sided Brownian motion $\mathcal{W}$ is premultiplied by the random Fisher information Γ_η, so this result involving mixed normal limit is under stable convergence. In order to use it in practice a studentization procedure is required, i.e. the quantity $n\vartheta_n^2(\hat{\tau}_n - \tau^*)$ has to be normalized by the inverse of Γ_η evaluated at the change point estimator $\hat{\tau}_n$. In this case, the limit above is more similar to the one of (9.6). The joint convergence of the normalized $\hat{\tau}_n$ and X_{τ^*} is indeed also proved in Iacus and Yoshida (2009).

9.1.6 A modified quasi-MLE

The quasi-likelihood approach does not consider the drift term in the estimation. For small sample sizes it is possible to construct a new contrast function

$$\Phi_n'(t; \theta_1, \theta_2) = \sum_{i=1}^{[nt/2T]} G_i'(\theta_1) + \sum_{i=[nt/2T]+1}^{n} G_i'(\theta_2), \qquad (9.11)$$

where the term $\Delta_i Y$ in $G_i(\theta)$ of (9.10) is replaced by

$$\tilde{\Delta} Y_i = \frac{Y_{2i+1} - 2Y_{2i} + Y_{2i-1}}{\sqrt{2}}$$

and

$$G_i'(\theta) = \log \det S(X_{t_{2i-1}}, \theta) + \Delta_n^{-1} (\tilde{\Delta}_i Y)' S(X_{t_{2i-1}}, \theta)^{-1} (\tilde{\Delta}_i Y). \qquad (9.12)$$

The change-point estimator has the same asymptotic properties of the one in previous section but it has better properties for finite samples because it compensates for the unknown drift. Indeed, the use of $\tilde{\Delta}_i Y$ has the effect of compensating for the unknown drift and eliminates the initial condition Y_0. Indeed, the term $\tilde{\Delta}_i Y$ reduces to

$$\tilde{\Delta}_i Y = \frac{\left(\int\limits_{t_{2i}}^{t_{2i+1}} b_s ds - \int\limits_{t_{2i-1}}^{t_{2i}} b_s ds \right) + \left(\int\limits_{t_{2i}}^{t_{2i+1}} \sigma(X_s, \theta) dW_s - \int\limits_{t_{2i-1}}^{t_{2i}} \sigma(X_s, \theta) dW_s \right)}{\sqrt{2}}.$$

9.1.7   First- and second-stage estimators

In the previous construction, it is supposed that θ_1 and θ_2 are known and it was anticipated that the results still hold if consistent estimators exist. Indeed, it is possible to construct consistent estimators of θ_k, $k = 1, 2$, using the data over time interval $[0, a_n]$ for $k = 1$ and the one over $[T - a_n, T]$ for $k = 2$, respectively, for some sequence a_n tending to zero. Although the details can be found in the original paper, the intuition is that, in order to have consistent estimators, the sequence a_n must be such that $na_n \to \infty$ because this will guarantee that we have a sufficient number of observations to get the asymptotic result to work. It is also possible to consider two separate sequences $a_n^1 \to 0$ for the estimation of θ_1 and $a_n^2 \to 0$ for the estimation of θ_2. The Quasi-MLE estimators are obtained using the contrast function

$$\hat{\theta}_{1,n} = \arg\min_{\theta_1} \sum_{i=1}^{[na_n]} G_i(\theta_1) \quad \text{and} \quad \hat{\theta}_{2,n} = \arg\min_{\theta_2} \sum_{i=[n(1-a_n)]}^{n} G_i(\theta_2)$$

Once this first-stage estimators are obtained, it is possible to construct the first-stage estimator $\hat{\tau}_n$ of the change point as follows:

$$\hat{\tau}_n = \arg\min_t \Phi_n(t; \hat{\theta}_{1,n}, \hat{\theta}_{2,n}).$$

It is now possible to iterate the procedure to obtain second-stage estimators. Indeed, with the initial first-stage change point estimator $\hat{\tau}_n$ we can estimate θ_1 and θ_2 more accurately as follows:

$$\check{\theta}_{1,n} = \arg\min_{\theta_1} \sum_{i=1}^{[n\hat{\tau}_n]} G_i(\theta_1) \quad \text{and} \quad \check{\theta}_{2,n} = \arg\min_{\theta_2} \sum_{i=[n\hat{\tau}_n]+1}^{n} G_i(\theta_2)$$

and finally obtain the second-stage estimator of τ, i.e.

$$\check{\tau}_n = \arg\min_t \Phi_n(t; \check{\theta}_{1,n}, \check{\theta}_{2,n}).$$

The same procedure is possible with the modified contrast function Φ' in (9.11).

9.1.8   Numerical example

The change point statistics based on the contrast function (9.9) is implemented in package **yuima** via the function `CPoint`. We now show an example of use with the two-steps approach. Let us consider the two-dimensional diffusion process $X_t = (X_t^1, X_t^2)$ solution to the stochastic differential equation

$$\begin{pmatrix} dX_t^1 \\ dX_t^2 \end{pmatrix} = \begin{pmatrix} 1 - X_t^1 \\ 3 - X_t^2 \end{pmatrix} dt + \begin{bmatrix} \theta_{1.1} \cdot X_t^1 & 0 \cdot X_t^1 \\ 0 \cdot X_t^2 & \theta_{1.2} \cdot X_t^2 \end{bmatrix} \begin{pmatrix} dW_t^1 \\ dW_t^2 \end{pmatrix}$$

$$X^1(0) = 1, \quad X^2(0) = 1.$$

We first describe this model in R using the **yuima** package

```
R> library(yuima)
R> diff.matrix <- matrix(c("theta1.1*x1", "0*x2", "0*x1",
    "theta1.2*x2"),
+       2, 2)
R> drift.c <- c("1-x1", "3-x2")
R> drift.matrix <- matrix(drift.c, 2, 1)
R> ymodel <- setModel(drift = drift.matrix, diffusion =
    diff.matrix,
+       time.variable = "t", state.variable = c("x1", "x2"),
    solve.variable = c("x1",
+           "x2"))

R> require(yuima)
R> diff.matrix <- matrix(c("theta1.1*x1", "0*x2", "0*x1",
    "theta1.2*x2"),
+       2, 2)
R> drift.c <- c("1-x1", "3-x2")
R> drift.matrix <- matrix(drift.c, 2, 1)
R> ymodel <- setModel(drift = drift.matrix, diffusion
    = diff.matrix,
+       time.variable = "t", state.variable = c("x1", "x2"),
    solve.variable = c("x1",
+           "x2"))
```

and then simulate two trajectories. One up to the change point $\tau = 4$ with parameters $\theta_{1.1} = 0.1$ and $\theta_{1.1} = 0.2$, and a second trajectory with parameters $\theta_{1.1} = 0.6$ and $\theta_{1.2} = 0.6$. For the second trajectory, the initial value is set to the last value of the first trajectory.

```
R> n <- 1000
R> set.seed(123)
R> t1 <- list(theta1.1 = 0.1, theta1.2 = 0.2)
R> t2 <- list(theta1.1 = 0.6, theta1.2 = 0.6)
R> tau <- 0.4
R> ysamp1 <- setSampling(n = tau * n, Initial = 0, delta = 0.01)

YUIMA: 'Terminal' (re)defined.

R> yuima1 <- setYuima(model = ymodel, sampling = ysamp1)
R> yuima1 <- simulate(yuima1, xinit = c(1, 1),
    true.parameter = t1)
R> x1 <- yuima1@data@zoo.data[[1]]
R> x1 <- as.numeric(x1[length(x1)])
R> x2 <- yuima1@data@zoo.data[[2]]
R> x2 <- as.numeric(x2[length(x2)])
R> ysamp2 <- setSampling(Initial = n * tau * 0.01, n = n * (1 -
+    tau), delta = 0.01)

YUIMA: 'Terminal' (re)defined.

R> yuima2 <- setYuima(model = ymodel, sampling = ysamp2)
R> yuima2 <- simulate(yuima2, xinit = c(x1, x2),
    true.parameter = t2)
R> yuima <- yuima1
R> yuima@data@zoo.data[[1]] <- c(yuima1@data@zoo.data[[1]],
    yuima2@data@zoo.data[[1]][-1])
R> yuima@data@zoo.data[[2]] <- c(yuima1@data@zoo.data[[2]],
    yuima2@data@zoo.data[[2]][-1])
```

The composed trajectory is visible in Figure 9.3. We first test the ability of the change point estimator to identify τ when for given true values

```
R> t.est <- CPoint(yuima, param1 = t1, param2 = t2, plot = TRUE)
R> t.est$tau

[1] 3.99
```

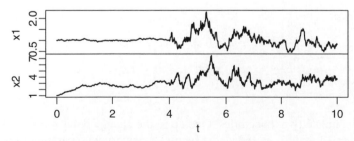

Figure 9.3 An example of two-dimensional trajectory with change point at $\tau = 4$.

Now we proceed with a two-stage estimation approach. We first estimate the parameters $\theta_{1.1}$ and $\theta_{1.2}$ before and after the change point τ using the observations in the tails

```
R> low <- list(theta1.1 = 0, theta1.2 = 0)
R> tmp1 <- qmleL(yuima, start = list(theta1.1 = 0.3,
   theta1.2 = 0.5),
+     t = 1.5, lower = low, method = "L-BFGS-B")
R> tmp2 <- qmleR(yuima, start = list(theta1.1 = 0.3,
   theta1.2 = 0.5),
+     t = 8.5, lower = low, method = "L-BFGS-B")

R> coef(tmp1)

 theta1.1  theta1.2
0.0946981 0.1913319

R> coef(tmp2)

 theta1.1  theta1.2
0.7429437 0.8549956
```

and obtain the first-stage change point estimator

```
R> t.est1 <- CPoint(yuima, param1
   = coef(tmp1), param2 = coef(tmp2))
R> t.est1$tau

[1] 3.99
```

With this first change point estimator, we estimate again the parameters in the diffusion matrix

```
R> tmp11 <- qmleL(yuima, start = as.list(coef(tmp1)),
   t = t.est1$tau -
+      0.1, lower = low, method = "L-BFGS-B")

YUIMA: attempting to coerce 'grid' to a list, unexpected
   results may occur!

R> coef(tmp11)

  theta1.1    theta1.2
0.09653777 0.20041300

R> tmp21 <- qmleR(yuima, start = as.list(coef(tmp2)),
   t = t.est1$tau +
+      0.1, lower = low, method = "L-BFGS-B")

YUIMA: attempting to coerce 'grid' to a list, unexpected
   results may occur!
```

R> coef(tmp21)

```
theta1.1   theta1.2
0.7829717 0.7924929
```

and finally calculate the second-stage estimator of the change point using the second-stage estimators of the parameters

```
R> t.est2 <- CPoint(yuima, param1 =
    coef(tmp11), param2 = coef(tmp21))
R> t.est2$tau
```

```
[1] 3.99
```

The same analysis can be performed using the modified constrast function (9.12) in **yuima** using the option symmetrized = TRUE in the function CPoint.

9.2   Asynchronous covariation estimation

Suppose that two Itô processes are observed only at discrete times in an asynchronous manner as usually happens in practice with high frequency data. We are interested in estimating the covariance of the two processes accurately in such a situation. Let $T \in (0, \infty)$ be a terminal time for possible observations. We consider a two-dimensional Itô process (X^1, X^2) satisfying the stochastic differential equations

$$
\mathrm{d}X_t^i = \mu_t^i \mathrm{d}t + \sigma_t^i \mathrm{d}W_t^i, \quad t \in [0, T]
$$
$$
X_0^i = x_0^i
$$

for $i = 1, 2$. Here W^i denote standard Wiener processes with a progressively measurable correlation process $\mathrm{d}\langle W_1, W_2 \rangle_t = \rho_t \mathrm{d}t$, μ_t^i and σ_t^i are progressively measurable processes, and x_0^i are initial random variables independent of (W^1, W^2). Estimation of covariation between two diffusion processes is our goal, but the formulation in the above terms allows for more sophisticated stochastic structures and the theory presented in this section has been developed in a series of papers by Hayashi and Yoshida (2005, 2008) and still under development to handle microstructure noise issues.

The process X^i is supposed to be observed over the increasing sequence of times T^{ik} $(k = 0, 1, \ldots)$ starting at 0, up to time T. Thus, the observable quantities are $(T^{ik}, X^{i,k})$ with $T^{ik} \leq T$. Each T^{ik} is allowed to be a stopping time, so possibly depends on the history of (X^1, X^2) as well as the precedent stopping times. Two sequences of stopping times T^{1k} and T^{2j} are *asynchronous*, and irregularly spaced, in general. Figure 9.4 shows a typical example of irregular sampling where the intervals I^k and J^j are the intervals determined respectively by subsequent elements of the sequences of random times T^{1k} and T^{2j}.

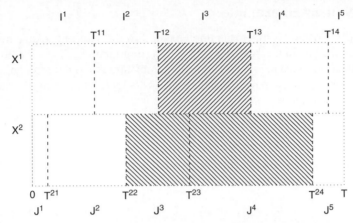

Figure 9.4   Example of asynchronous random sampling for a two-dimensional Itô process.

The parameter of interest is the quadratic covariation between X^1 and X^2 defined as follows:

$$\theta = \langle X^1, X^2 \rangle_T = \int_0^T \sigma_t^1 \sigma_t^2 \rho_t \mathrm{d}t. \qquad (9.13)$$

The target variable θ is random in the general setup but in our examples it will be a constant quantity. This quantity can be estimated with the *asynchronous covariance estimator* also called the Hayashi-Yoshida estimator defined as follows:

$$U_n = \sum_{i,j:T^{1i} \leq T, T^{2j} \leq T} (X_{T^{1i}}^1 - X_{T^{1\{i-1\}}}^1)(X_{T^{2j}}^2 - X_{T^{2\{j-1\}}}^2)\mathbf{1}_{\{I^i \cap J^j \neq \emptyset\}}. \qquad (9.14)$$

That is, the product of any pair of increments $(X_{T^{1i}}^1 - X_{T^{1\{i-1\}}}^1)$ and $(X_{T^{2j}}^2 - X_{T^{2\{j-1\}}}^2)$ will make a contribution to the sum only when the respective observation intervals $I^i = (T^{1\{i-1\}}, T^{1,i}]$ and $J^j = (T^{2,\{j-1\}}, T^{2,j}]$ are overlapping with each other as in Figure 9.4. The estimator U_n is consistent and possesses an asymptotically mixed normal distribution as $n \to \infty$ if the maximum length between two consecutive observing times tends to 0. This is an important result for covariance estimator because, before this result was proved, the usual covariance estimator suffered from the so-called Epps effect (Epps 1979). Epps showed that stock return correlations decrease as the sampling frequency of data increases. Since his discovery the phenomenon has been detected in several studies of different stock markets and foreign exchange markets. With the usual covariance estimator in high resolution data the correlations are significantly smaller than their asymptotic value as observed on daily data. Hayashi and Yoshida (2005) show indeed that even under very mild assumptions, the covariance estimator is biased and inconsistent. In fact, under certain assumptions, it converges to zero. The Hayashi-Yoshida estimator is not affected by this Epps effect.

9.2.1   Numerical example

The Hayashi-Yoshida estimator is implemented in the **yuima** package in the `cce` function as well as in the **realized** package in the function `rc.hy`. We now apply the `cce` function to asynchronous high-frequency simulated data. As an example, consider a two-dimensional stochastic process (X^1_t, X^2_t) satisfying the stochastic differential equation

$$\begin{aligned} dX^1_t &= \sigma_{1,t} dB^1_t, \\ dX^2_t &= \sigma_{2,t} dB^2_t. \end{aligned} \tag{9.15}$$

Here B^1_t and B^2_t denote two standard Wiener processes; however we take them correlated in the following way:

$$B^1_t = W^1_t, \tag{9.16}$$

$$B^2_t = \int_0^t \rho_s dW^1_s + \int_0^t \sqrt{1 - \rho_s^2} dW^2_s, \tag{9.17}$$

where W^1_t and W^2_t are independent Wiener processes, and ρ_t is the correlation function between B^1_t and B^2_t. We consider $\sigma_{i,t}, i = 1, 2$ and ρ_t of the following form in this example:

$$\sigma_{1,t} = \sqrt{1 + t}, \quad \sigma_{2,t} = \sqrt{1 + t^2}, \quad \rho_t = \frac{1}{\sqrt{2}}.$$

The parameter we want to estimate is the quadratic covariation between X^1 and X^2:

$$\theta = \langle X_1, X_2 \rangle_T = \int_0^T \sigma_{1,t} \sigma_{2,t} \rho_t dt = 1. \tag{9.18}$$

So we first build the model within the **yuima** package

```
R> diff1 <- function(t, x1 = 0, x2 = 0) sqrt(1 + t)
R> diff2 <- function(t, x1 = 0, x2 = 0) sqrt(1 + t^2)
R> rho <- function(t, x1 = 0, x2 = 0) sqrt(1/2)
R> diff.matrix <- matrix(c("diff1(t,x1,x2)", "diff2(t,x1,x2)
     * rho(t,x1,x2)",
+       "", "diff2(t,x1,x2) * sqrt(1-rho(t,x1,x2)^2)"), 2, 2)
R> cor.mod <- setModel(drift = c("", ""), diffusion = diff.matrix,
+       solve.variable = c("x1", "x2"))
```

and prepare a function `true.theta` to calculate numerically the true value of θ

```
R> true.theta <- function(T, sigma1, sigma2, rho) {
+       f <- function(t) {
+           sigma1(t) * sigma2(t) * rho(t)
+       }
+       integrate(f, 0, T)
+ }
```

so that it is possible to play with the model to test different situations. For the sampling scheme, we will consider two independent Poisson sampling. That is, each configuration of the sampling times T^{ik} is realized as the Poisson random measure with intensity np_i, and the two random measures are independent each other as well as the stochastic processes. Under this particular random sampling scheme, it is known from Hayashi and Yoshida (2005, 2008) that

$$n^{1/2}(U_n - \theta) \to N(0, c),\tag{9.19}$$

as $n \to \infty$, where

$$c = \left(\frac{2}{p_1} + \frac{2}{p_2}\right) \int_0^T \left(\sigma_{1,t}\sigma_{2,t}\right)^2 dt$$
$$+ \left(\frac{2}{p_1} + \frac{2}{p_2} - \frac{2}{p_1 + p_2}\right) \int_0^T \left(\sigma_{1,t}\sigma_{2,t}\rho_t\right)^2 dt$$

$$\tag{9.20}$$

and hence it is possible to estimate the asymptotic variance of the covariance estimator. So we simulate the model

```
R> set.seed(123)
R> Terminal <- 1
R> n <- 1000
R> yuima.samp <- setSampling(Terminal = Terminal, n = n)

YUIMA: 'delta' (re)defined.

R> yuima <- setYuima(model = cor.mod, sampling = yuima.samp)
R> X <- simulate(yuima)
R> theta <- true.theta(T = Terminal, sigma1 =
     diff1, sigma2 = diff2,
+        rho = rho)$value
R> theta

[1] 0.9995767
```

We calculate the covariance from the complete synchronous series

```
R> cce(X)$covmat[1, 2]

[1] 1.086078
```

and now we apply the random sampling using the two Poisson processes. We first construct two grids of random sampling

```
R> p1 <- 0.2
R> p2 <- 0.3
R> newsamp <- setSampling(random = list(rdist = c(function(x)
     rexp(x,
```

```
+      rate = p1 * n/Terminal), function(x) rexp(x, rate = p1 *
+      n/Terminal))))
```

Now `newsamp` contains information about Poisson random sampling so we can apply the `subsampling` function to obtain the asynchronous series

```
R> Y <- subsampling(X, sampling = newsamp)
```

The result is visible in Figure 9.5. We can now estimate the covariance on the new path `Y`

```
R> cce(Y)$covmat[1, 2]
```

```
[1] 1.070313
```

and we get a reasonable good estimate $\hat{\theta}$. Using (9.20) we also obtain the asymptotic variance of $\hat{\theta}$

```
R> var.c <- function(T, p1, p2, sigma1, sigma2, rho) {
+      tmp_integrand1 <- function(t) (sigma1(t) * sigma2(t))^2
+      i1 <- integrate(tmp_integrand1, 0, T)
+      tmp_integrand2 <- function(t) (sigma1(t) * sigma2(t) *
   rho(t))^2
+      i2 <- integrate(tmp_integrand2, 0, T)
+      2 * (1/p1 + 1/p2) * i1$value + 2 * (1/p1 + 1/p2 - 1/(p1 +
+          p2)) * i2$value
+ }
```

and calculate the approximate standard deviation

```
R> vc <- var.c(T = Terminal, p1, p2, diff1, diff2, rho)
R> sqrt(vc/n)
```

```
[1] 0.2188988
```

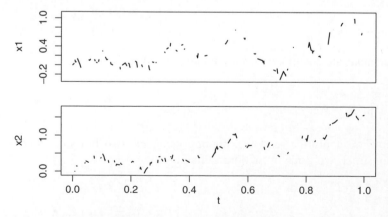

Figure 9.5   Simulated path of a two-dimensional asynchronous diffusion model.

9.3   LASSO model selection

The least absolute shrinkage and selection operator (LASSO) is a useful and well studied approach to the problem of model selection and its major advantage is the simultaneous execution of both parameter estimation and variable selection Efron *et al.* (2004), Knight and Fu (2000), Tibshirani (1996). This is realized by the fact that the dimension of the parameter space does not change (while it does with the information criteria approach, e.g. in AIC, BIC, etc.), because the LASSO method only sets some parameters to zero to eliminate them from the model. The LASSO method usually consists in the minimization of an L^2 norm under L^1 norm constraints on the parameters. Thus it usually implies least squares or maximum likelihood approach plus constraints.

Originally, the LASSO procedure was introduced for linear regression problems, but, in recent years, this approach has been applied to time series analysis by several authors mainly in the case of autoregressive models. For example, just to mention a few, Wang *et al.* 2007 consider the problem of shrinkage estimation of regressive and autoregressive coefficients, while Nardi and Rinaldo (2008) consider penalized order selection in an AR(p) model. The VAR case was considered in Hsu *et al.* (2008). Very recently Caner (2009) studied the LASSO method for general GMM estimator also in the case of time series.

Here we present the LASSO approach for discretely observed diffusion processes. For diffusion processes, the LASSO method requires some additional care because the rate of convergence of the parameters in the drift and the diffusion coefficient are different. We point out that, the usual model selection strategy based on AIC Uchida and Yoshida (2005) usually depends on the properties of the estimators but also on the method used to approximate the likelihood. Indeed, AIC requires the precise calculation of the likelihood Iacus (2008) while the LASSO approach presented here depends solely on the properties of the estimator and so the problem of likelihood approximation is not particularly compelling.

Let $\{X_t, t \geq 0\}$ be a d-dimensional diffusion process solution of the following stochastic differential equation

$$\mathrm{d}X_t = b(\alpha, X_t)\mathrm{d}t + \sigma(\beta, X_t)\mathrm{d}B_t \qquad (9.21)$$

where $\alpha = (\alpha_1, ..., \alpha_p) \in \Theta_p \subset \mathbb{R}^p$, $p \geq 1$, $\beta = (\beta_1, ..., \beta_q) \in \Theta_q \subset \mathbb{R}^q$, $q \geq 1$, $b : \Theta_p \times \mathbb{R}^d \to \mathbb{R}^d$, $\sigma : \Theta_q \times \mathbb{R}^d \to \mathbb{R}^d \times \mathbb{R}^d$ and B_t is a standard Brownian motion in $\mathbb{R}^d$. We assume that the functions b and σ are known up to the parameters α and β. We denote by $\theta = (\alpha, \beta) \in \Theta_p \times \Theta_q = \Theta$ the parametric vector and with $\theta_0 = (\alpha_0, \beta_0)$ its unknown true value. Let $\Sigma(\beta, x) = \sigma(\beta, x)^{\otimes 2}$. The sample path of X_t is observed only at $n + 1$ equidistant discrete times t_i, such that $t_i - t_{i-1} = \Delta_n < \infty$ for $1 \leq i \leq n$ (with $t_0 = 0$ and $t_{n+1} = t$). We denote by $\mathbf{X}_n = \{X_{t_i}\}_{0 \leq i \leq n}$ the sample observations. The asymptotic scheme adopted in this paper is the following: $n\Delta_n \to \infty$, $\Delta_n \to 0$ and $n\Delta_n^2 \to 0$ as $n \to \infty$. The process is supposed to be ergodic and the usual assumption on the regularity of the coefficients of the stochastic differential equations are supposed to hold. For details see the original paper of De Gregorio and Iacus (2010a).

In order to introduce the LASSO problem, we consider the negative quasi-loglikelihood function $\mathbb{H}_n : \mathbb{R}^{(n+1)\times d} \times \Theta \to \mathbb{R}$

$$
\mathbb{H}_n(\mathbf{X}_n, \theta) = \frac{1}{2} \sum_{i=1}^{n} \{ \log \det(\Sigma_{i-1}(\beta))
$$

$$
+ \frac{1}{\Delta_n}(\Delta X_i - \Delta_n b_{i-1}(\alpha))' \Sigma_{i-1}^{-1}(\beta)(\Delta X_i - \Delta_n b_{i-1}(\alpha)) \} \quad (9.22)
$$

where $\Delta X_i = X_{t_i} - X_{t_{i-1}}$, $\Sigma_i(\beta) = \Sigma(\beta, X_{t_i})$ and $b_i(\alpha) = b(\alpha, X_{t_i})$. This quasi-likelihood has been used by, e.g., Yoshida (1992), Genon-Catalot and Jacod (1993) and Kessler (1997) to estimate stochastic differential equations because the true transition probability density for $X_t, t \in [0, T]$, does not have a closed form expression. The function (9.22) is obtained by discretization of the continuous time stochastic differential equation (9.21) by Euler-Maruyama scheme, that is

$$
X_{t_i} - X_{t_{i-1}} = b(\alpha, X_{t_{i-1}})\Delta_n + \sigma(\beta, X_{t_{i-1}})(W_{t_i} - W_{t_{i-1}})
$$

and the increments $(X_{t_i} - X_{t_{i-1}})$ are conditionally independent Gaussian random variables for $i = 1, ..., n$.

We denote by $\dot{\mathbb{H}}_n(\mathbf{X}_n, \theta)$ the vector of the first derivatives with respect to θ and by $\ddot{\mathbb{H}}_n(\mathbf{X}_n, \theta)$ the Hessian matrix. Let $\tilde{\theta}_n : \mathbb{R}^{(n+1)\times d} \to \Theta$ be the quasi-maximum likelihood estimator of $\theta \in \Theta$, based on (9.22), that is

$$
\tilde{\theta}_n = (\tilde{\alpha}_n, \tilde{\beta}_n)' = \arg\min_\theta \mathbb{H}_n(\mathbf{X}_n, \theta).
$$

Let $\mathcal{I}(\theta)$ be the positive definite and invertible Fisher information matrix at θ given by

$$
\mathcal{I}(\theta) = \begin{pmatrix} \Gamma_\alpha = [\mathcal{I}_b^{kj}(\alpha)]_{k,j=1,...,p} & 0 \\ 0 & \Gamma_\beta = [\mathcal{I}_\sigma^{kj}(\beta)]_{k,j=1,...,q} \end{pmatrix}
$$

where

$$
\mathcal{I}_b^{kj}(\alpha) = \int \frac{1}{\sigma^2(\beta, x)} \frac{\partial b(\alpha, x)}{\partial \alpha_k} \frac{\partial b(\alpha, x)}{\partial \alpha_j} \mu_\theta(dx),
$$

$$
\mathcal{I}_\sigma^{kj}(\beta) = 2 \int \frac{1}{\sigma^2(\beta, x)} \frac{\partial \sigma(\beta, x)}{\partial \beta_k} \frac{\partial \sigma(\beta, x)}{\partial \beta_j} \mu_\theta(dx).
$$

Moreover, we consider the matrix

$$
\varphi(n) = \begin{pmatrix} \frac{1}{n\Delta_n}\mathbf{I}_p & 0 \\ 0 & \frac{1}{n}\mathbf{I}_q \end{pmatrix}
$$

where $\mathbf{I}_p$ and $\mathbf{I}_q$ are respectively the identity matrix of order p and q. Let $\Lambda_n(\theta) = \varphi(n)^{1/2}\ddot{\mathbb{H}}_n(\mathbf{X}_n, \theta)\varphi(n)^{1/2}$. Under the usual regularity conditions (De Gregorio and Iacus 2010a), we have that the following two properties hold true

i) $\Lambda_n(\theta_0) \xrightarrow{p} \mathcal{I}(\theta_0)$, $\displaystyle\sup_{||\theta||\leq\epsilon_n} |\Lambda_n(\theta + \theta_0) - \Lambda_n(\theta_0)| = o_p(1)$, for $\epsilon_n \to 0$ as $n \to \infty$;

ii) $\tilde{\theta}_n$ is a consistent estimator of θ_0 and asymptotically Gaussian, i.e.

$$\varphi(n)^{-1/2}(\tilde{\theta}_n - \theta_0) \xrightarrow{d} N(0, \mathcal{I}(\theta_0)^{-1}).$$

9.3.1 Modified LASSO objective function

The classical adaptive LASSO objective function, in this case, should be given by

$$\mathbb{H}_n(\mathbf{X}_n, \theta) + \sum_{j=1}^{p} \lambda_{n,j}|\alpha_j| + \sum_{k=1}^{q} \gamma_{n,k}|\beta_k| \qquad (9.23)$$

where $\lambda_{n,j}$ and $\gamma_{n,k}$ assume real positive values representing an adaptive amount of the shrinkage for each elements of α and β. The LASSO estimator is the minimizer of the objective function (9.23). Usually, this is a nonlinear optimization problem under L_1 constraints which might be numerically challenging to solve. Nevertheless, using the approach of Wang and Leng (2007), the minimization problem can be transformed into a quadratic minimization problem (under L_1 constraints) which is asymptotically equivalent to minimizing (9.23). Indeed, by means of a Taylor expansion of $\mathbb{H}_n(\mathbf{X}_n, \theta)$ at $\tilde{\theta}_n$, one has immediately that

$$\mathbb{H}_n(\mathbf{X}_n, \theta) = \mathbb{H}_n(\mathbf{X}_n, \tilde{\theta}_n) + (\theta - \tilde{\theta}_n)'\dot{\mathbb{H}}_n(\mathbf{X}_n, \tilde{\theta}_n)$$

$$+ \frac{1}{2}(\theta - \tilde{\theta}_n)'\ddot{\mathbb{H}}_n(\mathbf{X}_n, \tilde{\theta}_n)(\theta - \tilde{\theta}_n) + o_p(1)$$

$$= \mathbb{H}_n(\mathbf{X}_n, \tilde{\theta}_n) + \frac{1}{2}(\theta - \tilde{\theta}_n)'\ddot{\mathbb{H}}_n(\mathbf{X}_n, \tilde{\theta}_n)(\theta - \tilde{\theta}_n) + o_p(1)$$

Therefore, we define the LASSO-type estimator $\hat{\theta}_n : \mathbb{R}^{(n+1)\times d} \to \Theta$ as the solution of

$$\hat{\theta}_n = (\hat{\alpha}_n, \hat{\beta}_n)' = \arg\min_{\theta} \mathcal{F}(\theta) \qquad (9.24)$$

where

$$\mathcal{F}(\theta) = (\theta - \tilde{\theta}_n)'\ddot{\mathbb{H}}_n(\mathbf{X}_n, \tilde{\theta}_n)(\theta - \tilde{\theta}_n) + \sum_{j=1}^{p} \lambda_{n,j}|\alpha_j| + \sum_{k=1}^{q} \gamma_{n,k}|\beta_k|. \qquad (9.25)$$

Then, the least squares problem based on (9.25) is asymptotically equivalent to the original LASSO problem deriving from the objective function (9.23), but much easier to solve numerically. The function $\mathcal{F}(\theta)$ is a penalized quadratic form and has the advantage of providing a unified theoretical framework. Indeed, the objective function (9.23) allows us to perform correctly the LASSO procedure only if $\mathbb{H}_n$ is strictly convex and this fact restricts the choice of the possible contrast functions for the model (9.21). Then, the function (9.25) overcomes this criticality. We also point out that $\mathcal{F}(\theta)$ has two constraints, because the drift and diffusion parameters α_j and β_k are well separated with different rates of convergence. It is also possible to show that the LASSO estimators solutions to (9.25) satisfy the so-called oracle property, which means: 1) it identifies the right subset model (i.e. only parameters which are zero are effectively set to zero) and 2) it has the optimal estimation rate and converge to a Gaussian random variable $N(0, \Sigma)$ where Σ is the covariance matrix of the true subset model.

9.3.2   Adaptiveness of the method

Clearly, the theoretical and practical implications of our method rely on the specification of the tuning parameter $\lambda_{n,j}$ and $\gamma_{n,k}$. As observed in Wang and Leng (2007), these values could be obtained by means of some model selection criteria like generalized cross-validation, Akaike information criteria or Bayes information criteria. Unfortunately, this solution is computationally heavy and then impracticable. Therefore, we propose to choose the tuning parameters as in Zou (2006) in the following way

$$\lambda_{n,j} = \lambda_0 |\tilde{\alpha}_{n,j}|^{-\delta_1}, \qquad \gamma_{n,k} = \gamma_0 |\tilde{\beta}_{n,j}|^{-\delta_2} \qquad (9.26)$$

where $\tilde{\alpha}_{n,j}$ and $\tilde{\beta}_{n,k}$ are the unpenalized estimator of α_j and β_k respectively, $\delta_1, \delta_2 > 0$ and usually taken as unitary. The asymptotic results hold under the additional conditions

$$\frac{\lambda_0}{\sqrt{n\Delta_n}} \to 0, \quad (n\Delta_n)^{\frac{\delta_1-1}{2}}\lambda_0 \to \infty, \quad \text{and} \quad \frac{\gamma_0}{\sqrt{n}} \to 0, \quad n^{\frac{\delta_2-1}{2}}\gamma_0 \to \infty$$

as $n \to \infty$, as proved in De Gregorio and Iacus (2010a).

9.3.3   LASSO identification of the model
for term structure of interest rates

As an application of the LASSO approach we reanalyze the U.S. Interest Rates monthly data from 06/1964 to 12/1989 for a total of 307 observations. These data have been analyzed by many authors including Nowman (1997), Aït-Sahalia (1996), Yu and Phillips (2001) just to mention a few references. We do not pretend to give the definitive answer on the subject, but just to analyze the effect of the model selection via the LASSO in a real application. The data used for this application were taken from the R package **Ecdat** by Croissant (2010).

```
R> library(Ecdat)
R> library(sde)
R> data(Irates)
R> rates <- Irates[, "r1"]
R> plot(rates)
```

The different authors all try to fit to the data in Figure 9.6 a version of the so called CKLS model presented in Section 5.3. We remember that the CKLS process is the solution X_t of the following stochastic differential equation:

$$dX_t = (\alpha + \beta X_t)dt + \sigma X_t^\gamma dB_t.$$

This model encompasses several other models depending on the number of non-null parameters as Table 5.1 shows. This makes clear why the model selection on the CKLS model is quite appealing for this application. In this application, we estimate the parameters using quasi-likelihood method (QMLE in the tables) in the first stage, then set the penalties as in (9.26) and run the LASSO optimization. To this aim, we make use of the function `lasso` in the **yuima** package. We describe the model and the data for the `yuima` object

```
R> require(yuima)
R> X <- window(rates, start = 1964.471, end = 1989.333)
R> mod <- setModel(drift = "alpha+beta*x", diffusion =
    matrix("sigma*x^gamma",
+     1, 1))
R> yuima <- setYuima(data = setData(X), model = mod)
```

and then we let `lasso` estimate the CKLS parameters by using both quasi-maximum likelihood and LASSO method by first using mild penalties, i.e., $\lambda_0 = \gamma_0 = 1$ in (9.26). This is specified in a single list argument in `lasso`

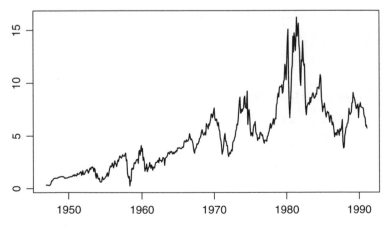

Figure 9.6   The U.S. Interest Rates monthly data from 06/1964 to 12/1989.

```
R> lambda1 <- list(alpha = 1, beta = 1, sigma = 1, gamma = 1)
R> start <- list(alpha = 1, beta = -0.1, sigma = 0.1, gamma = 1)
R> low <- list(alpha = -5, beta = -5, sigma = -5, gamma = -5)
R> upp <- list(alpha = 8, beta = 8, sigma = 8, gamma = 8)
R> lasso1 <- lasso(yuima, lambda1, start = start, lower = low,
   upper = upp,
+      method = "L-BFGS-B")

Looking for MLE estimates...

Performing LASSO estimation...
```

and we obtain the quasi-maximum likelihood estimates

```
R> round(lasso1$mle, 3)

 sigma  gamma  alpha   beta
 0.133  1.443  2.076 -0.263
```

and the LASSO estimates

```
R> round(lasso1$lasso, 3)

 sigma  gamma  alpha   beta
 0.131  1.449  1.486 -0.145
```

Further, we use strong penalties, i.e., $\lambda_0 = \gamma_0 = 10$

```
R> lambda10 <- list(alpha = 10, beta = 10, sigma = 10, gamma = 10)
R> lasso10 <- lasso(yuima, lambda10, start = start, lower = low,
+      upper = upp, method = "L-BFGS-B")

Looking for MLE estimates...

Performing LASSO estimation...
```

and check the results

```
R> round(lasso10$mle, 3)

 sigma  gamma  alpha   beta
 0.133  1.443  2.076 -0.263

R> round(lasso10$lasso, 3)

sigma gamma alpha  beta
0.117 1.503 0.591 0.000
```

Very strong penalties suggest that the model does not contain the term β and, in both cases, the LASSO estimation suggest $\gamma = 3/2$, therefore a model quite close to Cox *et al.* (1980). Being a shrinkage estimator, the LASSO estimates

Table 9.1   Model selection on the CKLS model for the U.S. interest rates data.
Table taken from Yu and Phillips (2001) and updated with LASSO results.
Standard errors in parenthesis when available.

Model	Estimation Method	α	β	σ	γ
Vasicek	MLE	4.1889	−0.6072	0.8096	–
CKLS	Nowman	2.4272	−0.3277	0.1741	1.3610
CKLS	Exact Gaussian	2.0069	−0.3330	0.1741	1.3610
		(0.5216)	(0.0677)		
CKLS	QMLE	2.0755	−0.2630	0.1325	1.4433
		(0.992)	(0.196)	(0.026)	(0.103)
CKLS	QMLE + LASSO	1.4863	−0.1454	0.1309	1.4493
	with mild penalization	(0.701)	(0.138)	(0.018)	(0.073)
CKLS	QMLE + LASSO	0.5914	0.0002	0.1168	1.5035
	with strong penalization	(0.211)	(0.005)	(0.018)	(0.073)

have very low standard error compared to the other cases. Our application of
the LASSO method is reported in Table 9.1 along with the results from Yu and
Phillips (2001) just for comparison.

We now perform a small example to prove the effectiveness of the LASSO
method also in the multdimensional case. We consider a two-dimensional stochastic differential equation of the form:

$$\begin{pmatrix} dX_t^1 \\ dX_t^2 \end{pmatrix} = \begin{pmatrix} -\theta_{2.1}X_t^1 - \theta_{2.2} \\ -\theta_{2.2}X_t^2 - \theta_{2.1} \end{pmatrix} dt + \begin{bmatrix} \theta_{1.1} & 1 \\ \theta_{1.2} & 1 \end{bmatrix} \begin{pmatrix} dW_t^1 \\ dW_t^2 \end{pmatrix}$$

$$X_0^1 = 1, \quad X_0^2 = 1$$

which we prepare into a `yuima` model

```
R> diff.matrix <- matrix(c("theta1.1", "theta1.2", "1", "1"), 2,
+     2)
R> drift.c <- c("-theta2.1*x1", "-theta2.2*x2", "-theta2.2",
    "-theta2.1")
R> drift.matrix <- matrix(drift.c, 2, 2)
R> ymodel <- setModel(drift = drift.matrix, diffusion =
    diff.matrix,
+     time.variable = "t", state.variable = c("x1", "x2"),
    solve.variable = c("x1",
+         "x2"))
R> n <- 100
R> ysamp <- setSampling(Terminal = (n)^(1/3), n = n)

YUIMA: 'delta' (re)defined.

R> yuima <- setYuima(model = ymodel, sampling = ysamp)
```

We set the true values for $\theta_{1.2}$ and $\theta_{2.2}$ to zero and simulate the trajectory

```
R> set.seed(123)
R> truep <- list(theta1.1 = 0.6, theta1.2 = 0, theta2.1 = 0.5,
   theta2.2 = 0)
R> yuima <- simulate(yuima, xinit = c(1, 1), true.parameter =
   truep)
```

we finally apply the LASSO method

```
R> est <- lasso(yuima, start = list(theta2.1 = 0.8,
   theta2.2 = 0.2,
+      theta1.1 = 0.7, theta1.2 = 0.1), lower =
   list(theta1.1 = 1e-10,
+      theta1.2 = 1e-10, theta2.1 = 0.1, theta2.2 = 1e-10),
   upper = list(theta1.1 = 4,
+      theta1.2 = 4, theta2.1 = 4, theta2.2 = 4), method =
   "L-BFGS-B")

Looking for MLE estimates...

Performing LASSO estimation...

R> unlist(truep)

theta1.1 theta1.2 theta2.1 theta2.2
     0.6      0.0      0.5      0.0

R> round(est$mle, 3)

theta1.1 theta1.2 theta2.1 theta2.2
   0.559    0.000    0.741    0.024

R> round(est$lasso, 3)

theta1.1 theta1.2 theta2.1 theta2.2
   0.558    0.000    0.670    0.000
```

and see that the LASSO method shrinks towards zero the two estimates of $\theta_{1.2}$ and $\theta_{2.2}$ as expected.

9.4   Clustering of financial time series

In recent years, there has been a lot of interest in mining financial time series data. Although many measures of dissimilarity are available in the literature (see e.g. Liao (2005), for a review) most of them ignore the underlying structure of the stochastic model which drives the data. Among the few measures which take into account the properties of the data generating model we can mention Hirukawa (2006) which considers non-Gaussian locally stationary sequences; Piccolo (1990) proposed an AR metrics and Otranto (2008) adapted it to GARCH models. Caiado *et al.* (2006) used an approach based on periodograms; Xiong and Yeung (2002) proposed a model based clustering for mixtures of ARMA models.

Kakizawa *et al.* (1998) and Alonso *et al.* (2006) performed clustering based on several information measures constructed on the estimated densities of the processes. In this section we consider a distance tailored to measure discrepancy of discretely observed diffusion processes and apply this distance in a simple cluster analysis. Cluster analysis is an explorative data analysis tool that, starting from a matrix of dissimilarities between couple of observations in a sample, groups the observations into clusters (subgroups) according to some rule. Roughly speaking, rules can be of agglomerative type (individual observations are put together) or of a divisive type (the observations which are more dissimilar are put away from the initial group of all units). Once the first groups are formed, other rules are necessary to decide, e.g. in the agglomerative case, how to aggregate more units to the formed groups or groups together, etc. An interesting review on cluster methods can be found in Kaufman and Rousseeuw (1990) and the corresponding implementation in R is available through the **cluster** package.

9.4.1   The Markov operator distance

This new dissimilarity De Gregorio and Iacus (2010b) is based on a new application of the results by Hansen *et al.* (1998) on identification of diffusion processes observed at discrete time when the time mesh Δ between observations is not necessarily shrinking to zero. The theory proposed in Hansen *et al.* (1998) has been used in Kessler and Sørensen (1999) and Gobet *et al.* (2004) in parametric and nonparametric estimation of diffusion processes respectively. The theory is based on the fact that, when the process is not observed at high frequency, i.e. $\Delta \not\to 0$, the observed data form a true Markov process for which it is possible to identify the Markov operator P_Δ. Consider now the regularly sampled data $X_i = X(i\Delta)$, $i = 0, \ldots, N$, from the sample path of $\{X_t, 0 \le t \le T\}$, where $\Delta > 0$ and is not shrinking to 0 and such that $T = N\Delta$. The process $\mathbf{X} = \{X_i\}_{i=0,\ldots,N}$ is a Markov process and under mild regularity conditions, all the mathematical properties of the model are embodied in the transition operator defined as follows:

$$P_\Delta f(x) = \mathbb{E}\{f(X_i)|X_{i-1} = x\}.$$

Notice that P_Δ depends on the transition density from X_{i-1} to X_i, so we put explicitly the dependence on Δ in the notation. This operator is associated with the infinitesimal generator of the diffusion, namely the following operator on the space of continuous and twice differentiable functions $f(\cdot)$

$$L_{b,\sigma} f(x) = \frac{\sigma^2(x)}{2} f''(x) + b(x) f'(x).$$

We assume that, under condition (9.2), the invariant density $\mu = \mu_{b,\sigma}(\cdot)$ of the process X_t exists and its explicit form is:

$$\mu_{b,\sigma}(x) = \frac{m(x)}{C_0} = \frac{\exp\left\{2\int_{\bar{x}}^x \frac{b(y)}{\sigma^2(y)}dy\right\}}{C_0 \sigma^2(x)}.$$

Then, the operator is unbounded but self-adjoint negative on $L^2(\mu) = \{f : \int |f|^2 d\mu < \infty\}$ and the functional calculus gives the correspondence (in terms of operator notation)

$$P_\Delta = \exp\{\Delta L_\mu\}.$$

This relation has been first noticed by Hansen *et al.* (1998) and Chen *et al.* (1997). For a given L^2-orthonormal basis $\{\phi_j, j \in J\}$ of $L^2([l, r])$, where J is an index set, following Gobet *et al.* (2004) it is possible to obtain the matrix $\hat{\mathbf{P}}_\Delta(\mathbf{X}) = [(\hat{P}_\Delta)_{j,k}(\mathbf{X})]_{j,k \in J}$, which is an estimator of $< P_\Delta \phi_j, \phi_k >_{\mu_{b,\sigma}}$, where

$$(\hat{P}_\Delta)_{j,k}(\mathbf{X}) = \frac{1}{2N} \sum_{i=1}^{N} \{\phi_j(X_{i-1})\phi_k(X_i) + \phi_k(X_{i-1})\phi_j(X_i)\}, \quad j, k \in J.$$

$$(9.27)$$

The terms $(\hat{P}_\Delta)_{j,k}$ are approximations of $< P_\Delta \phi_j, \phi_k >_{\mu_{b,\sigma}}$, that is, the action of the transition operator on the state space with respect of the unknown scalar product $< \cdot, \cdot >_{\mu_{b,\sigma}}$ and hence can be used as 'proxy' of the probability structure of the model. Therefore, we introduce the following dissimilarity measure.

Definition 9.4.1 *Let* $\mathbf{X}$ *and* $\mathbf{Y}$ *be discrete time observations from two diffusion processes. The Markov Operator distance is defined as*

$$d_{MO}(\mathbf{X}, \mathbf{Y}) = \left\| \hat{P}_\Delta(\mathbf{X}) - \hat{P}_\Delta(\mathbf{Y}) \right\|_1 = \sum_{j,k \in J} \left| (\hat{P}_\Delta)_{j,k}(\mathbf{X}) - (\hat{P}_\Delta)_{j,k}(\mathbf{Y}) \right|, \quad (9.28)$$

where $(\hat{P}_\Delta)_{j,k}(\cdot)$ *is calculated as in (9.27) separately for* $\mathbf{X}$ *and* $\mathbf{Y}$.

Notice that $d_{MO}(\mathbf{X}, \mathbf{Y})$ is the element-wise L^1 distance for matrixes, not simply a dissimilarity measure (i.e. it also respects the triangular inequality).

Like the invariant density $\mu_{b,\sigma}$, the Markov operator itself cannot perfectly identify the underlying process, in the sense that, for some (b_1, σ_1) there might exist another couple (b_2, σ_2) such that $\mu_{b_1,\sigma_1}(x) = \mu_{b_2,\sigma_2}(x)$. The same considerations apply to the infinitesimal generator and hence to the Markov operator. Nevertheless, the distance d_{MO} helps in finding similarities between two (or more) processes in terms of the action of their Markov operators.

9.4.2 Application to real data

In this section we compare the Markov operator distance with few other distances. We denote by $\mathbf{X} = \{X_i, i = 1, \ldots, N\}$ and $\mathbf{Y} = \{Y_i, i = 1, \ldots, N\}$ two discretely observed data from continuous time diffusion processes. We compare the following distances.

9.4.2.1   The Markov-operator distance

The Markov operator distance d_{MO} is calculated using formula (9.28). As in Reiß (2003) we deal with a basis 50 orthonormal B-splines on a compact support of degree 10 (see Ramsay and Silverman (2005)). As compact support we consider the observed support of all simulated diffusion paths enlarged by 10 %. This function is implemented in the function MOdist in the package **sde**.

9.4.2.2   Short-time-series distance

Proposed by Möller-Levet *et al.* (1978) is based on the idea to consider each time series as a piecewise linear function and compare the slopes between all the interpolants. It reads as

$$d_{STS}(X, Y) = \sqrt{\sum_{i=1}^{N} \left(\frac{X_i - X_{i-1}}{\Delta} - \frac{Y_i - Y_{i-1}}{\Delta} \right)^2}.$$

This measure is essentially designed to discover similarities in the volatility between two time series regardless of the average level of the process, i.e. one process and a shifted version of it will have zero distance.

9.4.2.3   The Euclidean distance

The usual Euclidean distance

$$d_{EUC}(X, Y) = \sqrt{\sum_{i=1}^{N} (X_i - Y_i)^2}$$

is one of the most used in the applied literature. We use it only for comparison purposes.

9.4.2.4   Dynamic time warping distance

The Euclidean distance is very sensitive to distortion in time axis and may lead to poor results for sequences which are similar, but locally out of phase (Corduas 2007). The Dynamic Time Warping (DTW) distance was introduced originally in speech recognition analysis (Sakoe and Chiba (1978); Wang and Gasser (1997)). DTW allows for nonlinear alignments between time series not necessarily of the same length. Essentially, all shiftings between two time series are attempted and each time a cost function is applied (e.g. a weighted Euclidean distance between the shifted series). The minimum of the cost function over all possible shiftings is the dynamic time warping distance d_{DTW}. In our applications we use the

Euclidean distance in the cost function and the algorithm as implemented in the R package **dtw** (Giorgino 2009).

9.4.2.5   The data

We consider time series of daily closing quotes, from 01 March 2006 to 31 December 2007, for the following 20 financial assets: Microsoft Corporation (MSOFT in the plots), Advanced Micro Devices Inc. (AMD), Dell Inc. (DELL), Intel Corporation (INTEL), Hewlett-Packard Co. (HP), Sony Corp. (SONY), Motorola Inc. (MOTO), Nokia Corp. (NOKIA), Electronic Arts Inc. (EA), LG Display Co., Ltd. (LG), Borland Software Corp. (BORL), Koninklijke Philips Electronics NV (PHILIPS), Symantec Corporation (SYMATEC), JPMorgan Chase & Co (JMP), Merrill Lynch & Co., Inc. (MLINCH), Deutsche Bank AG (DB), Citigroup Inc. (CITI), Bank of America Corporation (BAC), Goldman Sachs Group Inc. (GSACHS) and Exxon Mobil Corp. (EXXON). Quotes come from NYSE/NASDAQ. Source Yahoo.com. Missing values (the same 19 festivity days over 520 daily data) have been linearly interpolated. These assets come from both electronic hardware, appliance and software vendors or producers, financial institutions of different type and a petrol company. These data are preloaded in the data set `quotes` of the **sde** package and plotted in Figure 9.7

```
R> require(sde)
R> data(quotes)
R> Series <- quotes
R> nSeries <- dim(Series)[2]
R> plot(Series, main = "", xlab = "")
```

As anticipated, we now cluster these financial data using the four distances d_{MO}, d_{EUC}, d_{STS} and d_{DTW}. For the Short-Time-Series distance d_{STS} we write our own code first

```
R> STSdist <- function(data) {
+      nSer <- NCOL(data)
+      d <- matrix(0, nSer, nSer)
+      colnames(d) <- colnames(data)
+      rownames(d) <- colnames(data)
+      DELTA <- deltat(data)
+      for (i in 1:(nSer - 1)) for (j in (i + 1):nSer) {
+          d[i, j] <- sqrt(sum((diff(data[, i])/DELTA - diff(data[,
+              j])/DELTA)^2))
+          d[j, i] <- d[i, j]
+      }
+      invisible(d)
+ }
```

We now use `MOdist` function from the **sde**, our function `EUCdist`, the standard R function `dist` for the Euclidean distance and finally make use of the **dtw**

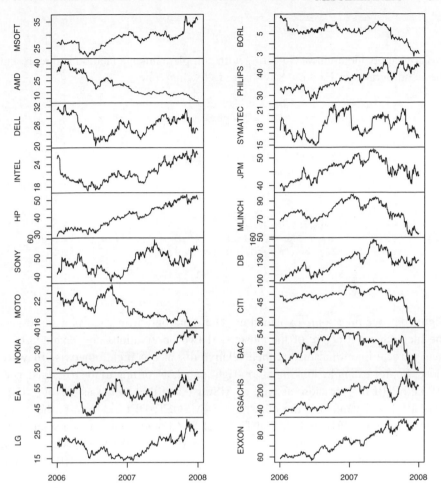

Figure 9.7   Paths of the 20 assets considered: from 01 March 2006 to 31 December 2007.

package. With the only aim of making the distances comparable, we normalize each by its maximum value, therefore all distances will be in the interval [0, 1]

```
R> dMO <- MOdist(Series)
R> dMO <- dMO/max(dMO)
R> dEUC <- dist(t(Series))
R> dEUC <- dEUC/max(dEUC)
R> dSTS <- STSdist(Series)
R> dSTS <- dSTS/max(dSTS)
R> require(dtw)
```

```
Loaded dtw v1.14-3. See ?dtw for help, citation("dtw") for usage
    conditions.
```

```
R> dDTW <- dist(t(Series), method = "dtw")
R> dDTW <- dDTW/max(dDTW)
```

We now apply hierarchical clustering with complete linkage method and represent the dendrograms in Figure 9.8 with the following code

```
R> cl <- hclust(dMO)
R> plot(cl, main = "Markov Operator Distance", xlab = "",
   ylim = c(0,
+      1))
R> rect.hclust(cl, k = 6, border = gray(0.5))
R> cl1 <- hclust(as.dist(dEUC))
R> plot(cl1, main = "Euclidean Distance", xlab = "", ylim = c(0,
+      1))
R> rect.hclust(cl1, k = 6, border = gray(0.5))
R> cl2 <- hclust(as.dist(dSTS))
R> plot(cl2, main = "STS Distance", xlab = "", ylim = c(0, 1))
R> rect.hclust(cl2, k = 6, border = gray(0.5))
R> cl3 <- hclust(as.dist(dDTW))
R> plot(cl3, main = "DTW Distance", xlab = "", ylim = c(0, 1))
R> rect.hclust(cl3, k = 6, border = gray(0.5))
```

Figure 9.8 requires some explanation. The dendrogram in each plot represents the hierarchical structure of the clusters. If two observations are separated by a long vertical line it means that their relative distance is high, otherwise they are close. So, for example, looking at top-right panel, according to the d_{MO} distance, HP and PHILIPS are close as well as MSOFT and DELL, but as a group, {HP, PHILIPS} is distant from the other group {MSOFT, DELL}. In turn, the group formed by {HP, PHILIPS, MSOFT, DELL} is homogenous and distant from the other homogeneous groups {AMF, SYMATEC, MOKIA, INTAL, MOTO, LG}. In particular, LG is separated by a long distance from, e.g. PHILIPS. And so forth.

In Figure 9.8 the dendrogram for the d_{MO} distance, identifies 5 or 6 groups and in particular isolates BORL and 'DB + GSACHS' into separates clusters very clearly (the difference between 5 and 6 groups is that in the 6 groups clustering 'MLINCH + EXXON' are put in a separate cluster). To isolate the BORL asset via the dendrograms of d_{EUC} and d_{DTW} we need to cut at least into 6 groups. The counter effect of this cutting is that DB an GSACHS go into different clusters for these metrics. The metric d_{STS} does not appear to give sharp indication on how to separate clusters. We have then decided to cut all the dendrograms into 6 groups, the result is represented by the gray boxes.

Different methods produce different groupings but with some overlap. In order to be more specific in the similarities of the results produced by different methods, we make use of a similarity measure proposed by Gravilov *et al.* (2000). Given two clustering $C = C_1, \ldots, C_K$ (the clusters formed by adopting one distance) and $C' = C'_1, \ldots, C'_{K'}$ (the clustering obtained using another

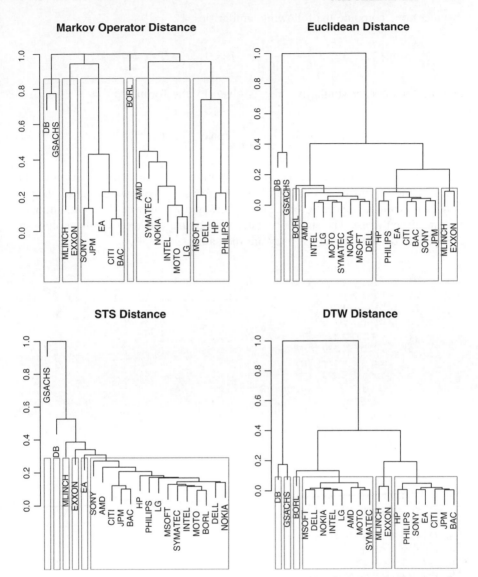

Figure 9.8   Clustering according to different distances. Distances normalized to 1 just for graphical representation. Although the markers of the terminal nodes go below the zero line (see e.g. right-bottom plot), the final nodes are obtained cutting the dendrogram above the zero line, which is represented as a dotted line just to help visualization.

distance), we compute the following similarities

$$\text{sim}(C_i, C_j') = 2\frac{|C_i \cap C_j'|}{|C_i| + |C_j'|}, \quad i = 1, \ldots, K, j = 1, \ldots, K',$$

and the final cluster similarity index is given by the formula

$$\text{Sim}(C, C') = \frac{1}{K} \sum_{i=1}^{K} \max_{j=1,\ldots,K'} \text{sim}(C_i, C_j'). \tag{9.29}$$

This index is not symmetric, so, we also apply the symmetrized version of the index, namely $(\text{Sim}(C, C') + \text{Sim}(C', C))/2$, because the real number of clusters is not known in advance. In formula (9.29) K and K' may be different, although it is not in our case. The similarity index will return 0 if the two clusterings are completely dissimilar and 1 if they are the same.

```
R> Sim <- function(g1, g2) {
+       G1 <- unique(g1)
+       G2 <- unique(g2)
+       l1 <- length(G1)
+       l2 <- length(G2)
+       sim <- matrix(, l1, l2)
+       for (i in 1:l1) {
+           idx <- which(g1 == i)
+           for (j in 1:l2) {
+               idx2 <- which(g2 == j)
+               sim[i, j] <- 2 * length(intersect(idx, idx2))/
    (length(idx) +
+                   length(idx2))
+           }
+       }
+       sum(apply(sim, 2, max))/l1
+ }
R> G <- cutree(cl, k = 6)
R> G1 <- cutree(cl1, k = 6)
R> G2 <- cutree(cl2, k = 6)
R> G3 <- cutree(cl3, k = 6)
R> A <- matrix(, 4, 4)
R> A[1, 1] <- Sim(G, G)
R> A[1, 2] <- Sim(G, G1)
R> A[1, 3] <- Sim(G, G2)
R> A[1, 4] <- Sim(G, G3)
R> A[2, 1] <- Sim(G1, G)
R> A[2, 2] <- Sim(G1, G1)
R> A[2, 3] <- Sim(G1, G2)
R> A[2, 4] <- Sim(G1, G3)
R> A[3, 1] <- Sim(G2, G)
R> A[3, 2] <- Sim(G2, G1)
R> A[3, 3] <- Sim(G2, G2)
R> A[3, 4] <- Sim(G2, G3)
R> A[4, 1] <- Sim(G3, G)
```

```
R> A[4, 2] <- Sim(G3, G1)
R> A[4, 3] <- Sim(G3, G2)
R> A[4, 4] <- Sim(G3, G3)
R> S <- (A + t(A))/2
```

The results are presented in Table 9.2. The similarity matrix in Table 9.2 shows that d_{EUC} and d_{DTW} form the same groups, i.e. they are essentially the same metric for this data set. The clustering made using d_{MO} is only partially in agreement with d_{EUC} and d_{DTW} (0.84). The difference is mainly in the placement of the subgroups 'HP + PHILIPS' and 'MSOFT + DELL'. Further, the d_{MO} distance considers 'DB + GSACHS' together, which makes sense for this distance probably because these two time series have the highest volatilities.

EA goes together with SONY in all dendrograms, which is not an unrealistic evidence in that the company essentially produces software for game consoles. Also for CITI, BAC and JPM the methods agree on their placement. To stress more on the comparisons in terms of levels, drift and volatilities of the 20 time series considered, in Figure 9.9 we also plot the data using the same vertical scale.

In summary, all but the d_{STS} distance provide similar evidence. Nevertheless, d_{MO} easily separates BORL (an outlier if we think at the levels and the volatility of times series, see Figure 9.9) and 'GSACHS + DB', while with the other two competitors, in order to separate BORL, we need to force an additional splitting which separates GSACHS and DB. This looks quite unfortunate from a substantial point of view. Of course, this is merely an exercise and the analysis cannot go deeper than this from a simple cluster analysis. In fact, other financial and economics considerations have to be done in analyzing the composition of the clusters obtained by any method.

9.4.3   Sensitivity to misspecification

As mentioned, one rarely knows if the number of clusters to select is the precise given number if the data we observed really follow the assumptions of Section 9.4.1. To test the robustness of the Markov operator distance against model misspecification we report an experiment taken from the original paper of

Table 9.2 Similarity matrix between the clusters formed by different metrics. Similarity calculated according to the similarity index defined in (9.29) (left table) and its symmetrized version (right table).

	d_{MO}	d_{EUC}	d_{STS}	d_{DTW}	d_{MO}	d_{EUC}	d_{STS}	d_{DTW}
d_{MO}	1	0.84	0.6	0.84	1	0.81	0.54	0.81
d_{EUC}	0.79	1	0.71	1		1	0.69	1
d_{STS}	0.48	0.67	1	0.67			1	0.69
d_{DTW}	0.79	1	0.71	1				1

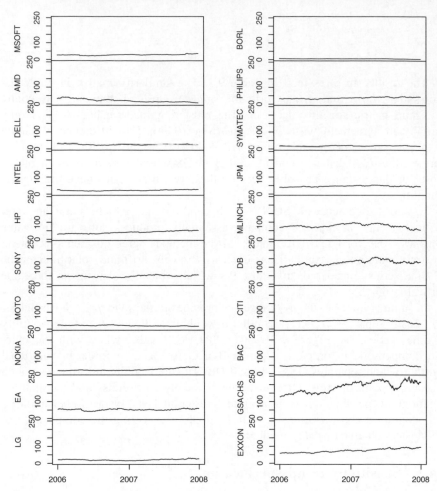

Figure 9.9   Paths of the 20 assets considered in Figure 9.7 represented on the same scale to put in evidence differences in the levels and volatilities.

De Gregorio and Iacus (2010b). This experiment simulates 23 paths according to the six different models M_j, $j = 1, \ldots, 6$, obtained via the combinations of drift b_k and diffusion coefficients σ_k, $k = 1, \ldots, 4$ presented in the following table:

	$\sigma_1(x)$	$\sigma_2(x)$	$\sigma_3(x)$	$\sigma_4(x)$
$b_1(x)$	M1		M4	
$b_2(x)$		M2	M3	
$b_3(x)$		M5		
$b_4(x)$				M6

where

$$b_1(x) = 1 - 2x, \quad b_2(x) = 1.5(0.9 - x), \quad b_3(x) = 1.5(0.5 - x),$$
$$b_4(x) = 5(0.05 - x)$$

and

$$\sigma_1(x) = 0.5 + 2x(1 - x), \quad \sigma_2(x) = \sqrt{0.55x(1 - x)},$$
$$\sigma_3(x) = \sqrt{0.1x(1 - x)}, \quad \sigma_4(x) = \sqrt{0.8x(1 - x)}.$$

For each model M_j a different number of n_j of trajectories has been simulated in order to have an unbalanced simulation design, i.e. $n_1 = 5$, $n_2 = 3$, $n_3 = 4$, $n_4 = 3$, $n_5 = 4$, $n_6 = 1$. Further, one trajectory generated with model M_1, say X^1, is reversed around the line $y = 1$, i.e. if $1 - X^1 = \tilde{X}^1$, hence $\tilde{X}^1$ has drift $-b_1(x)$ and the same quadratic variation of X^1. So it still belongs to the class M_1 with respect to volatility. Then, an additional trajectory was simulated using model M_1 but with different initial value. By the ergodic property of the simulated path, its invariant law still belongs to model M_1. Therefore, finally $n_1 = 7$.

Each path was simulated using (second) Milstein scheme (see e.g. Kloden and Platen (1999) or Iacus (2008)) with time lag $\delta = 1e - 3$. Observations have been then resampled at rate $\Delta = 0.01$ and observed paths of length $N = 500$ and $N = 1000$ have been used in the analysis in order to capture sample size effects. Due to the fact that the number of clusters is known in advance, i.e. $K = 6$, we can use the cluster similarity index (9.29) for the two clustering $C = C_1, \ldots, C_K$ (the real clusters formed by the six models) and $C' = C'_1, \ldots, C'_{K'}$ (the clustering obtained using one of the above distances).

Four different experiments were performed. In all cases hierarchical clustering with complete linkage method was used.

9.4.3.1   Experiment 1: Nonperturbed, correctly specified

Simulate according the above scheme 25 trajectories; calculate the distance matrixes d_{MO}, d_{STS}, d_{EUC} and d_{DTW} and run clustering. Cut the dendrograms into $K = K' = 6$ groups. Calculate the Sim index for each clustering solution.

9.4.3.2   Experiment 2: Nonperturbed, misspecified

Simulate according the above scheme 25 trajectories; calculate the four distances and run cluster analysis. Cut the dendrograms into $K' = 5$ groups, real number of groups $K = 6$. Calculate the Sim index for each clustering solution.

9.4.3.3   Experiment 3: Perturbed, correctly specified

Simulate according the above scheme 25 trajectories. Perturbate the experiment adding 2 trajectories from an ARIMA(1,0,1) process with mean 0.5 and AR coefficient 0.9, MA coefficient $= -0.22$, with Gaussian innovations $N(0, 0.01)$ (the parameters of the model are chosen in a way that the simulated trajectories looks qualitatively similar to the ones in the Experiment 1.) Calculate the four distances, use the same clustering approach as in Experiment 1, set $K = 7$ and cut the dendrograms into $K' = 7$ groups.

9.4.3.4   Experiment 4: Perturbed, misspecified

Proceed as in Experiment 3, set $K = 7$ and cut the dendrograms $K' = 6$ groups.

Each experiment is replicated only 100 times and the average value of the cluster similarity index Sim is reported in Table 9.3 for different sample sizes $N = 500$ (up) and $N = 1000$ (down). The number of replications is limited due to excessively long computational time of the DTW distance in dimension 23. To test the stability of the Monte Carlo results of the first few 100 replications, we drop d_{DTW} from the Monte Carlo analysis and replicate each of the four experiments 5000 times. Table 9.3 also reports in parenthesis the average values but calculated over the 5000 replications.

Experiment 3 corresponds to a perturbation of the diffusion setup with an ARIMA process, while Experiment 4 corresponds to a misspecified setting: there

Table 9.3   Results of the simulation experiments. Average values of the Sim index over 100 replications and, with the exclusion of the d_{DTW} distance, 5000 replications (in parentheses). $\Delta = 0.01$, sample size $N = 500$ (up) and $N = 1000$ (bottom).

Experiment: $N = 500$	d_{MO}	d_{EUC}	d_{STS}	d_{DTW}
Nonperturbed, correctly specified	0.84	0.49	0.27	0.69
	(0.83)	(0.49)	(0.27)	(–)
Nonperturbed, misspecified	070	0.44	0.24	0.60
	(0.69)	(0.43)	(0.24)	(–)
Perturbed, correctly specified	0.81	0.45	0.39	0.65
	(0.81)	(0.45)	(0.39)	(–)
Perturbed, misspecified	0.71	0.41	0.37	0.58
	(0.70)	(0.41)	(0.37)	(–)
Experiment: $N = 1000$	d_{MO}	d_{EUC}	d_{STS}	d_{DTW}
Nonperturbed, correctly specified	0.94	0.51	0.27	0.69
	(0.93)	(0.50)	(0.26)	(–)
Nonperturbed, misspecified	0.75	0.45	0.24	0.63
	(0.75)	(0.45)	(0.24)	(–)
Perturbed, correctly specified	0.91	0.47	0.39	0.67
	(0.90)	(0.46)	(0.39)	(–)
Perturbed, misspecified	0.78	0.43	0.37	0.59
	(0.78)	(0.42)	(0.37)	(–)

are $K = 7$ real clusters, but we induce misclassification selecting only $K' = 6$ groups. In Experiment 2 there is only misspecification where the number of real groups K is higher than the number of the groups generated with the cluster K'.

As emerges from the analysis of Table 9.3 we see that all methods perform better in Experiment 1, although a clear ordering–for all experiments–emerges in the different metrics to discover the correct groups: $d_{MO} \prec d_{DTW} \prec d_{EUC} \prec d_{STS}$, where $d_1 \prec d_2$ means: 'distance d_1 classifies *better* than d_2'. In the case of perturbation (Experiment 3) one should expect the Markov operator distance should fail to detect the ARIMA group, and instead should not expect any change in performance of the other metrics because they do not assume a particular stochastic structure of the model. But Table 9.3 shows that all methods are equally affected and d_{MO} looks quite robust. Although there is a decrease of performance of d_{MO} in the misspecified case (Experiment 4), the d_{MO} distance still performs much better that the other competitors. All methods increase performance as the number of observations N increases, but the enhancement of the d_{MO} is particularly remarkable. This is due to the property of the estimator of the Markov operator, which gets better and better as the sample size increases.

9.5   Bibliographical notes

The study of volatility, realized variance, and power variations has a long history in finance. Early studies are probably Andersen and Bollerslev (1998), Andersen *et al.* (2003) and Barndorff-Nielsen and Shephard (2004) from which it is possible to get all the relevant references. More recently the problem of estimation under nonsynchronicity was considered, for example, in Hayashi and Yoshida (2005, 2008). In parallel with the study of nonsynchronicity, the high frequency data from finance also revealed the microstructure noise effect and its impact on estimates of all the above quantities. Some early references on the subject are Aït-Sahalia *et al.* (2005), Bandi and Russell (2006) and Hansen and Lunde (2006). The problem of model selection was considered in Uchida and Yoshida (2005) and in the recent work on the Lasso method presented in this chapter. Both works are good sources of references in this direction. Finally, clustering and other methods for continuous time financial model have been considered in De Gregorio and Iacus (2010b) and references therein.

References

Aït-Sahalia, Y. (1996). Testing continuous-time models of the spot interest rate. *Rev. Financial Stud* **9**, 2, 385–426.

Aït-Sahalia, Y., Mykland, P., and Zhang, L. (2005). How often to sample a continuous-time process in the presence of market microstructure noise. *Review of Financial Studies* **18**, 351–416.

Alonso, A. M., Berrendero, J. R., Hernández, A., and Justel, A. (2006). Time series clustering based on forecast densities. *Computational Statistics & Data Analysis* **51**, 2, 762–776.

Andersen, T. G. and Bollerslev, T. (1998). Answering the skeptics: Yes, standard volatility models do provide accurate forecasts. *International Economic Review* **39**, 885–905.

Andersen, T. G., Bollerslev, T., Diebold, F. X., and Labys, P. (2003). Modeling and forecasting realized volatility. *Econometrica* **71**, 579–625.

Bai, J. (1994). Least squares estimation of a shift in linear processes. *Journal of Times Series Analysis* **15**, 453–472.

Bai, J. (1997). Estimation of a change point in multiple regression models. *The Review of Economics and Statistics* **79**, 551–563.

Bandi, F. and Phillips, P. (2003). Fully nonparametric estimation of scalar diffusion models. *Econometrica* **71**, 241–283.

Bandi, F. and Russell, J. R. (2006). Separating microstructure noise from volatility. *Journal of Financial Economics* **79**, 655–692.

Barndorff-Nielsen, O. E. and Shephard, N. (2004). Econometric analysis of realised covariation: High frequency based covariance, regression and correlation in financial economics. *Econometrica* **72**, 885–925.

Caiado, J., Crato, N., and Peña, D. (2006). A periodogram-based metric for time series classification. *Computational Statistics & Data Analysis* **50**, 10, 2668–2684.

Caner, M. (2009). Lasso-type gmm estimator. *Econometric Theory* **25**, 270–290.

Chen, G., Choi, Y., and Zhou, Y. (2005). Nonparametric estimation of structural change points in volatility models for time series. *Journal of Econometrics* **126**, 79–144.

Chen, X., Hansen, L., and Scheinkman, J. (1997). Shape preserving spectral approximation of diffusions. *Working paper*.

Corduas, M. (2007). Dissimilarity criteria for time series data mining. *Quaderni di Statistica* **9**, 107–129.

Cox, J., Ingersoll, J., and Ross, S. (1980). An analysis of variable rate loan contracts. *J. Finance* **35**, 2, 389–403.

Croissant, Y. (2010). *Ecdat: Data sets for econometrics*. R package version 0.1-6.

Csörgő, M. and Horváth, L. (1997). *Limit Theorems in Change-point Analysis*. John Wiley & Sons, Inc., New York.

De Gregorio, A. and Iacus, S. M. (2008). Least squares volatility change point estimation for partially observed diffusion processes. *Communications in Statistics, Theory and Methods* **37**, 15, 2342–2357.

De Gregorio, A. and Iacus, S. M. (2010a). Adaptive lasso-type estimation for ergodic diffusion processes. http://services.bepress.com/unimi/statistics/art50/.

De Gregorio, A. and Iacus, S. M. (2010b). Clustering of discretely observed diffusion processes. *Computational Statistics & Data Analysis* **54**, 598–606.

Efron, B., Hastie, T., Johnstone, I., and Tibshirani, R. (2004). Least angle regression. *The Annals of Statistics* **32**, 407–489.

Epps, T. (1979). Comovements in stock prices in the very short run. *Journal of the American Statistical Association* **74**, 291–298.

Florens-Zmirou, D. (1993). On estimating the diffusion coefficient from discrete observations. *J. App. Prob.* **30**, 790–804.

Genon-Catalot, V. and Jacod, J. (1993). On the estimation of the diffusion coefficient for multidimensional diffusion processes. *Ann. Inst. Henri Poincaré* **29**, 119–151.

Giorgino, T. (2009). Computing and visualizing dynamic time warping alignments in r: The dtw package. *Journal of Statistical Software* **31**, 1–24.

Gobet, E., Hoffmann, M., and Reiß, M. (2004). Nonparametric estimation of scalar diffusions based on low frequency data. *The Annals of Statistics* **32**, 2223–2253.

Gravilov, M., Anguelov, D., Indyk, P., and Motwani, R. (2000). Mining the stock market; which measure is best? *Proceedings of the 6th International Conference on Knowledge Discovery and Data Mining* 487–496.

Hansen, L., Scheinkman, J., and Touzi, N. (1998). Spectral methods for identifying scalar diffusions. *Journal of Econometrics* **86**, 1–32.

Hansen, P. R. and Lunde, A. (2006). Realized variance and market microstructure noise. *Journal of Business & Economic Statistics* **24**, 2, 127–161.

Hayashi, T. and Yoshida, N. (2005). On covariance estimation of nonsynchronously observed diffusion processes. *Bernoulli* **11**, 359–379.

Hayashi, T. and Yoshida, N. (2008). Asymptotic normality of a covariance estimator for nonsynchronously observed diffusion processes. *Annals of the Institute of Statistical Mathematics* **60**, 367–406.

Hinkley, D. (1971). Inference about the change-point from cumulative sum tests. *Biometrika* **58**, 509–523.

Hirukawa, J. (2006). Cluster analysis for non-Gaussian locally stationary processes. *Int. Journal of Theoretical and Applied Finance* **9**, 113–132.

Hsu, N., Hung, H., and Chang, Y. (2008). Subset selection for vector autoregressive processes using the lasso. *Computational Statistics & Data Analysis* **52**, 3645–3657.

Iacus, S. (2008). *Simulation and Inference for Stochastic Differential Equations. With R Examples*. Springer Series in Statistics, Springer, New York.

Iacus, S. and Yoshida, N. (2009). Estimation for the change point of the volatility in a stochastic differential equation. http://arxiv.org/abs/0906.3108.

Inclan, C. and Tiao, G. (1994). Use of cumulative sums of squares for retrospective detection of change of variance. *Journal of the American Statistical Association* **89**, 913–923.

Kakizawa, Y., Sumway, R. H., and Taniguchi, M. (1998). Discrimination and clustering for multivariate time series. *J. Amer. Statist. Assoc.* **93**, 328–340.

Karatzas, I. and Shreve, S. (1991). *Brownian Motion and Stochastic Calculus*. Springer-Verlag, New York.

Kaufman, L. and Rousseeuw, P. (1990). *Finding Groups in Data: An Introduction to Cluster Analysis*. John Wiley & Sons, Inc., New York.

Kessler, M. (1997). Estimation of an ergodic diffusion from discrete observations. *Scandinavian Journal of Statistics* **24**, 211–229.

Kessler, M. and Sørensen, M. (1999). Estimating equations based on eigenfunctions for a discretely observed diffusion process. *Bernoulli* **5**, 299–314.

Kim, S., Cho, S., and Lee, S. (2000). On the cusum test for parameter changes in garch(1,1) models. *Commun. Statist. Theory Methods* **29**, 445–462.

Kloden, P. and Platen, E. (1999). *Numerical Solution of Stochastic Differential Equations*. Springer, New York.

Knight, K. and Fu, W. (2000). Asymptotics for lasso-type estimators. *The Annals of Statistics* **28**, 1536–1378.

Kutoyants, Y. (1994). *Identification of Dynamical Systems with Small Noise*. Kluwer, Dordrecht.

Kutoyants, Y. (2004). *Statistical Inference for Ergodic Diffusion Processes*. Springer-Verlag, London.

Lee, S., Ha, J., Na, O., and Na, S. (2000). The cusum test for parameter change in time series models. *Scandinavian Journal of Statistics* **30**, 781–796.

Lee, S., Nishiyama, Y., and Yoshida, N. (2006). Test for parameter change in diffusion processes by cusum statistics based on one-step estimators. *Ann. Inst. Statist. Mat.* **58**, 211–222.

Liao, T. (2005). Clustering of time series data – a survey. *Pattern Recognition* **38**, 1857–1874.

Möller-Levet, C., Klawonn, F., Cho, K.-H., and Wolkenhauer, O. (1978). Dyanmic programming algorithm optimization for spoken work recognition. *IEEE Transactions on Acoustic, Speeach and Signal Processing* **26**, 143–165.

Nardi, Y. and Rinaldo, A. (2008). Autoregressive processes modeling via the lasso procedure. http://arxiv.org/pdf/0805.1179.

Nowman, K. (1997). Gaussian estimation of single-factor continuous time models of the term structure of interest rates. *Journal of Finance* **52**, 1695–1703.

Otranto, E. (2008). Clustering heteroskedastic time series by model-based procedures. *Computational Statistics & Data Analysis* **52**, 4685–4698.

Pham, D. (1981). Nonparametric estimation of the drift coefficient in the diffusion equation. *Statistics* **12**, 1, 61–73.

Piccolo, D. (1990). A distance measure for classifying arima models. *Journal of Time Series Analysis* **11**, 153–164.

Ramsay, J. O. and Silverman, B. W. (2005). *Functional Data Analysis*. Springer, New York.

Reiß, M. (2003). Simulation results for estimating the diffusion coefficient from discrete time observation. *Department of Mathematics, Humboldt University* Available at http://www.mathematik.hu-berlin.se/reiss/sim-diff-est.pdf.

Sakoe, H. and Chiba, S. (1978). Dyanmic programming algorithm optimization for spoken work recognition. *IEEE Transactions on Acoustic, Speeach and Signal Processing* **26**, 143–165.

Smaldone, L. (2009). *Analysis of the change point of the volatility in financial markets (in italian)*. Master Thesis in Finance, University of Milan.

Song, J. and Lee, S. (2009). Test for parameter change in discretely observed diffusion processes. *Statistical Inference for Stochastic Processes* **12**, 2, 165–183.

Stanton, R. (1997). A non parametric model of term structure dynamics and the market price of interest rate risk. *Journal of Finance* **52**, 1973–2002.

Tibshirani, R. (1996). Regression shrinkage and selection via the lasso. *J. Roy. Statist. Soc. Ser. B* **58**, 267–288.

Uchida, M. and Yoshida, N. (2005). Aic for ergodic diffusion processes from discrete observations. *Faculty of Mathematics, Kyushu University, Fukuoka, Japan* **MHF 2005-12, March 2005**, 1–20.

Wang, H. and Leng, C. (2007). Unified lasso estimation by least squares approximation. *J. Amer. Stat. Assoc.* **102**, 479, 1039–1048.

Wang, H., Li, G., and Tsai, C. (2007). Regression coefficient and autoregressive order shrinkage and selection via the lasso. *J. R. Statist. Soc. Series B* **169**, 1, 63–78.

Wang, K. and Gasser, T. (1997). Alignment of curves by dynamic time warping. *Annals of Statistics* **25**, 1251–1276.

Xiong, Y. and Yeung, D. (2002). Mixtures of arma models for model-based time series clustering. *Proceedings of the IEEE International Conference on Data Mining* 717–720.

Yoshida, N. (1992). Estimation for diffusion processes from discrete observation. *J. Multivar. Anal.* **41**, 2, 220–242.

Yu, J. and Phillips, P. (2001). Gaussian estimation of continuous time models of the short term interest rate. *Cowles Foundation Discussion Paper* cowles.econ.yale.edu/P/cd/d13a/d1309.pdf, 1309, 241–283.

Zou, H. (2006). The adaptive lasso and its oracle properties. *J. Amer. Stat. Assoc.* **101**, 476, 1418–1429.

Appendix A

'How to' guide to R

This appendix is a compact guide to the R language which focuses on special aspects of the language that are relevant to this book. A review of R packages useful in finance will be discussed in Appendix B. Along with these pages, the reader is also invited to read the quick guide called 'An Introduction to R' that comes with every installed version of R or the introductory book on the R environment by Dalgaard (2008).

A.1 Something to know first about R

All commands in R must be typed via the command-line interface, and graphical user interfaces (GUI) are very limited, with a few exceptions, such as the package Rcmdr by John Fox.

All the commands are given as inputs to R after the prompt > (R> in this book) and are analyzed by the R parser after the user presses the 'return'/'enter' key (or a newline character is encountered in the case of a script file).

```
R> cat("help me!")

help me!
```

R inputs can be multiline; hence, if the R parser thinks that the user did not complete some command (because of unbalanced parentheses or quotation marks), on the next line a + symbol will appear instead of a prompt.

```
R> cat("help me!"
+
```

This can be quite frustrating for novice users, so it is better to know how to exit from this impasse. Depending on the implementation of R usually pressing CTRL+C or ESC on the keyboard helps. Otherwise, for GUI versions of R, pushing

Option Pricing and Estimation of Financial Models with R, First Edition. Stefano M. Iacus.
© 2011 John Wiley & Sons, Ltd. Published 2011 by John Wiley & Sons, Ltd.

the 'stop' button of the R console will exit the parser. Of course, another solution is to complete the command (with a ')' in our example).

A.1.1   The workspace

Almost every command in R creates an object and not just text output, and objects live in the workspace. The workspace, or groups of objects, can be saved and loaded into R with the `save.image` (or `save`) and `load` commands, respectively. The user is prompted about saving workspace when exiting from R. This workspace is saved in the current directory as a hidden (on some operating systems) file named `.RData` and reloaded automatically the next time R is started.

A.1.2   Graphics

Usually R graphics are displayed on a device that corresponds to a window for a GUI version of R (for example, under MS-Windows, X11, or Mac OS X). Otherwise a Postscript file `Rplots.ps` is generated in the current working directory. Sometimes, in interactive uses of R, it is useful to use `par(ask = TRUE)` to pause R at each new plot or `par(ask = FALSE)` to avoid such pauses. We do not discuss the multiple R graphic systems here but the reader can refer to Murrel (2005) and Deepayan (2008).

A.1.3   Getting help

The casual user will find it very hard to get started without prior knowledge of which command is needed to perform a particular task. The help system is not that useful either. But the R system is thought in a way that every command has its help page and documentation always matches the actual implementation of the command. To get information about a particular command one should use the `help`, like `help(load)` or `?help`. For some special operators, the user should specify the argument like this: `?"for"`, `?"+"`, etc. When the documentation contains the section 'Examples' with R code inside, the code from that page can be executed automatically with `example(topic)`, where `topic` is the corresponding R command of interest, try e.g. `example(plot)`.

In case one wants to execute a fuzzy search on the help system, one can use the command `help.search("topic")` and R will return several options which partly match the word `topic`, try e.g. `help.search("regression")`. It is also possible to extend the search for a term or for more complicated queries to the Web using `RSiteSearch`, try e.g. `RSiteSearch("nonlinear regression")`. The search will be extended to all documentation pages for packages in the R repository and to all the pertinent mailing lists.[1] The web site for R mailing lists and related projects is http://stat.ethz.ch/mailman/listinfo.

---

[1] If you have a question, have a look at the mailing list archives first.

There is a rich repository of quick guides or electronic books on R and its use in different disciplines which can be found under the section 'Documentation/Contributed' on CRAN. The direct link to the page is http://cran.r-project.org/other-docs.html. Finally, we mention the 'Task Views', which are collections of R packages organized by macro areas, e.g. 'Finance'. These are again hosted on CRAN and the direct link is http://cran.r-project.org/web/views.

A.1.4 Installing packages

This book, like many others, requires several add-on packages which are not distributed with the basic R system. The main repository for R packages is called CRAN 'The Comprehensive R Archive Network' and its main address is http://cran.r-project.org/. To install a package in the R system, one should use the command `install.packages` with a package name as argument, i.e. `install.packages("sde")` to install the package sde. R GUIs usually offer some option to get the list of all packages at the repository (around 2000) and install those selected by point and click actions.

Another important source of R packages is the R-Forge repository. It is a repository mainly for developers but where users can also find pre-release of developer versions of packages already on CRAN or even packages not necessarily hosted on CRAN. The home page of R-Forge is http://r-forge.r-project.org. To install a package from R-Forge one can use a command like

```
R> install.packages("sde",repos="http://R-Forge.R-project.org")
```

i.e. adding to `install.packages` the option `repos` with a proper web address.

A.2   Objects

As mentioned, most functions in R return objects rather than text output. Clearly objects can be created anew with commands. We now describe how to create, inspect and manipulate objects.

A.2.1   Assignments

To create an object it is necessary to use the operator '<-' which has the meaning 'assign the right-hand side to the left-hand side', or use the more common operator ' = ', as in the following lines in which we create an object named x and assign the number 4 to it:

```
R> x <- 4
R> x = 4
```

Similarly, one can use the operator '->' which assigns the left-hand side to the right-hand side, i.e. x -> 4. The following command creates a more interesting vector `y` containing the numbers 2, 7, 4, and 1 concatenated in a single object using the function `c()`:

```
R> y <- c(2, 7, 4, 1)
R> y

[1] 2 7 4 1
```

A matrix can be created using the `matrix` command

```
R> z <- matrix(1:30, 5, 6)
R> z

     [,1] [,2] [,3] [,4] [,5] [,6]
[1,]    1    6   11   16   21   26
[2,]    2    7   12   17   22   27
[3,]    3    8   13   18   23   28
[4,]    4    9   14   19   24   29
[5,]    5   10   15   20   25   30
```

Where `1:30` produces a sequence from 1 to 30 by unitary step, i.e.

```
R> 1:30

 [1]  1  2  3  4  5  6  7  8  9 10 11 12 13 14 15 16 17 18 19 20
     21 22 23 24 25
[26] 26 27 28 29 30
```

The command `matrix` requires at least three arguments, where the second and third are the number of rows and columns and the first one is an object which is used recursively to fill the elements of the matrix. Of course, we can create an empty matrix with

```
R> matrix(, 5, 6)

     [,1] [,2] [,3] [,4] [,5] [,6]
[1,]   NA   NA   NA   NA   NA   NA
[2,]   NA   NA   NA   NA   NA   NA
[3,]   NA   NA   NA   NA   NA   NA
[4,]   NA   NA   NA   NA   NA   NA
[5,]   NA   NA   NA   NA   NA   NA
```

where NA is the R symbol for the missing values, or empty numerical vectors with `numeric`

```
R> numeric(4)

[1] 0 0 0 0
```

The command `ls()` shows the current content of the workspace

```
R> ls()
```

```
[1] "x" "y" "z"
```

Notice that all objects which are created but not assigned, are not kept in the workspace. Like `numeric`, there are several command to allocate objects for the different data types available in R, i.e. we have functions like `integer`, `character`, etc. or use the function `vector` as follows:

```
R> vector(mode = "numeric", 4)
```

```
[1] 0 0 0 0
```

is equivalent to `numeric(4)`. Objects can also have length zero

```
R> w1 <- numeric(0)
R> w1
```

```
numeric(0)
```

or can be initialized as NULL

```
R> w2 <- NULL
R> w2
```

```
NULL
```

which is useful if one wants to enlarge these objects later in subsequent tasks. In the above, `w1` and `w2` are objects of different types and, in particular, `w1` is an object of class `numeric` while `w2` is not. It is also possible to use the command `assign` to create objects and this is sometimes useful when the name of the object has to be created dynamically in the R code. The following is an example of use in which `o1` is created as before and `o2` is created via `assign`

```
R> O1 <- 1:4
R> O1
```

```
[1] 1 2 3 4
```

```
R> ls()
```

```
[1] "O1" "w1" "w2" "x"   "y"   "z"
```

```
R> assign("O2", 5:8)
R> ls()
```

```
[1] "O1" "O2" "w1" "w2" "x"   "y"   "z"
```

```
R> O2
```

```
[1] 5 6 7 8
```

A.2.2   Basic object types

Objects classes can be created from scratch in the R language, and this is usually the case for many R packages, but the basic classes are integer, numeric, complex, character, etc. which can be aggregated in vectors, matrixes, arrays or lists. While vectors, arrays and lists contain elements all of the same type, the lists are more general and can contain objects of different size and type but in addition can also be nested. For example, the following code loads a data set and estimates a linear model via lm and assign the result to an object mod. The statistical analysis per se is not relevant here, we just notice that an estimated regression model in R is not just an output of coefficients with their significance, but an object

```
R> data(cars)
R> mod <- lm(dist ~ speed, data = cars)
R> mod

Call:
lm(formula = dist ~ speed, data = cars)

Coefficients:
(Intercept)        speed
    -17.579        3.932
```

We now look at the structure of the object created by the linear regression using the command str which inspects the structure of the object.

```
R> str(mod)

List of 12
 $ coefficients : Named num [1:2] -17.58 3.93
  ..- attr(*, "names")= chr [1:2] "(Intercept)" "speed"
 $ residuals    : Named num [1:50] 3.85 11.85 -5.95 12.05 2.12 ...
  ..- attr(*, "names")= chr [1:50] "1" "2" "3" "4" ...
 $ effects      : Named num [1:50] -303.914 145.552 -8.115
   9.885 0.194 ...
  ..- attr(*, "names")= chr [1:50] "(Intercept)" "speed" "" "" ...
 $ rank         : int 2
 $ fitted.values: Named num [1:50] -1.85 -1.85 9.95 9.95 13.88 ...
  ..- attr(*, "names")= chr [1:50] "1" "2" "3" "4" ...
 $ assign       : int [1:2] 0 1
 $ qr           :List of 5
  ..$ qr   : num [1:50, 1:2] -7.071 0.141 0.141 0.141 0.141 ...
  .. ..- attr(*, "dimnames")=List of 2
  .. .. ..$ : chr [1:50] "1" "2" "3" "4" ...
  .. .. ..$ : chr [1:2] "(Intercept)" "speed"
  .. ..- attr(*, "assign")= int [1:2] 0 1
  ..$ qraux: num [1:2] 1.14 1.27
  ..$ pivot: int [1:2] 1 2
  ..$ tol  : num 1e-07
  ..$ rank : int 2
  ..- attr(*, "class")= chr "qr"
```

```
$ df.residual  : int 48
$ xlevels      : list()
$ call         : language lm(formula = dist ~ speed, data = cars)
$ terms        :Classes 'terms', 'formula' length 3 dist ~ speed
 .. ..- attr(*, "variables")= language list(dist, speed)
 .. ..- attr(*, "factors")= int [1:2, 1] 0 1
 .. .. ..- attr(*, "dimnames")=List of 2
 .. .. .. ..$ : chr [1:2] "dist" "speed"
 .. .. .. ..$ : chr "speed"
 .. ..- attr(*, "term.labels")= chr "speed"
 .. ..- attr(*, "order")= int 1
 .. ..- attr(*, "intercept")= int 1
 .. ..- attr(*, "response")= int 1
 .. ..- attr(*, ".Environment")=<environment: R_GlobalEnv>
 .. ..- attr(*, "predvars")= language list(dist, speed)
 .. ..- attr(*, "dataClasses")= Named chr [1:2] "numeric"
  "numeric"
 .. .. ..- attr(*, "names")= chr [1:2] "dist" "speed"
$ model        :'data.frame':   50 obs. of  2 variables:
 ..$ dist : num [1:50] 2 10 4 22 16 10 18 26 34 17 ...
 ..$ speed: num [1:50] 4 4 7 7 8 9 10 10 10 11 ...
 ..- attr(*, "terms")=Classes 'terms', 'formula' length 3
  dist ~ speed
 .. .. ..- attr(*, "variables")= language list(dist, speed)
 .. .. ..- attr(*, "factors")= int [1:2, 1] 0 1
 .. .. .. ..- attr(*, "dimnames")=List of 2
 .. .. .. .. ..$ : chr [1:2] "dist" "speed"
 .. .. .. .. ..$ : chr "speed"
 .. .. ..- attr(*, "term.labels")= chr "speed"
 .. .. ..- attr(*, "order")= int 1
 .. .. ..- attr(*, "intercept")= int 1
 .. .. ..- attr(*, "response")= int 1
 .. .. ..- attr(*, ".Environment")=<environment: R_GlobalEnv>
 .. .. ..- attr(*, "predvars")= language list(dist, speed)
 .. .. ..- attr(*, "dataClasses")= Named chr [1:2] "numeric"
  "numeric"
 .. .. .. ..- attr(*, "names")= chr [1:2] "dist" "speed"
 - attr(*, "class")= chr "lm"
```

For the above we see that mod is essentially a list object of 12 elements and it is of class 'lm' (for linear models). For example, the first one is called coefficients and can be accessed using the symbol $ as follows:

```
R> mod$coefficients
```

```
(Intercept)       speed
 -17.579095    3.932409
```

```
R> str(mod$coefficients)
```

```
 Named num [1:2] -17.58 3.93
 - attr(*, "names")= chr [1:2] "(Intercept)" "speed"
```

The vector `coefficients` is a 'named vector'. One can obtain or change the names of the elements of a vector with

```
R> names(mod$coefficients)

[1] "(Intercept)" "speed"
```

or change them with

```
R> names(mod$coefficients) <- c("alpha", "beta")
R> mod$coefficients

    alpha        beta
-17.579095    3.932409
```

Similarly, one can assign or get the names of the rows or the columns of an R matrix

```
R> z

     [,1] [,2] [,3] [,4] [,5] [,6]
[1,]    1    6   11   16   21   26
[2,]    2    7   12   17   22   27
[3,]    3    8   13   18   23   28
[4,]    4    9   14   19   24   29
[5,]    5   10   15   20   25   30

R> rownames(z) <- c("a", "b", "c", "d", "e")
R> colnames(z) <- c("A", "B", "C", "D", "E", "F")
R> z

  A  B  C  D  E  F
a 1  6 11 16 21 26
b 2  7 12 17 22 27
c 3  8 13 18 23 28
d 4  9 14 19 24 29
e 5 10 15 20 25 30
```

As anticipated, lists can be nested. For example, the object `model` inside `mod` is itself a list

```
R> str(mod$model)

'data.frame':   50 obs. of  2 variables:
 $ dist : num  2 10 4 22 16 10 18 26 34 17 ...
 $ speed: num  4 4 7 7 8 9 10 10 10 11 ...
 - attr(*, "terms")=Classes 'terms', 'formula' length 3
   dist ~ speed
   .. ..- attr(*, "variables")= language list(dist, speed)
   .. ..- attr(*, "factors")= int [1:2, 1] 0 1
   .. .. ..- attr(*, "dimnames")=List of 2
   .. .. .. ..$ : chr [1:2] "dist" "speed"
```

```
.. .. .. ..$ : chr "speed"
.. ..- attr(*, "term.labels")= chr "speed"
.. ..- attr(*, "order")= int 1
.. ..- attr(*, "intercept")= int 1
.. ..- attr(*, "response")= int 1
.. ..- attr(*, ".Environment")=<environment: R_GlobalEnv>
.. ..- attr(*, "predvars")= language list(dist, speed)
.. ..- attr(*, "dataClasses")= Named chr [1:2] "numeric"
  "numeric"
.. .. ..- attr(*, "names")= chr [1:2] "dist" "speed"
```

or, more precisely a data.frame which is essentially a list with the property
that all the elements have the same length. The data.frame object is used to
store data sets, like the cars data set

```
R> str(cars)

'data.frame':   50 obs. of  2 variables:
 $ speed: num  4 4 7 7 8 9 10 10 10 11 ...
 $ dist : num  2 10 4 22 16 10 18 26 34 17 ...
```

and it is assumed that the elements of a data.frame correspond to variables,
while the length of each object is the same as the sample size.

A.2.3   Accessing objects and subsetting

We have seen that $ can be used to access the elements of a list and hence of
a data.frame, but R also offer operators for enhanced subsetting. The first one
is [which returns an object of the same type of the original object

```
R> y

[1] 2 7 4 1

R> y[2:3]

[1] 7 4

R> str(y)

 num [1:4] 2 7 4 1

R> str(y[2:3])

 num [1:2] 7 4
```

or, for matrix-like objects

```
R> z

  A  B  C  D  E  F
a 1  6 11 16 21 26
```

```
b 2   7 12 17 22 27
c 3   8 13 18 23 28
d 4   9 14 19 24 29
e 5 10 15 20 25 30
```

```
R> z[1:2, 5:6]
```

```
   E  F
a 21 26
b 22 27
```

and subsetting can occur also on nonconsecutive indexes

```
R> z[1:2, c(1, 3, 6)]
```

```
  A  C  F
a 1 11 26
b 2 12 27
```

or in different order

```
R> z[1:2, c(6, 5, 4)]
```

```
   F  E  D
a 26 21 16
b 27 22 17
```

One can subset objects also using names, e.g.

```
R> z[c("a", "c"), "D"]
```

```
 a  c
16 18
```

We can also use a syntax like

```
R> z["c", ]
```

```
 A  B  C  D  E  F
 3  8 13 18 23 28
```

leaving one argument out to mean 'run all the elements' for that index. Further, R allows for negative indexes which are used to exclude indexes

```
R> z[c(-1, -3), ]
```

```
  A  B  C  D  E  F
b 2  7 12 17 22 27
d 4  9 14 19 24 29
e 5 10 15 20 25 30
```

but positive and negative indexes cannot be mixed.

The subsetting operator [also works for lists

```
R> a <- mod[1:2]
R> str(a)

List of 2
 $ coefficients: Named num [1:2] -17.58 3.93
  ..- attr(*, "names")= chr [1:2] "alpha" "beta"
 $ residuals   : Named num [1:50] 3.85 11.85 -5.95 12.05 2.12 ...
  ..- attr(*, "names")= chr [1:50] "1" "2" "3" "4" ...
```

where we have extracted the first two elements of the list mod using mod[1:2]. We can use names as well and the commands below return the same objects

```
R> str(mod["coefficients"])

List of 1
 $ coefficients: Named num [1:2] -17.58 3.93
  ..- attr(*, "names")= chr [1:2] "alpha" "beta"

R> str(mod[1])

List of 1
 $ coefficients: Named num [1:2] -17.58 3.93
  ..- attr(*, "names")= chr [1:2] "alpha" "beta"
```

Notice that mod[1] returns a list with one element but not just the element inside the list. For this purpose one should use the subsetting operator [[. The next group of commands returns the element inside the list

```
R> str(mod[[1]])

 Named num [1:2] -17.58 3.93
 - attr(*, "names")= chr [1:2] "alpha" "beta"

R> str(mod[["coefficients"]])

 Named num [1:2] -17.58 3.93
 - attr(*, "names")= chr [1:2] "alpha" "beta"

R> str(mod$coefficients)

 Named num [1:2] -17.58 3.93
 - attr(*, "names")= chr [1:2] "alpha" "beta"
```

We have mentioned that a data.frame looks like a particular list, but with more structure and used to store data sets. The latter are always thought as matrixes and indeed it is possible to access the elements of a data.frame using the subsetting rules for matrixes, i.e.

```
R> cars[, 1]
```

```
[1]    4   4   7   7   8   9  10  10  10  11  11  12  12  12  12  13  13  13  13  14
    14  14  14  15  15
[26]  15  16  16  17  17  17  18  18  18  18  19  19  19  20  20  20  20  20  22  23
    24  24  24  24  25
```

```
[1]    4   4   7   7   8   9  10  10  10  11  11  12  12  12  12  13  13  13  13  14
    14  14  14  15  15
[26]  15  16  16  17  17  17  18  18  18  18  19  19  19  20  20  20  20  20  22  23
    24  24  24  24  25
```

which is equivalent to the following

```
R> cars$speed
```

```
[1]    4   4   7   7   8   9  10  10  10  11  11  12  12  12  12  13  13  13  13  14
    14  14  14  15  15
[26]  15  16  16  17  17  17  18  18  18  18  19  19  19  20  20  20  20  20  22  23
    24  24  24  24  25
```

```
R> cars[, 1]
```

```
[1]    4   4   7   7   8   9  10  10  10  11  11  12  12  12  12  13  13  13  13  14
    14  14  14  15  15
[26]  15  16  16  17  17  17  18  18  18  18  19  19  19  20  20  20  20  20  22  23
    24  24  24  24  25
```

Notice that only the output is not a data.frame while[2]

```
R> str(cars[1])
```

```
'data.frame':    50 obs. of  1 variable:
 $ speed: num  4 4 7 7 8 9 10 10 10 11 ...
```

```
R> head(cars[1])
```

```
  speed
1     4
2     4
3     7
4     7
5     8
6     9
```

is a proper (sub) data.frame although the matrix-like subsetting operator as a different behaviour if used on columns or rows: cars[,1] returns the element but, for example,

```
R> cars[1:3, ]
```

```
  speed dist
1     4    2
2     4   10
3     7    4
```

returns a data.frame with the selected number of rows and all columns.

---

[2] The commands head and tail show the first and last rows of a data.frame respectively.

A.2.4   Coercion between data types

Functions like `names`, `colnames`, but also `levels`, `attributes`, etc. are used to
retrieve and set properties of objects and are called accessor functions. Objects
can be transformed from one type to another using functions with names `as.*`.
For example, `as.integer` transforms an object into an integer whenever possible
or eventually return a missing value

```
R> pi
```

```
[1] 3.141593
```

```
R> as.integer(pi)
```

```
[1] 3
```

```
R> as.integer("3.14")
```

```
[1] 3
```

```
R> as.integer("a")
```

```
[1] NA
```

Other examples are `as.data.frame` to transform `matrix` objects into true
`data.frame` objects and vice versa. For more complex classes one can also try
the generic function `as`.

A.3   S4 objects

We have used several times the term 'class' for R objects. This is because each
object in R belongs to some class and for each class there exist generic functions
called methods which perform some task on that object. For example, the function
`summary` provide summary statistics which are appropriate for some object

```
R> summary(cars)
```

```
     speed           dist
 Min.   : 4.0   Min.   :  2.00
 1st Qu.:12.0   1st Qu.: 26.00
 Median :15.0   Median : 36.00
 Mean   :15.4   Mean   : 42.98
 3rd Qu.:19.0   3rd Qu.: 56.00
 Max.   :25.0   Max.   :120.00
```

```
R> summary(mod)
```

```
Call:
lm(formula = dist ~ speed, data = cars)

Residuals:
```

```
     Min        1Q  Median        3Q      Max
 -29.069    -9.525  -2.272    9.215   43.201

Coefficients:
        Estimate Std. Error t value Pr(>|t|)
alpha -17.5791      6.7584  -2.601    0.0123 *
beta    3.9324      0.4155   9.464 1.49e-12 ***
---
Signif. codes:  0 '***' 0.001 '**' 0.01 '*' 0.05 '.' 0.1 ' ' 1

Residual standard error: 15.38 on 48 degrees of freedom
Multiple R-squared: 0.6511, Adjusted R-squared: 0.6438
F-statistic: 89.57 on 1 and 48 DF,  p-value: 1.490e-12
```

The standard set of classes and methods in R is called S3. In this framework, a method for an object of some class is simply an R function named method.class, e.g. summary.lm is the function which is called by R when the function summary is called with an argument which is an object of class lm. In R methods like summary are very generic and the function methods provides a list of specific methods (which apply to specific types of objects) for some particular method. For example

```
R> methods(summary)

 [1] summary.aov           summary.aovlist       summary.aspell*
 [4] summary.connection    summary.data.frame    summary.Date
 [7] summary.default       summary.ecdf*         summary.factor
[10] summary.glm           summary.infl          summary.lm
[13] summary.loess*        summary.manova        summary.matrix
[16] summary.mlm           summary.nls*          summary.
                                                    packageStatus*
[19] summary.POSIXct       summary.POSIXlt       summary.ppr*
[22] summary.prcomp*       summary.princomp*     summary.stepfun
[25] summary.stl*          summary.table         summary.
                                                    tukeysmooth*

   Non-visible functions are asterisked
```

The dot '.' naming convention is quite unhappy because one can artificially create functions which are not proper methods, for example the t.test function is not the method t for objects of class test but it is just an R function which performs ordinary two-samples t test. The new system of classes and methods which is now fully implemented in R is called S4. Objects of class S4 apparently behave like all other objects in R but they possess properties called 'slots', which can be accessed differently from other R objects. The next code estimates the maximum likelihood estimator for the mean of a Gaussian law. It uses the function mle from the package stats4 which is an S4 package as the name suggests. Again, we are interested in the statistical part of this example

```
R> require(stats4)
R> set.seed(123)
```

```
R> y <- rnorm(100, mean = 1.5)
R> f <- function(theta = 0) -sum(dnorm(x = y, mean = theta,
    log = TRUE))
R> fit <- mle(f)
R> fit

Call:
mle(minuslogl = f)

Coefficients:
    theta
1.590406
```

we now have a look at the object `fit` returned by the `mle` function

```
R> str(fit)

Formal class 'mle' [package "stats4"] with 8 slots
  ..@ call     : language mle(minuslogl = f)
  ..@ coef     : Named num 1.59
  .. ..- attr(*, "names")= chr "theta"
  ..@ fullcoef : Named num 1.59
  .. ..- attr(*, "names")= chr "theta"
  ..@ vcov     : num [1, 1] 0.01
  .. ..- attr(*, "dimnames")=List of 2
  .. .. ..$ : chr "theta"
  .. .. ..$ : chr "theta"
  ..@ min      : num 133
  ..@ details  :List of 6
  .. ..$ par      : Named num 1.59
  .. .. ..- attr(*, "names")= chr "theta"
  .. ..$ value    : num 133
  .. ..$ counts   : Named int [1:2] 6 3
  .. .. ..- attr(*, "names")= chr [1:2] "function" "gradient"
  .. ..$ convergence: int 0
  .. ..$ message  : NULL
  .. ..$ hessian  : num [1, 1] 100
  .. .. ..- attr(*, "dimnames")=List of 2
  .. .. .. ..$ : chr "theta"
  .. .. .. ..$ : chr "theta"
  ..@ minuslogl:function (theta = 0)
  ..@ method   : chr "BFGS"
```

We now see that this is an S4 objects with slots that, as the structure suggests, can be accessed using the symbol @ instead of $. For example,

```
R> fit@coef

    theta
1.590406
```

To get the list of methods for S4 objects one should use the function `showMethods`
```
R> showMethods(summary)
```

```
Function: summary (package base)
object="ANY"
object="mle"
```

A.4   Functions

In the previous section we have created a new function called f to define the
log-likelihood of the data. In R functions are created with the command function
followed by a list of arguments and the body of the function (if longer than one
line) has to be contained within '{' and '}' like in the next example in which
we define the payoff function of a call option

```
R> g <- function(x, K = 110) {
+     max(x - K, 0)
+ }
```

The function returns the last calculation unless the command return is used.
By default, in the function g we have set the strike price K = 100 and x is the
argument which represents the price of the underlying asset.

```
R> g(120)

[1] 10

R> g(99)

[1] 0

R> g(115, 120)

[1] 0
```

In R arguments are always named, so the function can be called with arguments
in any order if named, e.g.

```
R> g(150, 120)

[1] 30

R> g(K = 120, x = 150)

[1] 30
```

In the definition of g we have fixed a default value for the argument K to 100,
so if it is missing in a call, it is replaced by R with its default value. The
argument x cannot be omitted, therefore a call like g(K = 120) will produce
an error.

A.5   Vectorization

Most of R functions are vectorized, which means that if a vector is passed to a function, the function is applied to each element of the function and a vector of results is returned as in the next example:

```
R> set.seed(123)
R> x <- runif(5, 90, 150)
R> x
```

```
[1] 107.2547 137.2983 114.5386 142.9810 146.4280
```

```
R> sin(x)
```

```
[1]   0.4263927 -0.8026760  0.9916244 -0.9992559  0.9414204
```

But functions should be prepared to be vectorized. For example, our function g is not vectorized:

```
R> g(x)
```

```
[1] 36.42804
```

Indeed, in the body of g the function max is used and it operates as follows: first x-K is calculated:

```
R> x - 100
```

```
[1]   7.254651 37.298308 14.538615 42.981044 46.428037
```

and then the max calculates the maximum of the vector c(x-100, 0). To vectorize it we can use the function sapply as follows:

```
R> g1 <- function(x, K = 110) {
+       sapply(x, function(x) max(x - K, 0))
+ }
R> x
```

```
[1] 107.2547 137.2983 114.5386 142.9810 146.4280
```

```
R> g1(x)
```

```
[1]   0.000000 27.298308  4.538615 32.981044 36.428037
```

and we get five different payoffs. The functions of class *apply are designed to work iteratively on different objects. The function sapply iterates the vector in the first argument and applies the functions in the second argument. The function apply works on arrays (e.g. matrixes), lapply iterates over list's, etc.

The usual `for` and `while` constructs exist in R as well, but their use should be limited to real iterative tasks which cannot be parallelized as in our example. A `for` version of the function `g` can be the following:

```
R> g2 <- function(x, K = 110) {
+      n <- length(x)
+      val <- numeric(n)
+      for (i in 1:n) val[i] <- max(x[i] - K, 0)
+      val
+ }
```

or, in a more R-like fashion, as follows:

```
R> g3 <- function(x, K = 110) {
+      val <- NULL
+      for (u in x) val <- c(val, max(u - K, 0))
+      val
+ }
R> g1(x)
```

```
[1]   0.000000 27.298308   4.538615 32.981044 36.428037
```

```
R> g2(x)
```

```
[1]   0.000000 27.298308   4.538615 32.981044 36.428037
```

```
R> g3(x)
```

```
[1]   0.000000 27.298308   4.538615 32.981044 36.428037
```

The vectorized versions are usually faster then the ones iterated using `for` loops:

```
R> y <- runif(10000, 90, 150)
R> system.time(g1(y))
```

```
   user   system  elapsed
  0.034    0.001    0.035
```

```
R> system.time(g2(y))
```

```
   user   system  elapsed
  0.051    0.003    0.054
```

```
R> system.time(g3(y))
```

```
   user   system  elapsed
  0.261    0.068    0.344
```

Notice that the function `g3` is particularly inefficient because instead of allocating and assigning the results, it grows the vector `val` dynamically.

A.6 Parallel computing in R

There are several options for parallel computing in R. As usual, each solution has pro and cons, and here we present some options which appear to be the simplest to be used. The first option for the casual user is the package snow (Simple Network of Workstations) by Luke Tierney. This package allows the creation of cross-platform clusters (i.e., the nodes of the cluster may be on different platforms) very easily. The very first application of the package is to exploit the power of today's CPUs which are usually multicore. We assume a dual core machine in the next example, and start a cluster with two nodes, one for each core in our CPU.

```
R> library(snow)
R> cl <- makeSOCKcluster(c("localhost", "localhost"))
```

The cluster has been created over sockets (which are interprocess communication between tasks in operating systems) hence, not particularly efficient and localhost means 'this machine'. It is also possible to start a cluster with

```
R> makeCluster(2)
```

or

```
R> makeCluster(2, type = "SOCK")
```

in an interactive session because, usually, the sockets connection is the standard connection type, so all the above are essentially equivalent. The next code splits the replications of 100 simulated paths of an Ornstein-Uhlembeck process on the two nodes. For this we prepare a function f which simulates a number x of trajectories on each node

```
R> f <- function(x) {
+     require(sde)
+     sde.sim(model = "OU", theta = c(1, 1, 1), M = x)
+ }
```

and apply this function to the nodes of the cluster in this way:

```
R> tmp <- parLapply(cl, c(50, 50), f)
R> str(tmp)

List of 2
 $ : mts [1:101, 1:50] 1 1.136 1.038 0.865 0.724 ...
 ..- attr(*, "dimnames")=List of 2
 .. ..$ : NULL
 .. ..$ : chr [1:50] "X1" "X2" "X3" "X4" ...
 ..- attr(*, "tsp")= num [1:3] 0 1 100
 ..- attr(*, "class")= chr [1:2] "mts" "ts"
 $ : mts [1:101, 1:50] 1 0.861 0.693 0.653 0.705 ...
```

```
..- attr(*, "dimnames")=List of 2
.. ..$ : NULL
.. ..$ : chr [1:50] "X1" "X2" "X3" "X4" ...
..- attr(*, "tsp")= num [1:3] 0 1 100
..- attr(*, "class")= chr [1:2] "mts" "ts"
```

and we should not forget to stop the cluster

```
R> stopCluster(cl)
```

Compare the result with a single call

```
R> tmp <- f(100)
```

To check the errata corrige of the book, type
vignette("sde.errata")

```
R> str(tmp)
```

```
 mts [1:101, 1:100] 1 0.963 0.88 0.93 0.912 ...
 - attr(*, "dimnames")=List of 2
  ..$ : NULL
  ..$ : chr [1:100] "X1" "X2" "X3" "X4" ...
 - attr(*, "tsp")= num [1:3] 0 1 100
 - attr(*, "class")= chr [1:2] "mts" "ts"
```

When doing parallel computing it is necessary to take care that the random number generators provide independent streams of numbers on each node, otherwise a typical Monte Carlo analysis can be invalidated. The package rlecuyer can provide such functionality to R and can be used directly in a cluster created with snow via the function clusterSetupRNG. The following is an example of use:

```
R> cl <- makeSOCKcluster(c("localhost", "localhost"))
R> clusterSetupRNG(cl, seed = rep(123, 2))

[1] "RNGstream"

R> tmp <- parLapply(cl, c(50, 50), f)
R> stopCluster(cl)
```

The socket method is not very efficient if there is the need to pass big amount of data. The package snow is able to run smoothly clusters using other methods like MPI (Message-Passing Interface) via the package Rmpi which supports MPICH, MPICH2, LAM-MPI, Deino MPI and Open MPI, or PVM (Parallel Virtual Machine) via the package rpvm or the NWS (NetWorkSpaces) via the nws package. All these options are more powerful than socket connections, but require a bit of fine tuning of the hardware which cannot be discussed here. More details can be found in Rossini *et al.* (2007, 2008). The packages Rmpi, rpvm and nws all work independently of the package snow. The package snowfall although based on snow, has some additional functionalities which make

scripting of parallelized R code easier and also allow us to write the same code
with or without parallelization using a simple switch. It is also better in that
it creates global variables that are passed through all nodes when the cluster
is started and offer more effective debug tools. We provide an example of use
of previous code. The cluster is started and stopped using sfInit and sfStop
respectively.

```
R> require(snowfall)
R> sfInit(parallel = TRUE, cpus = 2)

R Version:   R version 2.10.1 (2009-12-14)

R> cl2 <- sfGetCluster()
R> clusterSetupRNG(cl2, seed = rep(123, 2))

[1] "RNGstream"

R> tmp <- parLapply(cl2, c(50, 50), f)
R> sfStop()
R> str(tmp)
List of 2
 $ : mts [1:101, 1:50] 1 0.985 1.044 1.118 1.104 ...
  ..- attr(*, "dimnames")=List of 2
  .. ..$ : NULL
  .. ..$ : chr [1:50] "X1" "X2" "X3" "X4" ...
  ..- attr(*, "tsp")= num [1:3] 0 1 100
  ..- attr(*, "class")= chr [1:2] "mts" "ts"
 $ : mts [1:101, 1:50] 1 1.052 1.001 0.994 0.983 ...
  ..- attr(*, "dimnames")=List of 2
  .. ..$ : NULL
  .. ..$ : chr [1:50] "X1" "X2" "X3" "X4" ...
  ..- attr(*, "tsp")= num [1:3] 0 1 100
  ..- attr(*, "class")= chr [1:2] "mts" "ts"
```

Notice that the code to execute commands on the node of the cluster is the same
as in snow. Other packages implement implicit parallelization in which the code
is automatically distributed to the different nodes of the cluster (or different CPUs
like the multicore package). For an updated review we suggest you look at the
Task View on CRAN named 'HighPerformanceComputing'.

A.6.1   The foreach approach

A particular attention merits the foreach package. This package automatically
distributes the parallelized tasks to the nodes of a cluster in a way that makes
writing of the code quite simple. The advantage is also that, when the cluster
is not available, the code still works but runs sequentially. We start by
writing a simple code which makes use of the foreach command and the
dopar operator

```
R> require(foreach)
R> set.seed(123)
R> tmp <- foreach(i = rep(50, 2)) %dopar% f(i)
R> str(tmp)

List of 2
 $ : mts [1:101, 1:50] 1 0.944 0.97 0.9 0.979 ...
  ..- attr(*, "dimnames")=List of 2
  .. ..$ : NULL
  .. ..$ : chr [1:50] "X1" "X2" "X3" "X4" ...
  ..- attr(*, "tsp")= num [1:3] 0 1 100
  ..- attr(*, "class")= chr [1:2] "mts" "ts"
 $ : mts [1:101, 1:50] 1 0.951 0.69 0.718 0.658 ...
  ..- attr(*, "dimnames")=List of 2
  .. ..$ : NULL
  .. ..$ : chr [1:50] "X1" "X2" "X3" "X4" ...
  ..- attr(*, "tsp")= num [1:3] 0 1 100
  ..- attr(*, "class")= chr [1:2] "mts" "ts"

R> getDoParWorkers()

[1] 1
```

in the above case, the `foreach` command executes several times the function `f` with argument `i` varying in the set `c(50,50)`. So it calls, sequentially two times `f`. The function `getDoParWorkers` tells us the number of nodes in the cluster. With the `foreach` approach, a cluster can be a rather generic one as in the **snow** package. As before, we first set up a cluster using the **snowfall** package:

```
R> require(snowfall)
R> sfInit(parallel = TRUE, cpus = 2)
R> cl2 <- sfGetCluster()
R> clusterSetupRNG(cl2, seed = rep(123, 2))

[1] "RNGstream"
```

Then, we load the `foreach` package and and `doSNOW` package which is used to inform the **foreach** package which parallel back end should be used by `dopar`. This is done using the function `registerDoSNOW`

```
R> require(foreach)
R> require(doSNOW)
R> registerDoSNOW(cl2)
```

For other cluster structures of back ends, the user need to provide their registration functions. Fortunately, there are already several ready to use solutions. We will show an example later. Let us continue and make use of the `foreach` functionality with the new parallel back end

```
R> tmp <- foreach(i = rep(50, 2)) %dopar% f(i)
R> str(tmp)
```

```
List of 2
 $ : mts [1:101, 1:50] 1 0.985 1.044 1.118 1.104 ...
 ..- attr(*, "dimnames")=List of 2
 .. ..$ : NULL
 .. ..$ : chr [1:50] "X1" "X2" "X3" "X4" ...
 ..- attr(*, "tsp")= num [1:3] 0 1 100
 ..- attr(*, "class")= chr [1:2] "mts" "ts"
 $ : mts [1:101, 1:50] 1 1.052 1.001 0.994 0.983 ...
 ..- attr(*, "dimnames")=List of 2
 .. ..$ : NULL
 .. ..$ : chr [1:50] "X1" "X2" "X3" "X4" ...
 ..- attr(*, "tsp")= num [1:3] 0 1 100
 ..- attr(*, "class")= chr [1:2] "mts" "ts"
```

```
R> getDoParWorkers()
```

```
[1] 2
```

Now we need to stop the cluster with `sfStop` but also inform `foreach` that parallel executing is no longer possible. We do this by registering the sequential back end with command `registerDoSEQ`

```
R> sfStop()
R> registerDoSEQ()
R> getDoParWorkers()
```

```
[1] 1
```

Now, if we run again the `foreach` statement, it will be executed sequentially.

```
R> tmp <- foreach(i = rep(50, 2)) %dopar% f(i)
```

A similar approach can be done using a cluster created with the **multicore** package along with the **doMC** package to teach `foreach` that a new cluster is in place:

```
R> require(doMC)
R> registerDoMC()
R> options(cores = 2)
R> tmp <- foreach(i = rep(50, 2)) %dopar% f(i)
R> str(tmp)
```

```
List of 2
 $ : mts [1:101, 1:50] 1 0.878 0.855 0.724 0.798 ...
 ..- attr(*, "dimnames")=List of 2
 .. ..$ : NULL
 .. ..$ : chr [1:50] "X1" "X2" "X3" "X4" ...
 ..- attr(*, "tsp")= num [1:3] 0 1 100
 ..- attr(*, "class")= chr [1:2] "mts" "ts"
 $ : mts [1:101, 1:50] 1 1.136 1.03 0.989 0.903 ...
 ..- attr(*, "dimnames")=List of 2
 .. ..$ : NULL
```

```
.. ..$ : chr [1:50] "X1" "X2" "X3" "X4" ...
..- attr(*, "tsp")= num [1:3] 0 1 100
..- attr(*, "class")= chr [1:2] "mts" "ts"

R> getDoParWorkers()

[1] 2

R> registerDoSEQ()
R> getDoParWorkers()

[1] 1
```

Notice that all the above codes have in common the `foreach` statement. So writing code in this way makes the software ready to run against most cluster structures out of the box, provided we have prepared it correctly.

A.6.2   A note of warning on the multicore package

The **multicore** package uses the 'fork' system call to spawn a copy of the current process which performs the computations in parallel. Modern operating systems use the copy-on-write approach which makes this very appealing for parallel computation since only objects modified during the computation will be actually copied and all other memory is directly shared.

However, the copy shares everything including any user interface elements (windows, menus, etc.). This may cause the above example to execute rather slowly if you try it from the R GUI. So the preferred use of **multicore** package is in command line scripts. This appears a trivial statement because intensive computation is usually done in a batch environment, but it is still good to know if you don't want to be disappointed by using **multicore** package in the wrong environment.

A.7   Bibliographical notes

There are many basic books apart from the one mentioned earlier (Dalgaard 2008), such as Crawley (2007), which cover the basic functionalities of the R language. A simple search with the keyword R in on-line book stores will return hundreds of titles. For advanced programming techniques on the standard S language we should mention Chambers (2004) and Venables and Ripley (2000). For S4 programming some recent references are Chambers (2008) and Gentleman (2008). For advanced graphics one should not miss the books of Murrel (2005) and Deepayan (2008).

References

Chambers, J. (2004). *Programming with Data: A Guide to the S Language (revised version)*. Springer, New York.

Chambers, J. (2008). *Software for Data Analysis: Programming with R*. Springer, New York.

Crawley, M. (2007). *The R Book*. Wiley, Chichester.

Dalgaard, P. (2008). *Introductory Statistics with R (2nd Ed.)*. Springer, New York.

Deepayan, S. (2008). *Lattice: Multivariate Data Visualization with R*. Springer, New York.

Gentleman, R. (2008). *R Programming for Bioinformatics*. Chapman & Hall/CRC, Boca Raton.

Murrel, P. (2005). *R Graphics*. Chapman & Hall/CRC, Boca Raton.

Rossini, A., Tierney, L., and Li, N. (2007). Simple parallel statistical computing in R. *Journal of Computational and Graphical Statistics* **16**, 399–420.

Rossini, A., Tierney, L., and Li, N. (2008). Snow: A parallel computing framework for the R system. *International Journal of Parallel Programming* **37**, 78–90.

Venables, W. N. and Ripley, B. D. (2000). *S Programming*. Springer, New York.

Appendix B

R in finance

Finance includes many subfields. In this book we considered only option pricing, econometric estimation and simulation of financial models and analysis of financial data. But finance also includes important fields like trading and portfolio optimization. The family of R packages offers several opportunities in this direction.

For example, the **fPortfolio** package from **Rmetrics** implements Markowitz Portfolio Theory, Mean-Variance Frontiers, Mean-CVaR, etc., but also offers a framework for backtesting analysis. The **fPortfolio** package comes with an additional ebook Würtz *et al.* (2009), which is worth reading. The package **portfolio** focuses on equity portfolio strategies and also implements matching portfolios for benchmark comparisons. Another interesting solution in this direction is the **backtest** package. We should also mention **PerformanceAnalytics** which is a library of functions designed for evaluating the performance and risk characteristics of financial assets or funds. Another growing library of packages is dedicated to trading. We mention just a few: **fTrading** for basic trading analysis; **TTR** to construct technical trading rules and **ttrTests** for testing these rules; **IBrokers** which is a set of API to interact with the Interactive Brokers Trader Workstation. For risk management analysis, apart from a combination of the above mentioned packages, one can check the **VaR** package for Value-at-Risk analysis and the **CreditMetrics** which implements the CreditMetrics risk model functionalities. For more information on other R tools for finance, we suggest looking at the TaskView on Finance available at `http://cran.r-project.org/web/views/Finance.html`. We now focus on what is strictly related to this book.

B.1 Overview of existing R frameworks

Although the R community has moved only recently to finance, there is a growing number of tools appearing every day. We briefly describe here two large collections which originate from different approaches.

Option Pricing and Estimation of Financial Models with R, First Edition. Stefano M. Iacus.
© 2011 John Wiley & Sons, Ltd. Published 2011 by John Wiley & Sons, Ltd.

B.1.1   Rmetrics

The first large suite of tools is the **Rmetrics** library which we have already used several times in the book, for example in Chapters 6 and 7. The Rmetrics project is now supported by the nonprofit Rmetrics Association which also publishes several ebooks on, but not limited to, finance. The packages of this collection were initially designed for teaching purposes but later evolved in the direction of being a more operational suite of tools. Of direct interest to the reader of this book are the packages:

- **fBasics**: which is a collection of functions to explore and investigate basic properties of financial returns and related quantities. The fields covered include techniques of explorative data analysis and the investigation of distributional properties, including parameter estimation and hypothesis testing;

- **fOptions**: a library of function for the pricing of basic European and American put and call options;

- **fAsianOptions**: includes different approximation methods for the pricing of Asian option in the Black and Scholes model;

- **fExoticOptions**: standard Asian option pricing, as well as barrier options, binary options, lookback options, etc.

- **timeDate**: a framework for chronological and calendar objects which we will discuss below.

This small list is only a part of the suite specifically focused on option pricing.

B.1.2   RQuantLib

The **RQuantLib** is an R interface to the QuantLib C++ library from the homonymous project. The open source QuantLib project[1] aims to provide a comprehensive software framework for quantitative finance. It is a low level framework written in C++. The loading of the **RQuantLib** package for the average R user is not easy, because it requires a working GCC compiler, the preliminary installation of the boost framework and finally the installation of the quantlib library. These steps are made easier for users of Debian OS, but in general they require the usual skills of installing applications from source code. For this reason, we haven't made use of these functionalities in this book. Nevertheless, we think it is worth signaling the **RQuantLib** package because it contains several interesting functionalities, in particular for European, American and Asian option pricing as well as implied volatility analysis. For more informations, we suggest checking the developer's web page at http://dirk.eddelbuettel.com/code/rquantlib.html.

---

[1] http://www.quantlib.org

B.1.3   The `quantmod` package

The package **quantmod** is another framework for quick analysis of financial data. This package implements several methods for plotting data and importing them. It is strictly related to the **xts** package, discussed later, which is one of the time series classes in R and is also related to the **TTR** package for doing on-the-fly calculations of indexes to be plotted on the charts. We will discuss in more detail in Section B.5 the data import functionalities but here we mention a few of the graphical capabilities. The main function is the `chartSeries` which can plot `xts` object in a very effective way. We have used the basic plot several times in this book (see, e.g., Chapter 6). We start by getting the data for the AAPL symbol

```
R> require(quantmod)
R> getSymbols("AAPL")

[1] "AAPL"
```

The function `getSymbols` creates an object of class `xts` in the R workspace, with the same name as the symbol. Figure B.1 represents the basic plot obtained with the following basic call to `chartSeries`:

```
R> chartSeries(AAPL, theme = "white", TA = NULL)
```

To this graph one can add several features like the Bolliger bands, the volume of exchanges, as Figure B.2, generated by the next code, shows.

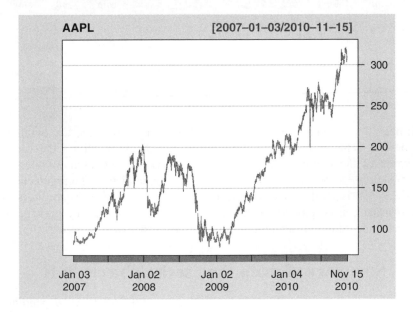

Figure B.1   Basic `chartSeries` plot.

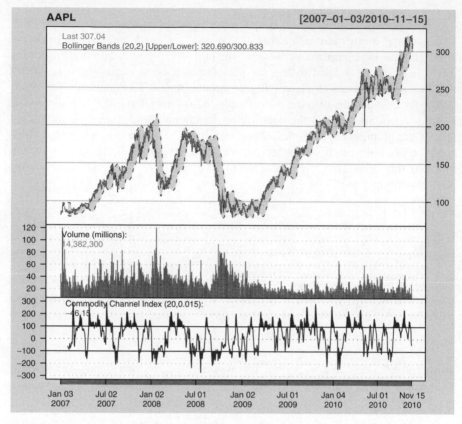

Figure B.2   Advanced `chartSeries` *plot.*

```
R> chartSeries(AAPL, theme = "white", TA = "addVo();addBBands();
   addCCI()")
```

Bollinger Bands, which are added using function `addBBands`, are obtained as the present value of the assets plus or minus two (or sometimes the) standard deviation of the moving average of the last, say, 20 quotations of the assets. The function `addBBands` can be configured in several ways. The argument `TA` in `chartSeries` is used to add technical indicators to the plot. There are several indexes that can be plotted against a `chartSeries` plot and the package **TTR** also provide additional functionalities. The user can write his own `TA` function.

B.2   Summary of main time series objects in R

Financial time series in R can be handled in different ways depending on their nature.

B.2.1   The `ts` class

The basic class of time series objects is the `ts` class. This class has an extension called `mts` for multidimensional times series but, apart from the dimensionality of the data, they share the same properties. The `ts` class is meant for regular time series where observations have a given `frequency` (e.g. 12 for monthly data, 7 for daily data, etc.) and a given time distance between observations `deltat`. When an object of this class is created, one should specify the `start` date and/or the final date `end`. For example, if we want to create a time series of quarterly data starting from the second quarter of 1959 we write something like

```
R> X <- ts(1:10, frequency = 4, start = c(1959, 2))
R> X
```

```
     Qtr1 Qtr2 Qtr3 Qtr4
1959         1    2    3
1960    4    5    6    7
1961    8    9   10
```

If we want to create monthly data starting from July 1954, we write something like

```
R> set.seed(123)
R> X <- ts(cumsum(1 + round(rnorm(100), 2)), start = c(1954, 7),
+       frequency = 12)
R> X
```

	Jan	Feb	Mar	Apr	May	Jun	Jul	Aug	Sep	Oct
1954							0.44	1.21	3.77	4.84
1955	10.15	9.88	10.19	10.74	12.96	14.32	15.72	16.83	17.27	20.06
1956	22.29	22.82	22.75	23.53	23.50	23.77	24.14	23.45	25.29	26.44
1957	29.98	30.68	32.58	34.46	36.28	37.97	39.52	40.46	41.15	41.77
1958	42.60	45.77	47.98	47.86	48.46	48.99	50.77	51.69	52.94	53.91
1959	58.01	60.53	59.98	61.56	62.68	63.90	65.28	65.78	66.45	66.43
1960	69.11	70.16	72.08	75.13	75.64	74.33	76.34	76.63	76.94	78.97
1961	80.65	81.51	82.52	83.91	84.54	86.18	86.96	88.29	90.39	91.83
1962	96.64	98.19	99.43	99.80	102.16	102.56	105.75	108.28	109.04	109.01
	Nov	Dec								
1954	5.97	8.69								
1955	21.56	20.59								
1956	26.30	28.55								
1957	42.08	42.87								
1958	54.87	57.24								
1959	66.36	67.66								
1960	79.69	79.47								
1961	92.50	94.65								
1962										

There are several accessory functions to obtain information from a `ts` object. In particular, `time` extracts the vector of time instants of each observation of the time series; `deltat` returns the Δt between observations; `end` and `start` returns initial and final date and `frequency` the frequency of the time series.

```
R> time(X)[1:10]
```

```
 [1] 1954.500 1954.583 1954.667 1954.750 1954.833 1954.917
     1955.000 1955.083
 [9] 1955.167 1955.250
```

```
R> deltat(X)
```

```
[1] 0.08333333
```

```
R> start(X)
```

```
[1] 1954      7
```

```
R> end(X)
```

```
[1] 1962     10
```

```
R> frequency(X)
```

```
[1] 12
```

These accessory functions are also available for the other classes presented below, eventually with some specificity. In addition to the accessory functions, one can extract a subseries using the `window` function. For example, if instead of monthly data we want to extract quarterly data from x above, we can do the following:

```
R> window(X, frequency = 4)
```

	Qtr1	Qtr2	Qtr3	Qtr4
1954			0.44	4.84
1955	10.15	10.74	15.72	20.06
1956	22.29	23.53	24.14	26.44
1957	29.98	34.46	39.52	41.77
1958	42.60	47.86	50.77	53.91
1959	58.01	61.56	65.28	66.43
1960	69.11	75.13	76.34	78.97
1961	80.65	83.91	86.96	91.83
1962	96.64	99.80	105.75	109.01

B.2.2   The zoo class

The `zoo` class can host time series in a very abstract way. Indeed, `zoo` objects are objects indexed by an abstract set of indexes we can put in relation with the set of integer numbers $\mathbb{Z}$ (hence the name 'zoo'). To use `zoo`, objects are defined in the **zoo** package. When a `zoo` object is created, if the set of indexes is not created, by default an increasing sequence is used.

```
R> require(zoo)
R> X <- zoo(rnorm(10))
R> X
```

```
        1              2              3              4              5
-0.71040656   0.25688371  -0.24669188  -0.34754260  -0.95161857
        6              7              8              9             10
-0.04502772  -0.78490447  -1.66794194  -0.38022652   0.91899661

R> str(X)

'zoo' series from 1 to 10
  Data: num [1:10] -0.71 0.257 -0.247 -0.348 -0.952 ...
  Index:  int [1:10] 1 2 3 4 5 6 7 8 9 10
```

To access or modify the indexes one can use either time or, better, index

```
R> index(X)

[1]  1  2  3  4  5  6  7  8  9 10
```

The advantage of the zoo indexing is that it can host irregular time series. For example, if we generate 10 random times from the exponential distribution we can create a time series under Poisson random sampling as follows:

```
R> X <- zoo(rnorm(10), order.by = cumsum(rexp(10, rate = 0.1)))
R> X

     8.2326       12.9902       47.6261       60.3664       71.1813
-1.02412879   0.11764660  -0.94747461  -0.49055744  -0.25609219
    74.1509       74.9925      104.5265      124.209      130.8476
 1.84386201  -0.65194990   0.23538657   0.07796085  -0.96185663

R> str(X)

'zoo' series from 8.23260500985098 to 130.847554074331
  Data: num [1:10] -1.024 0.118 -0.947 -0.491 -0.256 ...
  Index:  num [1:10] 8.23 12.99 47.63 60.37 71.18 ...
```

If one wants to use an approach similar to ts to create regularly space time series, one should use explicitly the zooreg function:

```
R> Xreg <- zooreg(cumsum(1 + round(rnorm(100), 2)), start = c(1954,
+       7), frequency = 12)
R> time(Xreg)[1:10]

 [1] 1954.500 1954.583 1954.667 1954.750 1954.833 1954.917
     1955.000 1955.083
 [9] 1955.167 1955.250
```

It is possible to convert ts object to zoo object without problems, but the contrary is possible only if the time series is regularly space, otherwise times are completely ruined as in the next example:

```
R> Y <- as.ts(X)
R> time(X)
```

```
[1]    8.232605  12.990195  47.626143  60.366421  71.181272
       74.150864
[7]    74.992471 104.526480 124.209003 130.847554
```

```
R> time(Y)
```

```
Time Series:
Start = 1
End = 10
Frequency = 1
 [1]  1  2  3  4  5  6  7  8  9 10
```

B.2.3 The **xts** class

The xts class, where the 'x' stands for 'extensible', extends the functionality of the zoo class specifically to handle time and dates. It also extends the object in the sense that it allows for the inclusion of meta data like 'last data update' or similar. It is also written entirely in low level language to gain in speed when accessing or subsetting the object which may be slow in some cases for the other classes of objects above. The **xts** is required to use this new class. The xts function does not assign an index to the object, so one has to explicitly create the index of times

```
R> require(xts)
R> X <- xts(rnorm(10), order.by = as.Date(cumsum(rexp(10,
           rate = 0.1))))
R> X
```

```
                   [,1]
1970-01-16 -1.0155926
1970-01-22  1.9552940
1970-01-31 -0.0903196
1970-02-04  0.2145388
1970-02-05 -0.7385277
1970-02-08 -0.5743887
1970-02-21 -1.3170161
1970-03-15 -0.1829254
1970-03-23  0.4189824
1970-03-23  0.3243043
```

```
R> str(X)
```

```
An 'xts' object from 1970-01-16 to 1970-03-23 containing:
  Data: num [1:10, 1] -1.0156 1.9553 -0.0903 0.2145 -0.7385 ...
  Indexed by objects of class: [Date] TZ:
  xts Attributes:
 NULL
```

and in the above we have used as.Date to transform a vector of indexes in times. With as.xts it is possible to convert objects from zoo and ts to xts only if the time indexes are transformed first in true time/class objects or if an additional argument order.by is specified appropriately. The package **xts**

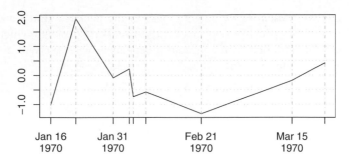

Figure B.3   Result of the plot method for objects of class xts.

redefines the plot method for objects of class xts to put in evidence the irregular time spaces. The next code produces the graph in Figure B.3.

```
R> plot(X)
```

B.2.4   The irts class

Another framework for irregular times series is provided by the package **tseries** in the class irts. The use of irts is similar to zoo but the arguments are reversed: the first argument is the vector of times and the second argument is the vector of values of the time series

```
R> require(tseries)
R> X <- irts(cumsum(rexp(10, rate = 0.1)), rnorm(10))
R> X

1970-01-01 00:00:05 GMT -1.388
1970-01-01 00:00:16 GMT -0.2646
1970-01-01 00:00:28 GMT -0.9473
1970-01-01 00:00:36 GMT 0.7395
1970-01-01 00:01:00 GMT 0.8968
1970-01-01 00:01:11 GMT -0.346
1970-01-01 00:01:30 GMT -1.782
1970-01-01 00:01:30 GMT 0.4649
1970-01-01 00:01:30 GMT -1.951
1970-01-01 00:01:33 GMT -0.5161

R> str(X)

List of 2
 $ time : POSIXct[1:10], format: "1970-01-01 09:00:05"
   "1970-01-01 09:00:16" ...
 $ value: num [1:10] -1.388 -0.265 -0.947 0.74 0.897 ...
 - attr(*, "class")= chr "irts"
```

As usual there exist functions to convert objects from one class to another.

B.2.5   The `timeSeries` class

The last class we present is the timeSeries in the package **timeSeries** of the Rmetrics suite. This package depends on the **timeDate** package which is used to store time and dates object in a system independent way. We will discuss the relevance of this issue in Section B.3.

```
R> require(timeSeries)
R> X <- timeSeries(rnorm(10), as.Date(cumsum(rexp(10,
    rate = 0.1))))
R> X
```

```
GMT
                            TS.1
1970-01-21 06:11:34   0.4890803
1970-01-22 15:29:47   0.9020247
1970-01-26 07:37:02   0.6403630
1970-01-29 15:10:12   0.9512089
1970-01-30 05:31:07  -0.5991232
1970-01-30 10:14:35  -1.3306950
1970-03-06 10:14:41  -0.5922097
1970-04-08 10:04:43   1.8010509
1970-04-09 19:31:46   1.0553386
1970-04-12 19:39:18  -0.2919208
```

```
R> str(X)
```

```
Time Series:
 Name:              object
Data Matrix:
 Dimension:         10 1
 Column Names:      TS.1
 Row Names:         1970-01-21 06:11:34  ...  1970-04-12 19:39:18
Positions:
 Start:             1970-01-21 06:11:34
 End:               1970-04-12 19:39:18
With:
 Format:            %Y-%m-%d %H:%M:%S
 FinCenter:         GMT
 Units:             TS.1
 Title:             Time Series Object
 Documentation:     Wed Nov 17 00:18:09 2010
```

B.3   Dates and time handling

Dates and time stamps are peculiar of time series and R supports many formats. Most of the time the user downloads data from a service and wants to keep or transform the time stamp of the data. Or, vice versa, after data have been simulated, one wants to attach the correct time information to the data. We now explain the basic concepts and see how different classes of time series handle these formats. We start with the POSIX formats. POSIX (Portable Operating System Interface) is the IEEE standard for a common interface in most UNIX-like operating systems now available on other platforms. There is a POSIX standard for dates but also for many other tasks. Here we focus on dates. Using the function ISOdate it is possible to create a data object very easily as follows:

```
R> d <- ISOdate(2006, 6, 9)
R> d
```

```
[1] "2006-06-09 12:00:00 GMT"
```

The function ISOdate accepts the following arguments as we see using the command args:

```
R> args(ISOdate)
```

```
function (year, month, day, hour = 12, min = 0, sec = 0, tz = "GMT")
NULL
```

All arguments are easy to understand. The most important one is tz, the time zone argument. By default the time zone is set equal to 'GMT' which corresponds to the Coordinated Universal Time (UTC), formerly called the Greenwich Mean Time. It means, that the time represented by our object d is referred to as Greenwich local time. Similarly 'CET' is the Central European Time which corresponds to UTC+1. This is the time adopted by countries in central Europe like Italy, France, Spain, Germany, etc. We can see that this object d is indeed a POSIX time and in particular it is of class POSIXct where ct stands for 'calendar time'.

```
R> class(d)
```

```
[1] "POSIXt"  "POSIXct"
```

Internally objects of type POSIXct are stored as the number of seconds since 1970 in the UTC time zone. A second representation is called POSIXlt which is internally stored as a list with the following entries:

```
R> names(as.POSIXlt(d))
```

```
[1] "sec"   "min"   "hour"  "mday"  "mon"  "year"  "wday"  "yday"
    "isdst"
```

```
R> unlist(as.POSIXlt(d))
```

```
  sec    min   hour   mday    mon   year   wday   yday  isdst
    0      0     12      9      5    106      5    159      0
```

It is possible to convert one format into another using as.POSIXlt and as. POSIXct coercing functions. Due to the fact that POSIX dates contain calendar information, it is possible to represent them in different formats. Next is an example of several representations using the format function without comments:

```
R> format(d, "%a")

[1] "Fri"

R> format(d, "%A")

[1] "Friday"

R> format(d, "%b")

[1] "Jun"

R> format(d, "%B")

[1] "June"

R> format(d, "%c")

[1] "Fri  9 Jun 12:00:00 2006"

R> format(d, "%D")

[1] "06/09/06"

R> format(d, "%T")

[1] "12:00:00"

R> format(d, "%A %B  %d %H:%M:%S %Y")

[1] "Friday June  09 12:00:00 2006"

R> format(d, "%A   %d/%m/%Y")

[1] "Friday   09/06/2006"

R> format(d, "%d/%m/%Y (%A)")

[1] "09/06/2006 (Friday)"
```

and so forth. For a complete set of conversion operators % one should read the man page of the command format. It is also possible to convert strings into real date objects with the function strptime. The next example shows a simple example of use:

```
R> x <- c("1jan1960", "2jan1960", "31mar1960", "30jul1960")
R> strptime(x, "%d%b%Y")

[1] "1960-01-01" "1960-01-02" "1960-03-31" "1960-07-30"
```

In this case 'jan, mar, jul' are interpreted correctly as 'january, march, july' but in different locales, e.g. Italian, 'jan' and 'jul' will not be recognized by the system and hence `strptime` returns a NA date. The user should check their environment carefully before attempting data manipulations like the above. R offers functions like `Sys.getlocale` and `Sys.setlocale` to inspect and change the current 'locale' setting. The next example temporarily sets the locale settings to Italian and then switches it back to UK English:[2]

```
R> Sys.getlocale()

[1] "en_GB.UTF-8/en_GB.UTF-8/C/C/en_GB.UTF-8/en_GB.UTF-8"

R> Sys.setlocale("LC_ALL", "it_it")

[1] "it_it/it_it/it_it/C/it_it/en_GB.UTF-8"

R> strptime(x, "%d%b%Y")

[1] NA             NA            "1960-03-31" NA

R> Sys.setlocale("LC_ALL", "en_GB")

[1] "en_GB/en_GB/en_GB/C/en_GB/en_GB.UTF-8"

R> strptime(x, "%d%b%Y")

[1] "1960-01-01" "1960-01-02" "1960-03-31" "1960-07-30"
```

When data are created without any time specification, by default `ISOdate` uses 12am as reference and `as.POSIXct` uses 12pm

```
R> format(ISOdate(2006, 6, 9), "%H:%M:%S")

[1] "12:00:00"

R> format(as.POSIXct("2006-06-09"), "%H:%M:%S")

[1] "00:00:00"
```

B.3.1   Dates manipulation

It is also possible to extract information which is of interest in empirical finance. For example, the package **timeDate** implements functions like `holiday*` which describe the dates of holidays of several financial centers:

```
R> holidayNYSE()

NewYork
[1] [2010-01-01] [2010-01-18] [2010-02-15] [2010-04-02]
    [2010-05-31]
```

---

[2] Note that you have to check exact naming of the locales on your system. In our example it is OS X.

```
[6]  [2010-07-05]  [2010-09-06]  [2010-11-25]  [2010-12-24]
```

```
R> holidayNERC()
```

```
Eastern
[1]  [2009-12-31 19:00:00]  [2010-05-30 20:00:00]
     [2010-07-04 20:00:00]
[4]  [2010-09-05 20:00:00]  [2010-11-24 19:00:00]
```

It is possible to make calculations with times such as the following:

```
R> ISOdate(2006, 6, 9) - ISOdate(2006, 3, 1)
```

```
Time difference of 100 days
```

Or, using the timeDate class

```
R> my.dates = timeDate(c("2001-01-05", "2001-02-15"))
R> my.dates[2] - my.dates[1]
```

```
Time difference of 41 days
```

Time zones are also important when one wants to synchronize data coming from different financial sources. We will see that, if data possesses correct time stamps, this is quite easy. The package **timeDate** contains the function listFinCenter which lists the names of known financial centers. This is a growing list, but we can get an idea of what this command produces for European countries:

```
R> listFinCenter("Europe*")
```

```
 [1] "Europe/Amsterdam"    "Europe/Andorra"      "Europe/Athens"
 [4] "Europe/Belgrade"     "Europe/Berlin"       "Europe/Bratislava"
 [7] "Europe/Brussels"     "Europe/Bucharest"    "Europe/Budapest"
[10] "Europe/Chisinau"     "Europe/Copenhagen"   "Europe/Dublin"
[13] "Europe/Gibraltar"    "Europe/Guernsey"     "Europe/Helsinki"
[16] "Europe/Isle_of_Man"  "Europe/Istanbul"     "Europe/Jersey"
[19] "Europe/Kaliningrad"  "Europe/Kiev"         "Europe/Lisbon"
[22] "Europe/Ljubljana"    "Europe/London"       "Europe/Luxembourg"
[25] "Europe/Madrid"       "Europe/Malta"        "Europe/Mariehamn"
[28] "Europe/Minsk"        "Europe/Monaco"       "Europe/Moscow"
[31] "Europe/Oslo"         "Europe/Paris"        "Europe/Podgorica"
[34] "Europe/Prague"       "Europe/Riga"         "Europe/Rome"
[37] "Europe/Samara"       "Europe/San_Marino"   "Europe/Sarajevo"
[40] "Europe/Simferopol"   "Europe/Skopje"       "Europe/Sofia"
[43] "Europe/Stockholm"    "Europe/Tallinn"      "Europe/Tirane"
[46] "Europe/Uzhgorod"     "Europe/Vaduz"        "Europe/Vatican"
[49] "Europe/Vienna"       "Europe/Vilnius"      "Europe/Volgograd"
[52] "Europe/Warsaw"       "Europe/Zagreb"       "Europe/Zaporozhye"
[55] "Europe/Zurich"
```

The information about the financial center can be passed when the date objects are created to fix the correct time zone and geographical information, e.g.

```
R> d1 <- timeDate("2001-01-05", Fin = "Europe/Paris")
R> d2 <- timeDate("2001-01-05", Fin = "America/New_York")
R> d1
```

```
Europe/Paris
[1] [2001-01-05 01:00:00]
```

```
R> d2
```

```
America/New_York
[1] [2001-01-04 19:00:00]
```

For further informations about date/time manipulation a suggested reading is the time/date FAQ Würtz *et al.* (2010) ebook.

B.3.2   Using date objects to index time series

We now see how to assign time and dates as index to time series. We consider only `zoo`, `xts` and `timeDate` objects. We first create some random data and create some string dates:

```
R> set.seed(123)
R> data <- rnorm(6)
R> charvec <- paste("2009-0", 1:6, "-01", sep = "")
R> charvec
```

```
[1] "2009-01-01" "2009-02-01" "2009-03-01" "2009-04-01"
    "2009-05-01"
[6] "2009-06-01"
```

then generate the corresponding objects with the different classes:

```
R> X <- zoo(data, as.Date(charvec))
R> Y <- xts(data, as.Date(charvec))
R> Z <- timeSeries(data, charvec)
```

and see how the look

```
R> X
```

```
 2009-01-01   2009-02-01   2009-03-01   2009-04-01   2009-05-01
-0.56047565  -0.23017749   1.55870831   0.07050839   0.12928774
 2009-06-01
 1.71506499
```

```
R> Y
```

```
                   [,1]
2009-01-01  -0.56047565
2009-02-01  -0.23017749
2009-03-01   1.55870831
```

```
2009-04-01  0.07050839
2009-05-01  0.12928774
2009-06-01  1.71506499

R> z

GMT
                TS.1
2009-01-01 -0.56047565
2009-02-01 -0.23017749
2009-03-01  1.55870831
2009-04-01  0.07050839
2009-05-01  0.12928774
2009-06-01  1.71506499
```

Similarly, we should have used one of the following approaches:

```
R> z1 <- zoo(data, as.POSIXct(charvec))
R> z2 <- zoo(data, ISOdatetime(2009, 1:6, 1, 0, 0, 0))
R> z3 <- zoo(data, ISOdate(2009, 1:6, 1, 0))

R> z1

 2009-01-01  2009-02-01  2009-03-01  2009-04-01  2009-05-01
-0.56047565 -0.23017749  1.55870831  0.07050839  0.12928774
 2009-06-01
 1.71506499
R> z2

 2009-01-01  2009-02-01  2009-03-01  2009-04-01  2009-05-01
-0.56047565 -0.23017749  1.55870831  0.07050839  0.12928774
 2009-06-01
 1.71506499

R> z3

 2009-01-01  2009-02-01  2009-03-01  2009-04-01  2009-05-01
-0.56047565 -0.23017749  1.55870831  0.07050839  0.12928774
 2009-06-01
 1.71506499
```

B.4   Binding of time series

Suppose we have two parts of the same time series collected in different periods
of time. It is possible to merge them by row, i.e. by date, using the rbind function.
If the time indexes do not overlap, all classes performs in the same way:

```
R> set.seed(123)
R> d1 <- rnorm(5)
R> d2 <- rnorm(7)
R> date1 <- ISOdate(2009, 1:5, 1)
R> date2 <- ISOdate(2009, 6:12, 1)
```

```
R> z1 <- zoo(d1, date1)
R> z2 <- zoo(d2, date2)
R> rbind(z1, z2)


2009-01-01 21:00:00 2009-02-01 21:00:00 2009-03-01 21:00:00
        -0.56047565         -0.23017749          1.55870831
2009-04-01 21:00:00
         0.07050839
2009-05-01 21:00:00 2009-06-01 21:00:00 2009-07-01 21:00:00
         0.12928774          1.71506499          0.46091621
2009-08-01 21:00:00
        -1.26506123
2009-09-01 21:00:00 2009-10-01 21:00:00 2009-11-01 21:00:00
        -0.68685285         -0.44566197          1.22408180
2009-12-01 21:00:00
         0.35981383


R> x1 <- xts(d1, date1)
R> x2 <- xts(d2, date2)
R> rbind(x1, x2)


                           [,1]
2009-01-01 21:00:00 -0.56047565
2009-02-01 21:00:00 -0.23017749
2009-03-01 21:00:00  1.55870831
2009-04-01 21:00:00  0.07050839
2009-05-01 21:00:00  0.12928774
2009-06-01 21:00:00  1.71506499
2009-07-01 21:00:00  0.46091621
2009-08-01 21:00:00 -1.26506123
2009-09-01 21:00:00 -0.68685285
2009-10-01 21:00:00 -0.44566197
2009-11-01 21:00:00  1.22408180
2009-12-01 21:00:00  0.35981383


R> s1 <- timeSeries(d1, date1)
R> s2 <- timeSeries(d2, date2)
R> rbind(s1, s2)


GMT
                    TS.1_TS.1
2009-01-01 12:00:00 -0.56047565
2009-02-01 12:00:00 -0.23017749
2009-03-01 12:00:00  1.55870831
2009-04-01 12:00:00  0.07050839
2009-05-01 12:00:00  0.12928774
2009-06-01 12:00:00  1.71506499
2009-07-01 12:00:00  0.46091621
2009-08-01 12:00:00 -1.26506123
2009-09-01 12:00:00 -0.68685285
2009-10-01 12:00:00 -0.44566197
2009-11-01 12:00:00  1.22408180
2009-12-01 12:00:00  0.35981383
```

but when indexes do overlap, some classes fail. In particular, zoo requires unique indexes. Thus, for example, if we create overlapping dates with this:

```
R> date2 <- ISOdate(2009, 4:10, 1)
R> z2 <- zoo(d2, date2)
```

the following code produces an error:

```
R> rbind(z1, z2)
```

```
Error in rbind(deparse.level,...) : indexes overlap
```

while the xts and timeSeries simply duplicate the entries:

```
R> x2 <- xts(d2, date2)
R> rbind(x1, x2)
```

```
                            [,1]
2009-01-01 21:00:00 -0.56047565
2009-02-01 21:00:00 -0.23017749
2009-03-01 21:00:00  1.55870831
2009-04-01 21:00:00  0.07050839
2009-04-01 21:00:00  1.71506499
2009-05-01 21:00:00  0.12928774
2009-05-01 21:00:00  0.46091621
2009-06-01 21:00:00 -1.26506123
2009-07-01 21:00:00 -0.68685285
2009-08-01 21:00:00 -0.44566197
2009-09-01 21:00:00  1.22408180
2009-10-01 21:00:00  0.35981383
```

```
R> s2 <- timeSeries(d2, date2)
R> rbind(s1, s2)
```

```
GMT
                       TS.1_TS.1
2009-01-01 12:00:00 -0.56047565
2009-02-01 12:00:00 -0.23017749
2009-03-01 12:00:00  1.55870831
2009-04-01 12:00:00  0.07050839
2009-05-01 12:00:00  0.12928774
2009-04-01 12:00:00  1.71506499
2009-05-01 12:00:00  0.46091621
2009-06-01 12:00:00 -1.26506123
2009-07-01 12:00:00 -0.68685285
2009-08-01 12:00:00 -0.44566197
2009-09-01 12:00:00  1.22408180
2009-10-01 12:00:00  0.35981383
```

A different approach is called merging which is analogous to the same functionalities for standard data.frame in the R system. Also in this case, the class zoo and xts perform similarly and timeSeries differentiates. Indeed, if we use the `merge` function both

```
R> merge(z1, z2)
```

```
                             z1          z2
2009-01-01 21:00:00 -0.56047565          NA
2009-02-01 21:00:00 -0.23017749          NA
2009-03-01 21:00:00  1.55870831          NA
2009-04-01 21:00:00  0.07050839   1.7150650
2009-05-01 21:00:00  0.12928774   0.4609162
2009-06-01 21:00:00          NA  -1.2650612
2009-07-01 21:00:00          NA  -0.6868529
2009-08-01 21:00:00          NA  -0.4456620
2009-09-01 21:00:00          NA   1.2240818
2009-10-01 21:00:00          NA   0.3598138
```

```
R> merge(x1, x2)
```

```
                             x1          x2
2009-01-01 21:00:00 -0.56047565          NA
2009-02-01 21:00:00 -0.23017749          NA
2009-03-01 21:00:00  1.55870831          NA
2009-04-01 21:00:00  0.07050839   1.7150650
2009-05-01 21:00:00  0.12928774   0.4609162
2009-06-01 21:00:00          NA  -1.2650612
2009-07-01 21:00:00          NA  -0.6868529
2009-08-01 21:00:00          NA  -0.4456620
2009-09-01 21:00:00          NA   1.2240818
2009-10-01 21:00:00          NA   0.3598138
```

produce a two-dimensional time series aligning the time indexes and setting to NA the missing observations in each time series. On the contrary, the `timeSeries` class produces one single one-dimensional time series keeping duplicates as in `rbind`:

```
R> merge(s1, s2)
```

```
GMT
                          TS.1
2009-01-01 12:00:00 -0.56047565
2009-02-01 12:00:00 -0.23017749
2009-03-01 12:00:00  1.55870831
2009-04-01 12:00:00  0.07050839
2009-04-01 12:00:00  1.71506499
2009-05-01 12:00:00  0.12928774
2009-05-01 12:00:00  0.46091621
2009-06-01 12:00:00 -1.26506123
2009-07-01 12:00:00 -0.68685285
2009-08-01 12:00:00 -0.44566197
2009-09-01 12:00:00  1.22408180
2009-10-01 12:00:00  0.35981383
```

To obtain a behaviour similar to `zoo` and `xts`, one should specify different time units. For example:

```
R> s2 <- timeSeries(d2, date2, units = "s2")
R> merge(s1, s2)
```

```
GMT
                           TS.1            s2
2009-01-01 12:00:00 -0.56047565          NA
2009-02-01 12:00:00 -0.23017749          NA
2009-03-01 12:00:00  1.55870831          NA
2009-04-01 12:00:00  0.07050839  1.7150650
2009-05-01 12:00:00  0.12928774  0.4609162
2009-06-01 12:00:00          NA -1.2650612
2009-07-01 12:00:00          NA -0.6868529
2009-08-01 12:00:00          NA -0.4456620
2009-09-01 12:00:00          NA  1.2240818
2009-10-01 12:00:00          NA  0.3598138
```

A final remark is that, while for `zoo` and `xts` the arguments in `rbind` are treated as symmetric, in `timeSeries` they are not. Hence, for example:

```
R> date1 <- ISOdate(2009, 1:5, 1)
R> date2 <- ISOdate(2009, 6:12, 1)
R> s1 <- timeSeries(d1, date1)
R> s2 <- timeSeries(d2, date2)
R> rbind(s1, s2)
```

```
GMT
                        TS.1_TS.1
2009-01-01 12:00:00 -0.56047565
2009-02-01 12:00:00 -0.23017749
2009-03-01 12:00:00  1.55870831
2009-04-01 12:00:00  0.07050839
2009-05-01 12:00:00  0.12928774
2009-06-01 12:00:00  1.71506499
2009-07-01 12:00:00  0.46091621
2009-08-01 12:00:00 -1.26506123
2009-09-01 12:00:00 -0.68685285
2009-10-01 12:00:00 -0.44566197
2009-11-01 12:00:00  1.22408180
2009-12-01 12:00:00  0.35981383
```

```
R> rbind(s2, s1)
```

```
GMT
                        TS.1_TS.1
2009-06-01 12:00:00  1.71506499
2009-07-01 12:00:00  0.46091621
2009-08-01 12:00:00 -1.26506123
2009-09-01 12:00:00 -0.68685285
2009-10-01 12:00:00 -0.44566197
2009-11-01 12:00:00  1.22408180
```

```
2009-12-01 12:00:00   0.35981383
2009-01-01 12:00:00  -0.56047565
2009-02-01 12:00:00  -0.23017749
2009-03-01 12:00:00   1.55870831
2009-04-01 12:00:00   0.07050839
2009-05-01 12:00:00   0.12928774
```

produce different ordering but one can still use, in all classes, the two functions
sort and rev:

```
R> sort(rbind(s2, s1))

GMT
                         TS.1_TS.1
2009-01-01 12:00:00  -0.56047565
2009-02-01 12:00:00  -0.23017749
2009-03-01 12:00:00   1.55870831
2009-04-01 12:00:00   0.07050839
2009-05-01 12:00:00   0.12928774
2009-06-01 12:00:00   1.71506499
2009-07-01 12:00:00   0.46091621
2009-08-01 12:00:00  -1.26506123
2009-09-01 12:00:00  -0.68685285
2009-10-01 12:00:00  -0.44566197
2009-11-01 12:00:00   1.22408180
2009-12-01 12:00:00   0.35981383


R> sort(rbind(s2, s1), decr = TRUE)

GMT
                         TS.1_TS.1
2009-12-01 12:00:00   0.35981383
2009-11-01 12:00:00   1.22408180
2009-10-01 12:00:00  -0.44566197
2009-09-01 12:00:00  -0.68685285
2009-08-01 12:00:00  -1.26506123
2009-07-01 12:00:00   0.46091621
2009-06-01 12:00:00   1.71506499
2009-05-01 12:00:00   0.12928774
2009-04-01 12:00:00   0.07050839
2009-03-01 12:00:00   1.55870831
2009-02-01 12:00:00  -0.23017749
2009-01-01 12:00:00  -0.56047565
```

to sort the dates in increasing or decreasing order or to revert the time stamps
of a time series when they are downloaded from external resources:

```
R> s2

GMT
                       TS.1
2009-06-01 12:00:00  1.7150650
2009-07-01 12:00:00  0.4609162
```

```
2009-08-01 12:00:00 -1.2650612
2009-09-01 12:00:00 -0.6868529
2009-10-01 12:00:00 -0.4456620
2009-11-01 12:00:00  1.2240818
2009-12-01 12:00:00  0.3598138

R> rev(s2)

GMT
                       TS.1
2009-12-01 12:00:00  0.3598138
2009-11-01 12:00:00  1.2240818
2009-10-01 12:00:00 -0.4456620
2009-09-01 12:00:00 -0.6868529
2009-08-01 12:00:00 -1.2650612
2009-07-01 12:00:00  0.4609162
2009-06-01 12:00:00  1.7150650
```

B.4.1   Subsetting of time series

Subsetting of time series is similar to indexing of `matrix` objects. For simplicity we make use of the data set `quotes` from the **sde** package which is stored in the `zoo` format.

```
R> require(sde)
```

```
To check the errata corrige of the book, type vignette
    ("sde.errata")
```

```
R> data(quotes)
R> str(quotes)
```

```
'zoo' series from 2006-01-03 to 2007-12-31
  Data: num [1:520, 1:20] 26.8 27 27 26.9 26.9 ...
 - attr(*, "dimnames")=List of 2
  ..$ : chr [1:520] "2006-01-03" "2006-01-04" "2006-01-05"
        "2006-01-06" ...
  ..$ : chr [1:20] "MSOFT" "AMD" "DELL" "INTEL" ...
  Index: Class 'Date'  num [1:520] 13151 13152 13153 13154
        13157 ...
```

We can see that the `Data` slot consists of a matrix with attributes for `colnames` and `rownames`, respectively the time series names and time stamps. We can access this object as follows:

```
R> quotes[1, 1:5]
```

```
          MSOFT   AMD  DELL INTEL    HP
2006-01-03 26.84 32.4 30.61 25.57 28.77
```

```
R> quotes[1:10, "MSOFT"]
```

```
2006-01-03 2006-01-04 2006-01-05 2006-01-06 2006-01-09 2006-01-10
     26.84       26.97       26.99       26.91       26.86       27.00
2006-01-11 2006-01-12 2006-01-13 2006-01-16
     27.29       27.14       27.19       27.04
```

but we can also use the data.frame-like access with $

```
R> quotes$MSOFT[1:10]
```

```
2006-01-03 2006-01-04 2006-01-05 2006-01-06 2006-01-09 2006-01-10
     26.84       26.97       26.99       26.91       26.86       27.00
2006-01-11 2006-01-12 2006-01-13 2006-01-16
     27.29       27.14       27.19       27.04
```

but we can also access data by dates. For example,

```
R> date <- as.Date(sprintf("2006-07-%.2d", 1:10))
R> date
```

```
 [1] "2006-07-01" "2006-07-02" "2006-07-03" "2006-07-04"
     "2006-07-05"
 [6] "2006-07-06" "2006-07-07" "2006-07-08" "2006-07-09"
     "2006-07-10"
```

```
R> quotes[date, 1:5]
```

```
             MSOFT    AMD   DELL   INTEL     HP
2006-07-03 23.700  24.60 24.590  19.360  32.51
2006-07-04 23.525  24.25 24.405  19.055  32.64
2006-07-05 23.350  23.90 24.220  18.750  32.77
2006-07-06 23.480  23.83 24.150  18.850  33.10
2006-07-07 23.300  23.56 23.870  18.560  32.85
2006-07-10 23.500  22.51 23.480  18.180  31.93
```

and we see that, in case of missing observations for some dates, these dates are ignored. Of course, because dates are objects, we can do selection on dates like this:

```
R> start <- as.Date("2006-06-25")
R> end <- as.Date("2006-07-10")
R> quotes[(time(quotes) >= start) & (time(quotes) <= end), 1:5]
```

```
             MSOFT    AMD   DELL   INTEL     HP
2006-06-26 22.820  24.66 23.840  18.280  32.49
2006-06-27 22.860  24.26 23.710  18.050  31.94
2006-06-28 23.160  23.89 23.850  18.660  31.59
2006-06-29 23.470  24.81 24.620  19.320  32.03
2006-06-30 23.300  24.42 24.460  19.000  31.68
2006-07-03 23.700  24.60 24.590  19.360  32.51
2006-07-04 23.525  24.25 24.405  19.055  32.64
2006-07-05 23.350  23.90 24.220  18.750  32.77
2006-07-06 23.480  23.83 24.150  18.850  33.10
2006-07-07 23.300  23.56 23.870  18.560  32.85
2006-07-10 23.500  22.51 23.480  18.180  31.93
```

B.5   Loading data from financial data servers

There are several ways to obtain data via http queries to famous financial data providers, local or remote data bases but also commercial services. For example, the package **quantmod** has a single function to get data from Yahoo! Finance[3], Google Finance[4], FRED[5] - Federal Reserve Bank of St. Louis or OANDA[6] but also from local MySQL data bases, csv files, or R data. We have already made use of the function getSymbols for downloading data from Yahoo! Finance with

```
R> getSymbols("AAPL")
```

```
[1] "AAPL"
```

```
R> attr(AAPL, "src")
```

```
[1] "yahoo"
```

but if we want to get the data from Google Finance we can specify the argument src as follows:

```
R> getSymbols("AAPL", src = "google")
```

```
[1] "AAPL"
```

```
R> attr(AAPL, "src")
```

```
[1] "google"
```

For exchange rates and currencies we can use both FRED or OANDA as follows:

```
R> getSymbols("DEXUSEU", src = "FRED")
```

```
[1] "DEXUSEU"
```

```
R> attr(DEXUSEU, "src")
```

```
[1] "FRED"
```

```
R> getSymbols("EUR/USD", src = "oanda")
```

```
[1] "EURUSD"
```

```
R> attr(EURUSD, "src")
```

```
[1] "oanda"
```

```
R> str(EURUSD)
```

---

[3] http://finance.yahoo.com
[4] http://finance.google.com
[5] http://research.stlouisfed.org/fred2
[6] http://www.oanda.com

```
An 'xts' object from 2009-07-06 to 2010-11-16 containing:
  Data: num [1:499, 1] 1.4 1.39 1.4 1.39 1.4 ...
 - attr(*, "dimnames")=List of 2
 ..$ : NULL
 ..$ : chr "EUR.USD"
 - attr(*, "src")= chr "oanda"
 - attr(*, "updated")= POSIXct[1:1], format: "2010-11-17 00:18:14"
  Indexed by objects of class: [Date] TZ:
  xts Attributes:
List of 2
 $ src     : chr "oanda"
 $ updated: POSIXct[1:1], format: "2010-11-17 00:18:14"
```

Notice that getSymbols returns an object of class xts in the R workspace with the same name of the symbol.

Another option is the **fImport** package which offers similar functionalities but returns objects of class timeSeries or fWEBDATA. For example, to get data from Yahoo! Finance, one can either use yahooSeries or yahooImport, e.g.

```
R> require(fImport)
R> X <- yahooSeries("AAPL")
R> str(X)

Time Series:
 Name:                  object
Data Matrix:
 Dimension:             252 6
 Column Names:          AAPL.Open AAPL.High AAPL.Low AAPL.Close
                        AAPL.Volume AAPL.Adj.Close
 Row Names:             2010-11-15  ...  2009-11-16
Positions:
 Start:                 2009-11-16
 End:                   2010-11-15
With:
 Format:                %Y-%m-%d
 FinCenter:             GMT
 Units:                 AAPL.Open AAPL.High AAPL.Low AAPL.Close
                        AAPL.Volume AAPL.Adj.Close
 Title:                 Time Series Object
 Documentation:         Wed Nov 17 00:18:19 2010
```

but

```
R> X <- yahooImport("AAPL")
R> str(X)

Formal class 'fWEBDATA' [package "fImport"] with 5 slots
  ..@ call      : language yahooImport(query = query,
                   file = file, source = source,
                   frequency = frequency, from = from, to = to,
                   nDaysBack = nDaysBack, save = save,  ...
  ..@ data      :Time Series:
```

```
Name:                   object
Data Matrix:
 Dimension:             6607 6
 Column Names:          Open High Low Close Volume Adj.Close
 Row Names:             2010-11-15  ...  1984-09-07
Positions:
 Start:                 1984-09-07
 End:                   2010-11-15
With:
 Format:                %Y-%m-%d
 FinCenter:             GMT
 Units:                 Open High Low Close Volume Adj.Close
 Title:                 Time Series Object
 Documentation:         Wed Nov 17 00:18:22 2010
 ..@ param       : Named chr [1:2] "AAPL" "daily"
 .. ..- attr(*, "names")= chr [1:2] "Instrument" "Frequency "
 ..@ title       : chr "Data Import from www.yahoo.com"
 ..@ description: chr "Wed Nov 17 00:18:22 2010 by user: "
```

and the class fWEBDATA stores the times series in a timeSeries object within the rich structure which describes more efficiently the source of the data. Similar functionalities exist for OANDA (fredSeries, fredImport) and FRED (oanda-Series, oandaImport). The **fOptions** package will soon be enhanced to allow for downloading data from additional resources not included in other packages.

Another option is the use of the function get.hist.quote from package **tseries** which downloads data either from Yahoo! Finance or OANDA and returns a zoo object. The use is as simple as follows:

```
R> x <- get.hist.quote("AAPL")

time series ends   2010-11-15

R> str(x)

'zoo' series from 1991-01-02 to 2010-11-15
  Data: num [1:5010, 1:4] 42.8 43.5 43 43 43.8 ...
 - attr(*, "dimnames")=List of 2
  ..$ : NULL
  ..$ : chr [1:4] "Open" "High" "Low" "Close"
  Index: Class 'Date'  num [1:5010] 7671 7672 7673 7676 7677 ...

R> x <- get.hist.quote(instrument = "EUR/USD", provider = "oanda",
+       start = Sys.Date() - 300)
R> str(x)

'zoo' series from 2010-01-21 to 2010-11-16
  Data: num [1:300] 1.42 1.41 1.41 1.41 1.41 ...
  Index: Class 'Date'  num [1:300] 14630 14631 14632 14633
          14634 ...
```

We mention finally that the **RBloomberg** allows for data fetching from Bloomberg. RBloomberg only works on a Bloomberg workstation, using the

Desktop COM API. The user also needs to install the **RDCOMClient** package or the **rcom** package for interprocess communications between R and the Bloomberg workstation.

B.6   Bibliographical notes

There is a growing interest of R in Finance, but still few specific publications are available at present. We already mentioned Würtz *et al.* (2009) and Würtz *et al.* (2010) from the Rmetrics group. Other books like Ruppert (2006), Franke *et al.* (2004) and Boland (2007) contain either code or have specialized support web sites with R examples from the books. General time series analysis books with applications to finance are Tsay (2005) and Carmona (2004).

References

Boland, P. J. (2007). *Statistical and Probabilistic Methods in Actuarial Science*. Chapman & Hall/CRC, Boca Raton, FL.

Carmona, R. (2004). *Statistical Analysis of Financial Data in S-Plus*. Springer, New York.

Franke, J., Härdle, W., and Hafner, C. (2004). *Statistics of Financial Markets: An Introduction*. Springer, New York, New York.

Ruppert, D. (2006). *Statistics and Finance: An Introduction*. Springer, New York.

Tsay, R. S. (2005). *Analysis of Financial Time Series. Second Edition*. John Wiley & Sons, Inc., Hoboken, N.J.

Würtz, D., Chalabi, Y., Chen, W., and Ellis, A. (2009). *Portfolio Optimization with R/Rmetrics*. Finance Online Publishing, Zurich.

Würtz, D., Chalabi, Y., Chen, W., and Ellis, A. (2010). *A Discussion of Time Series Objects for R in Finance*. Finance Online Publishing, Zurich.

Index

L^p integrability, 23
Γ, 27
χ^2, 28
δ, 249
δ-hedging, 249
δ-method, 70
γ greek, 259
$\mathcal{G}$-stable convergence, 103
ρ greek, 259
σ-algebra, 13
θ Greek, 258
φ-mixing, 102
:, 396
?, 394

absorbing state, 97
adapted, 81
addBBands, 422
additive, 194
AEAsian, 270
American option, 7, 285
AmericanPutExp, 291
AmericanPutImp, 296
apply, 409
arbitrage free, 223
arbitrage opportunity, 229
args, 429
as, 405
as.data.frame, 405
as.Date, 426
as.integer, 405
as.ts, 425
as.xts, 426
as.zoo, 425

asset price, 191
assign, 397
asymmetric double exponential, 210
asymptotically unbiased estimator, 58
asynchronous covariance estimator, 363
attributes, 405

backtest, 419
basket option, 8, 278
BAWAmericanApproxOption, 299
Bayes' rule, 16
benchmark, 341
Bermudan option, 286
Bernoulli
    sample, 79
Bessel function, 32
besselK, 180
best linear unbiased estimator, 59
Beta, 30
bias, 58
bilateral gamma, 179
binary, 252
binomial, 25
Borel, 51
Brown, 104
Brownian motion, 9, 104
    multidimensional, 145, 278
    translated, 232
    two-sided, 351

448 INDEX

BSAmericanApproxOption,
        300
Burkholder-Davis-Gundy
        inequality, 89, 127

c, 396
càd-làg process, 108
call option, 5
Cantelli, 51
cat, 393
Cauchy, 30
Cauchy functional equation, 135
Cauchy-Schwarz-Bunyakovsky
        inequality, 47
cce, 364
CGMY
    process, 209
change point, 350
Chapman-Kolomogorov equation,
        95, 98
character, 397
characteristic
    exponent, 38, 40
    function, 23
        de Finetti, 132
    triplet, 38
charSeries, 421
chartSeries, 274
Chebyshev inequality, 46
Chebyshev-Cantelli inequality, 46
Chebyshev-Markov inequality, 46
clusterSetupRNG, 412
coefficient
    diffusion, 128
    drift, 128
compensated Poisson random
        measure, 143
compensator, 87, 113
complementary, 13
complete additivity, 14
completeness, 223
complex conjugate, 23
complex, 398
compound Poisson process, 110
conditional

expectation, 55
probability, 15
confint, 70
consistent estimator, 60
contingent claim, 221
continuous
    mapping theorem, 50
    random variable, 19
    time
        Markov chain, 99
convergence
    almost sure, 49
    in r-th mean, 49
    in distribution, 48
    in probability, 48
    mean square, 49
    weak, 48
convolution, 35
COS, 329
counting process, 108
covariance, 22
    asynchronous estimator, 363
    function, 84
CPoint, 359
cpoint, 275, 350, 352
CRAN, 395
CreditMetrics, 419
cumulative distribution function, 19

data.frame, 401
dbeta, 31
dbinom, 25
dcauchy, 30
dcCIR, 201
dchisq, 29
de Finetti characteristic function,
        132
De Morgan's laws, 15
delta Greek, 249
deltat, 423
density
    function, 19
    method, 252
    transition, 107
dependence, 79

dexp, 27
dgamma, 28
dgh, 33
diffusion
    coefficient, 128
    equation, 108
    process, 128
digital option, 252
Dirac delta, 38, 117, 331
discrete
    Fourier transform, 43
    random variable, 19
dist, 378
distribution function, 19
dlnorm, 31
dmvnorm, 35
dnig, 32
dnorm, 28
Doob
    maximal L^2 inequality, 88
    maximal L^p inequality, 88
    maximal inequality, 88
Doob-Meyer decomposition, 87
dopar, 243, 413
doSNOW, 244, 414
doubling strategies, 229
dpois, 25
drift, 8, 192
    coefficient, 128
dsCIR, 201
dstable, 41
dt, 29
dtw, 378
dunif, 26

early exercise premium, 297
efficiency, 63
end, 423
Epps effect, 363
equivalent measure, 17, 261
Esscher transform, 315
estimating function, 66
estimator
    consistence, 60
EUCdist, 378

Euler's formula, 73
European option, 7, 222
events, 14
    elementary, 14
exercise price, 3
existence of solutions, 129
exotic option, 7
expected value, 21
expiry date, 3
exponential
    law, 26
    martingale, 86
extension space, 103

fair value, 4
family of experiments, 57
fAsianOptions, 272, 420
fast Fourier transform, 43
fBasics, 205, 420
fExoticOptions, 272, 420
filtration, 81
    natural, 81
fImport, 273
finite dimensional
    distribution, 80
first passage time, 89
first-order variation, 83
Fisher information, 62
fOptions, 240, 260, 299, 420
for, 410
foreach, 243, 413
format, 429
forward, 2
fPortfolio, 419
fraction
    optimal, 343
fractions, 342
FRED, 442
fredImport, 444
fredSeries, 444
frequency, 423
fTrading, 419
function, 408
future, 2
fWEBDATA, 443

Galton, 31
gamma, 179, 259
GBSCharacteristics, 260
GBSOption, 240, 321
GBSVolatility, 275
generalized
    hyperbolic
        process, 207
    tempered stable process, 209
generalized hyperbolic
    distribution, 32
generating function
    moment, 24
geometric
    Brownian motion, 8, 191, 210
        translated, 232
    telegraph process, 211
get.zoo.data, 178
getDoParWorkers, 244, 414
getSymbols, 421, 442
GH, 32
    process, 207
ghFit, 205
Girsanov, 130, 141
global Lipschitz condition, 129
Google, 442
grad, 165
Greek
    delta, 249
    gamma, 259
    kappa, 259
    rho, 259
    theta, 258
    vega, 259
growth
    optimal portfolio, 341
    rate, 342

Hölder inequality, 47
Hájek-Rényi
    inequality, 88
Hayashi-Yoshida estimator, 363
head, 404
heat equation, 108, 233

hedge ratio, 251
hedging, 6, 223
help, 394
help.search, 394
hessian, 165
Hessian matrix, 75
holder, 3, 223
homogeneous
    Poisson process, 110
hybrid diffusion system, 147
hypFit, 205

i.i.d., 24, 25, 57
IBrokers, 419
implied volatility, 275
independent, 16, 20
index, 425
index of stability, 40
inequality
    Burkholder-Davis-Gundy, 89,
        127
    Cauchy-Schwarz-Bunyakovsky,
        47
    Chebyshev, 46
    Chebyshev-Cantelli, 46
    Chebyshev-Markov, 46
    Doob
        maximal, 88
        maximal L^2, 88
        maximal L^p, 88
    Hölder, 47
    Hájek-Rényi, 88
    Jensen, 47, 57
    Kolmogorov, 47, 89
infinite activity, 138
infinitely divisible, 37
infinitesimal generator, 100
information, 19
install.packages, 395
instantaneous state, 101
integer, 397
integrable random variables, 21
intensity function, 110
invariant distribution, 97

inverse image, 18
inversion theorem, 43
irts, 427
ISOdate, 429
Itô
    formula, 124, 126, 146
    integral, 9, 119
    process, 123
Itô-Lévy
    decomposition, 137
    process, 143

Jensen's inequality, 47, 57
jump diffusion, 134

kappa, 259
Kolmogorov
    inequality, 47, 89
Kous process, 210

L-BFGS-B, 70
Lévy, 39
    exponent, 38
    jump diffusion, 134
    measure, 133, 136
    process, 133
Lévy-Khintchine formula, 38
Lamperti transform, 130
lapply, 409
large numbers
    strong law, 51
    weak law, 51
lasso, 371
law of total probability, 16
least squares method, 66
levels, 405
Lewis method, 170
likelihood, 61
Lindeberg's condition, 53
linear
    growth condition, 129
list, 399
listFinCenter, 432
lm, 398

load, 394
local martingale, 131
localizing sequence, 131
location, 40
log-likelihood, 61
log-normal, 31, 193
log-returns, 193
logLik, 69
Lorentz, 30
lower, 70
ls, 396
LSM, 310

makeCluster, 411
makeSOCKcluster, 411
marked point process, 109
market price of risk, 343
Markov
    chain, 92
        continuous time, 99
    operator, 375
    process, 91
    property
        strong, 96, 99
    switching, 147, 183
marks, 109
martingale, 84
    exponential, 86
    local, 131
    measure, 261
MASS, 35
matrix, 396
MatrixExp, 101
maximum likelihood estimator, 64
MCAsian, 268
MCdelta, 256
MCdelta2, 257
MCPrice, 242, 243
mean
    reverting, 200
    square error, 58
measurability, 18
measurable function, 18

measure
  equivalent, 17
  Lévy, 136
  martingale, 261
  of probability, 14
  random, 136
  risk neutral, 225, 261
Meixner
  distribution, 33
  process, 208
memoryless, 27
merge, 437
method, 70, 406
method of moments, 65
midori, 248
mixed moment, 22
mixing property, 101
MLE, 64
mle, 68, 169, 202, 406
modified Bessel function, 117, 331
MOdist, 377
moment, 21
  generating function, 24, 86
  mixed, 22
MSE, 58
msm, 101, 185
mts, 423
multicore, 244
multicore, 413
multidimensional
  Brownian motion, 145
  geometric Brownian motion, 278
mvrnorm, 35

Newton-Raphson, 167
NIG, 32, 179
  process, 208
nigFit, 205
nlm, 167
non-arbitrage, 6
nonhomogeneous
  Poisson process, 110
normal
  gamma, 179
  inverse Gaussian, 32, 179

normal inverse
  Gaussian
    process, 208
NULL, 397
numDeriv, 165
numeraire, 341
numeric, 396
nws, 412

OANDA, 442
oandaImport, 444
oandaSeries, 444
opefimor, 1
optim, 168
option, 2
  American, 7, 285
  basket, 8, 278
  Bermudan, 286
  binary, 252
  call, 5
  digital, 252
  European, 7, 222
  exotic, 7
  path-dependent, 7
  put, 5
  vanilla, 3
Ornstein-Uhlenbeck, 152

par, 394
path, 80
path-dependent option, 7
payoff, 5
  function, 221
pbeta, 31
pbinom, 25
pcauchy, 30
pcCIR, 201
pchisq, 29
PerformanceAnalytics, 419
pexp, 27
pgamma, 28
pgh, 33
plnorm, 31
pmvnorm, 35
pnig, 32

pnorm, 28
point process, 108
Poisson, 25
  process, 109
  random measure, 136
      compensated, 143
  random variable, 25
polyroot, 166
portfolio
  hedging, 229
  self-financing, 228
  strategy, 228
portfolio, 419
POSIXct, 429
POSIXlt, 429
ppois, 25
predictable, 83
probability, 14
  measure, 14
  space, 14
      extension, 103
  transition, 92, 98
Process
  Kou, 210
process
  Brownian, 104
  CGMY, 209
  continuous time, 80
  counting, 108
  diffusion, 128
  discrete time, 79
  gamma, 179
  generalized
      hyperbolic, 207
      tempered stable, 209
  Lévy, 133
  Markov, 91
      switching, 147, 183
  Meixner, 208
  normal inverse Gaussian, 208
  point, 108
      marked, 109
  Poisson, 109
  predictable, 83
  telegraph, 114

tempered stable, 207
  Variance Gamma, 179, 208, 323
  Wiener, 104
psCIR, 201
pstable, 41
pt, 29
punif, 26
put option, 5
put-call party, 239

qbeta, 31
qbinom, 25
qcauchy, 30
qcCIR, 201
qchisq, 29
qexp, 27
qgamma, 28
qgh, 33
qlnorm, 31
qmle, 198
qmvnorm, 35
qnig, 32
qnorm, 28
qpois, 25
qsCIR, 201
qstable, 41
qt, 29
quadratic
  variation, 83
quantile, 20
quantmod, 274
qunif, 26
quotes, 440

R-Forge, 395
Radon-Nikodým, 20, 130, 141
random
  experiment, 14
  measure, 136
  variable, 18
      Bernoulli, 24
      Beta, 30
      Binomial, 25
      Cauchy, 30
      Chi-square, 28

random (*Continued*)
    continuous, 19
    discrete, 19
    exponential, 26
    gamma, 27
    Gaussian, 28
    generalized hyperbolic, 32
    inverse Gaussian, 32
    Lévy, 39
    log-normal, 31
    Meixner, 33
    Poisson, 25
    Student's t, 29
    uniform, 26
  walk, 85, 92
rate
  Poisson process, 110
rates, 101
rbeta, 31
rbind, 434
rbinom, 25
RBloomberg, 444
rc.hy, 364
rcauchy, 30
rcCIR, 201
rchisq, 29
realized, 364
registerDoSEQ, 246, 415
registerDoSNOW, 414
return, 408
returns, 274
rev, 439
rexp, 27
rgamma, 28, 180
rgh, 33
rho, 259
Richardson's extrapolation, 164
risk
  free strategy, 6
  neutral measure, 225, 261
rlnorm, 31
Rmetrics, 240, 420
Rmpi, 412
rmvnorm, 35, 180
rnig, 32

rnorm, 28
RollGeskeWhaleyOption, 300
roundoff error, 163
rpois, 25
rpvm, 412
RQuantLib, 420
rsCIR, 201
RSiteSearch, 394
rstable, 41
rt, 29
runif, 26

sample
  space, 14
sapply, 409
save, 394
save.image, 394
scale, 40
score function, 62
sde, 175, 275, 377, 395
sde.sim, 175, 201
semimartingale, 131
setModel, 178
sfInit, 413
sfStop, 246, 413, 415
showMethods, 407
sigma-algebra, 13
sign function, 40
simMarkov, 184
simMSdiff, 185, 335
simulate, 178
skewness, 40
Slutsky's theorem, 50
snow, 411
snowfall, 244, 412
solution
  existence and uniqueness, 129
  strong, 128
  weak, 128
solve, 294
sort, 439
square integrability, 23
stable, 179
  convergence, 103

in distribution, 148
law, 38
stableFit, 205
standard
    error, 160
    normal, 28
start, 423
state space, 80
stats4, 68, 406
stochastic
    differential equation, 9, 128
    exponential, 326
    integral, 9, 119, 121
    process, 79
        covariance function, 84
        increments, 84
        quadratic variation, 83
        total variation, 83
stochastically continuous, 133
stopCluster, 412
stopping time, 89
str, 398
strike price, 3, 222
strong
    law of large numbers, 51
    Markov property, 96, 99
    solution, 128
strptime, 430
subsampling, 366
summary, 70
Sys.getlocale, 431
Sys.setlocale, 431

tail, 404
telegraph
    equation, 115
    process, 114
        geometric, 211
tempered stable, 179
    process, 207
theta, 258
thinning, 170
time, 423
timeDate, 420, 428
timeSeries, 274, 428

total variation, 83
trajectory of a process, 79
transform
    Esscher, 315
    Lamperti, 130
transient, 97
transition
    density, 107
    matrix, 92
    probability, 92, 98
translated
    Brownian motion, 232
    geometric Brownian motion, 232
trivial σ-algebra, 13
truncation error, 163
TS
    process, 207
ts, 423
tseries, 427
TTR, 419
ttrTests, 419
TurnbullWakemanAsian-
        ApproxOption, 272
two-sided Brownian motion, 351

unbiased estimator, 58
underlying asset, 3
uniform, 26
uniqueness of solutions, 129
uniroot, 166

vanilla, 3
VaR, 419
variance, 21
variance–covariance matrix, 22
variance gamma
    process, 208
vcov, 69
vector, 397
vega, 259
velocity process, 114
VG, 179, 208, 323
volatility, 4, 8, 192
    matrix, 278
    smile, 276

volatility
    implied, 275
Vsmc, 338

wave equation, 115
weak
    convergence, 48
    law of large numbers, 51
    solution, 128
while, 410
Wiener, 104
    process, 9
wild, 168

window, 424
writer, 3, 223

xts, 426

Yahoo, 442
yahooImport, 443
yahooSeries, 273, 443
yuima, 177, 270
yuima-data, 178

zoo, 178, 424
zooreg, 425